MW01406253

Science and Development *of* Muscular Strength

Timothy J. Suchomel, PhD, CSCS,*D, RSCC,*D, mISCP
University of Pittsburgh

HUMAN KINETICS

Library of Congress Cataloging-in-Publication Data

Names: Suchomel, Timothy J. author
Title: Science and development of muscular strength / Timothy J. Suchomel.
Description: Champaign, IL : Human Kinetics, [2026] | Includes bibliographical references and index.
Identifiers: LCCN 2025000589 (print) | LCCN 2025000590 (ebook) | ISBN 9781718223660 hardcover | ISBN 9781718223677 epub | ISBN 9781718223684 pdf
Subjects: LCSH: Muscle strength | Isometric exercise | Weight training
Classification: LCC QP321 .S918 2025 (print) | LCC QP321 (ebook) | DDC 612.7/4--dc23/eng/20250224
LC record available at https://lccn.loc.gov/2025000589
LC ebook record available at https://lccn.loc.gov/2025000590

ISBN: 978-1-7182-2366-0 (print)

Copyright © 2026 by Timothy J. Suchomel

Human Kinetics supports copyright. Copyright fuels scientific and artistic endeavor, encourages authors to create new works, and promotes free speech. Thank you for buying an authorized edition of this work and for complying with copyright laws by not reproducing, scanning, or distributing any part of it in any form without written permission from the publisher. You are supporting authors and allowing Human Kinetics to continue to publish works that increase the knowledge, enhance the performance, and improve the lives of people all over the world.

To report suspected copyright infringement of content published by Human Kinetics, contact us at **permissions@hkusa.com**. To request permission to legally reuse content published by Human Kinetics, please refer to the information at **https://US.HumanKinetics.com/pages/permissions-translations-faqs**.

The web addresses cited in this text were current as of November 2024, unless otherwise noted.

Senior Acquisitions Editor: Roger W. Earle; **Developmental Editor:** Amy Stahl; **Managing Editor:** Kim Kaufman; **Copyeditor:** Christina West; **Indexer:** Dan Connolly; **Permissions Manager:** Laurel Mitchell; **Senior Graphic Designer:** Joe Buck; **Cover Designer:** Keri Evans; **Cover Design Specialist:** Susan Rothermel Allen; **Photograph (cover):** Andy_Gin/iStock/Getty Images; **Photographs (interior):** © Timothy J. Suchomel, unless otherwise noted; **Photo Asset Manager:** Laura Fitch; **Photo Production Manager:** Jason Allen; **Senior Art Manager:** Kelly Hendren; **Illustrations:** © Human Kinetics; **Printer:** Versa Press

Human Kinetics books are available at special discounts for bulk purchase. Special editions or book excerpts can also be created to specification. For details, contact the Special Sales Manager at Human Kinetics.

Printed in the United States of America

10 9 8 7 6 5 4 3 2 1

The paper in this book is certified under a sustainable forestry program.

Human Kinetics
1607 N. Market Street
Champaign, IL 61820
USA

United States and International
Website: **US.HumanKinetics.com**
Email: info@hkusa.com
Phone: 1-800-747-4457

Canada
Website: **Canada.HumanKinetics.com**
Email: info@hkcanada.com

Human Kinetics' authorized representative for product safety in the EU is Mare Nostrum Group B.V., Mauritskade 21D, 1091 GC Amsterdam, The Netherlands.
Email: gpsr@mare-nostrum.co.uk

E9225

To Shana, for providing me with the strength to write a book about strength. I am beyond grateful that you could look past our first interaction at the isometric mid-thigh pull rack. I love you and cannot thank you enough for your support.

Contents

Acknowledgments   vii

Chapter 1   **Defining Strength**   1

    The History of Strength Competitions   1
    Definitions of Strength   2
    The Importance of Strength Within Sport   5
    Redefining Strength   8
    Strength Definition   9

Chapter 2   **Strength-Related Responses and Adaptations to Training**   13

    Neuromuscular System   13
    General Adaptation Syndrome   19
    Fitness–Fatigue Paradigm   19
    Timeline of Adaptation   20
    Adaptations to Training   20
    Influence of Training Stimuli on Strength Adaptations   24
    Impact of the Training Year   26
    Detraining   27
    External Stressors   28

Chapter 3   **Mechanisms of Strength**   29

    Genetics and Epigenetics (Nature and Nurture)   29
    Fiber Type   30
    Neuromuscular Factors   31
    Neuroendocrine System   34
    Muscle Architecture   36
    Motor Learning and Skill Acquisition   40

Chapter 4   **Principles and Organization of Training**   45

    Training Principles   45
    Organization of Training   51

| **Chapter 5** | **Training Methods** | **57** |

    Types of Training   57
    Additional Training Methods   69

| **Chapter 6** | **Advanced Training Methods** | **75** |

    Relative Strength Levels   75
    Potentiation Complexes   75
    Variable Resistance Training   80
    Accentuated Eccentric Loading   84

| **Chapter 7** | **Program Design** | **93** |

    Needs Analysis   93
    Programming Variables   94
    Training Age   109
    Training Year   110
    Overall Training Approach   111

| **Chapter 8** | **Concurrent Training** | **113** |

    The Interference Effect   113
    Conditioning Methods   116
    Concurrent Training Programming Considerations   120
    Practical Application   122

| **Chapter 9** | **Measuring Strength** | **125** |

    Rationale for Measuring Strength   125
    Strength Testing Considerations   126
    Types of Strength Testing   128
    Combined Assessment Methods   141
    Timing of Testing   144
    Integration of Sport Science and Evidence-Based Practice   145

Chapter 10 Monitoring and Adjusting Training Loads 147

Monitoring and Load Adjustment Methods 147
Motor Learning and Skill Acquisition Monitoring 158
Additional Considerations for Monitoring and Load Adjustment 159

Chapter 11 Nutritional Considerations for Muscular Strength 161

Energy Balance 161
Macronutrients 162
Micronutrients 167
Antioxidants 171
Hydration 171
Ergogenic Aids 172
Periodization of Nutrition 177
Monitoring Nutrition 178

Chapter 12 Recovery Considerations for Strength 181

Recovery Methods 182
Mental Fatigue and Recovery 189
The Placebo Effect 191
Periodization of Recovery 192
Measuring Recovery 192

References 195
Subject Index 279
Author Index 283
About the Author 309
Earn Continuing Education Credits/Units 310

Acknowledgments

I must first acknowledge my mentor, who truly is a pioneer in the field of sport science and strength and conditioning. A massive thank-you goes to Dr. Mike Stone for changing the way that I think about strength and for laying the foundation for so many of us in the field. You are one of the giants whose shoulders we stand on.

To Sophia Nimphius, for helping me think outside the box and opening my mind to new ideas in our field. You are brilliant, and I am incredibly grateful to be able to call you many things, including mentor, colleague, but most importantly, my friend.

To Katie Hirsch, for discussing the concepts covered within chapter 11 (Nutritional Considerations for Muscular Strength). You helped improve my level of thinking regarding nutrition and its practical applications, and for that I am forever grateful.

To Shona Halson, for providing your insight into the many methods of recovery and your outstanding contributions to our field. It means a lot to me that you were willing to take the time to discuss my ideas for chapter 12 (Recovery Considerations for Strength).

To my family, for their love and support. There are not enough words that I can type to say how grateful I am to have you in my life. Mom and Dad, you are my superheroes, and I cannot thank you enough for all the encouragement and inspiration you have given me. Thank you for being my sounding boards when I needed it. Mark and Christie, your knowledge and creativity leave me in awe and further my motivation. Thank you for being the best brother and sister anyone could ask for.

To my mentors, colleagues, and friends in the strength and conditioning field, including Meg Stone, Kimitake Sato, Paul Comfort, Travis Triplett, Jason Lake, John Wagle, and Bryan Mann. Thank you for the opportunities, conversations, advice, encouragement, and good times. We are all cogs in the sport science and strength and conditioning machine, but I believe that it is a lot easier and enjoyable moving things forward when working with outstanding individuals like yourselves.

To Roger Earle, for providing me with the opportunity to work on this project and for your continual guidance and availability throughout the entire publication process. The dedication that you show to your projects and the authors is invaluable.

To my past and present students, it has been an honor to work with you over the years. Thank you for the inspiration you have given me. Your passion cannot be matched, and I truly hope all of you go on to achieve everything that you seek in our field.

Last, but certainly not least, I must thank our two fur babies, Nubz and Keelo, for keeping me grounded during this process. Thank you for being my support animals during early morning and late evening writing sessions. The extra Cat Dad strength you provided helped me finish this book.

CHAPTER 1

Defining Strength

Muscular strength is a fitness characteristic that has often been used to distinguish superior and lesser athletes within competition. While the nature of competition varies, different types of strength can determine the winners and losers within specific events that occur within sport. Moreover, the winners of some sport competitions are determined purely on an athlete's strength characteristics (e.g., powerlifting and weightlifting). *Strength* is often defined by the maximum amount of weight an athlete can lift for a single repetition or multiple repetitions. Although this ability may be displayed in absolute or relative terms, this definition may be lacking additional context. The goal of this chapter is to provide readers with a more robust definition of strength and discuss its relationship to performance and its importance to the development of an athlete.

THE HISTORY OF STRENGTH COMPETITIONS

For thousands of years, the competitive nature of humans has included the question, "Who is the strongest?" The quest to be the strongest provides athletes with the notion that they are the superior being in that moment. The origins of lifting weights and feats of strength trace back to ancient China and Egypt (106). Chinese emperors as far back as 3600 and 3500 BC required their subjects to exercise daily. In addition, toward the end of the Chou dynasty (1122-249 BC), potential soldiers had to pass tests of strength before being allowed to serve. While these tests typically involved the use of heavy stones, dinglifting (lifting a bronze cauldron with three or four legs) became a primary test of strength in ancient China during the Warring States period (475-221 BC) (17, 76). This eventually evolved into competitions between individuals in which the winners of dinglifting contests were named the "Dinglifting Warrior" (109, 110). In Egypt, paintings in the Beni Hassan tomb dating back to 2040 BC show the sons of pharaohs running, wrestling, and lifting heavy objects, including sandbags, on a daily basis (78). Interestingly, the illustrations also depict the origins of the sport of weightlifting with movements relating to the one-hand snatch.

In Greece, the sixth century BC was known as the "Age of Strength" (106). Strength competitions were based on the ability of individuals to lift huge stones; although the first test of an individual's strength cannot be traced, an inscription on a 480kg block on the Greek island of Thera indicates that the stone had been lifted by Eumastas (78). Perhaps the most memorable feats of strength at the time were performed in Olympia, where athletes such as Bybon lifted a cylindrical stone weighing 143.5 kg over his head with one arm and Milo of Croton lifted and carried a 4-year-old heifer on his shoulders around the stadium (25). The Eumastas block and the Stone of Bybon may have been some of the first examples of "Manhood Stones." Berger and Weidt (9) wrote that young Scottish boys in the Middle Ages were compelled to place a minimum 100-kg stone onto another to be considered a man. In 1569, Italian physician Girolamo Mercuriale published *De Arte Gymnastica Libri Sex* (65), which presented the use of sandbags, stones, and other objects as tools to increase strength but also test an individual's willpower. Around the same time (16th and 17th centuries), there were contests of strength in Japan where individuals either lifted stones (*ishi*) or rice bales (*kome tawara*) (78). Other competitions in different countries began to follow suit; for example, Unspunnenfest in Interlaken, Switzerland, includes the Steinstossen (stone throwing) in which competitors lift the Unspunnen Stone (83 kg) to their chest or above their head before throwing it as far as possible. There are numerous other examples of lifting stones (e.g., Húsafell Stone, Nicol Stones, Odd Haugen Tombstone, and the Atlas Stones), but perhaps the most well-known lifting stones are the Stones of Dee or the two Dinnie Stones in Potarch, Aberdeenshire, Scotland (106). The Dinnie Stones have a combined weight of 332.5 kg, with the heavier stone weighing 188 kg and the lighter stone weighing 144.5 kg. The goal of the Dinnie Stone challenge is to lift the stones and carry them approximately 17.5 feet (5.22 meters) across the width of the Potarch Bridge. With additional movements and tasks, strength-based competitions have evolved from the lift-and-drop feats to include skills of unique strength demands across a variety of competitions.

In modern times, weightlifting, powerlifting, strongman, and highland games competitions serve as the primary settings for strength competitions. While entire books have been dedicated to each competition (2, 78, 107), the desire to be the strongest within the event pays homage to the origins of each sport and allows individuals to compete against historical figures and previous feats of strength. For example, British strongman Eddie Hall deadlifted 500 kg (1,102 lbs) at the 2016 Giants Live competition in Leeds, becoming the first human ever to do so while also breaking his own deadlift world record by 35 kg (465 kg was lifted earlier the same day). Four years later, Icelandic strongman Hafþór Björnsson deadlifted 501 kg (1,105 lbs) to best Hall and set a new world record. In the sport of weightlifting, Lasha Talakhadze of Georgia produced the highest total in history (492 kg: 225 kg snatch and 267 kg clean and jerk) at the 2021 International Weightlifting Federation World Championships. It should be noted that at the time of this writing, the greatest scaled performance (Sinclair total of 504) was produced by Turkish weightlifter Naim Süleymanoğlu at the 1988 Seoul Olympics, in which he totaled 342.5 kg (152.5 kg snatch and 190 kg clean and jerk) at a body weight of 59.7 kg (60 kg weight class), a feat of relative strength that has not been matched. Some individuals pursue feats of strength focused on lifting the heaviest weight within a given exercise, whereas others pride themselves on their ability to lift the heaviest weights in relation to their body mass or the ability to carry, pull, or throw a weight. Although the demands of each sport or event are unique, the desire to be the strongest individual remains.

DEFINITIONS OF STRENGTH

Strength has been defined in a variety of ways within the strength and conditioning field. While many of these definitions focus on the maximum weight an athlete can lift during a specific exercise, others focus on the performance of a specific task. The following paragraphs will provide readers with different definitions of strength, highlight the context in which they are used, and discuss how they may be useful in sport and the physical development of athletes. Although general information regarding the assessment of these qualities will be mentioned in each section, a thorough overview of testing muscular strength is provided in chapter 9.

ABSOLUTE STRENGTH

Humans are competitive and frequently compare themselves to others in a variety of ways, including appearance, personality, athletic ability, and so on. In fact, social media and fitness influencers have capitalized on the question, "How much do you bench?" While this question may not be viewed seriously by many within the strength and conditioning field, the basis of the question is still rooted in human competition. In this light, *absolute strength* refers to the maximum weight an athlete can lift for either a single or given number of repetitions. This measure of strength capacity is quite general in that it does not account for athlete body weight, size, or body composition. Although this type of comparison may be applicable when comparing athletes who play the same sport or position within a sport (e.g., American football linemen, rugby props), comparisons between athletes who differ significantly in body weight are less applicable and do not provide strength and conditioning professionals with additional context about an athlete's abilities. In general, heavier athletes have an absolute strength advantage over lighter athletes due to potentially greater magnitudes of muscle volume and, by extension, greater muscle cross-sectional area (34, 35, 70).

RELATIVE OR SCALED STRENGTH

Because heavier athletes may have an advantage in maximal strength, it is difficult to directly compare them to lighter athletes. To reduce the competitive advantage of heavier athletes, certain sports (e.g., weightlifting, judo, wrestling) have created weight classes in which athletes that are similar in size can compete with one another. Within these sports, those who are stronger hold a competitive advantage; in fact, the relative strength of an athlete may directly determine the final standing in competition in some sports (e.g., weightlifting and powerlifting). Because not all sports include weight classes, strength and conditioning professionals may require a method to directly compare the strength characteristics of athletes of different body sizes. This may be accomplished by calculating an athlete's *relative strength*, which refers to the maximum weight an athlete can lift for either a single repetition or multiple repetitions divided by their body mass (load lifted / body mass). It should be noted that there are several scaling methods, including *ratio scaling* (described previously), *allometric scaling* (load lifted or force produced / body mass$^{0.67}$), and *group-specific allometric scaling* (load lifted or force produced / body mass^a; a = fitted allometric scaling exponent derived from a log-log transformation specific to the population assessed) (13, 34, 35, 70, 95); however, it is important that individuals measuring performance ensure that the appropriate assumptions are met before using a given method (95).

CONCENTRIC STRENGTH

Concentric muscle actions are characterized by a shortening of the muscle fibers resulting from greater contractile forces produced over resistive forces. Therefore, *concentric strength* refers to the maximum amount of weight an individual can lift during the concentric phase of an exercise (e.g., upward phase of a squat, bench press). Although concentric-only 1 repetition maximum (1RM) values have been assessed previously (27, 32, 97, 98, 99, 103), the concentric phase of a traditional exercise that includes both eccentric and concentric muscle actions limits the athlete's strength potential in that lift because athletes are weaker during concentric actions compared to eccentric actions (72). Thus, strength and conditioning professionals likely use paired muscle actions when assessing an athlete's concentric strength characteristics (e.g., back squat, bench press). It should be noted that some sporting events such as cycling (7) and swimming consist almost exclusively of concentric muscle actions, which may provide support for performing concentric-only maximal strength tests with these athletes.

ECCENTRIC STRENGTH

While concentric actions shorten muscle length, *eccentric actions* are characterized by a lengthening of the muscle fibers due to greater resistive forces compared to contractile forces. Eccentric strength has been shown to be 20% to 60% (29) or approximately 40% (72) greater than concentric strength. Like concentric-only strength assessments, researchers have also used eccentric-only 1RMs to assess the eccentric strength characteristics of participants (29, 84). It should be noted that regarding different tasks performed in sport, eccentric muscle actions rarely occur in isolation. In fact, eccentric and concentric muscle actions are often performed in succession during actions such as jumping, sprinting, and change-of-direction tasks, a pairing termed the *stretch–shortening cycle*. This had led some researchers to assess eccentric strength qualities during stretch–shortening cycle actions (12); however, further research needs to be completed to determine whether this is a valid method.

ISOMETRIC STRENGTH

Isometric muscle actions are characterized by no demonstrable changes in muscle fiber length due to equal contractile and resistive forces. Although isometric muscle actions may occur in specific sporting situations (e.g., rugby scrum, competitive climbing, wrestling), these actions, however brief, also occur between the eccentric and concentric actions as part of the stretch–shortening cycle (termed the *amortization phase* within plyometric exercises). In general, an athlete's *isometric strength* will be greater than their concentric strength but less than their eccentric strength (18). Whereas concentric strength and eccentric strength are typically measured in pounds or kilograms, isometric strength is quantified in Newtons or kilograms of force and may be assessed during both single-joint (e.g., knee extension or flexion) (73, 77) and multi-joint (e.g., mid-thigh pull, squat, bench press) tests (4, 14, 68). Although isometric strength may not seem to relate to dynamic strength based on appearance (i.e., no dynamic movement), researchers have shown positive relationships between isometric and dynamic strength (58, 59).

STRENGTH–ENDURANCE

Strength–endurance refers to the ability of an athlete to perform a large number of exercise repetitions at a given weight or percentage of their maximum. Although they are related to an athlete's endurance performance, it is important to note that the terms *strength–endurance*, *muscular endurance*, and *cardiorespiratory endurance* are not synonymous. For example, strength–endurance may be defined as performing 8 to 12 repetitions using moderate to moderately heavy loads, with the goal of improving work capacity; muscular endurance involves performing a large number of repetitions (≥12) using a submaximal load that may not contribute to maximal strength adaptations; cardiorespiratory endurance is an aerobic characteristic that refers to the ability to use oxygen for energy production during longer-duration, submaximal activity (e.g., running, cycling). Although it is not ideal due to a wide spectrum of body sizes and relative strength, the National Football League (NFL) Combine requires athletes to perform bench press repetitions at 225 lb (102 kg) until failure (i.e., referred to as the *NFL-225 test*) (56). As of 2023, the official NFL Combine record is 49 repetitions, which ultimately makes this a test of muscle endurance rather than maximal strength for many athletes. This notion is supported by the fact that the NFL-225 bench press test may not serve as a good test for tracking maximal strength in collegiate football players (55).

STRENGTH–SPEED

Strength–speed is a characterization of specific training methods or exercises that focus on the intent to move heavy loads quickly. Examples of exercises that may be classified as strength–speed exercises are weightlifting movements and their derivatives (e.g., clean, clean grip hang pull, snatch pull from the floor) (89) as well as heavily loaded jumps (e.g., hexagonal barbell jumps) (91, 92). While the prescription of strength–speed exercises may vary based on the individual needs of each athlete, these exercises may be featured in both general and absolute strength phases but also during strength–power and taper phases to enhance early rapid force production characteristics (93).

SPEED–STRENGTH

In contrast to strength–speed, *speed–strength* training methods and exercises are characterized by the intent to move light loads quickly. Exercises that fall under the speed–strength category are typically ballistic in nature and are implemented using relatively light loads, with the goal of improving rapid force production and power output characteristics (e.g., plyometric exercises, unloaded and lightly loaded jumps, bench press medicine ball throws) (91-93). During strength–power phases of training, strength and conditioning professionals may also pair strength exercises with speed–strength exercises to form a potentiation complex (see chapter 6 for more detail) and potentially augment the training stimulus an athlete receives (8, 80, 90, 100). However, like strength–speed exercises, speed–strength exercises should be prescribed based on the needs of the athlete during phases that complement their overall training goals.

REACTIVE STRENGTH

The *reactive strength* characteristics of an athlete refer to their ability to quickly transition between eccentric and concentric muscle actions (111). Researchers have indicated that reactive strength is strongly associated with maximal strength, especially eccentric strength (5). The reactive strength characteristics of an individual are often assessed during plyometric exercises such as a drop jump and are quantified as a reactive strength index (jump height / ground contact time) (63). While researchers have examined the modified reactive strength index during a countermovement jump (jump height / time to takeoff) (19, 60, 88), it should be noted that each variant may reflect different athlete qualities (50). This is likely due to the differences in braking and propulsion phase duration between drop jumps and countermovement jumps, because the latter is less reactive in nature. It should be noted that although upper-body drop "jumps" can be performed (96), further research is needed to determine whether these actions are truly reactive based on their phase durations.

ISOKINETIC STRENGTH

Isokinetic muscle actions are those in which the velocity of the movement is constant. The isokinetic strength qualities of an athlete are typically assessed using an isokinetic dynamometer and are quantified as a torque (Newtons × meters). Beyond some correlation research (57, 77, 101), isokinetic strength qualities are typically not assessed in healthy athletes. For example, athletes returning from an anterior cruciate ligament tear may complete isokinetic testing to determine the torque produced from both the knee flexors and extensors (102). The ability to test injured athletes in this manner may provide rehabilitation professionals and strength and conditioning professionals with information regarding the strength characteristics of an athlete performed under controlled conditions (i.e., velocity) as part of a return-to-play protocol. However, isokinetic testing may not be frequently completed in real-world scenarios with healthy athletes, owing to the need for specialized equipment, testing space, tester competency, and assessment of joints in isolation. Regarding isolated joint assessment, researchers have shown that changes in dynamic (1RM) and isokinetic strength are not equivalent (24); thus, the applicability of isokinetic strength and performance may be limited in some capacity.

THE IMPORTANCE OF STRENGTH WITHIN SPORT

Muscular strength can have a significant impact on the performance capacity of an athlete. Moreover, stronger athletes may be considered more resilient and may be less likely to get injured. For these reasons, developing the strength characteristics of an athlete starting at an early age may be of utmost importance. The following paragraphs will provide an overview of how improvements in an athlete's strength may affect their general and specific performance characteristics, the potential for injury, long-term athlete development, and the differences in the literature between stronger and weaker individuals.

> **Key Point**
>
> Developing the strength characteristics of an athlete starting at an early age may be of utmost importance.

PERFORMANCE

The authors of a 2016 review provided a thorough overview of the relationships between maximal strength (absolute and relative) and various general (e.g., jump, sprint, and change of direction) and specific sport performances (figure 1.1) (94). Since the previous review, researchers have continued to show meaningful relationships between various measures of strength and sport performance measures such as punch impact in boxers (49), sprint kayak finishing times (51), snatch and clean improvements in weightlifters (113), and the start and overall performance in sprint swimmers (40). Although correlation does not necessarily equal causation, strength and conditioning professionals cannot ignore the collective trend of data spanning nearly 50 years that supports the importance of strength and its influence on performance. Furthermore, it should be acknowledged that while increasing an athlete's maximal strength in certain lifts does not guarantee improvements in athletic success, strength may serve as the foundation upon which other characteristics are built (94).

> **Key Point**
>
> While increasing an athlete's maximal strength in certain lifts does not guarantee improvements in athletic success, strength may serve as the foundation upon which other characteristics are built.

INJURY RISK

A frequently mentioned reason to resistance train is to build resilience and decrease the risk of injury. While injuries can never be fully prevented, researchers have indicated that strength training reduced acute sports injuries to less than a third

Figure 1.1

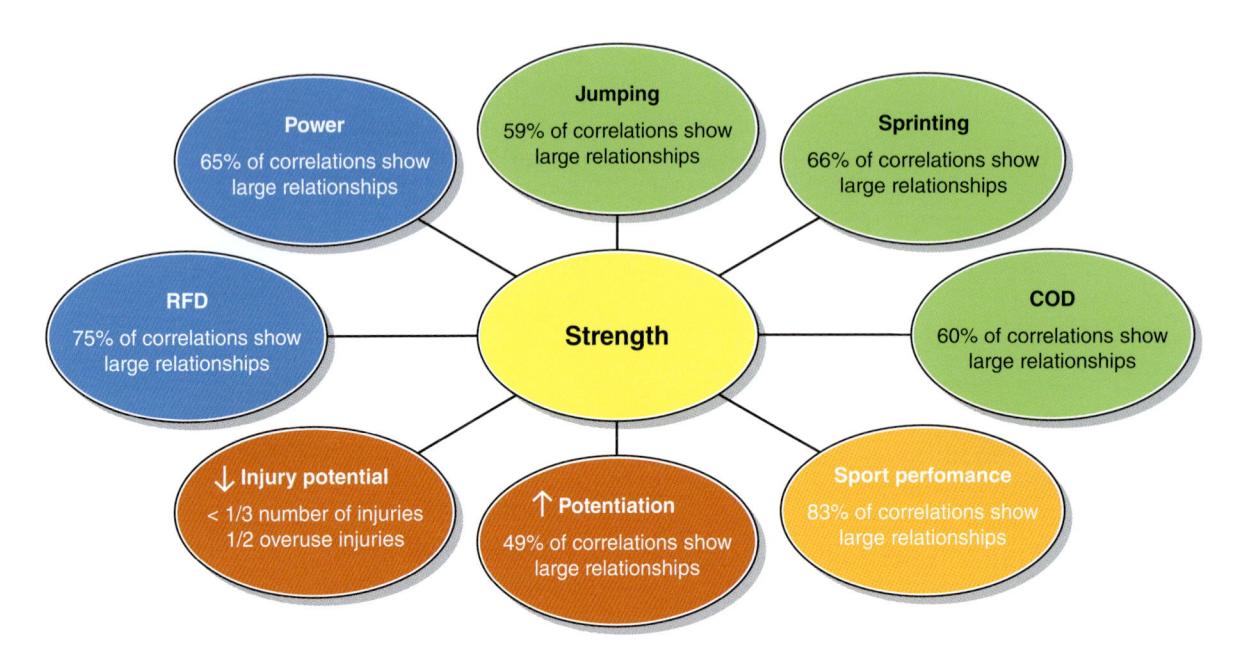

- **Strength** (center)
  - **Power**: 65% of correlations show large relationships
  - **Jumping**: 59% of correlations show large relationships
  - **Sprinting**: 66% of correlations show large relationships
  - **RFD**: 75% of correlations show large relationships
  - **COD**: 60% of correlations show large relationships
  - **↓ Injury potential**: < 1/3 number of injuries, 1/2 overuse injuries
  - **↑ Potentiation**: 49% of correlations show large relationships
  - **Sport performance**: 83% of correlations show large relationships

FIGURE 1.1 Influence of strength.
Based on Suchomel et al. (94).

and overuse injuries by approximately 50% (45). Furthermore, it was concluded that a progressive increase in resistance training volume of just 10% may decrease injury occurrence by 4.3% (44). This is likely due to chronic resistance training serving as a stimulus to strengthen not only the muscles but also cartilage, tendons, and bones (52), thus making the athlete more robust by increasing their ability to tolerate external loads. Other researchers have shown that lower levels of relative strength served as an indicator for higher injury risk (54), and they have also discussed specific relative strength standards related to injury occurrences (11). However, while improving the relative strength of an athlete may reduce their injury risk to an extent, the practice of chasing specific relative strength ratios may be discouraged. Specific to return-to-play protocols, it should be noted that large relationships between pre- and postoperative quadriceps strength have been observed (46) and that preoperative rehabilitation may improve postoperative knee-related function and muscle strength (1). Collectively, it appears that improving an athlete's muscular strength may not only reduce the potential risk of injury, but it may also, to an extent, reduce the return-to-play timeline. Finally, strength and conditioning professionals are encouraged to promote the use of full range-of-motion exercises because this has been shown to be an effective strategy for reducing injury risk (74).

Key Point

Improving an athlete's muscular strength may not only reduce the potential risk of injury, but it may also, to an extent, reduce the return-to-play timeline.

LONG-TERM ATHLETE DEVELOPMENT

A topic of frequent discussion and debate is whether it is appropriate for youth and adolescent athletes to resistance train. For example, it has been stated without evidence that "you have to be a teenager before lifting weights," "lifting will stunt a kid's growth," and "lifting weights is dangerous for kids." An abundance of research over the last decade has sought to dispel these myths. For example, the belief that individuals must wait until they are a specific age to start resistance training is not supported by evidence. In fact, Behringer and colleagues (6) concluded that resistance training may be an effective method for enhancing motor skills such as jumping, running, and throwing during childhood and youth years. Moreover, motor skill development and muscular strength has been noted as 1 of the 10 pillars of long-term athlete development (47), and children between ages 10 and 14 years, especially females, may benefit from such training (105). In fact, broad resistance training recommendations have been provided to aid strength and conditioning professionals in developing youth athletes (20, 47, 48).

The belief that lifting weights at a young age will damage growth plates and stunt growth is also not supported by the scientific literature. In fact, the opposite is true; it has been suggested that the mechanical stress from resistance training or other high-strain sports (e.g., weightlifting or gymnastics) may actually serve as an effective stimulus for bone formation and growth (10, 15, 21, 22, 53, 104). Regarding the belief that lifting weights is dangerous for children and adolescents, the findings of previous research demonstrate that there is a relatively low risk of injury during resistance training if individuals follow age-appropriate guidelines and receive qualified supervision and instruction (21, 28, 53, 69). Moreover, most of the sustained injuries have been deemed accidental (69).

Pichardo and colleagues (75) indicated that a structured training program that combined resistance training and plyometric exercises may reduce injury risk factors associated with jump landings in adolescent boys. This is likely because supervised resistance training in the weight room provides a controlled environment compared to sport, in which the environment is less controlled and more chaotic. Additional researchers have shown that the strength characteristics of young girls also relate to their jumping and sprinting performance as well as their movement skill during a back squat (82). It should be noted that the relative squat strength of elite youth and adolescent soccer players is strongly related to their squat jump, countermovement, and 30-meter sprint performances as well (37). Keiner and colleagues (38) showed that long-term (2 years) structured resistance training can produce relatively strong youth and adolescent athletes for their ages. Beyond the metrics provided earlier, the same research group also showed that change-of-direction performance can also improve simultaneously with an increase in back squat and front squat relative strength over a long-term period (2 years) in elite youth soccer players (39).

Key Point
The belief that lifting weights at a young age will damage growth plates and stunt growth is not supported by the scientific literature.

COMPARISONS BETWEEN STRONGER AND WEAKER INDIVIDUALS

Although increases in maximal strength may not directly transfer to an enhanced performance (94), the relationship between greater levels of strength and athletic performance cannot be ignored. Stronger individuals have been shown to jump higher and farther (3, 42, 81), sprint faster (16, 30, 64, 82, 108), and perform change-of-direction tasks better (71, 112) compared to weaker individuals. Furthermore, stronger athletes performed better than their weaker counterparts during sport-specific tasks, such as standing and three-step running throw velocity in handball (26), 100-meter sprinting (64), sprint time to first and second base in softball (71), tackling ability in rugby (83), and 25-meter track cycling performance (87).

Key Point
Stronger individuals have been shown to jump higher and farther, sprint faster, and perform change-of-direction tasks better compared to weaker individuals.

Beyond performance in sporting events, stronger individuals may benefit more from different types of training compared to those who are weaker. For example, researchers have shown that stronger individuals may acutely enhance their performance during potentiation complexes to a greater extent than those who are weaker (79, 98). Furthermore, Miyamoto and colleagues (66) indicated that individuals may increase their magnitude of potentiation after gaining additional strength. James and colleagues (33) also showed unique training adaptations of stronger individuals in responses to weightlifting, plyometric, and ballistic training. Collectively, these findings suggest that relatively stronger individuals may require novel training methods to further enhance their performance. An in-depth discussion of several advanced training tactics is provided in chapter 6.

REDEFINING STRENGTH

Despite the many definitions of strength provided earlier in this chapter, it is important that strength and conditioning professionals do not limit their definition of strength to how much weight an athlete can lift. Because force production is the foundation of movement, an understanding of the laws of motion and the impulse–momentum theorem—and how they relate to strength—is necessary to provide a more thorough definition.

Key Point

Strength and conditioning professionals should not limit their definition of strength to how much weight an athlete can lift.

NEWTON'S LAWS OF MOTION

While an athlete's absolute and relative strength characteristics are important to their performance capacity (94), the ability to lift heavier weights does not always translate to an improved performance. Thus, it may be viewed as shortsighted to limit the definition of strength to how much an athlete can lift in either an absolute or relative capacity. Therefore, to provide a more robust definition of strength, it is important for strength and conditioning professionals to understand how motion is created (i.e., how athletes move and how weights are lifted). Isaac Newton's *laws of motion* and the *impulse–momentum theorem* provide additional context and are described in table 1.1.

IMPULSE

An *impulse* is characterized by the amount of force applied over a given duration. However, when it comes to producing force in a task-specific context, *net impulse* (i.e., the impulse produced above an athlete's body mass or the combined weight of the athlete and load they are training with or moving) is the most important thing to consider. For example, propulsive net impulse directly determines an athlete's vertical jump height (41) but may also influence performance in competitive weightlifting (23).

Because net impulse influences athlete movement, strength conditioning professionals should understand how force production characteristics affect the shape of the impulse produced. Simply, two athletes can produce the same propulsive net impulse; however, the impulse may be produced in significantly different ways (figure 1.2). While the shape of the impulse created by an athlete may be determined by their ability to rapidly produce force, it may also be determined by the strategy used during the task itself. For example, McMahon and colleagues (62) showed that the shape of the braking or propulsion phases of a countermovement

TABLE 1.1 Newton's Laws of Motion and Impulse–Momentum Theorem as They Relate to Muscular Strength

Law or theorem	Description	Deadlifting example
Law of inertia	An object at rest stays at rest and a body in motion stays in motion until acted upon by an external force.	A barbell will remain in a motionless state until an athlete produces enough force to lift the barbell.
Law of acceleration	The acceleration of an object is directly proportional to the net forces that act upon it.	If an athlete produces a larger magnitude of force, the acceleration of the barbell will be higher.
Law of action–reaction	For every action, there is an equal, opposite, and simultaneous reaction.	The forces produced into the ground are received by the athlete in the opposite direction.
Impulse–momentum theorem	The impulse experienced by an object is equal to its change in momentum.	The impulse produced by an athlete on the barbell will produce an equal change barbell momentum.

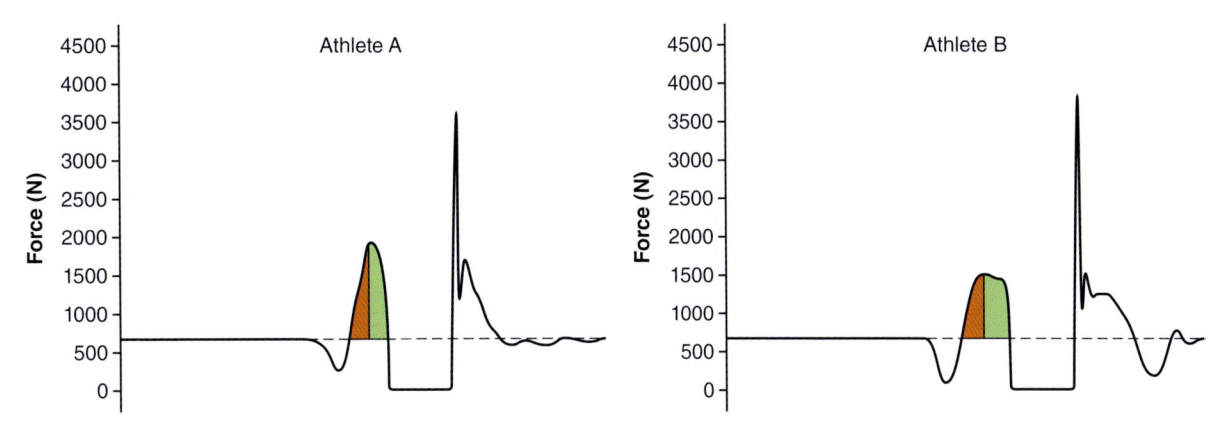

FIGURE 1.2 Comparison of impulse shape produced above body mass (i.e., net impulse) between two athletes during countermovement jumps. Athletes A and B both achieved the same jump height; however, the braking (red) and propulsion (green) impulse shapes were more peaked and shorter in duration during athlete A's jump compared to athlete B.

jump may either be taller and skinnier—produced either by a rapid unweighting phase or enhanced force production characteristics—or shorter and wider, indicating a less rapid unweighting phase or an inability to decelerate their downward momentum effectively. Restated, stronger athletes may have the capacity to produce taller and skinnier impulses based on their magnitude and rapid force production characteristics. This notion is supported by findings that showed greater impulses produced during countermovement jump braking and propulsion phases by senior-level rugby athletes compared to academy-level athletes (61). Further research has shown that differences in sprinting acceleration follow suit, because athletes who produce greater propulsive impulses accelerate better (31, 36, 43, 67). In fact, Morin and colleagues (67) displayed net impulses that were taller and skinnier produced by world-class sprinters compared to high-level sprinters.

STRENGTH DEFINITION

The common factor in Newton's laws of motion and the impulse–momentum theorem is the production of force. Because this athlete characteristic may be improved with progressive training prescription, the *production of force* may be defined as an ability (85, 86). Moreover, force may have to be produced against gravity, another opponent, or an implement used within the sport (94) and, thus, the necessary force production characteristics for each performance task are unique (figures 1.3 and 1.4). Therefore, for the purposes of this book, *strength* is defined as the ability to produce force against an external resistance within a task-specific context. It should be noted that within this definition, there is a magnitude of force, a rate at which the force is produced, and a duration over which force is applied, all of which relate to the unique demands of each performance task.

Key Point

Strength is defined as the ability to produce force against an external resistance within a task-specific context.

Given the unique demands of each athlete, strength and conditioning professionals are often told by sport coaches that their athletes need to develop "sport-specific strength." However, based on the definition of strength just provided, it could be argued that there is no such thing as sport-specific strength. Perhaps the more appropriate term for sport-specific strength is *task-specific strength*, because there is a specific amount of force that must be produced over a given duration to complete each task. Moreover, to perform the task effectively and gain a competitive advantage, it would be advantageous to produce more force over the same or a

10 Science and Development of Muscular Strength

	Maximum velocity sprint ground contact*	45-degree change of direction*	Bilateral countermovement jump	Hang power clean with 80% 1RM	Back squat with 70% 1RM
Phase duration	44 ms / 39 ms	113 ms / 24 ms	189 ms / 286 ms	176 ms / 262 ms	743 ms / 849 ms
Net impulse	42 Ns / 28 Ns	122 Ns / 26 Ns	130 Ns / 284 Ns	71 Ns / 240 Ns	125 Ns / 187 Ns

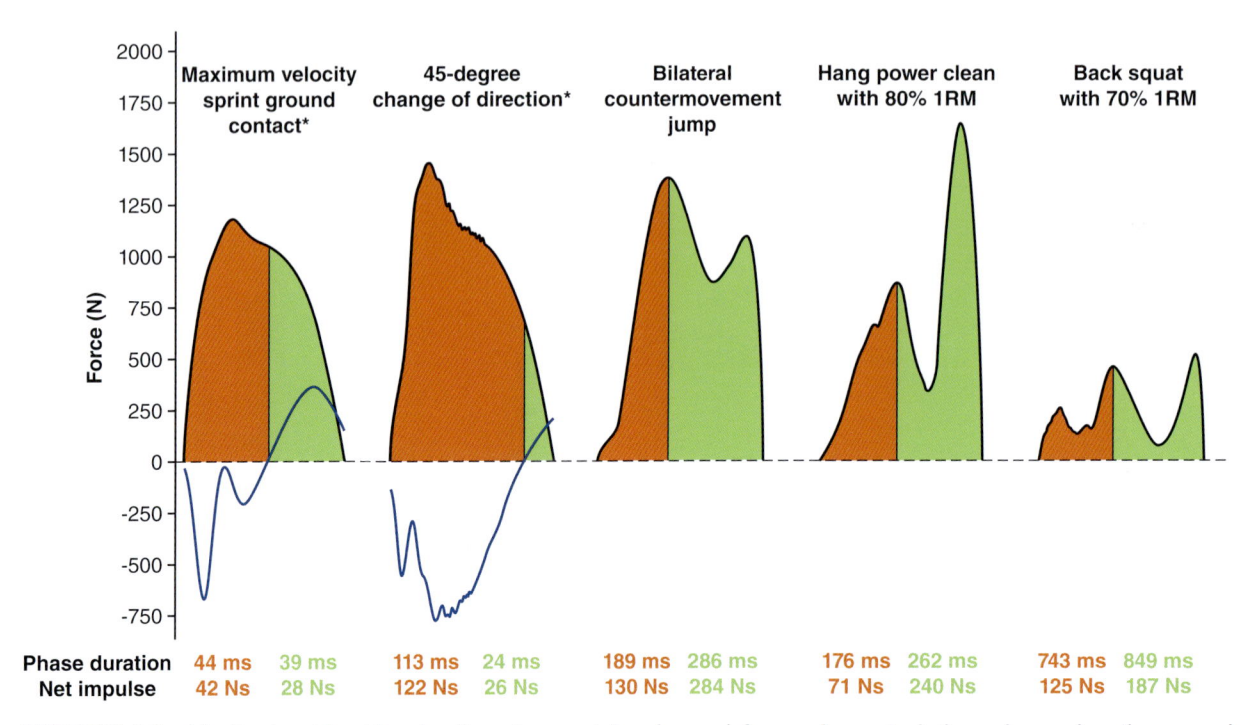

FIGURE 1.3 Vertical net braking (red) and propulsion (green) force characteristics, phase durations, and net impulses of a maximum velocity sprint ground contact, 45° change of direction, bilateral countermovement jump, hang power clean with 80% 1RM, and back squat with 70% of the 1RM. The asterisk indicates a unilateral task, the dashed black line indicates system weight, and the blue lines indicate horizontal force production.

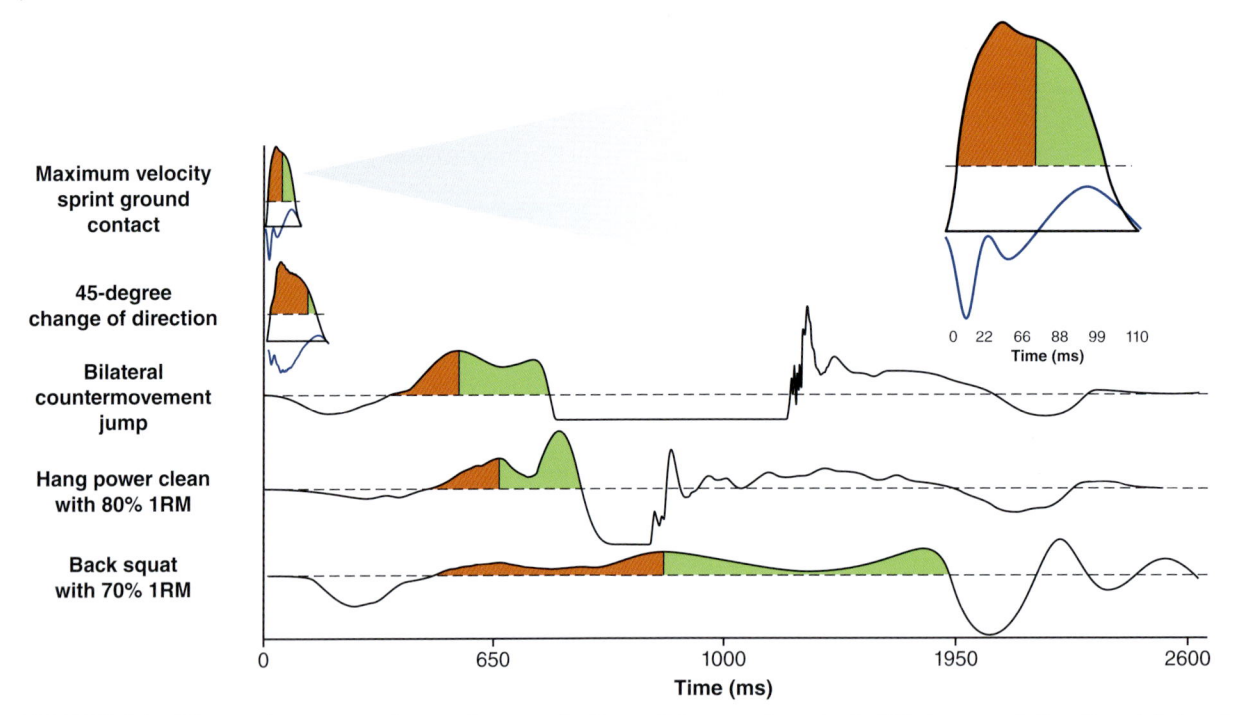

FIGURE 1.4 The performance of a maximum velocity sprint ground contact, 45° change of direction, bilateral countermovement jump, hang power clean with 80% 1RM, and back squat with 70% 1RM with respect to time. *Note:* The countermovement jump, hang power clean, and back squat repetitions lasted approximately the same duration from the initiation of the movement (unweighting) to stabilization after the repetition. Solid black lines indicate vertical force, dashed black lines indicate system weight, blue lines indicate horizontal force, red indicates the braking phase, and green indicates the propulsion phase.

shorter duration than the opponent. Thus, although individualization regarding training prescription is crucial, there are many athletes across a spectrum of sports who can be trained using similar training strategies that enhance their force production characteristics.

> **Key Point**
>
> *Task-specific strength* may be a more appropriate term than *sport-specific strength*, because there is a specific amount of force that must be produced over a given duration to complete each task.

Take-Home Points

- Beyond lifting competitions, strength has been shown to be an important aspect of athlete performance because it appears to enhance the ability to perform various performance tasks, such as jumping, sprinting, change-of-direction, and sport-specific movements.
- Enhancing an athlete's strength helps create a more resilient and robust athlete, possibly reducing the potential for injury.
- Because of its impact on long-term athlete development and athletic performance, muscular strength serves as the foundation upon which other abilities may be built.
- Newton's laws of motion and the impulse–momentum theorem underpin the ability of an athlete to produce force and perform various athletic movements.
- *Strength* is the ability to produce force against an external resistance within a task-specific context and consists of a magnitude of force, rate at which force is produced, and duration over which force is applied.

CHAPTER 2

Strength-Related Responses and Adaptations to Training

It is important that strength and conditioning professionals have a foundational knowledge of how an athlete's body may respond and adapt to specific training stimuli. To fully appreciate how various stimuli affect the strength characteristics of an athlete, a description of the neuromuscular system and the physiological processes that result in force production within the muscle is needed. In addition, information regarding how an adaptation takes place, the timeline of adaptations, and the factors that may influence an adaptation should also be considered. Thus, the aim of this chapter is to provide readers with an overview of the structure and function of the neuromuscular system, the responses and adaptations that may occur as the result of training stimuli aimed at improving muscular strength characteristics, and the factors that may impact targeted training adaptations.

NEUROMUSCULAR SYSTEM

Before discussing the responses and adaptations to training stimuli, a basic understanding of the *neuromuscular system* is needed to understand the physiological changes that underpin muscular strength adaptations. The neuromuscular system consists of the nervous system and the musculoskeletal system, in which nerve cells (i.e., neurons) innervate muscle cells (i.e., myocytes or muscle fibers) and stimulate them to generate force. Because muscle fibers are coupled with the skeletal system, the force generated within each muscle may be used to create movement. Muscle can be divided into smooth and striated (striped) muscle, with striated muscle being further subdivided into skeletal muscle and heart (cardiac) muscle. Although each muscle type possesses the same basic properties (table 2.1), the following discussion will focus on skeletal muscle, specifically its structure, function, and characteristics that relate to muscular strength development. Because a thorough discussion of neuromuscular physiology is beyond the scope of this book, those interested in learning more are encouraged to seek out several recommended resources (26, 118, 196).

STRUCTURE AND FUNCTION

Skeletal muscles exist in a variety of shapes and sizes, with the architecture of a given muscle being relatively complex, which may be a primary factor in defining its function (119). In general, skeletal muscles consist of several layers of connective tissue that differ in size and orientation (figure 2.1). The thick outermost layer, termed the *epimysium*, covers the entire surface of the muscle and is made up of collagen fibers that are woven into tight packets that connect with additional sublayers (specifically, the *perimysium*) (196). Deep to the *epimysium* are bundles of muscle fibers called *fasciculi*, each of which is surrounded by another connective tissue called the perimysium. Each muscle fascicle typically contains 100 to 150 muscle fibers; however, muscles that produce small or very fine movements possess a smaller number of fibers (59), whereas those that produce large magnitudes of force may have more muscle fibers within a fascicle (112, 119).

TABLE 2.1 Muscle Characteristics

Muscle property	Description
Adaptability	Ability to adapt and alter physiology or size based on the volume, intensity, and time frame of a training stimulus
Conductivity	Ability to conduct an action potential
Contractility	Ability to increase tension and shorten
Distensibility	Ability to be stretched by an external force
Elasticity	Ability to resist elongation; ability to return to original resting position after active or passive elongation
Irritability	Ability to react when stimulated
Relaxation	Ability to return to a resting state following a muscle action

Properties based on McComas (119) and MacIntosh et al. (112).

Each individual muscle fiber is covered by another collagen layer called the *endomysium*. Like the previously discussed muscle layers, the endomysium fibers connect with the perimysium but also to the basement membrane, thus providing additional stability to the overall structure of the muscle (112, 119). Each muscle fiber consists of rod-like organelles called myofibrils, which are made up of various protein filaments (e.g., actin, myosin, titin, nebulin) that form the functional units of the muscle cell termed *sarcomeres* (figure 2.2). The striated appearance of skeletal muscle is due to the parallel (stacked) orientation of sarcomeres, which may allow for greater magnitudes of force to be produced (222, 223). Further discussion on the events leading to force generation is provided later in this chapter, but readers should note that the interaction (i.e., crossbridge) between actin and myosin proteins within each sarcomere results in force being produced. While often viewed in a two-dimensional perspective, it is important to note that six actin filaments surround each myosin, whereas each actin filament is surrounded by three myosin filaments (figure 2.3), thus providing ample opportunities for interaction and subsequent force production. Beyond actin and myosin, the large filamentous proteins titin (77, 214) and nebulin (97) may also contribute to the rigidity of the sarcomere

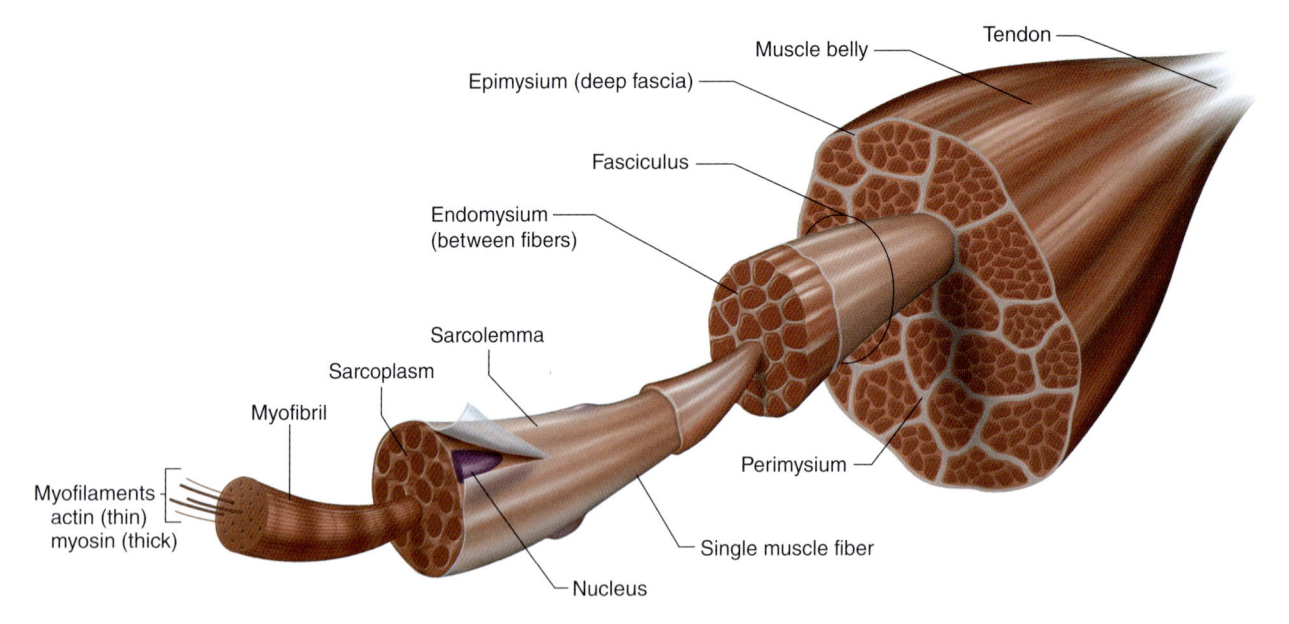

FIGURE 2.1 Schematic drawing of a muscle illustrating three types of connective tissue: epimysium (the outer layer), perimysium (surrounding each fasciculus, or group of fibers), and endomysium (surrounding individual fibers).

Reprinted by permission from N.T. Triplett, "Structure and Function of Body Systems," in *Essentials of Strength Training and Conditioning*, 4th ed., edited by G.G. Haff and N.T. Triplett (Champaign, IL: Human Kinetics, 2016), 4.

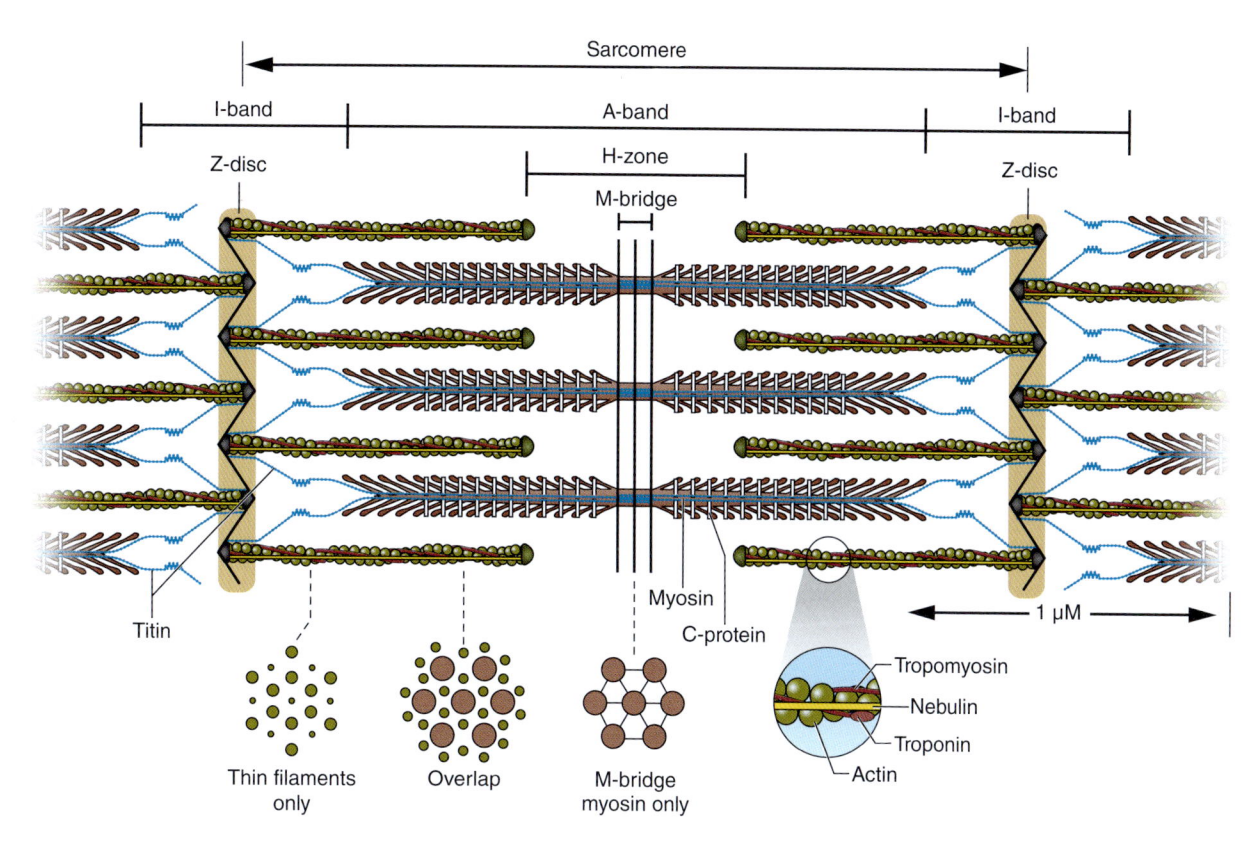

FIGURE 2.2 The structure of a sarcomere.

structure to allow for greater force transmission. In summary, although force is generated at the deepest level of the muscle, the organization of a muscle allows force to be effectively transmitted across its structural layers, through the tendon, and eventually to the bone to allow movement to occur.

Muscle fiber orientation is another structural consideration that may modify both the magnitude and rate of force production. Muscle fibers are arranged in either a fusiform or pennate format in which the fibers run parallel or obliquely (angled) to the longitudinal axis of the muscle, respectively. Whereas fusiform muscle fibers may run parallel, converge, or form a circular arrangement, pennate muscles can display unipennate, bipennate, or multipennate orientations of muscle fibers (figure 2.4). Based on the arrangement of the muscle fibers, the sarcomeres may be organized in a manner that may allow for either greater force production or fiber shortening velocity (222, 223). For example, greater force can be produced within pennate muscles because more fibers exist within an area of the muscle (i.e., fiber packing) (56, 158). Moreover, the pennation angle of a muscle can change following training (1), indicating that fiber arrangement can have a significant effect on strength adaptations. Further details on this concept are discussed in chapter 3.

MOTOR UNIT

A *motor unit* consists of an α motor neuron and all the muscle fibers that it innervates (181). Individual motor neurons consist of a cell body (soma), axon, and dendrites (figure 2.5); briefly, the dendrites of a neuron receive either stimulatory or inhibitory signals from the nervous system, all of which are summed at the *axon hillock* (i.e., conical projection of the neuron cell body that connects to the axon). If the sum of the signals exceeds a limiting threshold, an action potential is triggered and sent down the axon. Thus, when an action potential is transmitted by the α motor neuron, all the muscle fibers are stimulated to produce force—this is termed the *all-or-none principle*.

Another aspect of motor units that should be considered is the *innervation ratio*, which refers to the number of muscle fibers innervated by a single motor neuron. Innervation ratios may vary considerably between intraocular muscles of the eye and small muscles, ranging from 1:3 to 1:150,

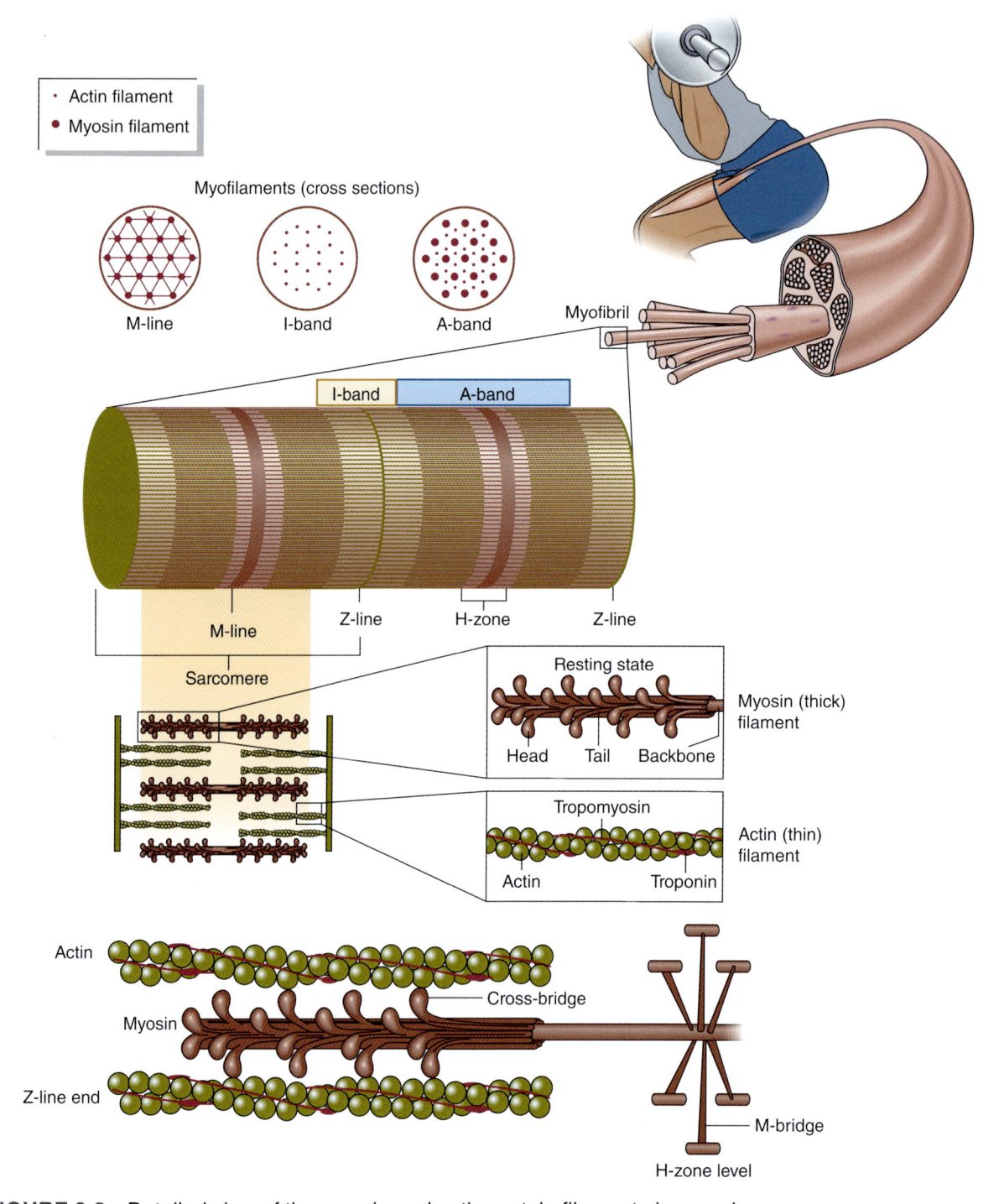

16 Science and Development of Muscular Strength

FIGURE 2.3 Detailed view of the myosin and actin protein filaments in muscle.

Reprinted by permission from N.T. Triplett, "Structure and Function of Body Systems," in *Essentials of Strength Training and Conditioning,* 4th ed., edited by G.G. Haff and N.T. Triplett (Champaign, IL: Human Kinetics, 2016), 6.

whereas larger muscles may have ratios as large as 1:2,000 (59, 119). Furthermore, these ratios typically determine the type of movements that the motor units are associated with; for example, small and large ratios are associated with fine motor control or movements that require large magnitudes of force with less consideration of fine movement, respectively. Within this spectrum, Type II motor units (particularly Type IIx) typically have very large innervation ratios, thus allowing them to produce larger and more rapid magnitudes of force; however, there may be considerable variation in motor unit number per muscle and size within the same muscle (94, 112, 119).

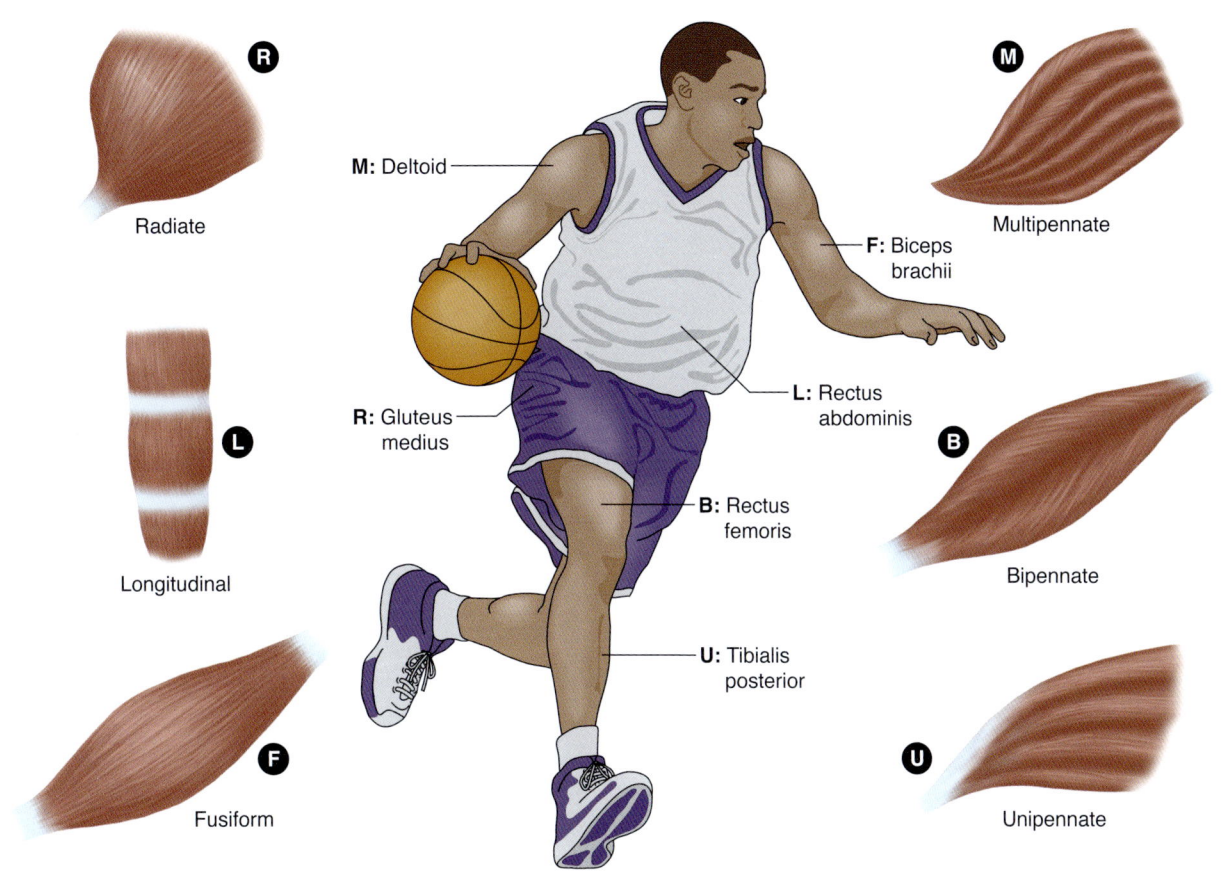

FIGURE 2.4 Muscle fiber orientation.

Reprinted by permission from NSCA, *Essentials of Strength Training and Conditioning,* 3rd ed., edited by T.R. Baechle and R.W. Earle (Human Kinetics, 2008), 76.

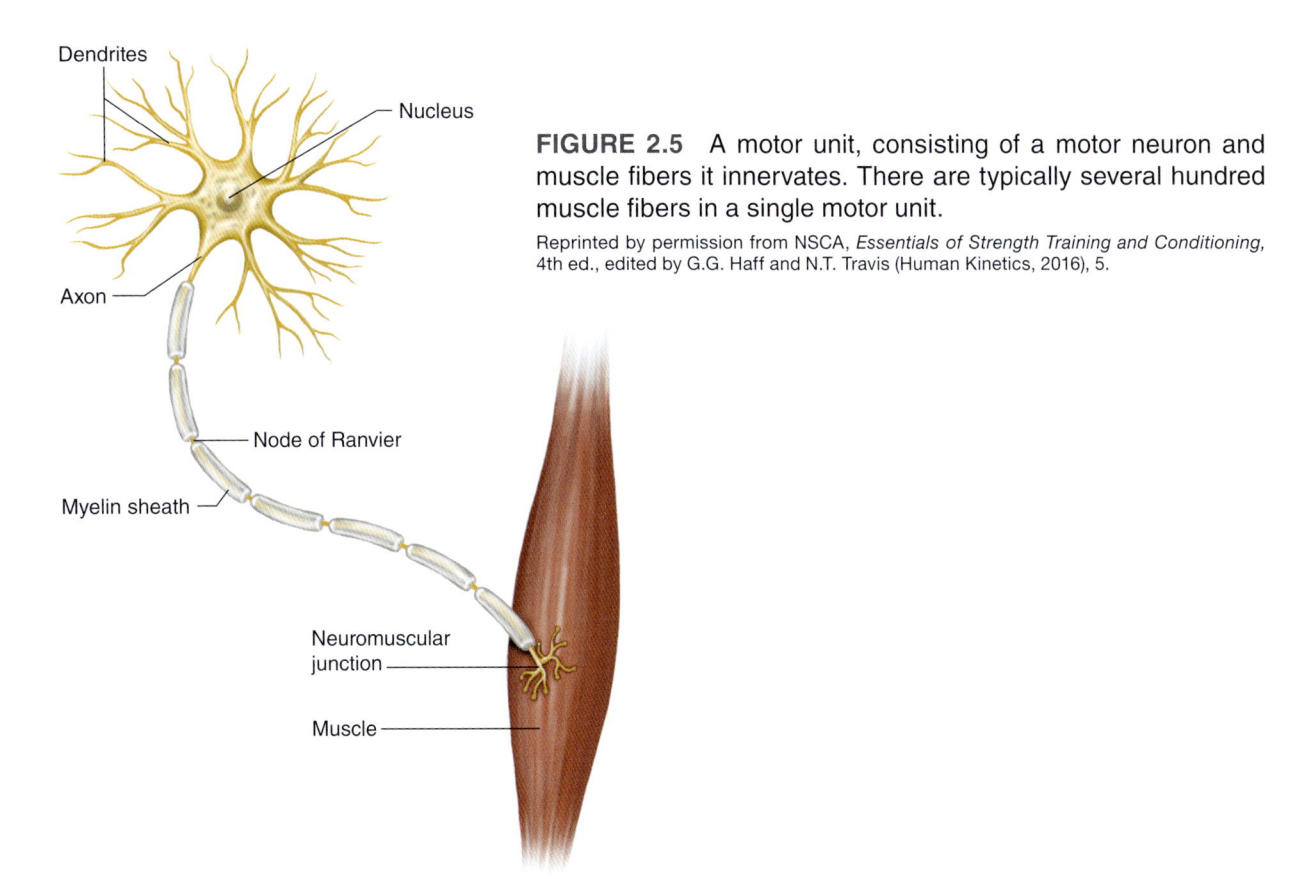

FIGURE 2.5 A motor unit, consisting of a motor neuron and muscle fibers it innervates. There are typically several hundred muscle fibers in a single motor unit.

Reprinted by permission from NSCA, *Essentials of Strength Training and Conditioning,* 4th ed., edited by G.G. Haff and N.T. Travis (Human Kinetics, 2016), 5.

SLIDING FILAMENT THEORY

The shortening and the lengthening of a muscle are caused by the sliding of actin filaments past the myosin filaments; this has been widely discussed as the *sliding filament theory* (85, 86, 87). Prior to the movement of the actin filament, a series of physiological events must take place for a voluntary muscle action to occur. The initiation of voluntary movement commences in the central nervous system when a stimulatory signal is sent down a chain of neurons starting within the motor cortex. Upon reaching the dendrites of the α motor neurons that innervate the muscle fibers needed to perform the movement, stimulatory and inhibitory signals are summed at the axon hillock. Assuming the summed signals surpass the limiting threshold, an action potential is sent down the axon to the motor end plate. Here, the action potential is transferred to the sarcolemma of a muscle fiber through the release of the neurotransmitter acetylcholine into the neuromuscular junction positioned between the motor end plate and sarcolemma. The acetylcholine molecules bind to receptors on the sarcolemma, which then trigger a depolarization event due to the influx of sodium ions (Na+). The depolarization signal moves along the sarcolemma until it travels down a transverse tubule (T-tubule) that leads to the interior of the muscle fiber (55, 112). Upon reaching the terminal cisternae that surround the T-tubules, the action potential activates a voltage-dependent dihydropyridine channel that, in turn, activates ryanodine receptors whose purpose is to mediate the release of calcium ions (Ca+2) that are stored in the sarcoplasmic reticulum (46). The released Ca+2 then binds to the troponin complex, which initiates a transformation change of tropomyosin resulting in the exposure of active binding sites on the actin myofilament. At this point, the myosin heads attach to the exposed binding sites forming a crossbridge, and force is produced within the muscle.

During concentric muscle actions, an action known as a *power stroke* occurs, where the myosin heads rachet upon attaching to actin, pulling the actin filament past the myosin filament, thus shortening the sarcomere and the overall muscle. Assuming enough adenosine triphosphate is present, the myosin detaches, recocks the myosin head, and reattaches to another binding site farther along the actin filament to form a new crossbridge. The continuous attachment and detachment of myosin to and from actin is known as *crossbridge cycling*.

During eccentric muscle actions, the muscle is actively resisting the external force by stretching the crossbridges formed between actin and myosin. Furthermore, it has been proposed that there is a decreased rate of crossbridge cycling during eccentric muscle actions, thus allowing a greater number of crossbridges to remain attached and tension to increase (79). Although not often discussed, it should be noted that both titin (77, 78, 80, 127, 151) and nebulin (10, 84, 97, 142, 143, 228) may play a significant role in force production during various muscle actions; a further discussion is provided in chapter 3.

FIBER TYPE CONTINUUM

The different muscle fiber types possess unique physiological characteristics and exist on a continuum. Although several methods exist that may be used to classify muscle fiber types (15, 28, 187, 197) and hybrid fiber types (124, 149), the three most identified fiber types in humans are Type I, Type IIa, and Type IIx fibers. Each fiber type varies in size, force production ability, fatigability, and other physiological characteristics (see chapter 3 for more detail). As noted previously, each fiber is innervated by an α motor neuron as part of a motor unit; however, the size of the fiber innervated is directly proportional to the size of the α motor neuron and its force production capacity. This is an important consideration because motor units are recruited in a specific order based on the force production needs of a given task (i.e., smaller motor units are recruited prior to larger motor units); the order of recruitment has been termed the *size principle* (75, 76). Fiber type can vary by both sex (64, 140, 186) and race (30) but can also play a role in athletic success. Elite and international-level sprint athletes (32, 210), weightlifters (54, 68, 178), and shot-put throwers (17) possess greater percentages of Type II versus Type I fibers. This information not only provides insight into the demands specific to sports or competitions, but it also sheds light on how training can affect fiber type characteristics. Researchers have summarized the existing evidence on fiber type transitions (150); a brief discussion of how training may stimulate fiber type transitions and the impact on strength characteristics is provided later in this chapter.

GENERAL ADAPTATION SYNDROME

The *general adaptation syndrome* (GAS) model was initially introduced and developed by Hans Selye during the 1930s to 1950s (171, 172, 173, 174, 175). Although his initial work focused on the responses to "diverse nocuous agents" (171), which included alarm, resistance, and exhaustion stages, the model was further developed to include the stress response related to homeostasis and subsequent adaptations (175). In later writings, Selye indicated that the GAS may be applied to promote adaptations and avoid exhaustion, including those produced following exercise (176). Briefly, when a stimulus (stress) is introduced, there is an initial negative response (alarm phase) that results in a reduction in performance from an individual's baseline level. This is followed by a resistance to the stress introduced, which allows an individual's performance to return to baseline levels using what Selye termed "adaptation energy" (resistance phase) (172).

It should be noted that a supercompensation effect, whereby an individual's performance exceeds their original baseline level, may also occur during the resistance phase; this concept was originally described within the stimulus–fatigue–recovery–adaptation model presented by Yakovlev (225, 226). However, as mentioned earlier, adaptation beyond baseline levels of performance may be based on homeostatic mechanisms and the individual's homeostatic set points of body temperature, blood pressure, hormone concentrations, and so on. Although these set points are controlled by positive or negative feedback loops, it should be noted that they may also be altered by consistent training (39, 41, 185).

Finally, the exhaustion phase occurs if the stimulus either exceeds the abilities of an athlete to adapt or is insufficient to promote an adaptation. Practically speaking, there will likely be multiple instances of alarm and resistance phases due to multiple training sessions performed during a given week plus general variation in training stimuli that alter both volume and intensity. For a thorough overview of the GAS model, readers are directed to Cunanan and colleagues (35).

FITNESS–FATIGUE PARADIGM

The *fitness–fatigue model* (or *fitness–fatigue paradigm*) is an extension of the GAS model and was introduced by Banister in the 1970s (14). The fitness–fatigue model focuses on potential changes in performance based on the interaction between fitness and fatigue characteristics following the introduction of a training stimulus. Briefly, *fitness* characteristics may be described as the underlying factors that contribute to a positive performance or adaptation. From a strength standpoint, these characteristics may include genetic characteristics, fiber type, neural drive (i.e., motor unit recruitment, firing frequency, synchronization), muscle cross-sectional area, and others (see chapter 3 for more information). In contrast to fitness characteristics, *fatigue* within the fitness–fatigue model simply refers to the accumulated fatigue that is produced following the completion of a given training stimulus.

The interaction between fitness and fatigue characteristics refers to a performance state called *preparedness*, which represents the potential of an athlete to perform well. However, both fitness and fatigue characteristics contribute to the residual effects of a training stimulus and may thus affect subsequent adaptations. Simply, if fitness characteristics dominate and fatigue decreases at a faster rate, the athlete has a greater level of preparedness. Using variations in volume and intensity, strength and conditioning professionals can plan for athletes to maximize their preparedness to peak at specific time periods (11, 12, 13). In contrast, if large levels of fatigue are present, they may mask the benefits of the fitness characteristics, ultimately leading to a lower level of physical preparedness. In this regard, larger volumes of training may negatively affect rapid force production characteristics due to the amount of fatigue that it may induce (114, 198).

Despite existing for nearly 50 years, the fitness–fatigue model has stood the test of time and may provide strength and conditioning professionals with valuable insight regarding the interplay between training stimuli and performance adaptations. However, researchers have continued to examine its validity as it relates to training load (217) and its interaction with a machine learning approach (88). With the continued growth of technology and artificial intelligence, more research in this area is expected.

> **Key Point**
>
> The interaction between fitness and fatigue characteristics refers to a performance state called *preparedness*, which represents the potential of an athlete to perform well.

It should be noted that the fitness–fatigue paradigm may also be viewed on an acute basis within potentiation complexes. These complexes use a conditioning activity to acutely enhance the fitness characteristics that may benefit the performance of a subsequent task; for example, researchers have investigated the effect of heavy squatting variations on subsequent jump performance (25, 34, 205, 206). However, strength and conditioning professionals should note that both athlete characteristics (e.g., relative strength, fiber type) and the design of the potentiation complex (e.g., mode, volume, intensity, rest interval, range of motion) may have a significant impact on subsequent performances (201). The timeline of potentiation was previously illustrated by both Sale (161) and Tillin and colleagues (209) to show that potentiation and fatigue co-exist following the performance of a conditioning stimulus. Because athletes may possess different capacities of fatigue resistance to a conditioning activity, the residual effects of both potentiation and fatigue may dissipate at different rates, which is why multiple windows of potentiation may be realized following various rest intervals (209). This presents a unique challenge to strength and conditioning professionals hoping to use this method as a training strategy. For more information on the use of potentiation complexes for strength development, readers are referred to chapters 5 and 6.

TIMELINE OF ADAPTATION

There are several underlying factors that may contribute to improvements in muscular strength, including genetics, epigenetics, muscle fiber type, neuroendocrine factors, neural factors, muscle architecture, and motor learning (see chapter 3 for more detail). Despite the contributions from each factor, it is commonly accepted that initial improvements in strength are attributed to neural changes. In fact, researchers have shown that improvements in coordination and learning (160), as well as increases in muscle activation and firing frequency (9, 67, 69, 129, 134), contribute to early improvements in muscle strength to a greater extent compared to muscle hypertrophy. However, it is important to note that later improvements in strength may be attributed to increases in muscle hypertrophy (162) and may be based on the long-term development of an athlete. For example, an inexperienced youth athlete must first develop physical literacy characteristics and basic exercise technique before they physically mature and benefit from the influence of neuroendocrine factors that contribute to muscle hypertrophy and strength adaptations.

A caveat to the sequence of underlying factor contributions to strength may be based on competition in weight class–based sports. For example, an individual competing in a specific weight class must make weight to be eligible to compete; thus, continual increases in muscle mass (i.e., hypertrophy) may not be possible. However, if these individuals continue to compete in the same weight class with a similar body composition and gain strength, it would seem logical to suggest that these improvements may be attributed to neural contributions (193, 194). Thus, there appear to be multiple windows of adaptation that may occur during the development of muscular strength characteristics. It is important that strength and conditioning professionals understand the contributions of underlying factors to muscle strength improvements during the long-term development of their athletes.

ADAPTATIONS TO TRAINING

As noted previously, there are many adaptations that may occur when an athlete is exposed to various stimuli aimed at improving their force production characteristics. The following paragraphs will discuss the impact of training on motor learning, fiber type, neural, neuroendocrine, and muscle architecture adaptations.

MOTOR LEARNING

The development of motor skills is a key aspect to improving athletic performance. For example, strength training may serve as a dynamical model for motor development, in which the time frames of developing peak performances and learning motor skills are similar when training or skill practice is distributed over longer training periods (130). Ideally, an athlete adapts to a given training stimulus and then recalibrates to perform a motor task more efficiently and effectively (93). To develop efficient and effective movement or to produce skilled movers, athletes should be exposed to varied practice so that they enhance adaptability (38, 93) to exhibit greater effectiveness under different task constraints (135); this may be accomplished by providing an athlete with various organism (e.g., physical, physiological, biomechanical, or psychological), environment (e.g., temperature, altitude), or task (e.g., goal of the task) constraints. However, strength and conditioning professionals should note that athletes may revert to previously learned, and possibly less efficient, motor patterns if the constraints of the task do not change in practice, leaving them unable to adapt to situations (139). When teaching exercise technique, athletes should be guided toward more effective movement with different constraints to allow them to search for an optimal solution to the task instead of restricting them to a single movement pattern (42). This, in turn, may allow athletes to learn how to use their "newfound strength" to produce and apply force more effectively (93, 202, 204).

Key Point

To develop efficient and effective movement or to produce skilled movers, athletes should be exposed to varied practice so that they enhance adaptability to exhibit greater effectiveness under different task constraints.

FIBER TYPE

As discussed earlier, fiber types exist on a spectrum, and their qualities may significantly affect the force production characteristics of an athlete (24, 125, 145, 147). In fact, researchers have shown greater percentages of Type I (slow-twitch) or Type II (fast-twitch) fibers in elite and international-level distance runners (33) or sprinters (32, 210), weightlifters (54, 68, 178), and shot-put throwers (17), respectively. Although these athletes may have been born with specific percentages of Type I, Type IIa, and Type IIx fibers, researchers suggest that it may be possible to shift existing fiber types (150). For example, hybrid fiber types, such as those that exist within the Type II fiber continuum, may serve as intermediates to different muscle fiber phenotypes (124, 149). Researchers have shown that hybrid fibers may disappear based on specific volume and intensities prescribed during training (148, 163, 197, 212); however, transitions between slow-twitch and fast-twitch fibers may be the result of changes in neuromuscular activity, mechanical loading, or conditions such as hypothyroidism or hyperthyroidism (148, 150).

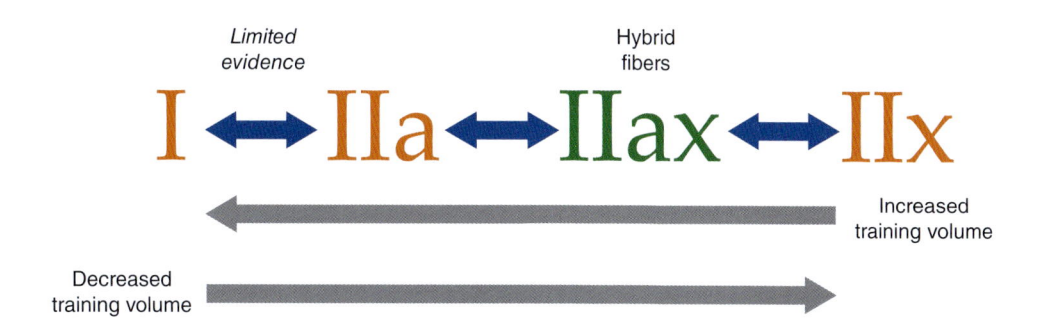

FIGURE 2.6 Fiber type transitions.

It should be noted that higher training volumes may shift fibers toward slower fiber types, whereas higher-intensity, lower-volume training may shift fibers toward faster fiber types (figure 2.6). However, the manipulation of volume and intensity can be used to optimize fiber type composition at specific times. For example, following a 6-week resistance training program that included a planned overreach and step taper, powerlifters showed a significant decrease in Type I and an increase in Type IIa single fiber phenotype percentage (212).

Finally, due to its rapid force production requirements, sprint training may preferentially recruit Type II muscle fibers, potentially leading to fiber type shifts toward faster phenotypes. It should be noted, however, that there is still some debate on whether complete fiber transformation (e.g., Type I to Type II) can occur in humans.

NEURAL

Neural adaptations to training exist across a wide spectrum and may include increases in motor unit recruitment, firing frequency, and synchronization and decreased neuromuscular inhibition. While some of these adaptations may be linked to motor program development changes, it is important to note that neural adaptations may be specific to the prescribed training stimulus. For example, ballistic exercise may lower the recruitment threshold of high-threshold motor units, given that larger magnitudes and rates of force production are needed to complete the task (113, 216). Additional research has shown that ballistic training using weightlifting movements and their derivatives (110) and sprinting (164) may improve motor unit firing frequency. From a training timeline standpoint, researchers have shown that ballistic training completed over the course of 12 weeks may lead to improvements in firing frequency (216). Like motor unit recruitment and firing frequency, motor unit synchronization has been enhanced using heavy resistance training (4) and rapid muscle actions (177). Finally, neuromuscular inhibition has been decreased after heavy resistance training (2, 3, 4), leading to enhanced maximal and rapid force production characteristics.

Beyond resistance training, sprinting may also lead to both central and peripheral nervous system adaptations that improve neural drive within the recruited muscle fibers (157). Given the existing research, strength and conditioning professionals should consider implementing heavy or ballistic-type exercises, or training methods that are ballistic in nature, to facilitate improvements in neural contributions to muscular strength characteristics. While the specific changes to motor neurons (e.g., functional changes to the neuromuscular synapse, protein synthesis, gene expression, and axon transport alterations and various biophysical properties) are beyond the scope of this chapter, readers are referred to Gardiner and colleagues (57) for a more in-depth discussion.

Key Point

Strength and conditioning professionals should consider implementing heavy or ballistic-type exercises, or training methods that are ballistic in nature, to facilitate improvements in neural contributions to muscular strength characteristics.

NEUROENDOCRINE

The elevated levels of both catecholamine (48, 53, 152) and testosterone (65, 70, 188) hormones that occur following training have been associated with improvements in strength characteristics, whereas cortisol has not (221). Moreover, elevated hormone levels may be associated with the type of training, volume, and intensity prescribed; for example, high-force and high-power training was shown to significantly increase plasma catecholamine levels following a resistance training session (102). Additional research has shown that higher-intensity anaerobic exercise may increase catecholamine levels up to 15-fold (96, 121).

A significant amount of research supports the notion that elevations in testosterone are directly related to the volume, intensity, and size of the involved musculature during aerobic (91, 224) and anaerobic exercise (90, 96, 99, 101, 122, 170, 220). Further research has shown that acute elevations in testosterone may be produced with large-muscle, multi-joint exercises as well as heavy resistances (85%-95% of the 1RM), moderate to high volumes of training, short rest intervals, and consistent resistance training (45, 51, 98, 99). Similarly, researchers have shown that both training volume and intensity can have a significant impact on elevated cortisol levels (49, 96, 144). However, further research indicated that lower volumes may significantly reduce resting cortisol levels in different types of athletes (21, 83, 144).

It is important to note that although acute exercise may affect the levels of these hormones, chronic, long-term training may produce different effects. For example, researchers have reported that significant reductions in catecholamine responses may be shown with chronic, long-term training (109), which suggests that there may be accommodation effects (47). Similar results related to cortisol were shown with chronic aerobic (92, 179, 208) and resistance (121, 191) training. Chronic changes in testosterone are unique because they may be based on different stages of physical development. For example, younger men (i.e., aged 30 years) were shown to display higher concentrations of free testosterone following resistance training compared to older men (i.e., aged 62 years) (100). Although much smaller in magnitude, it may also be possible for women in their 20s to display elevated levels of testosterone following resistance training (138, 218). Finally, the type of training may have a significant impact on long-term testosterone responses; for example, resistance-trained athletes have shown greater elevations compared to those who are endurance trained (213) or untrained (5). In addition, researchers have shown that strength–power athletes have higher circulating testosterone levels compared to endurance athletes (61, 89).

MUSCLE ARCHITECTURE

Various changes to the architecture of a muscle can be made following the introduction of different training stimuli. Perhaps the most well-documented muscle architectural changes related to the strength characteristics of an athlete are changes in muscle size (i.e., hypertrophy) (29, 66, 190). As discussed later in chapter 3, increases in hypertrophy may enhance the potential for greater force production due to an increased number of crossbridge formations between actin and myosin filaments within sarcomeres. Researchers have shown that similar magnitudes of muscle hypertrophy may be produced using light loads performed to failure and heavy loads (165, 166, 167, 168); however, it is important that strength and conditioning professionals distinguish between sarcoplasmic (i.e., noncontractile elements) and myofibrillar (i.e., contractile elements) hypertrophy, because the type may affect the magnitude and rate of force development as well as task-specific adaptations (211). Moreover, the type of training may also dictate the magnitude of hypertrophy produced; for example, researchers have shown that eccentric or paired eccentric and concentric muscle actions may produce greater hypertrophy compared to concentric-only actions (31, 219). Another aspect of hypertrophy to be considered is the way in which sarcomeres are added (i.e., in parallel or series). In theory, high-force training may lead to more sarcomeres being added in parallel, whereas high-velocity training may add sarcomeres in series, which may lead to greater force production via fiber packing (56, 158) or greater shortening velocities (222, 223), respectively.

Enhancements in both maximal and rapid force production can also be brought about by changes in pennation angle (1, 95), tendon stiffness (20), and other microstructures within the sarcomere (e.g., actin, myosin, titin) (113, 155). Regarding pennation angle, researchers have shown simultaneous increases in muscle thickness and pennation angle following resistance training (1, 95). Furthermore, increases in pennation during near-maximal or maximal dynamic muscle actions have been shown to increase both magnitude and rate of force development (7, 44). Additionally, researchers have shown that muscle fascicle length may increase despite no changes in pennation angle following high-velocity, light-load (<60% of the 1RM) training (6, 19), which may allow for faster muscle shortening velocities to occur (200). Additional enhancements to force production may be brought about by improvements in tendon stiffness. Researchers have shown that both isometric training (104, 106) and plyometric training (105) may lead to improvements in tendon stiffness, which may enhance force transmission, leading to more efficient movements performed within a timely manner. Finally, along with the addition of more sarcomeres via hypertrophy, there may be increases in passive tension within the sarcomeres from the addition of titin filaments (81), especially during eccentric muscle actions (78).

INFLUENCE OF TRAINING STIMULI ON STRENGTH ADAPTATIONS

The specific adaptations outlined earlier are likely specific to the training prescribed to the athlete. Thus, it is important that strength and conditioning professionals target the relevant physiological adaptations using stimuli that are specific to each athlete. In this regard, the program variables discussed in chapter 7 (i.e., exercise selection, frequency, exercise order, volume, set structure, intensity, and rest intervals) should be considered when designing training programs. The following paragraphs will discuss how the previously outlined variables affect both maximal and rapid force production adaptations.

EXERCISE SELECTION

The prescribed exercises can have a significant impact on the maximal and rapid force production characteristics of an athlete. For example, if an athlete lacks maximal strength, it is important that strength and conditioning professionals choose an exercise that permits the use of heavier loads to improve peak force production. In this example, the use of the front squat may improve peak force production of the lower body to a greater extent compared to a goblet squat. To target rapid force production, exercises that are more ballistic in nature may serve as better alternatives (82); it should be noted, however, that exercises that are traditionally nonballistic can be performed with ballistic intent (i.e., as fast as possible) to improve the force production characteristics of the exercise (16, 207, 227). In addition, it is important to prescribe exercises that may transfer to the specific movements performed in the athlete's sport or event. Thus, an emphasis on single-joint over multi-joint exercises in healthy athletes may be questionable, because it has the potential to hinder their potential strength adaptations (8, 18, 141, 189).

Key Point

The prescribed exercises can have a significant impact on the maximal and rapid force production characteristics of an athlete.

FREQUENCY

The frequency of training sessions may vary based on the training status of the athlete, the goals of different phases, the period of the training year, and the number of competitions within a given timeframe. The impact of different training periods on strength adaptations is discussed later in this chapter, but it is important to discuss the frequency of training in a dose–response context. A novice athlete may benefit from a lower frequency of training sessions before progressing to more training sessions (123, 180); this, in turn, may allow the athlete to develop their work capacity and tolerance to higher training intensities. It should be noted that training frequency may not affect strength adaptations to a large extent when volume is matched (36, 60, 154); however, in previous studies when volume was not matched, greater improvements were shown with greater training frequencies (60). Regarding the goals of individual training phases, greater frequencies may expose the athlete to a larger overall stimulus, which may benefit a variety of strength characteristics. It is important to note, however, that greater frequency may result in larger training volumes and thus greater fatigue. Therefore, strength and conditioning professionals may benefit strength–endurance and rapid force production characteristics if the training frequency is increased or decreased, respectively.

EXERCISE ORDER

The order of exercises prescribed within a training session may affect the ability of an athlete to perform subsequent repetitions of other exercises, especially when similar muscle groups are used (182). Thus, strength and conditioning professionals should program the exercises with the greatest neuromuscular demand at the beginning of a training session; this order may allow for the initial exercises to receive the greatest strength benefits (43, 183, 184). However, it should be noted that the training method used may dictate exercise order as well; for example, potentiation complexes may use either high-force (205, 206) or high-velocity exercises (27) to acutely improve the force production characteristics of an athlete during a subsequent exercise. Although either exercise order may produce positive adaptations (117), the adaptations may be highly individualized based on the design of the protocol and the athlete characteristics as well (201).

VOLUME

The impact of volume on strength adaptations may be based on the dose–response relationship (146, 153). While the maximal strength characteristics of athletes may initially benefit from additional training volume (103), greater volumes may develop more fatigue (229) and negatively affect rapid force production characteristics (115, 198); however, this may also be dependent on the individual and relative training intensity. It is important that novice athletes are eventually exposed to higher training volumes to not only improve their work capacity but also to solidify their technique with more repetitions. As noted previously, increases in training frequency may also affect the overall training volume that an athlete is exposed to. Thus, it is important that strength and conditioning professionals promote quality of training over quantity of training, especially as the training age of an athlete increases. This, in turn, may decrease the potential for nonfunctional overreaching, overtraining, and injury while promoting positive strength adaptations.

Key Point

While the maximal strength characteristics of athletes may initially benefit from additional training volume, greater volumes may develop more fatigue and negatively affect rapid force production characteristics.

SET STRUCTURE

Sets of exercise can be prescribed in a variety of ways, including traditional or various cluster set configurations (133). Although the use of traditional sets that require athletes to perform all the prescribed repetitions in succession may help develop the force production characteristics of an athlete, this type of set is not without its limitations. Traditional sets may be characterized by a decline in movement velocity (58, 215) and a breakdown of technique (72), especially when a larger number of repetitions are performed. Because this may negatively affect the overall training stimulus, strength and conditioning professionals may consider breaking the larger set into smaller sets, termed *cluster sets*. The primary benefits of cluster sets include the ability to use heavier loads, maintain higher movement velocities, and improve the overall quality of work (63, 215), all of which may improve the stimulus for both maximal and rapid force production characteristics.

INTENSITY

There is little doubt that heavier training loads can benefit the maximal strength characteristics of an athlete. While this may lead some strength and conditioning professionals to prescribe heavy loads performed to failure, it is important to note that this method of prescription may not provide any additional strength benefits compared to submaximal loading not performed to failure (40, 146). However, the literature does encourage the use of both heavy and light loading with a variety of strategies to promote the development of both maximal and rapid force production characteristics (62, 136, 137, 199). It is important that strength and conditioning professionals also consider the speed and phase of sprinting (132) as well as the angle (74) and entrance velocity (37) of change-of-direction tasks as measures of training intensity, because these methods of training are integral to the development of sport-specific skills and the strength characteristics of an athlete.

Key Point

Training to failure may not provide any additional strength benefits compared to submaximal loading not performed to failure.

REST INTERVALS

As mentioned previously, the quality of training is of utmost importance when it comes to developing the force production characteristics of an athlete. In this light, researchers have suggested that longer *interset rest intervals* (i.e., rest between sets) should be used when training to improve strength (156, 169). It is important to note that although the previous research was completed on traditional sets of training, *intraset rest intervals* may also be used during cluster sets to benefit the maximal and rapid force production stimulus an athlete receives. For example, longer *interrepetition rest intervals* (i.e., rest in between each repetition) may improve the overall quality of work performed (73) and allow the athlete to maintain their technique to a greater extent (72) and potentially lower their perceived exertion (71). Collectively, longer rest intervals performed during traditional or cluster sets may promote superior strength adaptations compared to those that are shorter.

IMPACT OF THE TRAINING YEAR

The training year may dictate the extent an athlete may be able to develop and express their force production characteristics. In general, a training year may be broken into off-season, preseason, in-season, and postseason training phases. Each phase should have a primary emphasis regarding the development of specific physiological adaptations that contribute to the force production characteristics of an athlete. However, the prescribed training stimulus and the associated volume of sport practice or competitions may affect the ability of an athlete to realize specific strength adaptations. The following paragraphs discuss how the different training seasons may affect the force production adaptations of athletes.

Key Point

The prescribed training stimulus and the associated volume of sport practice or competitions may affect the ability of an athlete to realize specific strength adaptations.

OFF-SEASON

The largest improvements in performance are typically made during the off-season phase of training due to limited amounts of sport-specific training, allowing a greater emphasis to be placed on the physical development of the athlete. From a strength development perspective, the off-season serves as the time in which the underlying mechanisms of force production can be developed; this may include changes in muscle architecture (e.g., developing greater magnitudes of muscle mass) and high-intensity training to improve the maximal force production capabilities of the athlete (203). In general, off-season training is characterized by higher volumes of training that may coincide with greater accumulated fatigue. While this may negatively affect rapid force production characteristics (115, 198), these training effects may be temporary and are resolved as the volume is decreased and intensity is increased as an athlete enters the preseason phase of training. However, it should be noted that reductions in rapid force production may also be mitigated by prescribing task-specific training that may benefit the underlying characteristics of force production to a greater extent (211).

PRESEASON

The primary goal of the preseason training phase is to build on what was gained during the off-season to prepare the athlete for competition. For example, if the emphasis in the off-season was to increase the muscle mass of an athlete, the preseason phase may serve as the period to realize the potential to increase maximal force production (203). Beyond maximal force production, it is important that strength and conditioning professionals emphasize the development of rapid force production characteristics during this phase to increase both the specificity of training and the likelihood of transfer to sport performance tasks. This is typically accomplished by decreasing training volume and increasing intensity during training sessions but also by providing a combination of training stimuli that emphasize both characteristics (62, 199). An additional benefit of these training manipulations is that they allow strength and conditioning professionals to manage fatigue as the volume of sport-specific training increases before competition. With individual-based sports that have a single major competition, decreases in volume or frequency and increases or maintenance of intensity may allow for a more effective taper and superior performance (23, 131, 212).

IN-SEASON

The primary goals of in-season training are, at a minimum, to maintain the force production characteristics developed during the off-season and preseason phases of training, to achieve peak performance for the most important event or events of the competitive season, and to manage fatigue. In-season training phases are typically characterized by lower volumes of training, including lower frequencies, while prescribing training methods or exercises that allow an athlete to continue producing high magnitudes and rates of force development. Because the number of competitions within a season (e.g., team vs. individual sports) or specific timeframe of the season may vary (e.g., tournaments, schedule changes), strength and conditioning professionals must adapt training programs to prescribe the most effective training stimulus. In other words, it is important to emphasize the biggest "bang for your buck" training stimuli without adding in additional volume that may contribute to fatigue. For example, although unilateral exercises may increase muscle activation of the involved musculature to a greater extent than bilateral training (120), these exercises may be limited in their capacity to be loaded heavily and may require additional volume because both limbs need to be trained (203) (e.g., lunges), which may also affect exercise efficiency and total session duration.

POSTSEASON

The competitive season can be both physically and mentally demanding; thus, postseason training requires a reduction in the overall training volume and intensity to allow the athlete to recover. Researchers have shown that the cessation of training can significantly affect the muscular strength characteristics of individuals (22); however, those with a higher training status displayed smaller reductions in performance, demonstrating an enhanced ability to retain these characteristics. While training may not be completely removed during the postseason training period, strength and conditioning professionals should de-emphasize the development of maximal and rapid force production during the immediate postseason and promote recovery and the performance of basic movement patterns. Failure to do so may instead promote nonfunctional overreaching and potentially overtraining (50, 52, 192, 195), because the athlete may not have fully recovered from the impact of the competitive season.

DETRAINING

The process of detraining is underpinned by a specific training principle termed *reversibility*, whereby an athlete may experience a deterioration or loss of their force production capabilities; this principle is discussed in greater detail in chapter 4. There are a variety of scenarios in which an athlete may lose their ability to produce force (e.g., removal of a training stimulus, injury, monotonous training). Therefore, it is important that the strength and conditioning professional is aware of the issue, can organize a plan to mitigate potential detriments, and can provide a training stimulus that eventually promotes the development of both maximal and rapid force production.

As mentioned previously, an injury is a common scenario that may lead to the detraining of specific strength characteristics. Following an injury, sports medicine staff must implement strategies to re-establish motor patterns to perform basic functions prior to sport-specific movements and skills (e.g., knee flexion or extension before squatting and jumping). However, after an athlete can perform basic movement patterns again, it is important to re-establish their ability to produce and apply force in addition to improving their motor skill coordination (159, 160). Return-to-play protocols exist on a continuum based on the limitations of the athlete. While early-stage rehabilitation protocols may include high-repetition sets with low to very low loads (e.g., 3 sets of 10 repetitions with body weight), athletes should be progressively loaded to expose previously injured tissue to heavier loads, thus allowing them to redevelop and improve their force production characteristics. Researchers have concluded that strength training may reduce injuries to less than one-third and halve overuse injuries (108) and that a progressive increase in resistance training volume of 10% may also decrease injury occurrence (107). Based on scope of practice, sports medicine staff should be the first individuals to provide the injured athlete with an initial training stimulus; however, it is the exception, rather than the norm, that sports medicine staff understand performance concepts to the same extent as they do the rehabilitation of injured tissue. Thus, it is important that there is a collaborative effort between sports medicine and strength and conditioning professionals so that a well-rounded return-to-play protocol is developed that promotes the development of an athlete's strength characteristics. For example, if an athlete is limited in their range of motion, a strength and conditioning professional may load the athlete within their limitations while still providing an effective training stimulus. It is important that whoever is predominantly designing the return-to-play training program understands the biomechanical characteristics of the sport skills and tasks performed by the athlete during competition, and that they subsequently design the program in a manner that enables a redevelopment of effective motor patterns for each skill or task (126).

> **Key Point**
>
> It is important that there is a collaborative effort between sports medicine and strength and conditioning professionals so that a well-rounded return-to-play protocol is developed that promotes the development of an athlete's strength characteristics.

EXTERNAL STRESSORS

While strength and conditioning professionals may be able to control the dynamics within various training facilities or venues, it may not be possible to account for the external stressors that affect an athlete. For example, academic stress or relationship issues may negatively affect the ability of an athlete to effectively train (111). Moreover, researchers have shown that periods of high academic stress increase the odds of an injury restriction within NCAA Division I football players regardless of whether the athlete consistently plays (116). Additional researchers showed that while initial improvements in strength characteristics were displayed during the first phase of a collegiate semester, decreases in performance were shown after additional conditioning stress was added during the second half of the semester (128). If these results are combined with the additional academic stress that may occur at the end of a typical semester due to final examinations, the ability of an athlete to train and improve their strength characteristics may be negatively affected. However, additional factors such as poor sleeping habits and nutrition may also worsen these issues. Thus, strength and conditioning professionals must develop effective athlete assessment protocols and relationships with their athletes to monitor their progress (or regression) while combating the impact that external stressors can have on performance within their scope of practice. Additional information on strength assessment and recovery considerations is provided in chapters 9 and 12, respectively.

Take-Home Points

- When exposed to a novel stimulus, athletes may experience changes in their force production characteristics as a product of motor learning as well as changes in fiber type, motor unit activation and firing characteristics, neuroendocrine influences, or muscle architecture.
- Early strength adaptations are primarily underpinned by neural adaptations, such as increased motor unit recruitment, firing frequency, and synchronization and decreased neural inhibition.
- Exposing athletes to various task constraints can help them become more efficient in their movements by allowing them to find an optimal solution. This, in turn, may help them develop an ability to use their "newfound strength" to produce and apply force more effectively.
- The magnitude and timing of the experienced adaptations may be based on the stimulus that an athlete receives, which may be modified based on exercise selection, training frequency, exercise order, volume, set structure, intensity, and rest interval duration.
- The time of the training year, the injury status of an athlete, and external stressors may also affect how an athlete adapts to various training stimuli.
- An interdisciplinary approach should be taken when developing return-to-play protocols for injured athletes and should include both sports medicine and strength and conditioning staff.
- Strength and conditioning professionals should not discount the impact of external stressors unrelated to physical training on the force production characteristics of their athletes.

CHAPTER 3

Mechanisms of Strength

There are several factors that may influence the human ability to produce force (i.e., strength). This chapter will highlight and discuss several physiological, biomechanical, and motor learning considerations that may affect the development of muscular strength. It is, however, important to note that the mechanisms that improve the force production capabilities of an athlete are multifactorial and may be influenced by their strength levels (14), training status (34), and genetics. To effectively prescribe different training strategies to improve the maximal and rapid force production characteristics of their athletes, it is important that strength and conditioning professionals understand the underpinning mechanisms of strength. Therefore, the goal of this chapter is to highlight many of the factors that influence the force production capabilities of an athlete, which may therefore modify their strength abilities.

GENETICS AND EPIGENETICS (NATURE AND NURTURE)

British mathematician, philosopher, and logician Bertrand Russell once wrote, "choose your parents wisely" (223). While the context of this quote did not refer to the performance capabilities of an athlete, it is important to recognize that there may be no substitute for innate talent (i.e., genetics) regarding athletic success and the ability to adapt and recover (252). Athletes may inherit specific characteristics such as a more mesomorphic somatotype (205, 239), higher testosterone concentrations, and specific muscle fiber types and metabolic enzyme functionality (68, 150, 195), all of which may benefit their force production characteristics. Thus, apart from being a biological parent, strength and conditioning professionals will not be able to influence the genetic characteristics inherited by an athlete; however, they do have the ability to dramatically influence athlete adaptation through various means.

Key Point

There may be no substitute for innate talent (i.e., genetics) regarding athletic success and the ability to adapt and recover.

Epigenetic characteristics represent heritable changes to an individual's DNA that do not alter the genetic sequence itself but may influence gene expression patterns (63, 88, 289). These characteristics are individual specific and may represent crucial control centers that dictate an athlete's performance capacity (254). It should be noted that several factors can influence the epigenetic characteristics of an athlete; these include training stimuli, the coach–athlete interaction, sleep, nutrition, work, travel, social life, injuries, and more (88, 254). This, in turn, may significantly affect the recovery and adaptation of an athlete and, ultimately, their expression of force. This notion brings up a subject of frequent debate: nature (genetics) versus nurture (training). Georgiades and colleagues (88) summarized this relationship as follows:

> Despite this complexity, the overwhelming and accumulating evidence, amounted through experimental research spanning almost two centuries, tips the balance in favour of nature in the "nature" and "nurture" debate. In other words, truly elite-level athletes are built—but only from those born with innate ability (p. 835).

With this in mind, it is important to point out that genetically gifted athletes may still have successful performances in sports even if their training did not lead to improved preparedness (224). However, strength and conditioning professionals must also consider the "window of adaptation," because each athlete may respond differently to a given training stimulus (46, 68, 113, 221). Thus, while epigenetic factors may be under genetic control (18, 63, 88), consideration must also be given to the fact that the magnitude of adaptation becomes smaller as athletes near their genetic ceiling (8). Given that an athlete's genetics, training age, and commitment may all influence their adaptation to training (90), it is important that strength and conditioning professionals implement strategies to maximize the potential for adaptation (e.g., individualized training programs, athlete monitoring, athlete education) while also recognizing the impact that external factors can have on performance (161).

To better understand the influence of genetics on sport performance, large, multination genomic projects have commenced (212). Two genetic polymorphisms, angiotensin I–converting enzyme insertion/deletion (ACE I/D) and α-actinin-3 gene (ACTN3) R577X, have received considerable attention in the strength and conditioning literature. Whereas ACE I/D has commonly been associated with endurance-type events (44, 162, 269), ACTN3 has been linked with strength–power events (67, 162, 163, 202) and improved strength characteristics (43, 211, 276, 280, 293). However, it should be noted that conflicting results also exist (10, 191, 194, 218, 222). While research focused on the influence of specific genetic characteristics on athlete performance is still in development, the findings of such research may influence athletes, parents, coaches, and national governing bodies to submit samples for genetic testing. However, consumers should first question how much influence each of these traits has on performance as well as be aware of the limitations of such testing (284). For a more thorough overview on the current state of DNA testing and its relationship with talent identification, readers are referred to a 2022 review by Varillas-Delgado and colleagues (274).

FIBER TYPE

As mentioned earlier, the genetic characteristics of athletes may allow them to possess the innate ability to produce greater magnitudes and rates of force production compared to those who do not have the same characteristics. Some of this may be explained by the inherited fiber type characteristics of the athlete (68, 150, 195). It should be noted that there are several ways to classify different motor unit and fiber types, such as contractile properties (35, 36), histochemical schemes based on myosin ATPase and enzymatic properties (16, 207), myofibrillar ATPase (33, 208, 243), and immunohistochemistry (identification of myosin heavy and light chains) (45, 257). While this may vary the number of muscle fiber types, the current discussion will focus on three primary fiber types: Type I, Type IIa, and Type IIx (table 3.1).

The fiber type composition of an athlete directly influences their maximal and rapid force production capabilities (30, 178, 206, 208). Briefly, Type IIx fibers produce the greatest forces, rates of force development, and power output, followed in order by Type IIa and Type I fibers. Therefore, athletes with greater percentages of Type II versus Type I fibers have a greater capacity to produce larger magnitudes and rates of force production. Researchers support this notion, because elite and international-level sprint athletes (47, 263), weightlifters (82, 98, 237), and shot-put throwers (20) have been shown to possess greater percentages of Type II compared to Type I fibers. Similar findings were shown in national-level powerlifters compared to untrained subjects (83). Researchers have also found that men, in general, may possess greater quantities of Type II fibers than women (95, 193, 242).

Key Point

The fiber type composition of an athlete directly influences their maximal and rapid force production capabilities.

Although athletes may be born with a specific percentage of Type I, Type IIa, and Type IIx fiber types, it may be possible to significantly shift existing fiber types (e.g., from Type I to Type IIa) to produce more favorable force production characteristics. This may be attributable, in part, to the existence of hybrid fibers (e.g., Type IIax) within skeletal muscle. These hybrid fibers serve as intermediates during fiber type transitions and provide a functional continuum of muscle fiber phenotypes (175, 210). Shifts in fiber type may be due to differences in myosin heavy chain composition along the length of single fibers. There is evidence to suggest

TABLE 3.1 Major Characteristics of Muscle Fiber Types

Characteristic	FIBER TYPES		
	Type I	Type IIa	Type IIx
Motor neuron size	Small	Large	Large
Recruitment threshold	Low	Intermediate/High	High
Nerve conduction velocity	Slow	Fast	Fast
Contraction speed	Slow	Fast	Fast
Relaxation speed	Slow	Fast	Fast
Fatigue resistance	High	Intermediate/Low	Low
Endurance	High	Intermediate/Low	Low
Force production	Low	Intermediate	High
Power output	Low	Intermediate/High	High
Aerobic enzyme content	High	Intermediate/Low	Low
Anaerobic enzyme content	Low	High	High
Sarcoplasmic reticulum complexity	Low	Intermediate/High	High
Capillary density	High	Intermediate	Low
Myoglobin content	High	Low	Low
Mitochondrial size, density	High	Intermediate	Low
Fiber diameter	Small	Intermediate	Large
Color	Red	White/Red	White

Reprinted by permission from NSCA, *Essentials of Strength Training and Conditioning,* 4th ed., edited by G.G. Haff and N.T. Travis (Human Kinetics, 2016), 10.

that these hybrid fibers may disappear based on the volume and intensities prescribed in training (209, 225, 257, 265). Briefly, fiber type transitions between fast and slow may be caused by modifications to neuromuscular activity, mechanical loading, and conditions, such as hypothyroidism or hyperthyroidism (209, 213). Practically speaking, higher volumes of training may tend to cause shifts in fiber type toward slower-type fibers, whereas reductions in volume may shift fibers in the opposite direction. For example, while investigating the differences between step and exponential tapers in powerlifters, Travis and colleagues (265) showed that fiber type transitions can be displayed in as little as 6 weeks. While most of the existing evidence supports the idea that fiber types can transition within the Type II fiber continuum, there is still debate over whether complete fiber transformation (i.e., Type I to Type II and vice versa) can take place in humans. A thorough discussion of the available evidence focused on fiber type transitions was provided by Plotkin and colleagues (213).

NEUROMUSCULAR FACTORS

There are a variety of neuromuscular factors that may contribute to the force production ability of an athlete, including motor unit recruitment, firing frequency, and synchronization as well as contributions from the stretch–shortening cycle and neuromuscular inhibition. The following paragraphs provide an overview of each of these factors and how they contribute to muscular strength.

MOTOR UNIT RECRUITMENT

The ability to recruit a greater number of motor units—and by extension, a greater number of muscle fibers—may lead to improvements in both maximal and rapid force production (111). Motor units are recruited based on their size (smallest to largest) in a sequenced manner, often termed

Henneman's size principle (see figure 3.1) (112). In this light, motor units are recruited based on the force production needs (magnitude and rate) of a given task, with slow-twitch Type I fibers being recruited with less demanding tasks and fast-twitch Type II fibers being recruited when tasks are more demanding. Researchers have shown that the order of recruitment holds true during slow, graded, isometric (180), and ballistic actions (58, 59). It should be noted, however, that although the size principle holds true during ballistic actions, motor unit recruitment thresholds may be lowered due to the need for greater magnitudes and rates of force production (164, 272). This serves as evidence that the demands of each task will dictate which motor units are recruited.

For a motor unit to adapt to a stimulus, it must be recruited. While all athletes are required to produce force during competition, the demands of their events will dictate the type of motor units recruited. For example, distance runners require high levels of muscular endurance (i.e., the ability to repeatedly produce submaximal forces for extended periods of time); thus, it is likely that these athletes will typically recruit low-threshold, slow-fatiguing (Type I) motor units. Unless the nature of their event changes and requires the runners to produce larger or more rapid magnitudes of force to sustain their performance, it is likely that high-threshold, fast-fatiguing (Type II) motor units may not need to be recruited. Weightlifters, on the other hand, require large magnitudes and rates of force development to be successful in both the snatch and clean and jerk lifts. Thus, it is likely that both low- and high-threshold motor units will be recruited based on the order of recruitment. In fact, researchers have suggested that the preferential recruitment of high-threshold motor units would benefit the rapid force production characteristics of athletes (62, 138). Based on the existing literature, it would appear beneficial to implement training strategies that allow athletes to recruit both Type I and Type II motor units on a conditional basis to enhance their force production characteristics.

FIRING FREQUENCY

Maximal and rapid force production may also be underpinned by the rate at which neural impulses are sent (i.e., "fired") to the muscle fibers by the α motor neuron within a given motor unit (see figure 3.2). This has been termed as either *firing frequency* or *rate coding*. Researchers have shown a significant increase in force production (300%-1,500%) when the firing frequency of motor units increases from their minimum to maximum potential (65). Due to the relationship between firing frequency and increased doublet charges (i.e., two consecutive motor unit discharges within a 5-ms interval), firing frequency has been linked to increases in rapid force production characteristics (272). This, in turn, may suggest that firing frequency underpins the ability to improve an athlete's strength characteristics during dynamic tasks that require large magnitudes of force within short durations. Researchers have shown that 12 weeks of ballistic training may improve motor unit firing frequency (272). Thus, training methods that include ballistic actions such as weightlifting movements and their derivatives (156) and sprinting (226) may benefit the firing frequency charac-

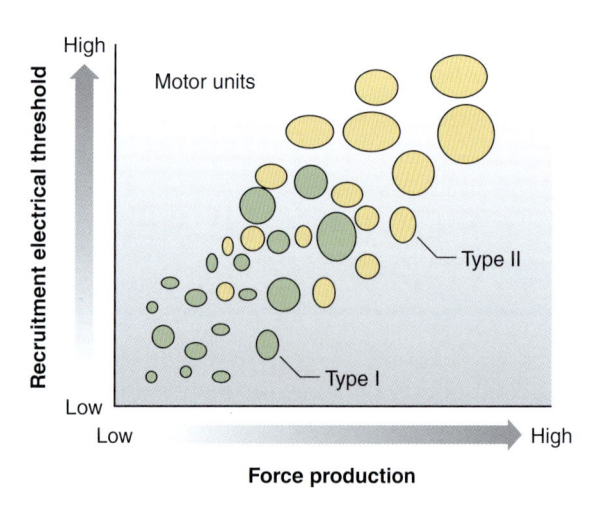

FIGURE 3.1 Graphic representation of the size principle, according to which motor units that contain Type I (slow-twitch) and Type II (fast-twitch) fibers are organized based on some "size" factor. Low-threshold motor units are recruited first and have lower force capabilities than higher-threshold motor units. Typically, to get to the high-threshold motor units, the body must first recruit the lower-threshold motor units. Exceptions exist, especially with respect to explosive, ballistic contractions that can selectively recruit high-threshold units to rapidly achieve more force and power.

Reprinted by permission from D. French, "Adaptations to Anaerobic Training Programs," in *Essentials of Strength Training and Conditioning*, 4th ed., edited for the National Strength and Conditioning Association by G.G. Haff and N.T. Triplett (Human Kinetics, 2016), 91.

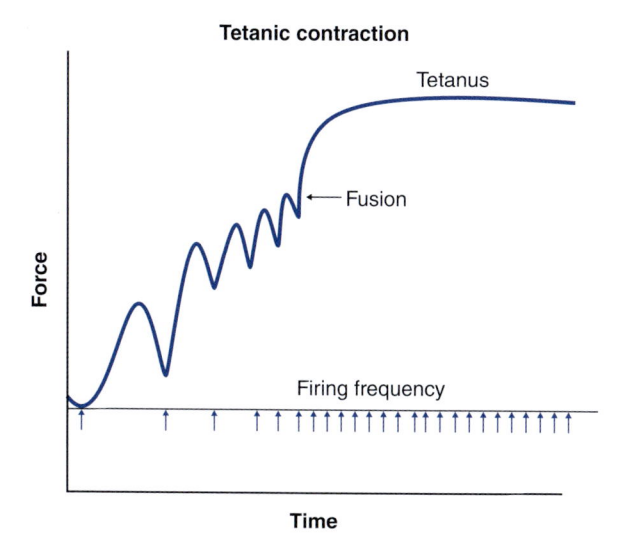

FIGURE 3.2 The influence of firing frequency on force production.

Adapted by permission from M. Stone, T. Suchomel, W. Hornsby, et al., *Strength and Conditioning in Sports:* From *Science to Practice* (Routledge, 2022), 27; permission conveyed through Copyright Clearance Center, Inc.

teristics of an athlete and, by extension, their force production characteristics.

MOTOR UNIT SYNCHRONIZATION

Motor unit synchronization refers to the simultaneous firing of multiple motor units. While this synchronization does not appear to positively affect maximal isometric force production (294), it may be more related to an individual's rapid force production characteristics (233). Resistance training serves as an effective tool to enhance motor unit synchronization (179). This notion is supported by the frequency of exposure to resistance training, as shown by the differences in synchronization in weightlifters compared to musicians and untrained individuals (236). Like other underpinning mechanisms of strength, motor unit synchronization may be enhanced by implementing heavy resistance training (4). It should be noted, however, that ballistic training did not enhance motor unit synchronization in one study (272) but it was enhanced during rapid muscle actions in another study (235). Due to mixed findings using ballistic training, further research is warranted. However, it does appear that both heavy and ballistic methods of training have the potential to produce favorable motor unit synchronization

adaptations to further enhance maximal and rapid force production characteristics.

CONTRIBUTION OF THE STRETCH–SHORTENING CYCLE

Stretch–shortening cycle actions occur when an eccentric (lengthening muscle) action precedes a concentric (shortening muscle) action in which the two are performed in immediate succession. Researchers have indicated that stretch–shortening cycles may benefit an athlete's force production characteristics (25, 51). Enhancement of the concentric phase is due to a combination of factors, including the use of stored elastic energy, a stretch reflex, and performance of muscle actions from an optimized muscle length with optimized muscle activation patterns (25, 26). Meechan and colleagues (176, 177) showed that force production characteristics were enhanced during weightlifting pulling derivatives when a countermovement was included.

NEUROMUSCULAR INHIBITION

Neuromuscular inhibition may be defined as a reduced neural drive of muscle groups that negatively affect force production due to neural feedback from the muscle and joint receptors during voluntary muscle actions (85). Restated, feedback sent from muscle and joint receptors results in a reduced amount of force production. Thus, if possible, a goal of strength training should be to reduce neuromuscular inhibition to allow for greater force production. Aagaard and colleagues (4) indicated that heavy resistance training may reduce inhibition and enhance force production by downregulating Ib afferent feedback to the spinal motor neuron pool. Additional researchers showed decreased neuromuscular inhibition (3) and enhanced rapid force production (2) following resistance training, along with simultaneous increases in spinal and supraspinal neural drive. Increases in maximal force production typically coincide with heavy resistance training. Based on the existing evidence, increases in maximal force production following heavy resistance training may partially be underpinned by reductions in neuromuscular inhibition.

NEUROENDOCRINE SYSTEM

The *neuroendocrine system* combines aspects of the nervous and endocrine systems to control homeostasis and other biological functions at the cellular level (135, 136). Several negative and positive feedback loops as well as external factors such as nutrition can significantly influence hormonal responses (135). It should be noted that the concentrations of each hormone prior to, during, or following training appears to be dictated by various aspects of the training program (e.g., exercise method, volume, intensity, rest periods) (147). Hormones regulate the demands of metabolism, manipulate fluid regulation, affect tissue growth, adapt protein synthesis and degradation, influence energy substrate use, and have the potential to alter an athlete's mood state (74). Because these effects and adaptations may have a direct impact on muscle hypertrophy and strength, a brief discussion is warranted. It is noted that hormones such as human growth hormone, insulin-like growth factors, insulin, and glucagon may significantly affect skeletal muscle hypertrophy and atrophy (74, 134, 228); however, they lack a direct influence on strength enhancement. Thus, the current discussion will focus on the effect that catecholamines, testosterone, and cortisol have on the strength characteristics of athletes.

CATECHOLAMINES

Catecholamines act as either neurotransmitters or circulating hormones (89, 92, 291), and their release may result in increased force production, rate of muscle contraction, blood pressure and flow, and energy availability but may also enhance the secretion rates of other hormones (e.g., testosterone) (134, 148). The rationale for this may be due to the anticipatory response prior to challenging training (75), which results in a heightened state of arousal and several physiological changes related to substrate utilization for energy (e.g., decreases cellular uptake of glucose, promotes glucagon secretion, and stimulates fat breakdown) and the distribution of blood to exercising muscles (74). Furthermore, catecholamines may act as central motor stimulators and peripheral vasodilators but also enhance enzyme function and calcium release in muscle (146), which may acutely enhance force production characteristics.

Circulating epinephrine may increase with exercise intensities as low as 40% of maximal oxygen consumption ($\dot{V}O_2$max) (174), but it will significantly increase with higher intensities (168). Researchers have shown that high-force and high-power resistance training results in significantly elevated plasma catecholamine levels following training (144). Although researchers have shown up to a 15-fold increase in catecholamines during anaerobic exercise (132, 172), post-exercise elevations in epinephrine have been correlated with the greatest improvements in strength (75, 215). Fry and colleagues (79) showed significant correlations between catecholamines (epinephrine and norepinephrine) and percentage change in strength. However, it should be noted that chronic training may reduce exercise-induced catecholamine responses (155), suggesting that there may be an accommodation phenomenon with training (74).

TESTOSTERONE

Testosterone is considered a primary anabolic hormone (279) and has been proposed as a physiological marker for the anabolic status of men and women (103, 151). While testosterone signals protein synthesis and may reduce protein breakdown—which may benefit muscle size (74)—it also interacts with neuron receptors, increases the amount of neurotransmitters, increases cell body size and dendrite characteristics (length and diameter), and influences structural protein changes also benefiting force production (23, 24, 81, 131, 185). Furthermore, researchers have shown increases in muscular strength along with fiber type transformations (i.e., Type IIx to IIa) with elevated serum testosterone (244), as well as associations with muscle hypertrophy, maximal and rapid force production, and power output (29, 101, 255). In fact, testosterone may augment the neural adaptations that occur with strength improvements in highly trained strength–power athletes (96, 104). These adaptations may be further enhanced when the resistance training program has the potential to induce elevated testosterone concentrations (106, 219).

Bhasin and colleagues (19) indicated that supraphysiological doses of testosterone (600 mg/week) produced a 17.7% increase in leg press strength despite a lack of resistance training stimulus. In males, testosterone may be elevated following both high-intensity aerobic endurance activity and resistance training (86, 145), with potentially small elevations also occurring in women following resistance training (190, 277). Testosterone increases appear to be proportional to relative intensity, volume, and the size of the involved musculature during both aerobic (126, 290) and anaerobic (125, 132, 140, 143, 173, 231, 285) exercise. Furthermore, it has been suggested that testosterone levels may be acutely elevated by prescribing one or multiple combinations of large-muscle, multi-joint exercises as well as heavy resistances (85%-95% of the 1RM), moderate-high volumes of training, short rest intervals, and consistent resistance training for at least 2 years (69, 78, 139, 140). Interestingly, greater levels of circulating testosterone may result from programs that are designed to improve morphological adaptations through high volumes of training compared to those focused on gaining strength through neural adaptations (49, 141, 157).

Resistance training may produce no change or a decrease in testosterone concentrations in sedentary and moderately trained young men (196, 256) and middle-aged men (188) but may increase resting testosterone in boys (270), resistance-trained men (244), middle-aged sedentary men (127), and young women (167). Despite testosterone being at its peak in the morning for men, there does not appear to be a difference in improvements in maximal strength with morning or afternoon resistance training sessions (232). In contrast, there is little variation in serum testosterone in women; however, the small increases that women experience following resistance training may benefit subsequent performances (277). It should be noted that younger men (i.e., aged 30 years) displayed higher concentrations of free testosterone after resistance training compared to those who were older (i.e., aged 62 years) (142). In addition, resistance-trained athletes showed greater testosterone responses compared to endurance-trained (266) and untrained (5) individuals. Moreover, those who compete in strength–power sports may have higher circulating testosterone compared to endurance athletes (94, 124), which may partially explain the force production capabilities of each athlete type; however, further research is needed on this front.

Kvorning and colleagues (152, 153) demonstrated that young men with normal testosterone concentrations improved strength and lean tissue mass to a greater extent than those who received luteinizing hormone blockers that produced very low testosterone but did not affect other anabolic signaling.

It is important to acknowledge the differences in resting testosterone between boys and girls as well as men and women. Vingren and colleagues (278) indicated that puberty produces large increases in resting testosterone in boys, whereas much smaller effects are seen in girls. Women in general have approximately 15- to 20-fold lower concentrations of testosterone compared to men (105). However, physically active women may have greater baseline concentrations of testosterone compared to those who are inactive (53). In addition, researchers have shown that small increases in total and free testosterone changes that resulted during strength training in women were related to their force production characteristics (104). A 2021 review by Hilton and Lundberg (118) concluded that testosterone suppression in transgender women may only trivially affect strength, lean body mass, muscle size, and bone density. However, the authors noted that minimal data on transgender female athletes have been collected. As the hot topic of transgender women competing with biological women remains, further research in this area is warranted.

CORTISOL

Cortisol, commonly referred to as a *stress hormone*, has a wide variety of functions related to metabolism. It is catabolic in nature and stimulates gluconeogenesis in the liver, reduced use of glucose within the cells, protein breakdown, increased amino acid availability, anti-inflammatory effects, and erythropoiesis (72, 279). Related to force production, cortisol has its greatest effect on Type II muscle fibers (49, 147) because it reduces protein synthesis and increases protein breakdown within these fibers. While protein degradation may still result from increased cortisol in Type I fibers, an increased amount of degradation may occur in Type II fibers due to greater overall amounts of protein within these fibers (227). Acutely, cortisol responses following resistance training may be crucial for the remodeling process, because muscle tissue needs to be disrupted before it can adapt to the stimulus placed upon it (147). For example, the increased magnitude of amino acid availability may contribute to protein synthesis and tissue regeneration (279). However, it should be noted that a net loss of contractile protein via catabolic processes or muscle atrophy may result in reductions in force production capability (70). Furthermore, cortisol may also reduce the production and release of anabolic hormones such as testosterone (50, 60, 134, 135), which, as noted earlier, may affect its anabolic effects on strength development.

The ratio between testosterone and cortisol (T:C) has been used by some to provide an indication of an individual's training status (103, 201). However, the use of T:C in this manner and as a predictor of strength and power characteristics has been questioned by others (147). Interestingly, Fry and Schilling (80) indicated that consistent training over several years can increase T:C and provide an indication of training tolerance. However, this may be an oversimplification because testosterone secretion remains relatively normal despite higher levels of cortisol (135). French (74) instead suggested that T:C may simply represent a gross indirect measure of anabolic and catabolic properties in skeletal muscle. Like testosterone, greater cortisol responses result from higher volumes of training with shorter rest periods compared to maximal strength workouts with heavier loads (48, 49, 137, 240). However, additional literature suggests that there is a dose-dependent training role for testosterone and cortisol that may affect the performance capabilities of athletes (50). In fact, subtle alterations in T:C within resistance training programs have been shown to have moderate to strong correlations with strength performance (7, 37, 77, 100, 102, 201). Stone and colleagues (253) showed inverse relationships between T:C and volume–load among national-level American weightlifters, displaying the sensitivity of T:C to training stress. Therefore, T:C may reflect a state of "preparedness" (121), with higher T:C ratios indicating an improved potential to perform well (201).

Key Point

Subtle alterations in the T:C ratio within resistance training programs have been shown to have moderate to strong correlations with strength performance.

There is limited information on how resistance training affects the cortisol levels of men and women. However, Vingren and colleagues (277) showed that glucocorticoid (catabolic) receptors were lower at rest and for up to 70 minutes following exercise in men compared to women. In contrast, women displayed an increase in androgen binding capacity following exercise, whereas men showed a continual reduction. Following the heavy 5RM protocol, women maintained their cortisol levels, whereas men displayed an increase. Researchers have also shown that training may initially increase serum cortisol, but longer-term training may allow levels to return to normal or slightly below baseline with aerobic (128, 238, 262) and resistance training (172, 246). Like catecholamines and testosterone, cortisol responses appear to be related to both training volume and intensity (76, 201). However, a reduction in training volume has been shown to reduce resting cortisol concentrations in different types of athletes as well (28, 121, 201).

MUSCLE ARCHITECTURE

Muscle architectural factors including hypertrophy (i.e., cross-sectional area), fiber arrangement, and structural proteins contributing to musculotendinous stiffness—such as titin and nebulin—may also contribute to the force production capabilities of an athlete. The following paragraphs highlight each of these factors and their impact on muscular strength.

HYPERTROPHY

Muscle hypertrophy, specifically in Type II muscle fibers, can significantly affect the force–velocity characteristics (38, 245) and, by extension, the peak and rapid force production characteristics of a whole muscle. Interestingly, researchers have shown that approximately 50% to 60% of changes in force production were attributed to an increase in muscle cross-sectional area and architectural alterations in untrained participants (186). Further research showed very large relationships (r = .70, R^2 = .49) between muscle cross-sectional area and force production in elite strength- and endurance-trained athletes and sprinters (97). From a physiological standpoint, increases in muscle hypertrophy improve the potential for greater force production due to an increase in crossbridge interactions between the previously and newly formed sarcomere myofilaments (actin and myosin). Furthermore, hypertrophied muscles may display greater pennation angles compared to normal muscles (130), which again may benefit the magnitude of force production due to a greater number of crossbridge interactions attributable to fiber packing with a unit area.

Despite the previously presented evidence of support, there are several factors that may influence the effect of muscle hypertrophy on an athlete's strength characteristics. First, the method of training or types of muscle actions may dictate whether sarcomeres would be added in series or in parallel (258). For example, eccentric-only or paired eccentric and concentric muscle actions have been shown to produce greater magnitudes of hypertrophy compared to concentric-only actions (41, 282). This may partially be explained by the preferential recruitment of Type II muscle fibers during eccentric muscle actions (61, 260), which may contribute to greater adaptations regarding the peak and rapid force production characteristics of an athlete.

A second explanation may be due to differences in the type of hypertrophy that results from training. Travis and colleagues (264) identified the differences between sarcoplasmic and myofibrillar hypertrophy; in particular, the authors discussed how the latter may contribute to greater maximal strength, rapid force production, and power output. A further discussion of the contributory effects of myofibrillar hypertrophy to maximal strength improvements was provided by Taber and colleagues (261). Briefly, *sarcoplasmic hypertrophy* refers to a chronic increase in sarcolemmal volume and noncontractile elements (e.g., mitochondria, transverse tubules) (216), whereas *myofibrillar hypertrophy* refers to an increase in the size or number of myofibrils (and sarcomeres) or protein abundance related to force production (110, 273).

Finally, variance in hypertrophy and subsequent strength improvements may exist due to differences in the timeline of when the former is measured and when strength is expressed during a given task (189, 249). Additionally, improvement in strength characteristics can be affected by other physiological or neural factors that contribute on an individualized basis (14).

Improvements in an athlete's strength characteristics may be explained by phase potentiation and residual training effects (181, 245, 259, 297). For example, researchers have shown that the residual training effects from previous training phases may carry over to subsequent training phases, thereby enhancing or potentiating them (107, 123). Thus, it could be argued that the greatest benefits from resistance training may follow a specific sequence or progression. This notion is supported by evidence suggesting that strength characteristics may benefit from phases of training that first increase the cross-sectional area of muscles (i.e., hypertrophy) and their force production capacity (i.e., work capacity) (181, 248, 297). Increases in muscle hypertrophy combined with concomitant or subsequent changes in a muscle's architecture, fiber type, and other neural factors may enhance the ability to benefit an athlete's maximum strength (181, 247, 297). While previous researchers have indicated that similar muscle hypertrophy may result from low-load, high-repetition and high-load, low-repetition training prescriptions (229), it should be noted that strength increases were maximized using the latter. Thus, further research is needed to determine the impact of phases of training focused on hypertrophy and how flipped set and repetition sequences (e.g., 10 sets of 3 repetitions vs. the traditional 3 sets of 10 repetitions) may affect an athlete's strength characteristics.

Key Point

Increases in muscle hypertrophy combined with concomitant or subsequent changes in a muscle's architecture, fiber type, and other neural factors may benefit an athlete's maximum strength.

Another aspect of muscle hypertrophy that may influence the force production characteristics of a muscle is the Type II:I fiber ratio. Campos and colleagues (38) indicated that an increase in muscle hypertrophy occurred simultaneously with a greater Type II:I ratio due to a greater rate of hypertrophy of Type II fibers. Moreover, strong relationships have been shown between improvements in squat jump power output and percentage change in Type II:I ratio following 8 weeks of resistance training (99). Conceptually, muscles with a greater Type II:I ratio have the capacity to produce greater magnitudes and rates of force production. However, it should be noted that prescription of training (method, volume–load) will have a considerable effect on the type of motor units (and muscle fibers) recruited and affected. Moreover, this may affect the way in which sarcomeres are added to the previously existing sarcomeres within the muscle. High-force training may promote the addition of sarcomeres in parallel, while high-velocity training may add sarcomeres in series (see figure 3.3). Although each method of adding sarcomeres may improve force production capacity due to the potential for more crossbridge interactions between actin and myosin, adding sarcomeres in parallel or in series tends to favor greater force production or shortening velocity, respectively. A greater number of sarcomeres in parallel allows for a greater potential of fiber packing density in which there are more crossbridge interactions per unit area (87, 220); when sarcomeres are added in series, the entire unit of muscle contracts in unison, thus allowing for greater shortening velocities (287, 288).

FIBER ARRANGEMENT

The differences in force production capacity between muscles that differ in fiber arrangement have been highlighted in numerous textbooks within the strength and conditioning field (136, 228, 250, 254). For example, it is well known that muscle fibers within a pennate or bipennate arrangement possess greater force production capacities compared to those in a parallel, circular, fusiform, and multipennate arrangement. Although muscle fibers may be the same size and possess the same anatomical cross-sectional area, pennate muscles differ in their force production capacity due to a greater physiological cross-sectional area (i.e., greatest distance across the muscle belly that is perpendicular to the fiber arrangement) compared to other fiber arrangements (see figure 3.4). Researchers have shown that strength training may produce simultaneous increases in muscle thickness and pennation angle (1, 130). These adaptations may allow for greater magnitudes of force to be produced due to the ability to pack a greater number of muscle fascicles (i.e., muscle fiber bundle) into a given area, ultimately resulting in more crossbridge interactions (122). However, further research has shown increases in muscle fascicle length despite no changes in pennation angle following high-velocity, light-load (<60% of the 1RM) training (6, 22). In this scenario, smaller pennation angles allow for greater muscle shortening velocities because the fascicles will be positioned more parallel to the muscle's aponeuroses (i.e., connective tissue that connects the muscle to bone or other tissue) (258).

Throughout a dynamic muscle action, the angle of pennation can vary and may change the functional parameters of some muscles (84, 197). For example,

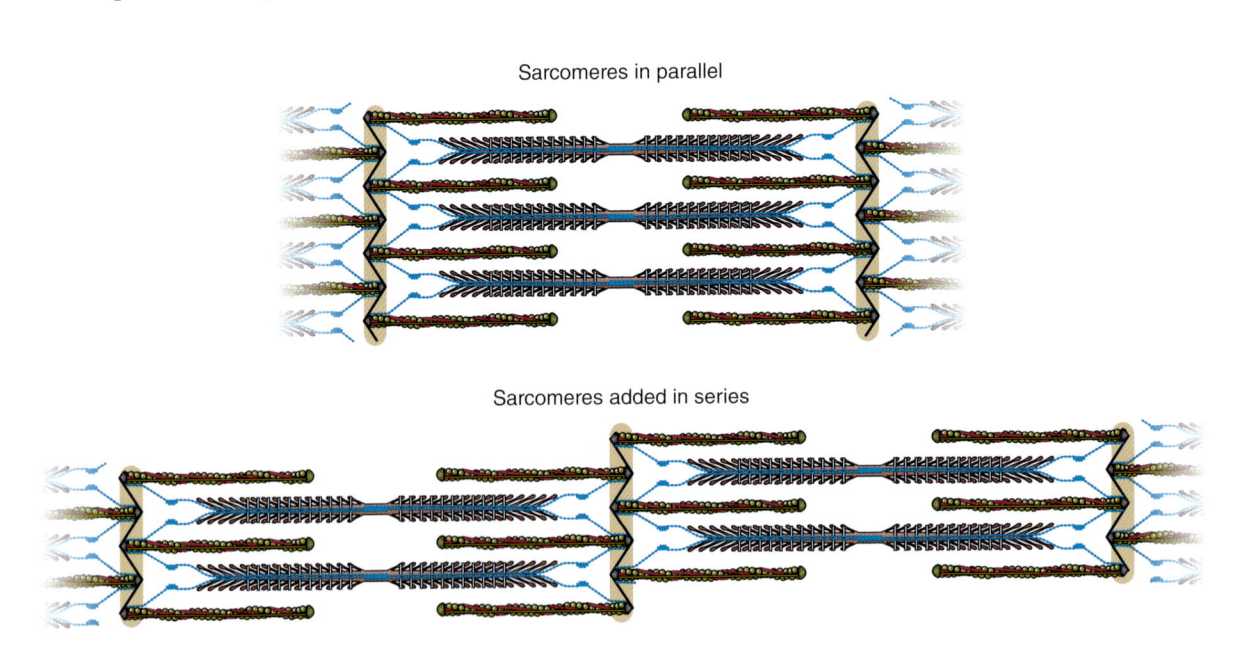

FIGURE 3.3 Sarcomeres in parallel and series.

increases in pennation angle during near-maximal or maximal dynamic actions may increase force and rate of force development during the later phases of a force–time curve (9, 66). This may result in a gearing effect (muscle fiber rotation) in which velocity is sacrificed to produce greater rates of force development and high forces (11, 31). However, the opposite may also occur in which pennation angles may decrease, resulting in velocity increases at the expense of force production. Lower (high-force) and higher (high-velocity) gear ratios within pennate muscles provide a unique mechanism by which force production is modified within the muscle during mechanically diverse actions (11). For example, changes in the shape of the muscle during movement allow pennate muscles to shift between and use low- and high-gearing effects, thus allowing effective force transmission when using both high (low-gearing) and low (high-gearing) loads.

As discussed previously, architectural changes specific to pennation angle may be specific to the muscle actions performed in training. For example, researchers have shown specific architectural adaptations to both eccentric and concentric training (21, 73). Moreover, weightlifters were shown to have longer fascicle lengths compared to track and field throwers (298). However, it should also be noted that muscle architectural changes may not be uniform throughout the muscle belly (64, 286). For example, the muscle action demands of different sport tasks may vary (e.g., sprint running or cycling), which may activate muscles in a unique manner (281). This is an important consideration for strength and conditioning professionals when selecting exercises to maximize the transfer of training adaptations to performance.

MUSCULOTENDINOUS STIFFNESS

Musculotendinous stiffness is a term used to describe the force transmission properties of the musculotendinous unit. Because muscular strength is based on the expression of force within a task-specific context, it is important to understand the spring-like behavior of tissues and their influence on force production. Simply, an increase in tissue stiffness may enhance the transmission of force from the muscle and tendon to the bone, thus allowing an athlete to move more effectively within a time-efficient manner. In fact, researchers have shown that tendon stiffness adaptations (27), as well as adaptations to other microstructures within sarcomeres (e.g., actin, myosin, titin, nebulin), can influence maximal force and rapid force production characteristics (164, 217) and power output (27, 149).

Titin

Titin is a large filamentous protein or viscoelastic spring that is often synonymous with the term *stiffness* (discussed earlier), and it is the largest protein in the human body that spans half the length of a sarcomere (114). From a structural standpoint, titin has a rigid association with myosin but also attaches to actin near its insertion point to the Z disc of the sarcomere, thus providing a link between actin and myosin and their crossbridge interaction (267, 268).

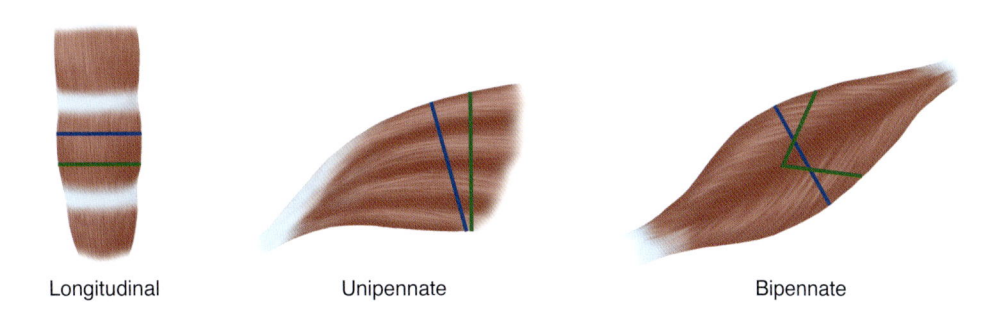

Longitudinal   Unipennate   Bipennate

FIGURE 3.4 Anatomical (blue lines) and physiological (green lines) cross-sectional areas. *Note:* The physiological cross-sectional area is greater than the anatomical cross-sectional area in unipennate and bipennate muscles compared to parallel muscles.

Powers and colleagues (214) showed decrements in force production within muscle sarcomeres after the removal of titin. In fact, researchers have suggested that titin may play a greater role in muscle function—namely, maintaining the rigidity of the sarcomere, thus allowing for enhanced force transmission (114, 267). This notion stems from findings that suggest titin may produce passive tension within the sarcomere (117), especially during eccentric muscle actions (115). In addition, several research groups have suggested that titin has a greater role within the sarcomere beyond serving as merely a structural tissue (116, 183, 217), with some suggesting that titin may serve as the third contractile microfilament, alongside actin and myosin (116). This is supported by additional research indicating that titin's characteristics may relate to muscle strength and power (82, 169). Although researchers are still investigating the role of titin during various muscle actions, its contributions to muscular strength and force transmission cannot be overlooked. Readers are referred to a review by Herzog (114) for a more thorough discussion on how titin contributes to force production.

Nebulin

Another large protein within the sarcomere that may contribute to force production is nebulin. Nebulin attaches to both the Z disc of the sarcomere and actin near the point where myosin heads attach during crossbridge formation (40). Specifically, nebulin attaches to tropomodulin at the pointed ends of actin (170), which may help regulate the length of the actin filament (40, 203). Beyond modifying the length of actin, nebulin may also factor into the regulation of force production (154, 198, 200). Like titin, the regulation of force may be through stiffness mechanisms where nebulin increases the stability of actin during muscle contraction (133). Additional findings suggest that the presence of nebulin may also increase isometric force, stiffness, and rate of force development by influencing crossbridge cycling and optimizing interactions between actin and myosin (15, 93). Research focusing on the properties and functions of nebulin is ongoing; however, further information on the influence of nebulin on force production can be found in several reviews (154, 199, 200, 296).

MOTOR LEARNING AND SKILL ACQUISITION

Briefly, motor skills require athletes to voluntarily move different parts of their body to achieve a specific task goal (165) and may be classified as gross, fine, continuous, discrete, open, or closed (182). To achieve the goal of a specific task, developing motor skills through practice would appear to be beneficial to performances within athletic competitions (54). However, it is important to consider that each independent component of an athlete's motor system (e.g., motor units, muscles, joints) serves as a degree of freedom; thus, the coordination of these components produces a specific movement of the body within a given motor task. Therefore, the improvement of existing coordination patterns and the development of new coordination patterns to meet the changing demands of training or competition may serve as a primary goal to improve the overall performance of an athlete (182).

To adequately complete various motor tasks in sport competitions, athletes will express movement during the performance of a motor skill within the constraints of their motor capacities (sometimes called *boundaries*) and their current skill developed through practice (129). Within this context, the force production characteristics of an athlete may be classified as a measure of motor capacity or motor capability (119). However, when considering practice of the motor skill (e.g., change of direction), it is important that strength and conditioning professionals expose their athletes to situations in which they must find movement task solutions to adequately develop a specific motor task (e.g., reacting to a stimulus, increasing the speed of a task). Specific to resistance training, learning how to squat (a motor task in itself) requires a progression using simple to complex variations (e.g., bodyweight squat before performing a goblet squat, front squat, back squat, and overhead squat) to develop the coordination pattern to perform the movement efficiently. Thus, introducing new variations in training may benefit the development of new coordination patterns (54), force production characteristics (184), and promotion of greater control under different task constraints (187).

> **Key Point**
>
> The force production characteristics of an athlete (i.e., strength) may be classified as a measure of motor capacity or motor capability.

Using a constraints-led approach, strength and conditioning professionals may influence the development of a particular motor skill. Constraints may take many forms, including those that are specific to the organism (e.g., physical, physiological, biomechanical, or psychological), environment (e.g., temperature, altitude), or task (e.g., goal of the task) (54). Using maximal sprinting as an example, an organismic trait that may influence the performance of the task is the strength of the athlete. Goodwin (91) indicated that insufficient strength may lead to overstriding at maximal velocity due to the inability to tolerate and generate the necessary vertical impulses during shorter stance phases. From an environmental standpoint, researchers have shown that wind and altitude (158), as well as terrain (204), can also significantly affect sprinting velocity. Finally, specific task constraints such as carrying a ball (17) or a field hockey stick (283) or sprinting with additional equipment (32) may negatively affect running velocity. Additional task constraints can come in the form of the instructions given to the athlete during a performance test; for example, Young and colleagues (295) showed that drop jump performance can be modified by instructing an athlete to jump as high as possible, minimize ground contact time, or a combination of both. Regardless of the constraints used in practice, a constraints-led approach provides athletes with an opportunity to find specific solutions that allow them to meet the goal of a given task (54). While strength and conditioning professionals may use a variety of constraints to develop effective and efficient movement, it is important that they understand how different constraints can affect the demand placed on the athlete (129, 182). This, in turn, may directly affect the movement formed by the athlete and their ability to produce and apply force during the task itself.

Another method used to enhance the performance of a motor task is through using a dynamical systems approach. Briefly, a *dynamical system* related to the performance of a motor task refers to the motor system of an athlete that is constantly changing and evolving over time (55); this system is characterized by the large number of degrees of freedom that are present within each task. Using a dynamical systems approach, an athlete forms coordinative structures (i.e., temporary organization and control of degrees of freedom) (271) through self-organization under constraint (230). In other words, an athlete may develop an effective movement pattern to complete a given motor task in response to the conditions presented to them. Thus, the coordinative structure of a given motor pattern may become relatively stable if the constraints of the task do not change and may allow athletes to return to this coordination pattern despite constraint changes (192).

As athletes continue to form different coordination patterns, they will eventually converge into what is termed an *attractor state* that is typically limited to several stable coordination patterns (230). Despite the existence of an attractor state, it has been proposed that practice should be used to guide athletes away from their initial attractor state to allow them to search for an optimal solution to the task rather than restricting athletes to a specific movement pattern (56). However, it is important that strength and conditioning professionals put these recommendations into context, because some exercises or movements are more complex than others and any technique that may promote injury should be avoided. In a 2023 article, Morrison and Newell (184) indicated that strength training may serve as a dynamical model for motor development. Specifically, the authors suggested that there appear to be compatible changes in strength training and motor skill learning due to common attractor states that arise from constraints from both practice and training. Moreover, the time frames of developing peak performances (250) and learning motor skills (12, 275) appear to be similar when training or skill practice is distributed over longer periods of time. While much of the previous discussion has referred to developing movement during a given exercise, it should be noted that a dynamical model may be used in the coordination of movements between two athletes during a team sport (13). Similar approaches have been discussed regarding the development of motor skills within agility tasks (39, 241); however, it appears that a dynamical systems approach can be used to develop and improve a motor skill, which includes the improvement of producing and applying force during the tasks. Evidence of this was provided by Haug and colleagues (109) who showed that through learning a complex exercise (i.e., hang power clean), an athlete can improve their ability to produce and apply force. An example of different steps that can be taken to aid the development of a motor skill (e.g., clean) is shown in figure 3.5.

The contribution that attentional focus, instructions or cueing, and feedback can have on movement should not be overlooked. For example, it is well established that an external attentional focus (i.e., focus on things external to the body and the outcome of a task) is broadly more effective for learning compared to an internal focus (i.e., focus on things within the body and movement during the task) (292). As mentioned previously, the instructions provided serve as a constraint on the athlete; thus, it is important to understand how athletes will respond to specific instructions and their phrasing to promote the best results. Finally, feedback to an athlete can come in many forms, including verbal, visual, and, when appropriate, tactile. How, when, and how often feedback is delivered may also affect skill acquisition. For example, an athlete learning how to perform a power clean may require significant feedback on their technique. Within this scenario, it is important not to place an overwhelming number of constraints on the athlete so that they can properly learn and complete the motor task, as noted earlier. However, given that feedback also serves as a temporary constraint (54), it should gradually be removed from practice so that the athlete does not become dependent on the feedback (52). While the measurement and monitoring of motor skills is discussed in chapter 9, readers interested in motor skill acquisition are referred to previous work by Moir (182).

Key Point

It is well established that an external attentional focus (i.e., focus on things external to the body and the outcome of a task) is broadly more effective for learning compared to an internal focus (i.e., focus on things within the body and movement during the task).

Assess intrinsic dynamics
- Athlete performs exercise without instructions
- Strength and conditioning professional assesses coordination dynamics

→ **Develop coordination dynamics**
- Athlete trains using the deadlift, front squat, and weightlifting pulling derivatives
- Strength and conditioning professional refines technique as necessary

→ **Practice structure**
- Athlete trains using blocked practice
- Strength and conditioning professional increases practice variability as an athlete becomes more advanced

↓

Alter task constraints
- Athlete performs the target exercise under various task constraints
- Strength and conditioning professional prescribes heavier loads, different starting positions, and complexes with other exercises

← **Simplification of the task**
- Athlete trains with lighter loads to execute the full exercise
- Strength and conditioning professional prescribes exercise regressions to improve the coordination of the target exercise

← **Instructions**
- Athlete is presented with an example of a skilled performer
- Strength and conditioning professional provides external or internal cues to the athlete if the basic coordination pattern can or cannot be performed, respectively

↓

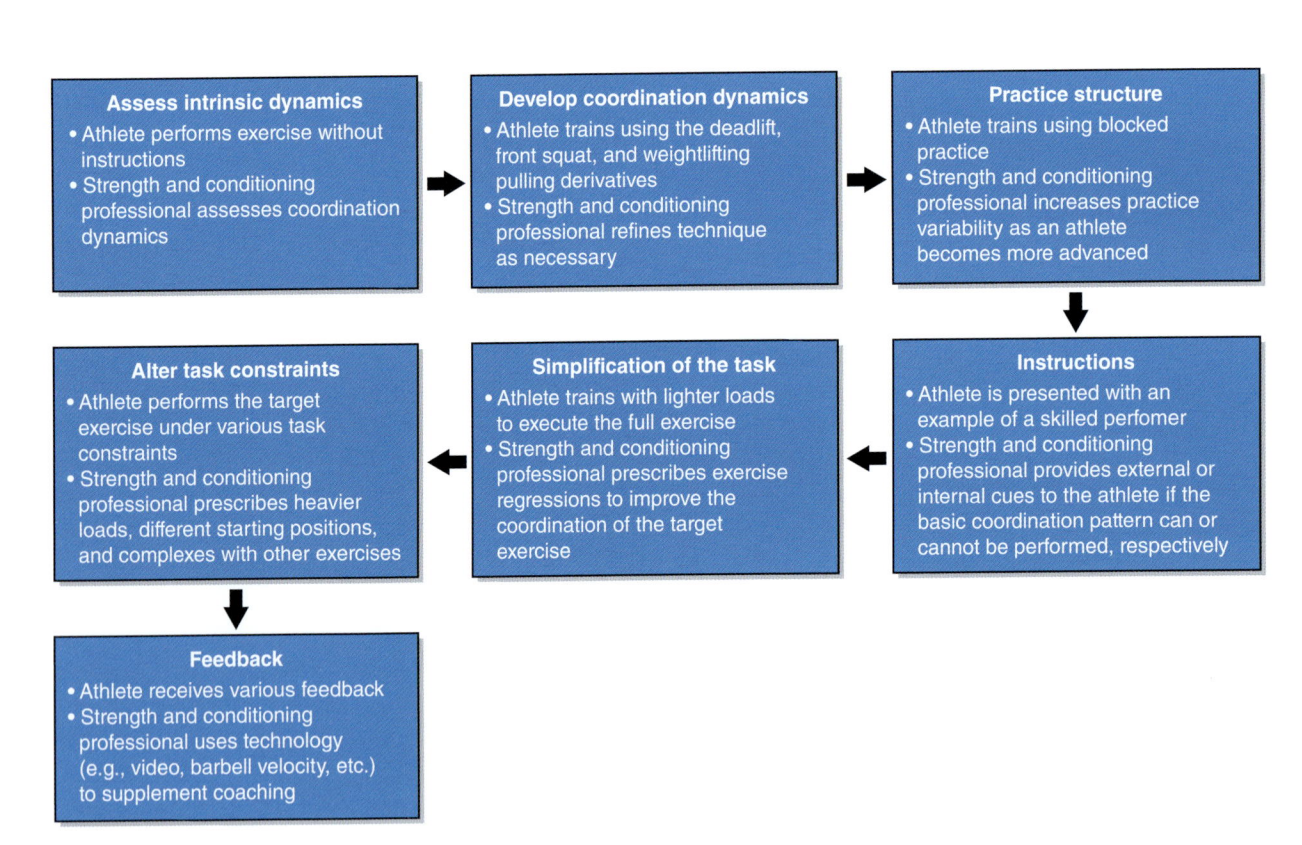

Feedback
- Athlete receives various feedback
- Strength and conditioning professional uses technology (e.g., video, barbell velocity, etc.) to supplement coaching

FIGURE 3.5 Developing a clean using a motor skill perspective.
Adapted by permission from G.L. Moir, *Strength and Conditioning: A Biomechanical Approach* (Jones & Bartlett Learning, 2016), 419.

TRAINING AGE

One of the 10 pillars of long-term athlete development focuses on motor skill and muscular strength development (159). Resistance training is safe for youth and adolescents if prescribed in a progressive manner that coincides with their abilities. Because *training age* is a general term that may have contextual definitions, strength and conditioning professionals should understand the differences between technical, tactical, and resistance exercise training age (RETA) (171). *Technical training age* relates to the ability of an athlete to make decisions and perform movements to gain a tactical advantage during sport contests. An athlete's *tactical training age* develops during sport practice under the guidance of the sport coach and improves their strategic problem-solving abilities during competitions. Finally, *resistance exercise training age* relates to the amount of resistance training experience an athlete has; of the terms just defined, RETA may be the most applicable to developing an athlete's muscular strength characteristics. Simply, an athlete's RETA influences their ability to adapt to specific training stimuli and may dictate programming decisions, which will be discussed in greater detail in chapter 7.

> **Key Point**
>
> Resistance training is safe for youth and adolescents if prescribed in a progressive manner that coincides with their abilities.

LEVERS

Muscles within the human body function as lever systems that comprise of fulcrums (joints), pulleys (tendons), and levers (perpendicular distances between points of muscular force or resistance and the fulcrum) (71). Within each lever system, force and resistance lever arms ultimately dictate the amount of force necessary to perform a given task. For example, a resistance arm may be altered based on the position of the external resistance relative to the joint or joints being used during a movement. However, it should be noted that the human body consists of different lever systems identified as a first-class, second-class, or third-class lever (71). An overview of the different lever systems is provided in table 3.2. Based on the type of lever system, an athlete may possess a mechanical advantage (i.e., less force is needed to complete a movement) or disadvantage (i.e., more force is needed to complete a movement).

ATHLETE ANTHROPOMETRICS

The overall size of an athlete (i.e., height and body mass) may directly contribute to the amount of force produced. For example, an athlete's muscle architecture, pennation angle, muscle insertion point, height, and moment arm may alter their mechanical advantage or disadvantage (discussed earlier) and may affect their ability to produce force. Regarding the impulse–momentum theory (chapter 1), larger athletes have the potential to create a greater impulse; therefore, they apply more force than smaller athletes due to a larger body mass. However, this often comes down to the ability of athletes to move at higher velocities as well. This may explain why larger athletes such as American football linemen may be assessed based on their ability to produce larger magnitudes of momentum rather than velocity (166). However, an athlete with a greater ratio of body mass to height may have an advantage over those with a lower ratio (234); that is, between two athletes with the same muscle mass, the shorter athlete may have a greater cross-sectional area and a potential strength advantage.

TABLE 3.2 Lever System Characteristics

Lever class	Characteristics	Example
First	Force and resistance arms are located equidistant from the joint	Atlanto-occipital joint
Second	Longer force arms than resistance arms; requires less force to complete a task relative to the resistance arm	Ankle joint
Third	Longer resistance arms than force arms; most common lever in the body; allows athletes to move with greater velocities	Elbow joint

The body dimensions of an athlete may also affect their ability to produce force within certain positions. Given that not all resistance training equipment is adjustable (e.g., bar height off the ground during exercises from the floor), strength and conditioning professionals must consider that athletes may not start in the same starting position during certain lower-body (42, 57, 160) and upper-body (108) exercises. Because a greater magnitude of work may be performed by taller athletes based on volume–load displacement (120, 251), the ability of taller athletes to develop strength within certain positions may present a greater challenge. This may, however, be dependent on the sport. As mentioned in chapter 1, former Turkish weightlifter Naim Süleymanoğlu snatched 152.5 kg and clean and jerked 190 kg in the 60 kg bodyweight category at the 1988 Seoul Olympics. Beyond the incredible ability of Süleymanoğlu to produce force efficiently during each lift, it is important to note that the displacement of each load was significantly smaller compared to those who competed in larger bodyweight categories. However, it should be noted that successful athletes may range across a wide spectrum of size despite playing the same sport (e.g., 2017 Major League Baseball American League Most Valuable Player [AL MVP] José Altuve is 5 feet, 6 inches [1.67 meters] tall and 2024 Major League Baseball AL MVP Aaron Judge is 6 feet, 7 inches [2.01 meters] tall).

Take-Home Points

- The magnitude and rate of force production may be influenced by many characteristics, including genetics and epigenetics, fiber type characteristics, neuromuscular factors, the neuroendocrine system, muscle architecture, motor learning, exercise technique, and various biomechanical factors.
- Although athletes are born with a specific fiber type composition, hybrid fibers may shift toward faster- or slower-type fibers based on neuromuscular activity, mechanical loading, and conditions, such as hypothyroidism or hyperthyroidism. Beyond genetic conditions, strength and conditioning professionals should understand how different training stimuli can influence potential shifts in fiber type.
- Heavy and ballistic-type training may positively increase motor unit recruitment, firing frequency, and synchronization while also reducing neuromuscular inhibition. This, in turn, may improve both the magnitude and rate of force development.
- Testosterone may augment neural adaptations associated with improvements in strength, which is why its elevation and ratio with cortisol have received so much attention. However, strength and conditioning professionals should be aware that the T:C ratio may simply represent a gross measure of the anabolic and catabolic properties of skeletal muscle rather than serve as an indicator of training status.
- Changes in muscle hypertrophy and, specifically, the Type II:I ratio can significantly affect the force production characteristics of an athlete. However, professionals should also be aware that the training stimulus provided may alter how hypertrophy adaptations can occur within the muscle (i.e., in series or parallel) and the type of hypertrophy produced (i.e., sarcoplasmic or myofibrillar).
- While actin and myosin have historically been discussed as the primary microfilaments that alter force production within the sarcomere, the large proteins titin and nebulin may play a substantial role in the rigidity of the sarcomere structure and force transmission.
- Success in sport requires athletes to perform various motor skills to achieve specific task goals. To increase an athlete's force production characteristics, various training stimuli should be provided to enhance existing coordination patterns and allow the athlete to produce and apply force more effectively.
- Some underpinning mechanisms of strength (e.g., genetics, anthropometrics) cannot be modified; therefore, strength and conditioning professionals should develop an understanding of how the remaining factors can affect force production.

CHAPTER 4

Principles and Organization of Training

There are five primary training principles that strength and conditioning professionals should consider when designing training programs for their athletes: overload, specificity, variation, reversibility, and individualization (see table 4.1). The ability to understand and manipulate the variables associated with each principle may have a significant impact on the development of both maximal and rapid force production characteristics (8, 97, 116, 117). Strength and conditioning professionals must also consider the organization of training through periodization to effectively integrate the five principles into the training process to optimize adaptations during specific time periods. This, in turn, may help manage fatigue and reduce the potential for nonfunctional overreaching and overtraining. The aim of this chapter is to provide readers with an overview of training principles and how periodization may be used to promote strength adaptations.

TRAINING PRINCIPLES

Overload, specificity, variation, reversibility, and individualization are the five primary training principles that serve as the foundation of exercise prescription. The following paragraphs will provide an overview of these training principles and how they may be used to drive the strength adaptations of each athlete.

OVERLOAD

An *overload* may be defined as any training stimulus that promotes an adaptation beyond the baseline abilities of an athlete. In general, overload stimuli are used to drive specific adaptations (e.g., physical, physiological, psychological) during different phases throughout the training year. Examples of overload stimuli may include either increases or decreases in volume, intensity (absolute or relative), range of motion, speed, or duration. However, the introduction of a new exercise or training method may also serve as a unique overload stimulus for athletes. Although some overload stimuli are more easily quantifiable (e.g., volume and duration), other stimuli may not be (e.g., variable resistance, eccentric training); moreover, a combined training stimulus of different methods (e.g., resistance training, speed development, change of direction, conditioning) may

TABLE 4.1 Primary Training Principles

Principle	Description
Overload	Training stimulus that promotes an adaptation beyond an athlete's baseline abilities
Specificity	Mechanical or metabolic similarity of an exercise or training method to a sport performance task
Variation	Manipulation of volume, intensity, set configuration, rest intervals, or exercise selection to remove linearity within a training program
Reversibility	Deterioration or loss of a previously gained fitness characteristic due to the removal or reduction of a training stimulus or involution from monotonous training
Individualization	Design of a training program based on an athlete's unique sport, event, or position as well as genetic characteristics and training age

further complicate the ability to quantify the overload stimulus. Thus, it is recommended that strength and conditioning professionals take a conservative approach when providing a progressive overload stimulus to their athletes to promote positive adaptations while managing fatigue.

From a strength development standpoint, it is important to provide a progressive overload stimulus that will benefit an athlete's force production characteristics during different training tasks. For example, researchers have shown that additional volume (71, 99), heavier loads (106), greater ranges of motion (10), faster running speeds (90), and sharper change-of-direction angles (107, 109) may provide a more intense stimulus for developing force production characteristics. However, it is important to put each variable into context, because the training status of the individual and the phase of training may dictate how the stimulus may affect an individual's strength characteristics.

First, while additional sets of exercise may benefit maximal strength adaptations, smaller versus larger doses may benefit novices and more experienced athletes to a greater extent, respectively (133). Moreover, it is important to note that reductions in training volume may also lead to improvements in strength characteristics, as shown within tapering strategies (87, 121). Similarly, heavier loads may increase maximal strength and peak force adaptations; however, decreases in load with certain exercises may also allow for the development of rapid force production characteristics (129, 130, 131). With range of motion, an increase may not always be possible, given an athlete's anthropometrics and joint mobility characteristics; however, decreases in range of motion may provide a unique overload stimulus, given the ability to use heavier loads (5) and ballistic intent (136) while providing athletes with sport-specific ranges of motion (100). Finally, from a running speed and change-of-direction angle standpoint, reductions in either may be required periodically to manage central fatigue, especially when these training methods are programmed with others. Conversely, reductions in running speed and cutting angles may allow athletes to focus more on developing different capacities, such as acceleration characteristics and the ability to maintain higher entrance velocities, respectively.

> **Key Point**
>
> Additional volume, heavier loads, greater ranges of motion, faster running speeds, and sharper change-of-direction angles may provide a more intense stimulus for developing force production characteristics.

As noted previously, the training status of an athlete may dictate how a strength and conditioning professional may prescribe a progressive overload stimulus to benefit an athlete's maximal and rapid force production characteristics. Simply, novice and less experienced athletes may not need as intense of a stimulus to achieve significant improvements in strength. In fact, researchers have shown that an emphasis on strength training alone may provide an effective stimulus for maximal and rapid force production compared to power or speed training (23, 24, 25, 58, 144). Moreover, the ability of an athlete to respond to and sustain an overload for extended periods of time may require a foundation of maximal strength (119). Such an emphasis may benefit tissue stiffness and tensile strength to enhance force transmission and reduce injury potential. Furthermore, greater mean, peak, and rapid force production may benefit velocity and power output, and enhance absolute (and possibly relative) endurance capacities—especially during high-intensity endurance activities—leading to a greater capacity to perform work (2, 12, 41, 110, 113, 134). However, as the strength characteristics of an athlete improve, strength and conditioning professionals are met with the challenge of continuing to provide unique overload stimuli to promote further adaptation and prevent performance stagnation. Within this scenario, the focus of training may shift from only improving maximal force production to providing a novel overload stimulus to emphasize rapid force production characteristics. A lack of change in training emphasis may lead to diminishing returns from training only to improve maximal strength (134). Researchers have shown that stronger individuals may adapt in a unique manner to stimuli focused more on rapid force production (25, 63). Furthermore, improvements in force production may be further enhanced when resistance training and sport training components (e.g., sprints, jumps, and change of direction) are combined (43, 68, 85, 101, 149).

SPECIFICITY

The principle of *specificity* is characterized by how similar an exercise or training method is to a given performance task regarding its mechanical or metabolic characteristics. Simply, greater program design specificity may lead to a greater transfer of training effect (i.e., degree of transfer to sport performance). *Mechanical specificity* refers to the similarity with the kinetic and kinematic characteristics of specific exercises and how they transfer to performances during athletic movement tasks: namely, the force, rate of force development, impulse, direction of force, range of motion, and movement pattern characteristics.

Related to the principle of specificity is the concept of dynamic correspondence, originally discussed by Yuri Verkhoshansky in the early 1990s (143). *Dynamic correspondence* provides a more nuanced version of specificity and includes five primary aspects: amplitude and direction of movements, accentuated regions of force production, dynamics of effort, rate and timing of maximum force production, and the arrangement of muscular work. While an overview of the individual aspects of dynamic correspondence is beyond the scope of this chapter, readers are directed to previous reviews that provide a thorough overview of each aspect (112, 122).

Apart from mechanical specificity, *metabolic specificity* refers to how closely a training stimulus matches the energy system characteristics (e.g., duration of activity and the magnitude and rate of energy utilization) of tasks performed in competition. For example, the dominant energy systems needed to be successful in baseball compared to running a marathon are considerably different, and they would thus require workouts that are more targeted to the demands of each sport or event. However, because the development of strength characteristics relies primarily on the phosphagen (i.e., adenosine triphosphate–phosphocreatine) system due to the focus of developing large magnitudes of force within shorter durations, the following will focus almost exclusively on mechanical specificity. Those interested in more information on metabolic specificity are directed to additional resources (11, 54, 119).

> **Key Point**
>
> Dynamic correspondence provides a more nuanced version of specificity and includes five primary aspects: amplitude and direction of movements, accentuated regions of force production, dynamics of effort, rate and timing of maximum force production, and the arrangement of muscular work.

Researchers have discussed mechanical specificity as it relates to the intramuscular (i.e., motor unit activation) and intermuscular (i.e., muscle activation patterns) organization of a given task (112, 122, 143), otherwise termed *task specificity*. Additional authors have suggested that task specificity may have a significant impact on strength–power and potentially endurance training (57, 65). These conclusions are based on the notion that task specificity may allow the motor cortex to organize motor unit activation and whole-muscle activation patterns (119). Thus, exercise stimuli that provide a greater degree of similarity with a given task increase the likelihood that the stimulus will transfer and lead to an improved performance (103, 116). Furthermore, intramuscular task specificity allows for only a specific motor unit task group to be activated for a defined activity (76, 103). This is evidenced by task-specific hypertrophy that has been displayed in sprinters (proximal thigh) (38, 56) versus track cyclists (distal thigh) (39) due to the biomechanical requirements needed to be successful in their events (118). Task specificity in this regard would then suggest that the back squat and front squat would be more specific to sprinters and track cyclists, respectively, owing to differences in muscle activation (22, 148). Thus, based on the underpinning mechanisms of strength discussed in chapter 3, these exercises would also serve as task-specific exercises that may benefit the force production characteristics for these athletes.

Based on the research just described, the specificity of exercise selection and mode of training becomes paramount to improving the transfer of training of strength characteristics to sport performance. Although different types of athletes possess similar movement patterns (e.g., sprinter vs. distance runner), the transfer of a given exercise may not necessarily provide the same benefits to each athlete because of the different needs of force production characteristics. For example, a power snatch or drop jump may better serve a sprinter compared to a distance runner, due to the need to develop large magnitudes to force rapidly during sprinting. In contrast, the demands of distance running typically require repeated submaximal forces to be produced during longer ground contact times (13).

Although some exercises may be chosen based on the perceived direction of force production and the involved musculature (21, 77), it is important that strength and conditioning professionals note that forces are being produced relative to the athlete rather than within a global frame (40). Regardless of the position of the athlete (e.g., standing upright, staggered stance with forward lean), vertical forces are being produced relative to the athlete's body. Thus, choosing exercises based on the force–vector theory may be questionable due to the concept of dynamic correspondence.

Finally, specificity of exercise selection may relate to similar force production characteristics being developed during training phases of different methods. For example, weightlifting derivatives performed from the floor, knee, or mid-thigh position may benefit the force production characteristics of athletes focused on improving the acceleration, transition, and maximal velocity phases of sprinting, respectively (29, 125). Thus, strength and conditioning professionals may consider exercise specificity when planning the phases of different training methods. The organization and planning of specific training phases will be discussed in greater detail later in this chapter.

VARIATION

Variation aims to remove the linearity of training programs by manipulating both overload and specificity characteristics to promote performance adaptations. In this regard, strength and conditioning professionals may manipulate a variety of training variables, including volume, intensity, set configuration, rest intervals, or exercise selection, to benefit an athlete's force production characteristics. Furthermore, planned variation may assist in preventing training monotony or strain and athlete staleness while also reducing fatigue and providing a wide spectrum of training stimuli (30).

Researchers have shown that more variation within training may produce greater training adaptations compared to programming constant repetitions with a higher or lower number of sets and training to failure or not to failure (82, 93, 120, 146, 147). An abundance of literature supports the idea of training using a heavy and light emphasis for the development of both maximal and rapid force production characteristics (50, 51, 91, 92, 124). Combined heavy and light variation may be introduced into the training process for athletes at both macro and micro levels (124). At the macro level, strength and conditioning professionals may provide variation to their athletes by planning out specific training phases used to develop specific fitness characteristics (e.g., hypertrophy, peak force, rapid force production, power output) through the periodization of the training program. Within this organization, athletes would be exposed to a combination of heavy and light loads while also developing the physiological characteristics that underpin improvements in force production. At the micro level, combined heavy and light loading stimuli may be provided within individual training phases, weeks, days, or exercises. A more thorough discussion of different programming strategies that can be used to improve both maximal and rapid force production characteristics is provided in chapter 7.

Key Point

Planned variation via volume, intensity, set configuration, rest intervals, or exercise selection may assist in preventing training monotony or strain and athlete staleness while also reducing fatigue and providing a wide spectrum of training stimuli.

Strength and conditioning professionals should note that the variation in periodization and programming may be based on the training status of the athlete and their needs (27, 111). For example, leading up to a competition, novice athletes may spend more time within an accumulation phase due to the need to develop various underpinning characteristics of force production. In contrast, advanced athletes may spend a greater portion of training time within the transmutation phase to fine-tune specific motor abilities before peaking their force production characteristics (60). Each scenario can be linked to specific windows of adaptation that an athlete may experience throughout their physical development (55, 80, 81). Thus, it appears that athletes with a greater training age may require more variation within their training programs compared to novices or those with less training experience. It should be noted that regardless of training status, athletes may not be able to tolerate consistently high-volume multi-joint exercises or heavy loading for extended periods, because this may lead to nonfunctional overreaching and potentially overtraining or injury (44, 45, 119). In fact, researchers have concluded that a similar training stimulus prescribed for longer than 6 weeks may lead to a plateau in maximal strength (138). Collectively, athletes with different training backgrounds may benefit from various degrees of variation within their training programs; this training principle may not only assist in maximizing force production adaptations, but it may also help manage fatigue and prevent training plateaus.

REVERSIBILITY

The principle of *reversibility* refers to the deterioration or loss of previously gained fitness characteristics, which may lead to a reduction in performance. These losses are typically caused by two factors: the removal or reduction of a stimulus that promoted the adaptation, or the involution that results from monotonous training or poor fatigue management (119). First, athletes may experience a detraining effect following the removal or reduction of a training stimulus; for example, a loss in rapid force production may occur during a strength–endurance phase that is characterized by higher training volumes (78, 121). In contrast, a loss in muscle size (i.e., muscle atrophy) may occur while training volume is reduced (3, 4, 139). Thus, strength and conditioning professionals may consider that reversibility is a product of programming variation because different training blocks or phases emphasize specific adaptations while others are de-emphasized.

It should be noted, however, that the loss of fitness characteristics may be mitigated—to an extent—through the use of retaining loads that comprise a small percentage of the overall training volume to maintain the desired adaptation (31). Yet it is important that strength and conditioning professionals distinguish between retaining loads that preserve previous adaptations and minimal effective doses that promote targeted training adaptations. When referring to strength improvement, the *minimal effective dose* denotes the smallest training stimulus that can be given to an athlete that improves maximal strength (46). Some researchers have indicated that a minimal effective dose for squat and bench press 1RM strength gain may include performing a single set of 6 to 12 repetitions using loads ranging 70% to 85% of the 1RM performed to failure two to three times per week for 8 to 12 weeks (1); however, the authors of the study noted that these adaptations may be suboptimal. Thus, strength and conditioning professionals should consider that a concentrated load focused on developing specific characteristics may produce residual training effects that persist into subsequent training phases. These concepts will be discussed in greater detail in the Periodization subsection later in this chapter.

Key Point

The losses in fitness characteristics due to reversibility are typically caused by two factors: the removal or reduction of a stimulus that promoted the adaptation, or the involution that results from monotonous training or poor fatigue management.

Specific to involution, a monotonous training program may result in a plateau or decrease in performance due to a lack of mechanical variation (42) or loading (132). As discussed earlier in this chapter, athletes can improve their ability to produce force during the programmed exercises, assuming the training program is appropriately designed and progressed with task specificity. However, it is possible that performing the same movements for a relatively long time (e.g., 12-16 weeks) without some variation may reduce the effectiveness of a given training stimulus and negatively affect the desired strength adaptations. In fact, researchers indicated that maximal strength adaptations may plateau if the same training program is prescribed for longer than 6 weeks despite small progressions in loading (138). Additional literature has suggested that continual increases in loading (i.e., linear programming) and consistent training to failure may also negatively affect training results, potentially leading to nonfunctional overreaching and overtraining (132). Similarly, involution related to poor fatigue management may occur when training programs promote nonfunctional overreaching without supplementing the program with adequate recovery periods (44, 45). For example, researchers showed that training to failure using RMs for 10 weeks resulted in moderate to large reductions in rate of force development (50 and 100 ms) compared with the implementation of set–repetition best loading (i.e., relative intensities based on a percentage of the best performance of a given set and repetition scheme) (17). The authors also concluded that training to failure produced significantly greater training strain during the final 7 weeks of training, indicating that this method of programming may ignore the potential ramifications of training involution.

INDIVIDUALIZATION

Using the principles discussed earlier, strength and conditioning professionals must also consider the importance of *individualization* when designing training programs. The needs of each athlete within their sport, event, or position are unique, and the genetic characteristics and training age of athletes suggest that they may all respond differently to specific training stimuli. Although some sport coaches may be reluctant to use strength and conditioning services based on their impression that all strength and conditioning programs are the same regardless of the sport or event, it is important to acknowledge that training programs across the sports spectrum are more similar than they are different. For example, the demands of baseball and volleyball athletes are considerably different, especially when different positions are taken into account; however, upon close inspection, each sport requires similar movement patterns, such as triple extension and flexion of the hip, knee, and ankle joints (e.g., jumping, landing, fielding or receiving position, diving), hip and torso rotation (e.g., baseball throwing and hitting vs. volleyball hitting, serving, and passing), accelerations, decelerations, and rapid changes of direction. Thus, many of the same exercises may be used by strength and conditioning professionals to develop these athletes. Although it is not advisable to create "cookie-cutter" programs in which all the exercises, loads, volumes, and so on are identical, exercise selection for the aforementioned athletes may be 80% to 90% similar, with 10% to 20% comprising the individualized portion.

Key Point

The needs of each athlete within their sport, event, or position are unique, and their genetic characteristics and training age suggest that they may all respond differently to specific training stimuli.

As mentioned previously, an athlete's training age can have a significant impact on the desired training outcomes. However, strength and conditioning professionals must also consider the exercise technique competency of their athletes. For example, it would be considered inappropriate to prescribe the snatch exercise to a novice athlete without demonstrating and coaching them through a step-by-step progression. Previous literature has provided technique guidelines for a wide range of traditional exercises, such as the squat (20), bench press (49), pull-up and lat pull-down (74), and deadlift (95). Additional authors have discussed the technique of several weightlifting movements and their derivatives (32, 33, 34, 35, 36, 126, 127, 128), but they also provide insight into how strength and conditioning professionals may progress and regress their athletes when teaching these technical exercises (19, 86, 123). While more complex variations of exercises may be an end goal, less complex variations may allow athletes to demonstrate similar or greater force production characteristics. For example, researchers have shown that athletes unfamiliar with the hang power clean exercise were able to improve their force production characteristics during vertical jumping within 4 weeks (52).

Researchers have investigated various assessments that have been used to guide training prescription, including the dynamic strength index (108), force–velocity profiles (104, 105), and eccentric utilization ratio (83). Although each method may help strength and conditioning professionals identify the weaknesses of an athlete, it is important to be aware of the shortcomings of these methods as they relate to training prescription for strength characteristics. For example, previous research has suggested that the dynamic strength index should not be taken at face value and should be examined within the context of a larger athlete monitoring testing battery (135). Second, researchers have shown that force–velocity profiles may not always be reliable (66, 67, 75) and may fail to appropriately identify a force or velocity imbalance (140). Finally, despite using various iterations of the eccentric utilization ratio, the metric does not appear to be strongly correlated to jump, sprint, or change-of-direction tasks (69); in addition, it may not provide accurate representation of jumping ability (70) and may not change despite improvements in jump performance (47, 53).

ORGANIZATION OF TRAINING

With differing competition seasons between sports, it is important for strength and conditioning professionals to logically organize the training programs of their athletes. By doing so, athletes can participate in a structured program that provides a timeline for when specific training goals should be attained, rather than unorganized programs that provide little direction and instead introduce chaos. Training plans can be organized over multiple timelines; for example, training may be based on a single year for collegiate athletes, whereas quadrennial plans may be used for those who hope to compete in the Olympics. Regardless of the training period duration, each year or cycle may be organized in an annual plan and further subdivided using various periodization strategies, both of which are discussed next.

ANNUAL PLAN

Following the conclusion of a competitive season, strength and conditioning professionals typically begin planning for the upcoming year by creating an annual plan. This plan includes several aspects related to the preparation of an athlete, such as all training, competition, athlete testing or monitoring sessions, and recovery sessions that are scheduled to take place (30). Thus, an annual plan serves as a blueprint that is used to guide the training process throughout the year in preparation for the competition period. Annual plans can be quite thorough and include the periodization model and the planned programming for the various training phases (119). Briefly, periodization includes the manipulation of training variables in a logical, phasic manner to maximize the potential of achieving specific training adaptations (117). Because periodization is rooted in nonlinearity (i.e., variation) (111), a periodization model may be used to organize an annual plan into different phases and timelines to optimize adaptations at specific times, manage fatigue, and reduce the likelihood of nonfunctional overreaching and overtraining during the training year (30, 31, 96). Although program design concepts are discussed in chapter 7, the concept of periodization will be discussed in greater detail next.

Key Point

Because periodization is rooted in nonlinearity (i.e., variation), a periodization model may be used to organize an annual plan into different phases and timelines to optimize adaptations at specific times, manage fatigue, and reduce the likelihood of nonfunctional overreaching and overtraining during the training year.

PERIODIZATION

Although the concepts of periodization and programming may be similar, it is important to identify them as separate entities within the training process. Simply, *periodization* refers to the planning, organization, and timing of training phases used to target specific training adaptations, whereas *programming* strategies drive the development of the adaptations during different training phases by using various exercise selection, volume, intensity, and rest period tactics (27). Researchers have shown that strength–power improvements were greater using a periodized approach compared to one that was nonperiodized (98, 145). Furthermore, the existing strength and conditioning literature and observational evidence suggests that most advanced or elite-level athletes use some model of periodization (111). The traditional model of periodization was formalized by Matveyev in the 1960s (72, 79) and featured broader general, specific, competition, and transition periods that were further subdivided into fitness phases and timelines including macrocycles, mesocycles, and microcycles. In addition, Matveyev noted the removal of linearity and inclusion of "wave oscillations" in loading dynamics (72).

A periodization model is used to divide the annual plan into various training phases throughout the year (e.g., preparatory, competitive, transition) to promote the physiological adaptations underpinning force production characteristics to maximize performance and manage fatigue at specific time points of the competitive season (6). Each phase of training can be further subdivided into different designated timelines used to define time periods to develop or emphasize specific performance characteristics; these timelines may include a multiyear plan, an annual training plan, macrocycles, mesocycles, microcycles, training days, and training sessions (table 4.2) (30). Briefly, *macrocycles* contain several mesocycles, ranging from several months to approximately 1 year, and may include general preparation, specific preparation, competition, taper, and active rest phases. *Mesocycles* include summated microcycles (i.e., a block of training weeks in which volume and intensity are modified to target a specific training adaptation via a concentrated load) that may range from 2 to 6 weeks to allow for the accumulation, transmutation, and realization of a training stimulus. *Microcycles* include several summated training sessions with a common training goal that are completed over a shorter duration (typically 1 week). *Training days* include all of the various training stimuli (e.g., resistance training, recovery, conditioning) that take place throughout a single day. Finally, a *training session* refers to a specific time period in which an athlete is training. It should be noted that the duration of the previously discussed timelines may be modified based on the competition calendar, athlete training status, or accumulated fatigue from the previous phase of training (111).

Although the traditional model of periodization provides a strong foundation of training organization, it is not without its limitations. First, the traditional model of periodization fails to account for the fact that athletes cannot maintain a performance peak for long periods of time. Issurin (60) noted that although maximum strength may be maintained for up to 30 days, sport performance is multifactorial and athletes may only maintain peak performance for approximately 5 to 8 days. This becomes especially complicated when there are many competitions within a short period of time (28, 73).

The second limitation of traditional periodization is its goal to develop multiple performance characteristics simultaneously, including those that may be considered noncompatible (e.g., cardiorespiratory endurance and rapid force production). Simply, increases in training volume may benefit the development of fitness characteristics; however, if multiple fitness characteristics are targeted simultaneously, a reduction in training volume may also negatively affect each characteristic (142). In addition, the simultaneous increase in overall work may make managing fatigue more difficult (14, 117), will likely favor endurance over rapid force production

TABLE 4.2 Periodization Cycles

Period	Duration	Description
Multiyear plan	2-4 years	A 4-year training plan is termed a quadrennial plan.
Annual training plan	1 year	The overall training plan can contain single or multiple macrocycles. Is subdivided into various periods of training including preparatory, competitive, and transition periods.
Macrocycle	Several months to a year	Some authors refer to this as an annual plan. Is divided into preparatory, competitive, and transition periods of training.
Mesocycle	2-6 weeks	Medium-sized training cycle, sometimes referred to as a block of training. The most common duration is 4 weeks. Consists of microcycles that are linked together.
Microcycle	Several days to 2 weeks	Small-sized training cycle; can range from several days to 2 weeks in duration; the most common duration is 1 week (7 days). Composed of multiple workouts.
Training day	1 day	One training day that can include multiple training sessions; is designed in the context of the particular microcycle it is in.
Training session	Several hours	Generally consists of several hours of training. If the workout includes >30 min of rest between bouts of training, it would comprise multiple sessions.

Reprinted by permission from G.G. Haff, "Periodization," in *Essentials of Strength Training and Conditioning*, 4th ed., edited for the National Strength and Conditioning Association by G.G. Haff and N.T. Triplett (Human Kinetics, 2016), 587. Adapted by permission from G.G. Haff and E.E. Haff, "Training Integration and Periodization," in *NSCA's Guide to Program Design*, edited for the National Strength and Conditioning Association by J. Hoffman (Human Kinetics, 2012), 220.

characteristics (37, 48, 102), and may impair the ability to learn new skills (9).

Finally, it may be difficult to implement traditional periodization within team sports. The original model of periodization focused on fewer major competitions per year, whereas team sports have increased the number of competitions performed within a season over the past several decades (62). To address the previous issues, several other models of periodization have been developed, including the conjugated successive system (142), block periodization (59), and phase potentiation (114, 115). Because each of the more contemporary models of periodization promotes the use of training blocks or phases, the following discussion will focus on block periodization. However, readers are referred to Stone and colleagues (111) for a thorough overview of periodization.

Block periodization is characterized by highly concentrated workloads (figure 4.1) that are focused on developing specific characteristics (i.e., blocks) and sequenced in a logical order to produce and benefit from residual training effects (111). The primary difference between block periodization and other models is the development of fitness characteristics throughout the training process (30, 31, 62, 117). As mentioned previously, the traditional model may aim to develop several performance characteristics simultaneously (59, 117). Although the traditional model of periodization may be effective compared to a nonperiodized training approach, it may not appropriately manage neuromuscular fatigue or address peaking periods during the competitive season (59). However, block periodization may be implemented in a single-targeted or multitargeted fashion (table 4.3) to address these modern issues and the needs of a given sport (62).

A single-targeted block periodization approach uses a concentrated load to emphasize the development of a single or very few related fitness characteristics (e.g., maximal and rapid force production) while other characteristics are de-emphasized (133, 142). Although this model allows for the use of summated microcycles and retaining loads to develop compatible characteristics and maintain the de-emphasized characteristics, respectively (30, 31), this approach may not be ideal for team sports that require the simultaneous development of multiple performance characteristics and technical factors (15, 72). However, it is important to note that if individual blocks are sequenced appropriately, the residual training effects may carry over and enhance

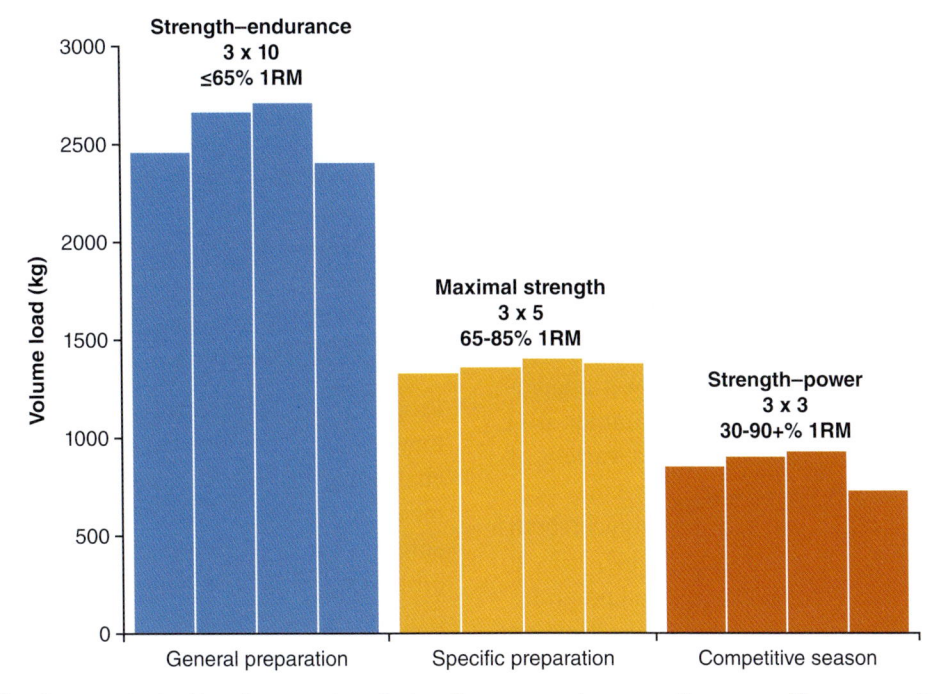

FIGURE 4.1 Concentrated load examples during the general preparation, specific preparation, and competitive season phases of training.

TABLE 4.3 Single-Targeted and Multi-Targeted Block Periodization

Descriptor	Single-targeted block periodization	Multi-targeted block periodization
Purpose	• Develop a single fitness characteristic • Maintain previously developed fitness characteristics	Develop multiple fitness characteristics concurrently
Rationale	• Useful in sports where relatively few tasks are developed, especially those developed simultaneously (e.g., track and field)	Necessary for sports that require multiple factors to be developed simultaneously (e.g., soccer, lacrosse, basketball, hockey)
Loading strategies	• Concentrated loads • Focused training volume of compatible factors • Minimizes volume of noncompatible factors • Summated microcycles • Retaining loads	Emphasis placed on training compatible fitness characteristics simultaneously Incompatible stimuli are ideally avoided during training
Additional benefits	• Superior delayed training effects following a period of restitution compared to multi-targeted block periodization • Phase potentiation effects	

Adapted by permission from T.J. Suchomel, S. Nimphius, C.R. Bellon, and M.H. Stone, "The Importance of Muscular Strength: Training Considerations," *Sports Medicine,* 48 (2018): 765-785.

(potentiate) subsequent blocks (64, 114, 115, 144). In contrast to a single-targeted approach, multitargeted block periodization aims to develop multiple compatible fitness characteristics (e.g., strength, power, and speed) simultaneously (61, 133). This method may be implemented more effectively with team sport athletes; however, it is important that strength and conditioning professionals take care in managing fatigue because the development of multiple characteristics may raise the overall training volume. In addition, it is crucial to avoid the development of noncompatible characteristics (e.g., speed and cardiorespiratory endurance) within an individual training block to avoid potential interference effects. The use of multitargeted training approaches may be viewed as a method of concurrent training; this concept and multiple approaches to training are discussed in chapter 8.

While it has been suggested that block periodization may result in greater strength–power adaptations (30) and an efficient training stimulus (94) compared to other models, there are a variety of programming methods (see chapter 7) that may benefit these characteristics and manage fatigue in both individual and team sport athletes (7, 141, 151). It should be noted that although the selected programming tactics may not have a considerable effect on the strength characteristics in novice athletes, these decisions become more important as the training age of athletes increases. Furthermore, sports with consistent competitions year-round (e.g., soccer, tennis) and nontraditional sports (e.g., snowboarding, BMX) have schedules that may require extensive modifications to the more traditional phases of training described earlier. However, it is important to recall that maximal strength may serve as a "vehicle" that may drive the improvement of other variables such as rapid force production and power output (134). Furthermore, because of the time constraints required to express peak force (e.g., ≥300 ms) often exceeding the timeframe needed to complete various sport skills (e.g., jumping, sprinting, change of direction), rapid force production should be viewed as an important quality that underpins sport performance (50, 113, 137).

SEQUENCING

Programming a proper *sequence* of training phases may lead to enhancements in various performance characteristics (16, 17, 51). Specific to the development of strength characteristics, it is important to enhance the underpinning characteristics of force production while using the residual training effects from different phases (see figure 4.2). Previous authors have suggested that the work capacity and

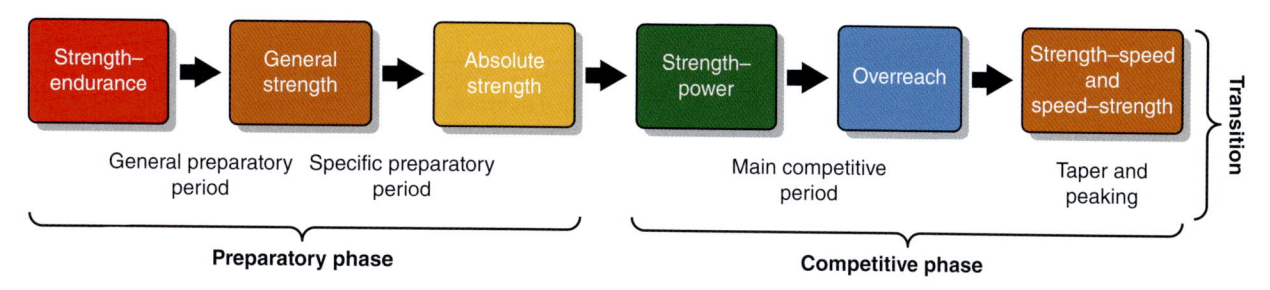

FIGURE 4.2 General sequence of training periods from a preparatory phase leading into a competitive phase.

Adapted by permission from T.J. Suchomel, S. Nimphius, C.R. Bellon, and M.H. Stone, "The Importance of Muscular Strength: Training Considerations," *Sports Medicine, 48 (2018): 765-785.*

muscle architectural adaptations that result from a strength–endurance phase may enhance an athlete's ability to improve their maximal strength (84, 114, 150). Additional literature supports the notion that developing maximal strength may increase the potential to improve rapid force production characteristics and power output (26, 133). Thus, a logical sequence of training phases would be moving from strength–endurance and general strength phases to a strength–power phase. This sequence of training phases may benefit from the training residuals that are produced in the previous phases and lead to enhancements in subsequent phases. Strength and conditioning professionals should be aware that the removal of a training stimulus that targets specific force production characteristics may lead to a loss in these characteristics based on the principle of reversibility. However, various authors have concluded that maximal strength may persist for relatively long periods of time (30 ± 5 days), whereas the loss of rapid force production characteristics happens quickly (5 ± 3 days) (59, 60, 88, 89, 111).

Key Point

Programming a proper sequence of training phases may lead to enhancements in various performance characteristics.

Although a sequence of strength–endurance, strength, and strength–power seems appealing for the development of strength characteristics, the training year requires more than three training phases because maximal strength may plateau in approximately 6 weeks if too similar of a training stimulus is prescribed (138). Thus, the wave-like variation in training loads described earlier may be required to use a long-term approach in developing both maximal and rapid force production characteristics. Interestingly, researchers have shown that performing a strength phase prior to a higher-volume phase focused on hypertrophy may lead to further benefits in both maximal strength and hypertrophy (18). Thus, strength and conditioning professionals may consider returning to a higher-volume training phase following a general strength phase to further benefit the muscle architectural characteristics that underpin force production. Upon returning to a previously completed phase of training, the use of different training tactics (e.g., methods, sets, repetitions, loads, set structure) may help provide a novel training stimulus to the athlete. However, it is important to note that small modifications to training phases may result in a significant change to the overall training stimulus; thus, a conservative approach is recommended when determining appropriate programming variables.

Take-Home Points

- Overload, specificity, variation, and reversibility are training principles that guide program design. However, the individualization of a training program may allow strength and conditioning professionals to target the weaknesses of an athlete and the long-term development of their strength characteristics.
- The concept of dynamic correspondence includes the amplitude and direction of movements, accentuated regions of force production, dynamics of effort, rate and timing of maximum force production, and the arrangement of muscular work and serves as the foundation of specificity. A greater understanding of dynamic correspondence may allow the selected training stimuli to more effectively transfer to sporting tasks.
- Before designing a training program, an annual plan that includes the planned phases of training, competitions, breaks, practices, and so forth should be developed. The planning and organization of an annual plan can be completed by using a method of periodization, which may allow an athlete to maximize their force production characteristics while their fatigue is effectively managed.
- The existing literature supports the use of block periodization whether it be a single-targeted or multitargeted approach, due to the ability to emphasize the development of specific characteristics within a set timeframe. Professionals should note that sports with year-round competitions (e.g., soccer, tennis) and nontraditional sports (e.g., snowboarding, BMX) may require more extensive programming modifications because competition schedules are not always consistent.
- Strength and conditioning professionals should note the impact of sequencing various training phases on the development of both maximal and rapid force production. Specifically, it is important to sequence training phases to use the training residuals produced from previous phases to enhance subsequent phases.

CHAPTER 5

Training Methods

Strength and conditioning professionals can prescribe a variety of training methods to enhance an athlete's strength characteristics. While the potential of each method to improve maximal or rapid force production may be dependent on the training age of an athlete, some methods may serve as superior alternatives and may transfer to motor skill performances during competition to a greater extent. In addition, other training methods may better serve athletes who have already developed a foundation of maximum strength and require a novel training stimulus for continued improvement. The aim of this chapter is to discuss a wide variety of training methods that strength and conditioning professionals may consider using to improve the strength characteristics of their athletes.

TYPES OF TRAINING

Many resistance training methods may be used to enhance an individual's strength characteristics. While each method has its own unique qualities, it is important for strength and conditioning professionals to understand the ability of each method to enhance an athlete's peak and rapid force production. In addition, practitioners should be aware of the neuromuscular demand of each method because several are often prescribed concurrently. This section will provide an overview of various resistance training methods and highlight their potential benefits and limitations. With this knowledge, strength and conditioning professionals should be able to make educated decisions on which methods may best address the desired strength qualities of their athletes.

BODYWEIGHT EXERCISE

The use of bodyweight exercise, also known as *calisthenics*, dates to ancient Greece and the words *kàlos* (meaning "beauty") and *sthénos* (meaning "strength") (217). Exercises that fall under this umbrella include bodyweight squats, push-ups, pull-ups, sit-ups, and many others. It should be noted that although plyometric exercises (discussed later in this chapter) are often performed using only an individual's body weight as resistance, they are prescribed for a different purpose and should thus not be grouped with traditional bodyweight exercises.

Bodyweight exercises are often closed-chain exercises that target multiple muscle groups and may increase the relative strength of individuals with less weight training experience (122). In addition, bodyweight exercises are often accessible and versatile in the sense that they may be performed in a variety of positions (e.g., traditional, wall, or feet-elevated push-ups). When bodyweight exercises are properly progressed, it may be possible to increase an individual's strength (156). Due to the simplicity and versatility of these exercises, it is not surprising that individuals in the fitness industry have consistently ranked bodyweight exercise as a top 10 fitness trend since 2013 (304). However, regarding the physical preparation of athletes, bodyweight exercises often serve as an introductory form of resistance training before progressing to loaded exercises and more complex movements. For example, athletes may perform bodyweight squats before progressing to a goblet squat, front squat, or back squat.

The primary limitation of bodyweight exercise is the inability to consistently provide an overload stimulus to enhance maximal strength and force production (122). Simply, bodyweight exercises do not provide the opportunity to apply large magnitudes of force against one's body mass. To increase or progress the intensity of a bodyweight exercise, one is limited to altering body position or increasing the number of repetitions. While the former may be effective with youth and adolescent athletes as well as those returning from injury, consistently increasing the number of repetitions may promote muscle endurance qualities rather than maximal and rapid force production characteristics. In fact, several studies have shown that prescribing more repetitions of the same exercise may not improve an individual's strength characteristics (31, 175, 250, 307). Recognizing the benefits and limitations of bodyweight training may allow practitioners to implement this form of exercise more effectively.

Key Point

Bodyweight exercises often serve as an introductory form of resistance training before progressing to loaded exercises and more complex movements.

MACHINE-BASED TRAINING

It is quite common to see a wide variety of exercise machines within a commercial fitness setting. From a client and practitioner standpoint, this variety provides opportunities to target many muscle groups and the ability to use a variety of machines within the same training session. Compared to free weight training, machines may provide a safer alternative because the added resistance moves within a specific plane of motion and can be modified by pulley systems (cams), elastic material, hydraulics, or pneumatic resistance (241). In addition, machine-based exercises may provide the opportunity to isolate specific muscle groups, which may benefit tissue capacity development and injury rehabilitation (285). Whereas some researchers have shown that training with machine-based exercises may improve maximal strength to a greater extent than free weight training (241, 252), others have shown contrasting findings (236).

Machine-based exercises can include single-joint (e.g., knee extension) or multi-joint (e.g., leg press) movements. While single-joint exercises may isolate specific muscle groups to a greater extent than multi-joint movements, their ability to benefit strength characteristics that transfer to sporting movements may be questioned because few single-joint movements are performed in competition (14, 22, 215, 266). For example, single-joint exercises may improve strength; however, they may lack the coordinative pattern of sporting movements to improve the sport task itself (e.g., isolated plantar flexion and vertical jump performance) (58, 266). In contrast, multi-joint movements that involve several muscle groups may serve as a superior alternative to develop strength characteristics that transfer to sport (8, 23, 97, 266). Despite the potential of multi-joint, machine-based exercise to enhance the force production characteristics of individuals, it should be noted that free weight exercises recruit stabilization musculature to a greater extent (111, 182, 240, 266) and may thus place greater coordination and muscle recruitment demands on an athlete (285). Combined with the ability to increase free testosterone (241), free weight exercise may serve as a superior alternative to machine-based movements when the focus is developing greater peak and rapid force production characteristics that may transfer to sport.

ISOMETRIC TRAINING

Isometric exercises have been implemented in resistance training programs for decades. This method includes performing a muscle action in which the involved muscle lengths and joint angles do not change during each effort. Despite no measurable work being performed—due to the lack of displacement—isometric exercises (figure 5.1) may be effectively used to enhance the maximal and rapid force production characteristics of athletes (168, 171, 213, 214).

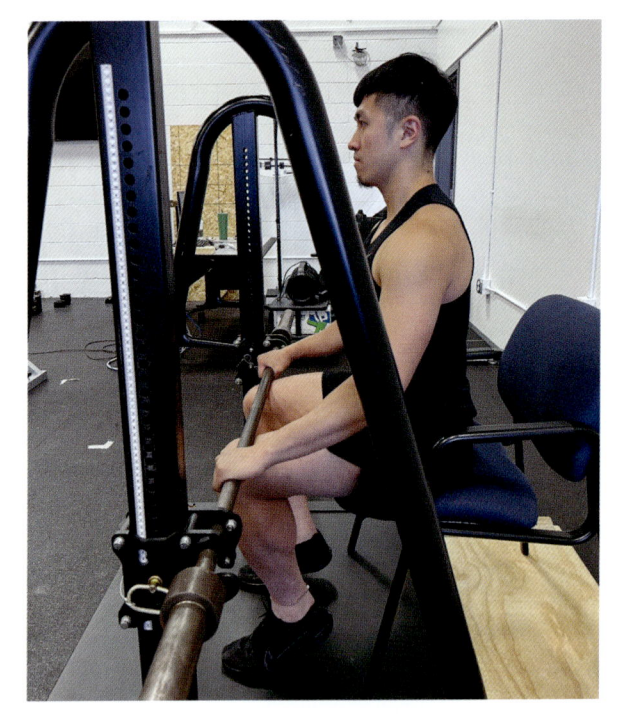

FIGURE 5.1 Isometric calf raise.
© Human Kinetics

Although single-joint isometric exercises such as knee extension (152), elbow flexion (59), and plantar flexion (168) have received attention within the scientific literature, several reviews (168, 171, 213, 214) and training studies (169, 170, 172) have highlighted various aspects of multi-joint isometric training. For example, researchers have shown that training with rapid, sustained isometric actions (3 seconds in duration) over 6 weeks produced greater isometric squat peak force adaptations compared with rapid, nonsustained isometric actions (1 second in duration) (169). The same research group also concluded that continuous inclusion of isometric squats for 24 weeks yielded greater maximal strength, sprint speed, and countermovement jump performance (172). Despite these findings, some may argue that isometric training lacks specificity when it comes to enhancing dynamic performance. However, it is important to consider that the force–time characteristics produced during isometric tasks may provide some insight into both upper- and lower-body dynamic performance. In this light, researchers concluded that peak force, as well as force produced at specific time intervals, during isometric tests may relate to various sport-specific performances, thus providing a rationale for their inclusion (171). The implementation of isometric exercises should not be overlooked due to the inclusion of isometric forces during a stretch–shortening cycle action (i.e., amortization phase). In theory, the ability to produce and sustain high magnitudes of force during this phase may allow athletes to "hold" higher braking (eccentric) forces and allow for greater early propulsion forces to be produced. Because of the potential to enhance peak force production and rate of force development characteristics (168, 171, 213, 214), practitioners may consider implementing said exercises during a resistance training phase that focuses on these qualities. However, it is important to consider the joint angle specificity of the isometric task and the positions in which an athlete may be during competition to promote effective transfer of training.

Strength and conditioning professionals should be aware that implementing isometric exercises with other exercises may provide a potentiation effect (see Potentiation Complexes later in this chapter). For example, various researchers have shown enhanced dynamic performances following different isometric exercise protocols (92, 96, 100, 110). Therefore, due to its simplicity and ability to train an athlete's strength characteristics, practitioners may consider implementing isometric training into resistance training programs.

Key Point

Despite no measurable work being performed—due to the lack of displacement—isometric exercises may be effectively used to enhance the maximal and rapid force production characteristics of athletes.

KETTLEBELL TRAINING

Originally used as a measure of weight in the 1700s, the term *kettlebell* originates from the Russian word *girya* and refers to a cast-iron weight in the shape of a cannonball and handle (50). Before becoming popular in commercial gyms and strength and conditioning settings around the world, kettlebells were primarily used by the Soviet military in their physical preparation training. From a strength and conditioning standpoint, kettlebell training serves as another alternative to traditional barbell and dumbbell exercises (30).

Common kettlebell exercises include two- or one-handed swings, modified weightlifting exercises (e.g., one-arm snatch or clean), and Turkish get-ups. Although researchers suggested that kettlebell training may improve various measures of maximal strength (139, 140, 159, 174, 216) and vertical jump performance (159, 216), additional research showed contrasting findings for vertical jump (140) and sprint (132) performance. To youth and adolescent athletes with lower training ages, kettlebell training may serve as an effective stimulus to enhance both peak and rapid force production. However, kettlebell training may also be limited in its capacity to consistently provide an overload stimulus. For example, researchers have shown that training with weightlifting movements (i.e., high pull and power clean combined with the back squat) produced greater back squat and power clean maximal strength and vertical jump improvements compared to training with kettlebells (216). Simply, the amount of weight an athlete can traditionally perform a clean with will exceed the amount of weight the same athlete can use during a kettlebell swing performed with good technique. Professionals should also consider the size of the kettlebell's handle and weight. For example, some commercial kettlebells weigh up to 109 kg (240 lb), and it is likely that the diameter of the handle will increase as the load increases. With the obvious limitation of grip for some athletes in this instance, strength and conditioning professionals must question whether there may be a more effective alternative.

Limited research has examined the long-term training effects of kettlebell training on an athlete's strength characteristics. While some have sought to improve the implementation of these exercises (123), it is important that practitioners understand the limitations of kettlebells and program their use based on the desired strength characteristics that are being pursued. Although they may not serve as the primary training stimulus for healthy athletes, kettlebells may serve as effective training tools for athletes recovering from an injury (25). Furthermore, kettlebell exercises may serve as part of learning progression or regression for athletes, but also as alternatives or supplemental movements to traditional exercises to prevent monotony in training and by extension, an athlete's desire to train.

BILATERAL AND UNILATERAL TRAINING

Bilateral exercises are those in which an external load is lifted with two limbs (e.g., back squat), whereas unilateral or partial-unilateral movements are defined as exercises in which the load is solely or primarily lifted by a single limb (figure 5.2) (183). While bilateral resistance is more commonly prescribed than unilateral training due to strong relationships between bilateral strength and general sport skills (e.g., jumping, sprinting, and change of direction [COD]) (286), the use of unilateral training often becomes a topic of debate within the strength and conditioning field. Proponents of the latter often cite that many movements in sport are performed unilaterally (e.g., sprinting, cutting); thus, some believe that training in this manner provides a greater degree of specificity and transfer. Interestingly, this may not be the case, because some researchers have shown greater adaptations to COD performance following bilateral versus unilateral training (11). These findings were supplemented by a 2021 meta-analysis that showed no differences between bilateral and unilateral training effects on short sprint and COD speed (192). Additional researchers indicated that similar improvements in bilateral and unilateral strength, sprint speed, and agility were shown after 5 weeks of either bilateral (back squat) and unilateral (rear-foot elevated split squat) training (260). Finally, Appleby and colleagues (10) showed that bilateral and unilateral training can improve movement-specific strength, but that each mode of training can transfer to the nontrained movement as well. Collectively, the existing literature supports the use of both bilateral and unilateral training; however, it has been suggested that strength and conditioning professionals should focus on the physiological stimulus rather than movement specificity when making programming decisions (11).

Key Point

Collectively, the existing literature supports the use of both bilateral and unilateral training; however, it has been suggested that strength and conditioning professionals should focus on the physiological stimulus rather than movement specificity when making programming decisions.

FIGURE 5.2 Rear-foot elevated split squat.
© Human Kinetics

An aspect of unilateral training that should not be overlooked when making programming decisions is the potential for increased muscle activation compared to bilateral movements. McCurdy and colleagues (184) showed greater gluteus medius, hamstring, and quadriceps muscle activation during a modified split squat compared to a traditional bilateral back squat. These findings are likely attributable to greater instability during unilateral exercises. While greater muscle activation may benefit muscle hypertrophy and strength, practitioners must consider the ability to load unilateral exercises safely. Simply, due to greater instability, a unilateral exercise cannot be prescribed with the same load as a bilateral exercise. As a result, bilateral exercises may provide a superior maximal and rapid force production stimulus due to their stability (19). Taking the research mentioned previously into account, strength and conditioning professionals should not exclude unilateral exercises but should instead prescribe them as assistance exercises to bilateral movements (227). Practitioners must also account for the additional volume that is associated with unilateral exercises because the number of repetitions may be doubled, assuming the athlete is performing the exercise with both limbs (e.g., 5 repetitions prescribed = 10 repetitions total). Due to the additional volume, unilateral exercises may be best implemented alongside bilateral exercises during preparatory phases of training that consist of higher training volumes. Obvious exceptions may include unilateral plyometric exercises during a strength–power phase of training.

STRONGMAN TRAINING

As discussed in chapter 1, strength competitions are rooted in the ability to lift various heavy objects. Strongman competitions have capitalized on this, and events include athletes lifting heavy loads for a maximum number of repetitions, carrying heavy loads for a given distance, or lifting a series of loads in the shortest amount of time possible (332). To prepare for the various events, strongman competitors may train using equipment such as loaded frames, stones, loaded sleds, logs, tires, sandbags, oversized dumbbells, and even some vehicles (331). While these pieces of equipment may allow for sport-specific training for strongman competitors, it should be noted that strength and conditioning professionals may use these objects as training tools with athletes who do not compete in strongman as well.

Researchers have shown that strongman training over a 7-week period may produce similar strength adaptations when compared to traditional resistance training (330). These findings may, in part, be due to the similar physiological demand placed on individuals during each method of training (121). It should, however, be noted that similar exercises performed using a strongman apparatus may not necessarily produce the same force production characteristics as a more traditional exercise. For example, Winwood and colleagues (328) showed that mean and peak vertical force production was similar between the strongman log exercise and the clean and jerk when performed at 70% of the participant's 1RM clean and jerk; however, mean and peak power output during the clean and jerk was significantly higher than the log lift. In another study, researchers showed that the barbell push press exercise produced significantly greater braking force, propulsion force, and impulse compared to both small and large log push presses (232). In addition, the back squat was shown to produce significantly greater vertical forces, but substantially lower anterior forces compared with a heavy sprint-style sled pull (327). While the aforementioned studies provide strength and conditioning professionals with just three examples of comparisons, further biomechanical analyses of strongman exercises such as the atlas stones (126), yoke walk (127), sprint-style sled pull (147), and farmer's walk (146) have also been completed. Because some strongman exercises may offer a novel training stimulus to athletes, additional research is warranted. Readers interested in further discussions of how to implement such exercises are directed to previous reviews (128, 329, 337).

BALLISTIC TRAINING

Exercises performed with ballistic intent (i.e., acceleration throughout the entire concentric phase of a movement), such as hexagonal barbell jumps and bench press throws, have been shown to produce greater magnitudes of power output than exercises that are nonballistic (272). Given that maximal strength (i.e., peak force production) and rapid force production underpin power output (298), it should not be surprising that ballistic exercise has been shown to produce greater maximal and rapid force, velocity, and power production as well as greater muscle activation compared to the same exercise performed quickly (160, 203, 292). Furthermore, researchers have shown that ballistic exercise may also allow for greater potentiation benefits to be realized compared to nonballistic exercise (246, 288, 289). This is likely due to the ability of ballistic exercises to lower the recruitment threshold of motor units (67, 308) and activate the entire motor neuron pool within milliseconds (78). As noted in chapter 3, the ability to recruit higher-threshold motor units favors large magnitudes of force being produced within a short duration, ultimately promoting favorable rapid force production adaptations. Therefore, it should come as no surprise that ballistic exercises are often prescribed for athletes throughout their careers; however, it may be important to consider the technical competency of individual athletes as well as their relative strength. For example, Cormie and colleagues (48) showed greater magnitudes of improvement in vertical jump performance with relatively stronger individuals following 10 weeks of training with jump squats. Further research showed greater improvements in jump squat velocity and net impulse in stronger participants compared to those that were weaker following combined weightlifting, plyometric, and ballistic training (137). In summary, ballistic exercises may serve as effective stimuli for desired strength adaptations; however, professionals must be selective when it comes to implementing these exercises with specific athletes and within resistance training phases throughout the year.

Key Point

Ballistic exercises may serve as effective stimuli for desired strength adaptations; however, professionals must be selective when it comes to implementing these exercises with specific athletes and within resistance training phases throughout the year.

PLYOMETRIC TRAINING

Plyometric exercises are rapid force production movements that include the stretch–shortening cycle in which an eccentric (lengthening) muscle action precedes a concentric (shortening) muscle action, with a brief amortization (i.e., isometric) period occurring between these phases. These exercises are unique in that they provide both a rapid braking stimulus and ballistic propulsion stimulus to the athlete. In fact, plyometric exercises may produce a series of adaptations benefiting an athlete's strength characteristics. In a series of meta-analyses, researchers concluded that plyometric training has the potential to improve maximal strength (66, 218, 229) as well as vertical jump (176, 218, 229), sprint (64, 218), and COD performance (13). In addition, plyometric training has also been shown to alter muscle thickness, pennation angle, and fascicle length, which may lead to increased tendon stiffness (229), thus modifying an athlete's force production characteristics. While much of the existing literature has focused on lower-body plyometric exercises, it is important to note that upper-body plyometric movements may also be implemented; however, they may not have as significant of an impact on an athlete's strength characteristics (254).

Although plyometric exercises are meant to benefit the rapid force production characteristics of an athlete, they may lack the ability to enhance maximal force. Some researchers have sought to combat this problem by having participants wear weighted vests during plyometric training (148). Although their results showed positive squat jump, countermovement jump, and five jump adaptations, prescribing greater loads during plyometric exercises may diminish the original plyometric stimulus. For example, greater loads upon landing during a plyometric exercise may lead to greater braking forces, longer durations between the eccentric and concentric muscle actions, or both. Therefore, professionals are encouraged to prescribe plyometrics in a periodized fashion by manipulating the intensity of the exercises (9, 84, 88, 90, 138, 141, 155) and the volume of contacts (85, 89). Moreover, it is important to consider the length of the training program, frequency of training sessions, and athlete characteristics (e.g., relative strength, exercise competency). Examples of different plyometric exercises at varying intensities are displayed in figure 5.3. It should be noted that there is no optimal method of implementing plyometric training (228). An important consideration is whether plyometric exercises are being implemented in isolation or are complementing other forms of training (e.g., traditional resistance training, sprinting, condition-

Depth/drop jumps (>30 cm)

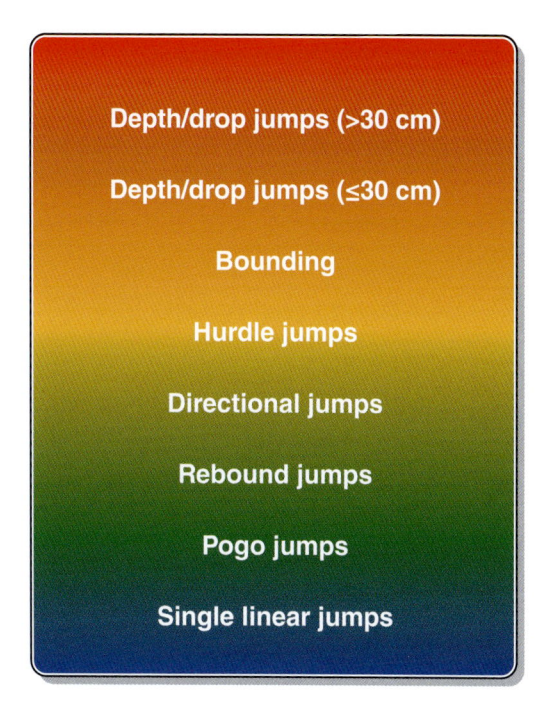

Depth/drop jumps (≤30 cm)

Bounding

Hurdle jumps

Directional jumps

Rebound jumps

Pogo jumps

Single linear jumps

FIGURE 5.3 Plyometric training intensity. Red indicates the highest intensity, while blue is the lowest intensity. *Note:* Unilateral variations of plyometric exercises have a higher relative intensity compared to bilateral variations.

ing). Although combined methods of training will be discussed later in this chapter, researchers have noted that plyometric training combined with traditional resistance training or weightlifting may lead to superior results compared to just one method on its own (65, 66).

WEIGHTLIFTING MOVEMENTS AND DERIVATIVES

Weightlifting movements such as the snatch, clean, and jerk and their derivatives (i.e., movements that are modified from the traditional competition lifts) have been consistently implemented by strength and conditioning professionals across a variety of sports and competitive levels (79, 82, 83, 86, 253). This is likely due to their ability to provide an effective overload stimulus during the rapid extension of the hip, knee, and ankle joints (i.e., triple extension). Moreover, these movements focus on moving light to heavy loads in a ballistic manner (273), likely producing favorable neuromuscular adaptations that benefit peak and rapid force production. Researchers have shown that weightlifting movements and their derivatives may provide a superior strength–power stimulus compared to traditional resistance training movements (12, 35, 36, 130), jump training (302, 306), and kettlebell training (216). A 2022 meta-analysis supports these findings for the improvement of an athlete's strength characteristics (195). Although the aforementioned research has primarily focused on propulsive strength characteristics, researchers have shown that weightlifting movements may provide a unique braking stimulus as well (191). However, it is important to note that a spectrum of weightlifting movements and derivatives exists and that the exercise–load combination of these exercises, as well as the technical competency of an athlete, may modify the propulsion or braking stimulus received.

Key Point

Weightlifting movements and their derivatives may provide a superior strength–power stimulus compared to traditional resistance training movements, jump training, and kettlebell training.

Although the traditional weightlifting competition lifts (i.e., snatch and clean and jerk) may be prescribed in an athlete's training program, derivatives (variations) of these lifts are often prescribed due to their decreased complexity (44). Weightlifting derivatives may be subdivided into catching, pulling, and overhead pressing derivative categories (table 5.1). Each is described next.

Catching derivatives modify either the starting position or catching position of the snatch or clean (43, 125, 269, 277, 299, 300, 301) and are often the most frequently prescribed weightlifting derivatives. As shown in the previous research, catching derivatives can be quite effective at improving an athlete's peak and rapid force production characteristics. However, as with all methods of training, it is important to note that exercises exist on a spectrum and may not always provide an optimal training stimulus for an athlete.

Weightlifting pulling derivatives may modify the starting position of either a snatch or clean but also remove the catch phase of the lift (274). By remov-

TABLE 5.1 Weightlifting Exercises and Derivatives*

Catching derivatives**	Pulling derivatives^	Overhead pressing derivatives
Midthigh clean/snatch	Midthigh pull	Push press
Countermovement clean/snatch	Countermovement shrug	Push jerk
Clean/snatch from the knee	Pull from the knee	Split jerk
Hang clean/snatch+	Hang pull+	Behind the neck push press/jerk^
Clean/snatch	Pull from the floor Hang hill pull Jump shrug	Behind the neck split jerk^

Reprinted by permission from P. Comfort, G.G. Haff, T.J. Suchomel, et al., "National Strength and Conditioning Association Position Statement on Weightlifting for Sports Performance," *Journal of Strength and Conditioning Research*, 37 (2023): 1163-1190.

*Variations from midthigh and knee can start with the barbell resting on blocks, or with the athlete holding the barbell and lowering to the start position and briefly pausing. Currently, there is minimal research comparing the kinetics or kinematics of these variations.

**All clean/snatch variations can be performed with a partial-squat (power) or full-squat catch.

^All derivatives may be performed with either clean or snatch grip.

+Starting with the legs extended, initiated by flexing the hips to perform a countermovement down to the knees (both above and below the knee commonly used), followed by the double knee bend and rapid triple extension.

ing the catch phase, athletes can focus on the rapid extension of the hip, knee, and ankle (plantarflexion) joints and potentially use loads in excess of a 1RM catching variation (45, 46, 112, 274, 287), thus promoting greater peak force adaptations. In addition, pulling derivatives such as the jump shrug (276) and hang high pull (275) may allow athletes to receive an effective rapid force production stimulus using light loads (<50% of the 1RM) with ballistic intent (150, 268, 270, 290, 291, 293, 296, 297). Interestingly, researchers have shown that individuals who train with weightlifting pulling derivatives may receive a similar (42) or greater (282, 283, 284, 291) strength–power propulsion stimulus compared to training with catching derivatives. Additional research has shown that the braking stimulus may be similar (47) or greater (280, 283) when training with pulling derivatives compared to catching derivatives; however, further research is needed in this area.

Finally, weightlifting overhead pressing derivatives are variations of the clean and jerk in which the clean is omitted and athletes typically lift the barbell off a squat rack, blocks, or stands before starting the movement (259). Less research has been completed on overhead pressing derivatives; however, these variations may provide athletes with an effective peak and rapid force production stimulus. Researchers have shown that athletes may be able to lift more weight during a jerk variation compared to a push press (256, 257); however, it is important to note that the technical competency of the athlete and the external load may alter the training stimulus (95, 258).

Collectively, weightlifting movements and their derivatives provide athletes with unique stimuli that may help them improve their peak and rapid force production characteristics. However, it is important to consider the training goals of an athlete and prescribe specific exercise–load combinations in order to achieve the desired adaptations (267).

POTENTIATION COMPLEXES

The term *potentiation* refers to the enhancement or increased potential of a given physical or physiological quality. Within a training method context, potentiation is typically realized as part of a complex (i.e., pairing) of two biomechanically similar exercises performed in succession. For example, a squat variation may be paired with a vertical jump (54, 288) or a bench press may be paired with a bench press throw (93, 149). Within these complexes, the goal is to enhance (potentiate) the performance of the latter exercise by using the former as a conditioning stimulus, often termed *postactivation performance enhancement* (PAPE) (56). An abundance of potentiation complexes have been examined within the scientific literature, with many researchers showing mixed findings (281). While the previous findings may be partially attributed to the design of the potentiation complex (e.g., exercise choice, volume–load, rest interval), researchers have shown that athlete characteristics—such as maximal strength—may have a significant impact on whether potentiation is realized (21, 37, 142, 190, 233, 244, 245, 289). Due to the influence that strength may have on PAPE, potentiation complexes have been classified as an advanced training tactic (285) and will therefore be discussed in greater detail in chapter 6.

VARIABLE RESISTANCE TRAINING

Large-muscle, multi-joint exercises such as the squat and bench press (and their respective variations) are commonly performed using eccentric and concentric muscle actions in which the load throughout the movement is constant. While prescribing exercises in this manner can certainly improve an athlete's strength characteristics, the mechanical advantage or disadvantage of each muscle group may change throughout the exercise. *Variable resistance training*, also known as *accommodated resistance training*, modifies the external resistance of an exercise to maximize force production throughout the entire range of motion (94), using chains or elastic bands.

Researchers have shown that variable resistance training modifies the loading profile of an exercise (136), which allows athletes to overcome mechanical disadvantages at various joint angles (87, 317) by matching changes in joint leverage (336). This may allow athletes to train through *sticking points*, in which an athlete typically demonstrates a diminished ability to produce force. Although the prospect of overcoming mechanical disadvantages may seem appealing for novice athletes, the change in resistance throughout the movement may create unnecessary variation in an athlete's technique, which may affect the athlete's ability to express force. While some variation is necessary for novices to enhance skill acquisition (i.e., varied practice) (202), adding too much variation, such as modifying the load throughout each repetition before exercise technique is solidified, may have a negative impact on the coordination of the movement. Thus, to ensure that the athlete develops movement competency without additional variation and to allow for its use as a novel training method later in the athlete's development, variable resistance training may also be classified as an advanced training method and will be discussed in greater detail in chapter 6.

ECCENTRIC TRAINING

Eccentric muscle actions are characterized by the forced lengthening of the musculotendinous unit due to greater resistive forces being applied compared to contractile forces being produced by the muscle (164). Published reviews have outlined the molecular and neural characteristics of eccentric muscle actions that may benefit an athlete's mechanical function (i.e., maximal strength, rate of force development, and power output), morphological adaptations (e.g., muscle fiber cross-sectional area and fascicle length), and neuromuscular characteristics (e.g., motor unit recruitment and firing rate) and performance (e.g., vertical jump, sprint, and COD) (77, 294). While greater detail on the underlying mechanisms of strength is provided in chapter 3, it should be noted that chronic eccentric training may produce similar or greater force production training effects compared to concentric, isometric, or combined eccentric and concentric training (76). This is likely because individuals are 20% to 60% (131) or approximately 40% (211) stronger during eccentric actions compared to concentric actions. This, in turn, has allowed researchers to investigate loads up to 150% 1RM during the eccentric phase of an exercise (80, 115, 133). Despite a growing body of literature, more information is needed to fully understand eccentric muscle actions and their contribution to an athlete's strength characteristics.

While training movements often consist of both eccentric and concentric muscle actions, an eccentric training stimulus places an emphasis on the eccentric phase through various means. For example, Mike and colleagues (189) discussed the use of the 2/1 technique (i.e., two limbs during the concentric phase, one limb during the eccentric phase), two-movement technique (i.e., compound exercise followed by isolation exercise), slow or superslow (i.e., tempo), and negatives with supramaximal loads (>100% of the 1RM). Additional details on the eccentric training stimuli potential and implementation of tempo, flywheel inertial training (FIT), accentuated eccentric loading (AEL), and plyometric training methods is provided in a two-part review

(294, 295). Finally, a 2022 review sought to classify different eccentric training methods by the speed of the eccentric muscle action (114). Although researchers have discussed the eccentric stimuli that may be produced with weightlifting catching and pulling derivatives (47, 191, 280, 283, 293), loaded jumps (161), eccentric cycling (41, 221), and COD drills (263), the following sections will focus on tempo, FIT, accelerated eccentrics, AEL, and plyometric training. The theoretical force production potential of each eccentric training method is shown in table 5.2.

Tempo

Perhaps one of the most common ways of providing an eccentric stimulus is by using tempo training, in which the duration of the lowering phase of a movement is increased (249). This method is often implemented using a submaximal load, whereby the concentric phase of the movement is performed with maximal intent following the prolonged eccentric phase. While the thought behind tempo training is to "overload" the eccentric phase of an exercise by increasing the time under tension, this method may be more suited for the development of muscular hypertrophy rather than strength (326). For example, researchers have shown that longer eccentric durations may lead to greater improvements in elbow flexor strength compared to shorter durations (151, 222); however, a much larger body of literature supports the notion that there are no favorable strength adaptations following longer eccentric tempos (15, 32, 99, 220, 243, 249, 265, 320, 323). It should be noted that most of the existing research has examined different eccentric durations using submaximal loading; thus, further research is needed using maximal or supramaximal loads when only the eccentric phase is performed, often termed *negatives* (189). Using tempo in this manner may allow for the recruitment of higher-threshold (type II) motor units and their associated fibers, whereas the use of submaximal loads may only stimulate Type I motor units, which may negatively affect the magnitude and rate of force production due to their slower contraction velocities (173).

Strength and conditioning professionals should be aware that longer tempos may negatively affect stretch–shortening cycle contributions to the magnitude and rate of force production. For example, Schilling and colleagues (237) demonstrated that purposefully slow durations reduce the potential for force production during resistance training movements. Moreover, researchers also showed reduced back squat (34) and bench press (34, 321) mean and peak barbell velocities during the concentric phase, decrements in countermovement jump height (243), and reductions in leg press rate of force development (265). Furthermore, the chronic use of longer eccentric tempos may increase perceived exertion (220), which may negatively affect overall training volume (98, 207, 322, 324) and intensity (325). Practically speaking, tempo training may be best implemented during resistance training phases that are characterized by higher training volumes and that emphasize the development of work capacity (295). Furthermore, while tempo training may affect muscle hypertrophy (15, 238, 326)—which can influence an athlete's force production characteristics—there does not appear to be much support for the use of long eccentric duration tempo training for improving maximal or rapid force production characteristics. However, as previously noted, further research examining the use of maximal and supramaximal eccentric training is needed.

TABLE 5.2 Theoretical Force Production Potential of Eccentric Training Methods

Eccentric training method	Peak force	Rapid force production*
Tempo eccentric training	++	+
Flywheel inertial training	+++	++
Accelerated eccentrics	++	++++
Accentuated eccentric loading	++++	++++
Plyometric training	+	+++++

Note: + = low potential; ++ = low to moderate potential; +++ = moderate potential; ++++ = moderate to high potential; +++++ = high potential.
*The technique and exercise may modify the force production characteristics.

Key Point

While tempo training may affect muscle hypertrophy, there is limited support for the use of long eccentric duration tempo training for improving maximal or rapid force production.

Flywheel Inertial Training

Flywheel inertial training (FIT) is another eccentric training method that has gained popularity. This method uses inertial resistance that is produced from different inertial discs, rotational speed, and the characteristics of the device itself (209). Briefly, through the concentric phase of a predetermined range of motion, an athlete accelerates the tether or cord that is wound around a portion of the device before the energy created by the athlete must be accepted during the eccentric phase and the tether or cord is rewound. Although FIT devices were originally investigated as a gravity-independent training tool in 1994 (20), an abundance of research projects have since been conducted regarding their physical performance benefits. As part of a 2022 umbrella review, de Keijzer and colleagues (61) sought to summarize the existing FIT literature as it relates to strength development. The authors concluded that all the included existing systematic reviews and meta-analyses (6, 177, 210, 225, 231, 313) suggested that FIT may be effective for improving muscular strength; however, it should be noted that many of the included studies used single-joint exercises, which may not transfer as effectively to sport performance as multi-joint tasks. Moreover, despite the large volume of FIT studies, very few have compared the training effects to other forms of resistance training; thus, it cannot be concluded that FIT provides a superior training stimulus (230).

A basic search of the literature reveals the frequent use of the term *eccentric overload*, referring to the potential to create greater eccentric force production characteristics with FIT compared to traditional training. While some researchers have shown that the impulse characteristics of each repetition increase when using larger flywheels (33), others indicated that no eccentric overload was found when examining force–time characteristics (197, 209). The latter findings are supported by the notion that FIT devices generally provide a closed (isolated) system in which the athlete must accept the eccentric impulse created during the concentric phase of the movement (law of conservation of energy). As noted in chapter 1, the shape of the impulse determines the stimulus that an individual receives, which, in relation to FIT, may be determined by the eccentric technique used (61, 294, 303). For example, following a maximum concentric effort, the forces produced during the eccentric phase of a flywheel exercise may be low, moderate, or high depending on the strategy of the individual (294). Higher mean forces would indicate that the athlete would be using a stiffer strategy to stop the momentum of the flywheel, whereas lower mean forces would suggest that a compliant strategy was used (figure 5.4). It should be noted that an athlete may also stop themselves rapidly without descending the same distance, ultimately creating a high-force, short-duration impulse, which would require high rates of force production. However, the ability to use a compliant strategy may also be based on the relative strength of the athlete (295), which in turn may modify the stimulus received. Although certain devices have since been developed that use motorized technology to provide additional eccentric forces, many devices fail to include such a stimulus. However, researchers have shown that when performing assisted squats (e.g., pushing with the arms to create additional concentric acceleration, as seen with a Hatfield squat), the eccentric stimulus could be increased (334). Another form of assisted flywheel exercise is through the use of the 2/1 technique (157), as described previously. Readers interested in how an eccentric overload stimulus may be provided with flywheel exercises are referred to a 2024 review by Martínez-Hernández (179).

An additional consideration with FIT is its feasibility in training and the ability to effectively prescribe training intensities. Researchers concluded that the biggest perceived barriers to FIT implementation among therapists and professional soccer practitioners are equipment cost or space, evidence of effectiveness, and scheduling (62, 63). Simply, athletes and strength and conditioning professionals are more likely to be familiar with traditional exercises using free weight equipment rather than FIT devices. In this light, FIT devices may serve as a novel training stimulus to athletes; however,

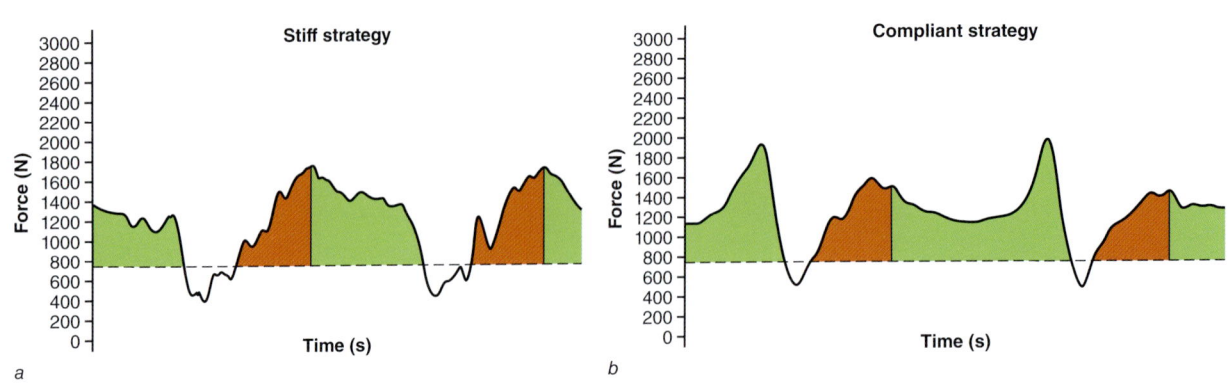

FIGURE 5.4 Stiff versus compliant flywheel strategies.

ecological validity is a valid concern, especially in a team setting. For the prescription of training intensities, it may be possible for strength and conditioning professionals to use movement velocity to track the intensity of FIT repetitions (33, 178). While this research is relatively new, it is important to investigate the potential differences between FIT exercises and traditional exercises using similar methods. Although the previous information outlined some of the existing literature on FIT, further information on the implementation of this training method can be found in several reviews (17, 18, 230).

Accelerated Eccentrics

Accelerated eccentric exercise requires athletes to perform the eccentric phase of a movement at an increased velocity that is the result of the eccentric execution strategy (114). Examples of these exercises may include both passive and active accelerated movements. Passive accelerated eccentrics require the athlete to perform the eccentric (braking) phase of a movement due to the downward acceleration caused from an external source (e.g., banded jumps). In contrast, active accelerated eccentrics are characterized by the athlete rapidly initiating the eccentric phase of an exercise before stopping (catching) the load in a certain position (e.g., hexagonal barbell drop catch; figure 5.5) (113).

The primary benefit of accelerated eccentrics is a rapid braking stimulus that may help athletes tolerate larger eccentric forces and benefit stretch–shortening cycle performance. Interestingly, despite being introduced as "accelerated powermetrics" by Verkhoshansky and Siff (312), minimal research has examined the effectiveness of this strategy. However, researchers have shown an increased eccentric demand (i.e., eccentric force, rate of force development, and impulse) during accelerated eccentric countermovement jumps (3) and drop jumps (1) compared to their traditional counterparts. It should be noted that propulsive phase characteristics were only improved during accelerated eccentric countermovement jumps, indicating that the difficulty of accelerated eccentric drop jumps may be too challenging for some athletes. This may be due to reflex inhibition in the presence of excessive eccentric loading (2). Van den Tillaar (309) demonstrated that higher forces are produced during a faster squat descent, thus providing the rationale for a loaded drop catch exercise. Unfortunately, no data exist regarding the longitudinal use of accelerated eccentric jumps or drop catches within training programs. Thus, although athletes may receive a rapid eccentric training stimulus acutely, the long-term training effects of accelerated eccentrics are unknown. Strength and conditioning professionals are therefore advised to interpret the existing findings with caution.

Accentuated Eccentric Loading

Another form of eccentric training has been termed *accentuated eccentric loading* or *AEL*. Although some have mistakenly classified FIT as AEL, this method requires that an exercise is performed using a heavier load during the eccentric phase compared to the concentric phase and that the movement is performed by pairing the eccentric and concentric phases with minimal disruption to the natural movement mechanics of the exercise (315). AEL has

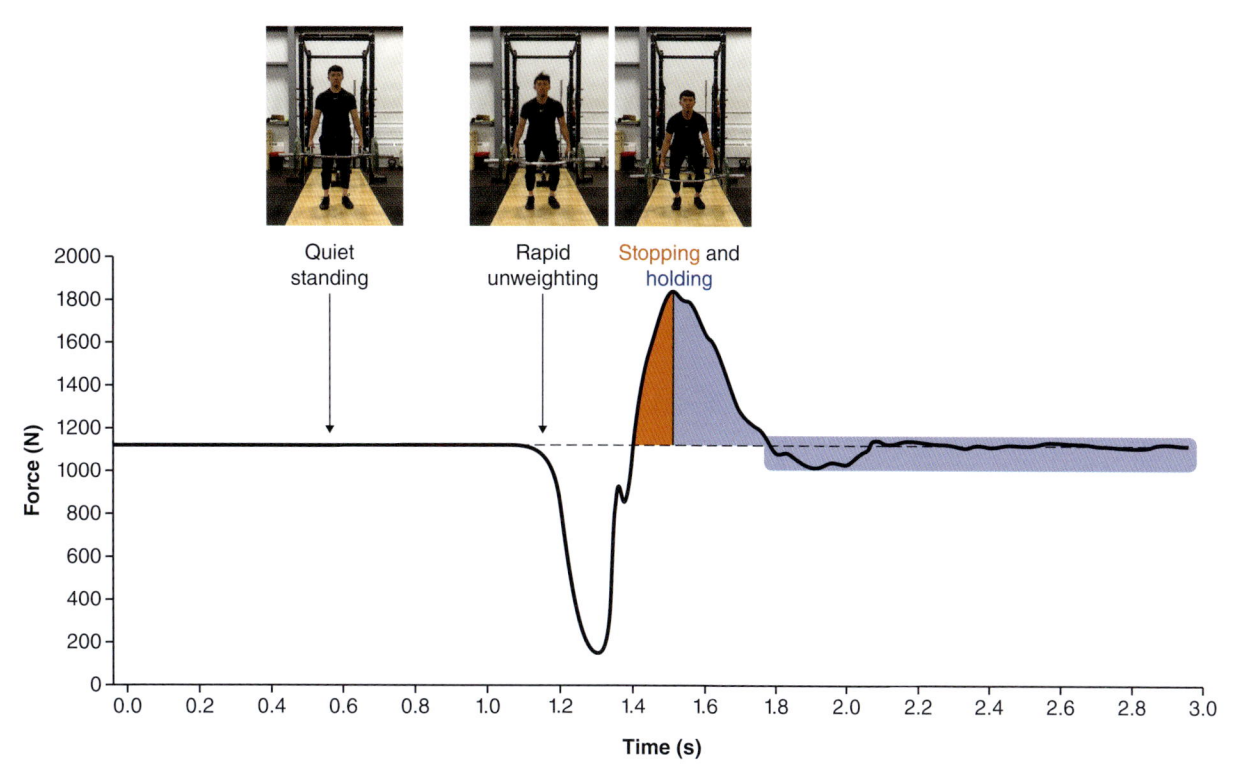

FIGURE 5.5 Hexagonal barbell drop catch.

become a popular topic of investigation due to its unique ability to produce favorable strength adaptations (316) and the potential to use submaximal (247), maximal, and supramaximal loading (187, 271) combinations. However, due to the changes in load during the movement, potential for very heavy load prescriptions, and the influence of relative strength (188), AEL has been classified as an advanced training method (295). Thus, a further discussion of AEL will be provided in chapter 6.

Plyometric Training as an Eccentric Stimulus

Although plyometric exercises were discussed earlier in this chapter, it is important to acknowledge their potential as an eccentric training stimulus as well. For example, these exercises include a unique eccentric stimulus due to the speed of the muscle action that occurs during the first phase of the stretch–shortening cycle (294). Specifically, tasks such as landing may provide athletes with a training stimulus that may help them accept the forces created upon ground contact (162). Because the speed at which the eccentric phase occurs directly affects the eccentric stimulus, strength and conditioning professionals must consider the impact of their coaching cues (278, 279, 335) and the intensity of the plyometric exercise (9, 84, 88, 90, 138, 141, 155) when implementing these exercises. Thus, these exercises should be properly progressed based on the desired volume and intensity (85, 89) and the abilities of the athlete (295) to mitigate the potential for injury during repetitive landings (4, 60). Furthermore, while less literature has focused on the intensity and implementation of upper-body plyometric exercises (254), it is logical to progress these exercises using similar principles.

ADDITIONAL TRAINING METHODS

Although the previously discussed training methods can serve as effective stimuli for improving muscular strength characteristics, additional methods that exist beyond the traditional exercises performed in a weight room setting should also be discussed. Because sprinting and COD tasks are used frequently in sport, the following paragraphs will discuss the ability to use them as strength stimuli. In addition, this section will conclude with a brief discussion regarding the potential use of isokinetic and blood-flow restriction (BFR) training to enhance the force production characteristics of athletes.

SPRINTING

Sprinting has been characterized as the net horizontal displacement of an athlete's center of mass during a maximal speed run lasting 15 seconds or less (235) and is most commonly viewed as a training outcome rather than a training stimulus. However, strength and conditioning professionals should note that sprinting may also serve as an effective training stimulus for both maximal and rapid force production. For example, elite sprinters may spend as little as 80 to 90 ms on the ground during each foot contact at top speed (318) while also producing forces between three and five times their body mass on a single limb (40). This rapid force production stimulus may induce both central and peripheral nervous system changes ultimately benefiting neural drive within the active muscle fibers (235). Moreover, sprinting may also target Type II muscle fibers and facilitate shifts to faster phenotypes, increase Type II muscle fiber cross-sectional area, and improve the Type II:I muscle fiber ratio (7, 308), thereby enhancing the potential for greater maximal and rapid force production characteristics.

It is important to note that the force production stimulus may change based on the phase of sprinting being trained. For example, Nagahara and colleagues (199) indicated that the vertical and horizontal forces produced during the acceleration, transition, and top speed phases may vary. Specifically, greater horizontal forces are produced during the acceleration phase compared to top speed, whereas vertical force production is favored during top speed sprinting (319). Because the ability to accelerate tends to be a universal skill in most sports, it should be noted that propulsive impulse characteristics appear to be a primary factor in enhancing this ability (135, 145, 194) as well as sprinting performance in general (200). Researchers expanded on this and noted that peak vertical force may serve as an indicator for achieving greater acceleration during the late acceleration phase (198). Because horizontal force is clearly important for acceleration performance (26, 186), sprint training strategies that allow athletes to produce high magnitudes of both horizontal and vertical forces would appear beneficial. The use of both inclined slopes (104) and resisted sled pulling (5, 28, 224) and pushing (29) may provide athletes with an effective stimulus that may benefit propulsive force production characteristics. Strength and conditioning professionals should be aware of the slope incline because it may modify the mechanics of the sprint and, by extension, the ability of an athlete to produce force in both the vertical and horizontal directions (212). Researchers indicated that there are biomechanical similarities between incline sprinting using 16% (approximately 9° slope angle) and 5% to 10.5% (approximately 3°-6° slope angle) gradients and early- and late-phase acceleration phases, respectively (212). During resisted sprints, some researchers suggest the optimal load is that which decreases maximal velocity by approximately 50% (approximately 80% body weight in adult men) (55); however, adaptations may be individualized (158, 193).

As athletes transition from acceleration to maximum velocity sprinting phases, ground contact time decreases while the vertical impulse demand increases (40, 318, 319). Thus, prescribing maximum velocity sprinting regularly to athletes may expose them to large magnitudes of force during much shorter ground contact times compared to the previous sprinting phases (39), thus providing an effective rapid force production stimulus. To enhance both maximal and rapid force production characteristics, strength and conditioning professionals may consider pairing sprinting phases with task-specific or mechanically specific exercises such as weightlifting movements and their derivatives (69). For detailed programs on how to pair sprinting progressions and resistance training strategies, readers are referred to works by DeWeese and colleagues (69, 70, 71, 72).

Key Point

To enhance both maximal and rapid force production characteristics, strength and conditioning professionals may consider pairing sprinting phases with task-specific or mechanically specific exercises such as weightlifting movements and their derivatives.

CHANGE-OF-DIRECTION TRAINING

Like sprinting, strength and conditioning professionals may view COD tasks as more of an outcome rather than a training stimulus; however, the ability to decelerate, slow or stop one's momentum, and accelerate in another direction may require an athlete to produce large magnitudes and rates of force development, depending on the velocity of the movement. Harper and colleagues (116) discussed the frequency of high-intensity (>2.5 meters per second) and very high–intensity (>3.5 meters per second) acceleration and deceleration demands in elite team sport. The authors reported that except for American football, each of the other analyzed sports, including Australian rules football, hockey, rugby league, rugby sevens, rugby union, and soccer, reported greater frequencies of high-intensity and very high–intensity decelerations compared to accelerations. As noted previously, propulsive force and impulse may underpin the ability to accelerate (135, 145, 194); however, to change direction effectively, it is also important to develop the ability of an athlete to decelerate. For the purposes of this discussion, *COD* will refer to a specific event where an athlete uses the necessary skills and abilities to change movement direction, velocity, or mode during both planned and agility conditions (68). To avoid confusion, *agility* is defined as skills and abilities needed to change movement direction, as opposed to a preplanned or predetermined COD (68).

As noted previously, a COD training stimulus is underpinned by both deceleration and acceleration demands. While the latter has been discussed in the context of sprinting, a greater understanding of deceleration as a training stimulus is needed. When examining maximal decelerations and accelerations, researchers have shown that deceleration forces may be approximately 2.7 times greater than accelerations (311). Moreover, these forces appear to be produced in a more "peaked" impulse shape indicating a rapid force production demand. Similar force–time profiles were also displayed during the preparatory deceleration steps (i.e., antepenultimate and penultimate) of 135° (201) and 180° (75) COD tasks. Strength and conditioning professionals should also note that the highest braking forces during COD tasks may occur during the late deceleration phase (120), thus stressing the importance of developing greater braking forces earlier during deceleration (75). For a thorough overview of deceleration, readers are directed to a 2022 review by Harper and colleagues (119).

Because the magnitude and the rate of force production are crucial for effective COD, the ability to generate such forces requires a focus on the physical preparation of the athlete. Researchers have highlighted the importance of developing eccentric (143, 262, 263), isometric (119, 261, 262, 263), and concentric (117, 118, 263, 338) strength characteristics to improve COD performance; however, it is important to note that the angle and velocity of the COD may alter the force production demands. For example, COD tasks ranging from 0° to 45°, from 45° to 60°, and from 60° to 180° may include limited, moderate, and substantial braking demands, respectively (74). Specifically, greater ground reaction forces are produced during sharper COD performances (49, 124, 239, 251) as well as those with a higher entrance velocity (57, 310). While strength and conditioning professionals should consider these conclusions when programming COD tasks for their athletes, it is important that a needs analysis be completed to determine movements that athletes will have to perform (68). COD training programs may use postural manipulations and progressive loading of "shallow" and "sharp" COD angles (73). Strength and conditioning professionals may also consider the importance of gradually progressing athletes by manipulating the constraints of the task (e.g., changes in time, loading) to increase the physical demand of the COD. This progression will eventually prepare the athlete for worst-case scenarios (i.e., introduction of an extreme internal response that threatens tissue integrity) to not only improve performance but also to minimize injury risk (144). Regarding the speed of the COD tasks, it is important to take an individualized approach before programming specific thresholds (181) while also considering the needs of the athlete's sport or event. Finally, it is important that strength and conditioning professionals effectively monitor the COD capabilities of their athletes during and following training interventions (204, 205, 206).

> **Key Point**
>
> Change-of-direction training programs may use postural manipulations and progressive loading of shallow and sharp COD angles.

ISOKINETIC TRAINING

As discussed in chapter 1, *isokinetic exercise* refers to performing a given movement (e.g., knee or elbow flexion or extension) in which the velocity of the movement is constant. First developed in the 1960s (129), isokinetic devices have been used in resistance training programs for rehabilitation of an injury (e.g., anterior cruciate ligament [ACL] rupture, chondromalacia, osteoarthritis) (16, 27, 51, 91, 314) and to address muscular size and strength imbalances (101, 102, 103). However, because the ability to implement multi-joint isokinetic training is limited (81)—and many isokinetic devices do not offer closed-chain exercise options—the effectiveness of isokinetic training may be limited to return-to-play protocols. Moreover, despite the importance of eccentric training for tendinopathy, hamstring strains, and ACL reconstruction within rehabilitation programs, it has been suggested that isokinetic training may be limited because the devices can be cost-prohibitive, require patients to be trained on their use, and often require a large footprint due to their size (167). Individuals with the means to purchase these devices and use them are referred to previous reviews that discuss different factors associated with isokinetic rehabilitation protocols (27, 91, 105, 106, 248).

Beyond injury rehabilitation, the use of isokinetic training for healthy individuals on various performance measures has been investigated. Researchers have shown that isokinetic training may improve vertical jump and sprinting performance (255), rapid force production characteristics (24), and peak torque during knee extension (163, 196) and flexion (196); however, other researchers showed no change in jumping performance (196). Strength and conditioning professionals should note, however, that isokinetic adaptations may be specific to the speed used within the training protocol (53, 163, 226). Moreover, readers should note that strength adaptations may be greater following traditional resistance training compared to isokinetic training (108, 109). Collectively, it is likely that the applicability of isokinetic training to strength adaptations and sport performance may be limited, thus questioning the use of this method beyond injury rehabilitation protocols.

BLOOD-FLOW RESTRICTION TRAINING

Although it is different from the other training methods described in this chapter, BFR has garnered attention within the resistance training literature. Briefly, BFR includes placing a tourniquet, inflatable cuff, or elastic wrap around the proximal portion of a limb to occlude distal blood flow and create a localized hypoxic environment (166). In theory, creating a hypoxic environment is thought to increase metabolic and hormonal responses to resistance training, potentially leading to greater strength and hypertrophy improvements (153, 154). While there is an abundance of BFR protocols, researchers have typically combined BFR using 40% to 80% of the pressure needed to occlude arterial blood flow with low-load resistance training (e.g., 20%-40% of the 1RM) performed with moderate training volumes (e.g., 4 sets performed to failure) and short rest periods (e.g., 30-60 seconds) (219).

The findings of several systematic reviews and meta-analyses have supported the use of BFR compared to traditional resistance training (107, 166, 333), whereas others have shown greater improvements in strength with high-load resistance training compared to BFR protocols (165, 223). It should be noted that additional BFR protocols that are both active (e.g., high-load resistance training, sprint or repeat sprints, sport training) and passive (e.g., rest or immobilization, or neuromuscular electrical stimulation) in nature are being developed (242). However, strength and conditioning professionals must consider the practicality of implementing BFR before choosing this method of training. For example, implementing BFR with athletes with rehabilitation or return-to-play protocols was shown to be effective (134); however, these individuals typically have specific restrictions, and the training environment is easier to control. In contrast, implementing BFR with large groups of athletes would require a lot more BFR equipment as well as the necessary supervision to ensure that it is being used appropriately; thus, using BFR in this setting may not be ideal. A variety of concerns, including development of blood clots (208), promotion of muscle damage (e.g., rhabdomyolysis) (38), and negative impacts on the cardiovascular system (264), have been raised with BFR training. Thus, it is important that strength and conditioning and sports medicine professionals consider different factors, such as cuff width, material, and athlete characteristics, that may affect the blood flow being restricted when using BFR protocols in training (180, 219).

Take-Home Points

- Strength and conditioning professionals have many different training methods at their disposal; however, it is important that they consider the potential of each method to improve the maximal and rapid force production characteristics of their athletes.
- Some training methods may be limited in their capacity to improve the maximal or rapid force production characteristics of an athlete (e.g., bodyweight exercise and machine-based training), whereas other methods may provide a superior training stimulus compared to traditional resistance training (e.g., weightlifting movements and derivatives and ballistic training). Moreover, the latter methods may transfer more effectively to movements performed in sport via dynamic correspondence.
- Training methods should be chosen based on the abilities of an athlete (e.g., training age, technique competency, maximal strength) in order to provide an effective training stimulus for the athlete's goals as well as their ability to implement the methods based on availability of equipment and team size.
- While some methods may be considered foundational (e.g., bodyweight and machine-based training), other modes of training are more complex in nature but may offer a unique stimulus for force production characteristics (e.g., weightlifting movements and derivatives, strongman exercises). It is also important to acknowledge that some training methods may be considered advanced (e.g., potentiation complexes, variable resistance training, and AEL).
- Strength and conditioning professionals should not overlook the potential of sprinting and COD as stimuli for developing force production characteristics. However, it is important to apply the overload, specificity, variation, reversibility, and individualization training principles to ensure that an effective training stimulus is provided while managing fatigue, given that these methods are typically combined with others.
- Although women are underrepresented within the strength and conditioning literature (52) and training prescriptions may differ between men and women (185), many of the same training methods can be implemented with male and female athletes to improve their force production characteristics (234, 305). Regardless of the sex of an athlete, strength and conditioning professionals should initially take a conservative approach when choosing training methods before progressing or regressing as appropriate.

CHAPTER 6

Advanced Training Methods

While chapter 5 discussed a variety of training methods that may enhance both maximal and rapid force production, it should be noted that not all training methods may be appropriate for all athletes. For example, inexperienced athletes require the necessary time to develop the appropriate technique for performing efficient exercise repetitions. Furthermore, adding variability to different tasks may add unnecessary complexity to the movement and may not allow effective transfer of training. Potentiation complexes, variable resistance training (VRT), and accentuated eccentric loading (AEL) have all been identified as advanced methods of training due to the neuromuscular demand they may place on an athlete (133). This chapter will discuss each of these methods in detail as well as provide readers with some insight into how they may be effectively implemented.

RELATIVE STRENGTH LEVELS

The relationship between muscular strength and various performance measures is well established (134). However, it should be noted that this relationship does not appear to be entirely linear in that the magnitude of performance improvements may decrease as relative strength continues to increase (54, 134), likely due to a smaller window of adaptation (75). As a result, several authors have suggested that it is important to introduce novel training methods that allow relatively stronger athletes to continue to benefit their force production characteristics (75, 97, 124). Interestingly, researchers have shown that individuals with greater levels of relative strength may respond in a unique way compared to weaker individuals when exposed to various stimuli such as weightlifting derivatives, plyometric exercise, and ballistic training (66, 67) as well as jump training (30) and potentiation complexes (68, 112, 136). Specifically, stronger individuals in those studies displayed greater improvements in net impulse, rapid force production, and both jump and sprint performance. Collectively, these findings support the notion that certain training stimuli may be more appropriate for stronger athletes compared to those who are weaker. For example, potentiation complexes, VRT, and AEL may place a considerable demand on the athlete's nervous system, challenge their technique stability, and expose them to supramaximal loads (i.e., >100% of the 1RM). Although researchers have compared the responses to various training stimuli using specific relative levels of strength (e.g., 2 times bodyweight squat) to differentiate the participants (67, 112, 136), it is important that strength and conditioning professionals look beyond the amount of weight an individual can lift relative to their body mass when deciding what training strategies to prescribe to their athletes. However, technique stability and chronic exposure to higher training intensities may be minimum prerequisites for those considering advanced training methods.

Key Point

Strength and conditioning professionals should look beyond the relative strength of an athlete and consider technique stability and chronic exposure to higher training intensities when deciding what training strategies to implement.

POTENTIATION COMPLEXES

As mentioned in the previous chapter, a *potentiation complex* refers to the pairing of two biomechanically similar exercises, with the goal of enhancing (i.e., potentiating) the performance of the subsequent exercise. Examples of such complexes may include performing a squat variation before a vertical jump (31, 135) or a bench press before a bench press throw (41, 71). Strength and conditioning professionals interested in potentiation complexes may have come across different terminology within the literature; for example, the terms *postactivation potentiation* (PAP) and *postactivation performance enhancement* (PAPE) have been commonly used to describe changes in performance. For clarity, PAP refers to an enhanced muscle contractile response for a given level of stimulation following an intense voluntary muscle action that includes a measurement of maximal twitch force evoked by supramaximal electrical stimulation (81). PAPE, on the other hand, refers to an enhancement in voluntary muscular performance following a high-intensity voluntary conditioning action (i.e., exercise stimulus) that occurs without confirmatory evidence of PAP (33).

Although the relationship between PAP and PAPE has been questioned due to potentially different underlying mechanisms (14, 168), it should be noted that relatively few studies have examined the presence of both PAP and PAPE. Several underlying mechanisms of PAP and PAPE, including increased phosphorylation of regulatory light chains, increased recruitment of higher-order motor units, increased muscle-tendon stiffness, pennation angle changes, and changes in muscle temperature, have been discussed previously (14, 148). While PAP may largely be explained by myosin light chain phosphorylation (14, 168), literature has indicated that PAPE may also be associated with the same mechanism (23). However, additional research may be needed to determine the mechanisms of PAPE. Researchers have proposed that PAP likely occurs within a shorter time frame compared to PAPE (14); however, it may be argued that PAP and PAPE may not be mutually exclusive if both are produced within the same time frame (120). By understanding the differences between PAP and PAPE, strength and conditioning professionals can focus on performance improvements and worry less about the physiological mechanisms that underpin PAP. To promote the potentiation of strength characteristics, the emphasis should be placed on acutely enhancing both maximal and rapid force production.

DESIGNING POTENTIATION COMPLEXES

An abundance of potentiation research has been published within the scientific literature, with many researchers showing mixed findings (113, 129, 148). However, the mixed findings may be attributable to the design of specific potentiation complexes or failure to account for the characteristics of the participants (129), both of which may affect the resulting magnitude and timing of potentiation. Regarding the design of potentiation complexes, strength and conditioning professionals must consider the potentiation stimulus (e.g., exercise meant to induce potentiation), the volume–load of the potentiation stimulus, the subsequent activity that is meant to be enhanced, and the rest interval in which an improved performance is to be realized. Failing to account for any of these factors may result in no change or a negative change in performance. First, the potentiation stimulus and subsequent exercise should be as biomechanically similar as possible. Results from previous research have shown that concentric-only quarter squats performed with 50% and 65% of the participant's 1RM quarter squat failed to enhance countermovement jump performance across several different time intervals (32). In contrast, participants who performed concentric-only half squats with 90% of their concentric-only half squat 1RM with ballistic intent enhanced squat jump performance 2 minutes post stimulus (135, 137), with potentiation being realized earlier in stronger participants (136). Collectively, these findings stress the importance of specificity within the design of the potentiation complex, including movement patterns, muscle actions, joint angles, and intent of the exercise (figure 6.1). These conclusions are supported by previous reviews that have highlighted the importance of using specific muscle actions and ranges of motion within potentiation complexes (113, 148).

Key Point

Regarding the design of potentiation complexes, strength and conditioning professionals must consider the potentiation stimulus (e.g., exercise meant to induce potentiation), the volume–load of the potentiation stimulus, the subsequent activity that is meant to be enhanced, and the rest interval in which an improved performance is to be realized.

FIGURE 6.1 *(a-b)* Partial squat and *(c-d)* squat jump paired within a potentiation complex.

Most potentiation studies have used complexes that use a high-force or power exercise to enhance the rapid force production characteristics of another high-power exercise. In contrast, relatively few studies have examined the effect of a potentiation complex on measures of strength (18, 40, 84, 107, 108). Rønnestad and colleagues investigated the acute effects of whole-body vibration on 1RM back squat (108) and half squat (107) performance. No differences in 1RM back squat performances were found; however, untrained and recreationally trained participants improved their 1RM half squat performance. Strength and conditioning professionals should, however, interpret these findings with caution, because the applicability and safety of performing squatting movements on vibration platforms may be questioned in a practical setting. Further research examined the effect of various plyometric exercises on acute 1RM back squat performance. Interestingly, protocols that included a single set of two, four, or six 33-centimeter (13 in) depth jumps (18) or tuck jumps and 43.2-centimeter (17 in) depth jumps (84) improved the 1RM of the participants. Although the previously discussed studies examined the acute impact on 1RM performance, none reported force production characteristics. However, Ebben and colleagues (40) examined the acute effect of a supramaximal quarter squat (120% 1RM) on the performance of 2 repetitions of the back squat performed with 80% 1RM. The researchers indicated that while there was no difference in peak force production, there was a significant improvement

(18.6%) in concentric rate of force development. Although most of the discussed studies focused on lower-body potentiation, a 2022 systematic review and meta-analysis indicated that upper-body potentiation complexes may also be effective (43).

As mentioned previously, the volume–load of the potentiation stimulus may also affect whether potentiation is realized. Tillin and Bishop (148) indicated that a given level of both potentiation and fatigue is present following a training stimulus, illustrating an acute version of the fitness–fatigue paradigm (167). The authors also highlighted that two windows of potentiation may exist; however, the type, volume, intensity, and duration of exercise may dictate whether fatigue or potentiation is realized over the other (164). If the primary goal of a potentiation complex is to enhance both maximal and rapid force production, it would appear beneficial to recruit higher-order motor units (i.e., Type II). As indicated in chapter 3, this can be accomplished by using exercises that can use heavy loads and be performed in a ballistic manner (38, 74, 82, 149). However, the volume of such a stimulus may need to be low to allow fatigue to dissipate and to allow potentiation to be realized earlier (129). In contrast, if the volume of a heavy, ballistic stimulus is higher, it is likely that greater fatigue will be present, ultimately delaying an athlete's ability to realize potentiation.

A final aspect of designing potentiation complexes is choosing the rest interval where potentiation can be realized and thereby acutely improve force production. As noted earlier, the interaction between potentiation and fatigue following a stimulus determines whether the subsequent performance will be positively or negatively impacted. While meta-analyses have indicated that the optimal rest interval for potentiation complexes ranges from 7 to 12 minutes (52, 164), the optimal rest interval may be specific to not only the potentiation complex (129) but also the individual athlete (13, 29). Another thing strength and conditioning professionals must consider regarding the rest interval of the potentiation complex is whether it is feasible in a weight room or competition setting. For example, much of the existing literature examined within the aforementioned meta-analyses required participants to complete the potentiating stimulus before resting quietly (e.g., sitting on a chair) prior to performing the subsequent exercise at various rest intervals. However, because the research suggests that optimal performance may be realized between 7 and 12 minutes, this does not appear to be reasonable or ecologically valid. Potentiation complexes may serve as an advanced training method due to the unique characteristics of the complex. Researchers have shown that stronger individuals potentiate faster and to a greater extent compared to those who are weaker (112, 114, 136). Beyond enhancing performance in a meaningful manner, potentiating faster adds to the ecological validity of using potentiation complexes within training. It should also be noted that stronger individuals have since developed greater fatigue resistance to heavier loads and may tolerate a higher-intensity stimulus. This, in turn, may allow them to enhance maximal and rapid force production characteristics within the appropriate training phases.

Key Point

Although meta-analyses suggest that optimal performances within potentiation complexes may be realized between 7 and 12 minutes, this does not appear to be practical.

ATHLETE CHARACTERISTIC CONSIDERATIONS

As mentioned previously, potentiation complexes are typically used to enhance the force production characteristics of an athlete—namely, rapid force production. Chapter 3 discusses the ability of different fiber types to affect both maximal and rapid force production and concludes that individuals who possess greater proportions of Type II compared to Type I muscle fibers may have an advantage at improving these strength characteristics. Similar conclusions can be made when comparing the magnitude and timing of potentiation following a conditioning stimulus of individuals with greater percentages of Type I versus Type II fibers. For example, several studies have concluded that individuals who are more Type II dominant may display greater PAP compared to those who are more Type I dominant (53, 56, 57, 101, 152, 153, 154). While the existence of PAP does not necessarily improve PAPE, Terzis and colleagues (145) indicated that the percentage of Type II fibers was strongly correlated with improvements in underhand shot-put throw performance following drop jumps. Collectively, the existing literature supports the notion that potentiation complexes may best serve individuals with greater proportions of Type II fibers, who may also possess greater levels of muscular strength (1, 85, 147). Given that endurance athletes typically possess a greater proportion of Type I fibers (24, 37, 146), it should not be surprising that the authors of a 2024 meta-analysis concluded that potentiation complexes do not appear to benefit endurance exercise performance (155). However, readers interested in the use of potentiation complexes within endurance sports are directed to a review by Boullosa and colleagues (15).

IMPLEMENTING POTENTIATION COMPLEXES IN TRAINING

The idea of implementing potentiation complexes within training may seem appealing to many strength and conditioning professionals; however, it is important to implement this training method within the appropriate phases. First, potentiation complexes are based on the fitness–fatigue paradigm (148), so it does not appear logical to implement potentiation complexes during a strength–endurance phase because the accumulated fatigue may mask any potentiation effects. Moreover, the goals of a strength–endurance phase (i.e., work capacity and muscle hypertrophy) (92, 133, 166) do not match the goals of a potentiation complex. As discussed earlier, many of the existing potentiation complexes show increases in performance underpinned by changes in rapid force production. Thus, it could be argued that potentiation complexes may be implemented more effectively in either general or absolute strength phases in which the goals are to increase the magnitude and rate of force production or during strength–speed and speed–strength phases that focus primarily on improving rapid force production and power output (133). Beyond the training goals, these phases are also characterized by lower volume–loads and may thus produce less accumulated fatigue, ultimately providing a favorable training environment for potentiation complexes. Table 6.1 presents examples of implementing potentiation complexes within various training blocks. It is important that strength and conditioning professionals recognize the global fatigue that multiple training methods can produce during resistance training sessions. Thus, when considering the implementation of potentiation complexes, strength and conditioning professionals must decide

1. if it is necessary to add them to the already existing stimuli,
2. if there are other (potentially simpler) training methods that may produce similar or greater training effects, and
3. if every athlete they are working with should use them or if they should be implemented on an individual basis.

IMPLEMENTING POTENTIATION COMPLEXES IN COMPETITION

While potentiation complexes within resistance training sessions are most common, it may also be possible to enhance performance using potentiation complexes during a competition. For example, Judge and colleagues (10, 69) showed that using overweighted implements during competition warm-ups may enhance weight throw performance at track meets. It should be noted that although the use of

TABLE 6.1 Implementing Potentiation Complexes Into a Strength–Power (3×3) Resistance Training Program Example

Day 1	Day 2	Day 3
Jerk	A) Hang clean pull	Hang power snatch
A) Back squat	B) Jump shrug	A) Back squat
B) Countermovement jump	Hex bar deadlift	B) Medicine ball vertical throw
A) Bench press	Pull-up	
B) Medicine ball throw		

Note: A and B exercises are paired within a potentiation complex.

overweighted implements may be feasible, these implements may not be readily available for every athlete. Moreover, implements that are heavier than the athlete is used to may lead to a negative performance due to changes in technique. Thus, athletes and sport or event coaches may elect to seek alternative methods that exclude additional equipment. Interestingly, researchers have concluded that both countermovement jumps and sprinting stimuli may acutely enhance shot-put performance (144). Additional researchers showed that depth jumps acutely improved sprinting performance (20). Therefore, because track and field competition venues have space for athletes to warm up, as well as various forms of seating for athletes and spectators that can be used for different jump efforts, the feasibility of using potentiation complexes within a track and field competition appears favorable.

As mentioned earlier, the design of potentiation complexes should emphasize specificity of the movements that are being paired. Kelekian and colleagues (70) indicated that weightlifters may enhance their maximal clean performance after performing clean pulls with either 85% or 120% of their clean 1RM. Weightlifting movements and derivatives are discussed in chapter 5, but it should be noted that specific pulling variations may allow for greater force production characteristics to be produced compared to catching variations (26, 27, 72, 73, 127, 138, 139, 141) and may thus provide an effective potentiation stimulus. From a competition standpoint, weightlifting coaches may consider using pulling variations within the warm-up protocols of their athletes in preparation for their lift attempts. Regarding sporting actions that may require multiple efforts such as swinging an implement, isometric efforts in various swing positions have been shown to acutely improve bat velocity in both baseball (63) and softball (50) players. In addition, researchers showed that performing countermovement jumps prior to golf driving performance may enhance golf club head speed (105). From an upper-body perspective, Finlay and colleagues (42) indicated that both punch-specific isometric efforts and elastic resistance acutely improved punch force and rate of force development. An advantage to isometric efforts and countermovement jumps is that no additional equipment is needed. In addition, an athlete may be able to use these strategies more than once during their respective competitions. While further research is needed, athletes, sport coaches, and strength and conditioning professionals may consider the aforementioned exercises as effective methods of implementing potentiation complexes within athletic competitions.

VARIABLE RESISTANCE TRAINING

In contrast to traditional resistance training where the load remains constant throughout the entire movement, VRT, also known as *accommodated resistance training* (8, 106), uses chains or elastic bands to modify the external resistance throughout the range of motion of an exercise (44). In doing so, the force production characteristics of the exercise are altered to allow for greater force, velocity, and power output to be produced during the early eccentric and late concentric phases of the lift (65), while also allowing athletes to match changes in joint leverage (167) and overcome mechanical disadvantages at various joint angles (39, 162). What makes this method advanced is the modified load throughout that motor task itself. If novice athletes cannot perform repetitions of a traditional resistance training movement without a modified load, this may lead to insufficient transfer to a task. Moreover, the implementation of VRT early in the development of an athlete may prevent strength and conditioning professionals from using VRT as a novel training method later. In fact, Rhea and colleagues (106) suggested that stronger athletes may benefit more from VRT compared to weaker athletes because of the potential to transfer their maximal strength into rapid force production movements.

Before discussing the potential benefits of using VRT to improve an athlete's strength characteristics, it is important to mention the different ways that this method has been investigated. Researchers have used a variety of loading protocols for VRT with both chains and bands, including relatively higher loading where the load at the bottom position of an exercise is equal between VRT and traditional loading (103, 109, 122), relatively lower loading where the load at the top position of an exercise is equal with traditional loading (9, 39, 51, 62, 98), and equated loading where the load for VRT is lower at the bottom and higher at the top compared to traditional loading (3, 46, 47, 76, 142, 162). Although it is not applicable to the use of chains, it should be noted that elastic band VRT has been examined in both an assisted and resisted manner (3). *Assisted VRT* refers to the attachment of the bands to the barbell and a position above the barbell, which may allow athletes to lift a greater load. In contrast, *resisted VRT* refers to the attachment of bands to the barbell and a position on the floor, which increases and decreases the overall load when they are stretched and relaxed, respectively. In line with the typical use of chains, the following discussion will primarily focus on resisted VRT.

ACUTE EFFECTS OF VARIABLE RESISTANCE TRAINING

A VRT stimulus aims to help athletes maximize force production throughout the entire range of motion (44), which may allow them to train through "sticking points" (150) and further allow them to increase force production throughout the end of a repetition (83, 109, 162). In a 2022 systematic review and meta-analysis, Shi and colleagues (117) indicated that VRT may be more effective at improving velocity and power output during resistance training repetitions compared to traditional loading. Although their analysis concluded that there were no significant differences in force production characteristics (i.e., mean force, peak force, and rate of force development), chains and bands may provide different loading characteristics (86). For example, increases in chain load increased peak force production (142) while peak force decreased as elastic band tension increased (46) during the deadlift exercise. However, peak force increased as elastic band tension increased during the back squat (162). From a rapid force production standpoint, it should also be noted that elastic bands produced significantly greater effects compared to chains (117). Galpin and colleagues (46) indicated that a greater percentage of load provided by bands (i.e., 35% of the total load) during heavy deadlifts (85% of the 1RM) resulted in a significantly improved rate of force development. In addition, Stevenson and colleagues (122) concluded that VRT produced a greater rate of force development during light back squats (55% of the 1RM) with 20% of the load accounted for by bands.

Key Point

Variable resistance training may be more effective at improving velocity and power output during resistance training repetitions compared to traditional loading.

VARIABLE RESISTANCE POTENTIATION COMPLEXES

Based on its force production characteristics, it should come as no surprise that researchers have investigated VRT as part of a potentiation complex to enhance different performances. In fact, researchers have shown that vertical jump (19, 80, 91, 104, 111), horizontal jump (114, 125), and short sprint (104, 165) performances were all enhanced following a VRT potentiation stimulus. Moreover, additional research showed 6.6% (90) and 7.7% (89) improvements in back squat maximal strength following a VRT potentiation protocol that included 2 sets of 3 repetitions performed at 85% of the 1RM with 35% of the resistance provided by chains and bands, respectively. Finally, researchers showed that countermovement jump force at peak power (111) as well as peak force (104) and rate of force development (91, 104) improved as part of VRT potentiation complexes. It should, however, be noted that similar to more traditional potentiation complexes, stronger athletes may potentiate to a greater extent than those who are weaker when using VRT potentiation complexes (114). These findings again support the notion that VRT may be classified as an advanced training method compared to other methods.

LONGITUDINAL ADAPTATIONS TO VARIABLE RESISTANCE TRAINING

Beyond an acute training stimulus, it is important to discuss the chronic use of VRT and its potential to enhance the strength characteristics of athletes. Several meta-analyses have been completed regarding the long-term maximal strength adaptations of VRT and have displayed mixed findings (4, 78, 99).

Whereas one of the previous analyses indicated that VRT may produce greater strength adaptations compared to traditional resistance training with a constant load (78), the other two concluded that there were no significant differences between training methods (4, 99). The different findings within each study may be explained by the study inclusion criteria, the wide variety of VRT protocols, and a combination of untrained and trained participants (117). Interestingly, Lin and colleagues (78) took a unique approach to determine how training status and load impacted maximal strength adaptations. Their results suggested that trained individuals may significantly improve their maximal strength with VRT compared to traditional loading when the total load was greater than 80% of the 1RM; however, no differences existed when the load was less than 80%. However, the opposite was true when examining the training effects of untrained individuals. As noted in chapter 3 and in additional research (55), initial changes in maximal strength are typically underpinned by neuromuscular changes (e.g., motor unit recruitment, firing frequency). These initial changes are also supported by studies that investigated the longitudinal use of VRT (5, 160). This may partially explain why U20 soccer players showed no differences in maximal strength, vertical jump, sprint, or change-of-direction performances following short-term (4-week) VRT or traditional loading programs (79). In another study, Rhea and colleagues (106) showed that 12 weeks of VRT back squat training significantly improved both 1RM back squat strength and power in NCAA Division I male athletes. The researchers concluded that VRT may be an effective form of training for both maximal strength as well as rate of force development. Furthermore, the authors suggested that VRT prescription may be best served for athletes with greater maximal strength and training experience.

Although it may be difficult to determine the effectiveness of longitudinal VRT using a meta-analysis due to differences in study inclusion criteria, VRT protocols with chains or bands, or study participants (117), the acute stimulus and potentiation benefits of VRT cannot be ignored. VRT appears to provide an effective neuromuscular stimulus that may benefit the maximal and rapid force production characteristics of an athlete. Strength and conditioning professionals should note, however, that the VRT stimulus may be based on the amount of total weight prescribed (78) as well as the percentage of total load that is accounted for by the chains or bands (46, 142, 162). Related, it should also be noted that the training status of the individual may affect the outcomes from VRT (78, 114); thus, it is important that VRT be implemented in a manner that benefits an athlete's strength characteristics as they relate to the goals of each training phase. Further research is needed on the ability of VRT to transfer to different sport movements as well as on the underlying mechanisms that produce specific training adaptations. Additionally, more research examining the combination of VRT and other training methods (e.g., plyometric exercises, weightlifting derivatives) is needed to improve the ecological validity of training programs.

IMPLEMENTING VARIABLE RESISTANCE TRAINING

Like other methods of training, VRT possesses several caveats that may affect the adaptations an athlete may receive. The following paragraphs will discuss how familiarization, exercise choice, loading, and frequency may modify a VRT training stimulus.

Familiarization

The change in load during VRT is gradual but it may cause instability and affect an athlete's exercise technique. This feeling may be intensified if the chains or bands are contributing a significant amount to the total load during initial familiarization. Strength and conditioning professionals should gradually introduce their athletes to VRT to ensure that their technique remains stable to promote an effective training stimulus. It has been suggested that VRT may provide the greatest training effects during the range of motion where the external load increases (123). This range of motion provides the greatest opportunity for acceleration during the propulsion phase of an exercise; with the gradual increase in load, it may allow for an effective rapid force production stimulus (162). It is important to match the athlete's characteristics (e.g., height, maximal strength) to provide an appropriate loading stimulus (e.g., chain size and length, band width), because inappropriate loading may hinder the acceleration throughout the movement, which may negatively affect the training effects received by the athlete.

Exercise Choice

When implementing VRT, the first thing strength and conditioning professionals should note is that although the use of chains and bands is possible with most exercises, they do not need to be incorporated with every exercise. For example, researchers previously investigated the use of VRT with both the snatch (25) and the clean (12) exercises and showed no differences in force production characteristics between VRT and traditional loading when performed at 80% and 85% of the 1RM. However, the authors noted that the weightlifters perceived the VRT condition to be more difficult due to the oscillation of the chains. Although the psychological effect of VRT is noteworthy, strength and conditioning professionals should be aware of the already technically complex nature of weightlifting movements before implementing VRT with these exercises. Indeed, the technical complexity of weightlifting catching variations such as the snatch and clean has been noted by previous authors (127, 128) and led to the investigation of weightlifting derivatives that simplify the exercises rather than complicate them further (28, 130, 131, 132). In line with the previous recommendations (161), it is therefore suggested that strength and conditioning professionals consider implementing VRT with multi-joint core exercises such as squatting, pressing, and pulling exercises (figures 6.2 and 6.3). The use of VRT with these exercises may provide athletes with effective maximal and rapid force production stimuli as evidenced by the existing literature.

Loading

Another factor strength and conditioning professionals must consider when implementing VRT is how the athletes are going to be loaded. Specifically, decisions need to be made regarding the method of

FIGURE 6.2 *(a-c)* Variable resistance training bench press.

FIGURE 6.3 *(a-c)* Variable resistance training back squat.

VRT (i.e., chains and bands) and what percentage of the total load is accounted for by both the chosen load on the barbell and the VRT method. While researchers have examined VRT using an absolute load for all participants (9, 65), a percentage of the participant's 1RM (122, 142), or a percentage of the load being tested (39, 162), the overall load chosen for a given exercise should coincide with the goals of a given training phase. Regarding peak force production, it would be beneficial to implement heavier loads (i.e., ≥80% of the 1RM) (78); however, based on the existing research, 20% to 35% of the total load should be accounted for by either chains or bands (46, 122). In contrast, the loading recommendations for rapid force production are less clear. However, strength and conditioning professionals should note that the existing findings on rapid force production may be based on the exercise examined and whether a stretch–shortening cycle occurs (e.g., back squat vs. deadlift). Rapid force production was shown to improve with loads as heavy as 85% of the 1RM (30% bands) during a deadlift (46) but as light as 55% of the 1RM (20% bands) during a back squat (122). Furthermore, strength and conditioning professionals should note that bands may benefit rapid force production characteristics to a greater extent compared to chains (117). Readers are directed to additional resources that provide insight into the amount of weight that specific chain sizes (11) and band thicknesses (119) may account for.

Key Point

Regarding peak force production adaptations during variable resistance training, heavier loads (i.e., ≥80% of the 1RM) would be beneficial with 20% to 35% of the total load accounted for by either chains or bands.

Frequency

Following familiarization, exercise choice, and loading, it is important to determine the frequency in which VRT is implemented during a training program. While VRT has typically been prescribed multiple times during any given week within training studies (4, 78, 99), the duration of the training programs has ranged from 1 week (6) to 24 weeks (118). However, as noted previously, the existing studies had a range of participants, including untrained, recreationally trained, and trained individuals as well as athletes and nonathletes.

From a training phase perspective, it would appear beneficial to implement VRT during low-volume training phases that allow for the development of both maximal and rapid force production without the accumulation of fatigue that may be produced during higher-volume training phases. However, based on the focus of either maximal or rapid force production, VRT may be implemented more frequently versus less frequently, respectively. For example, strength and conditioning professionals aiming to improve maximal force production may need to provide their athletes with a concentrated loading stimulus of VRT, which would include implementing VRT across several weeks during a training phase. This may also be the case when the aim is to increase rapid force production; however, it is important to be aware of the overall training volume and resulting fatigue, especially if using VRT within potentiation complexes (148). Strength and conditioning professionals must also determine whether it is necessary to use VRT with the same exercise multiple times per week. If athletes are exposed to VRT using the same exercise, it is important to modify the relative intensities to still provide the athletes with a light day to manage fatigue throughout a training week (34). Finally, if the desire is to prescribe VRT with multiple exercises during a single training session, strength and conditioning professionals must determine if it is feasible within the training session, whether the athlete can handle such a stimulus, if it is necessary to do so, and how to redistribute the overall volume–load of the other prescribed exercises throughout the rest of the week if the VRT exercises fatigue an athlete to the point where it negatively impacts their performance.

ACCENTUATED ECCENTRIC LOADING

As noted in chapter 5, AEL as a training method requires that an exercise is performed with a heavier eccentric load than concentric load and pairs eccentric and concentric muscle actions with minimal disruption to the natural movement mechanics of the exercise (158). Interestingly, AEL combines different aspects of both potentiation complexes and VRT. Whereas PAPE refers to the enhancement (potentiation) of a subsequent performance of a different exercise or motor task, AEL is unique in that it may be possible to realize within-set potentiation. For example, Merrigan and colleagues (88) indicated that concentric power output increased when using 120% and 65% during the eccentric and concentric phases of the bench press, respectively. Similar to VRT, AEL modifies the load that is being used during an exercise by removing a portion of the total load at the end of the eccentric (braking) phase (158). This can be accomplished by using weight releasers (102), spotters (16), or the athlete dropping part of the load (115). However, unlike VRT, AEL can be used with both strength and ballistic exercises, allowing for a greater range of force–velocity characteristics to be trained. Researchers have shown that AEL may be used to enhance maximal strength (16, 35, 159), rapid force production (2), and performance (2, 115). It should be noted that while previous literature has provided some insight regarding the implementation of AEL (87, 140), the goal of this section is to provide readers with additional insight on how best to use AEL to benefit an athlete's force production characteristics. Specifically, the following paragraphs will discuss familiarization, eccentric phase duration, loading, and frequency considerations. While this section will provide some insight on how to implement AEL within resistance training programs, more research needs to be completed.

SUBCATEGORIES OF ACCENTUATED ECCENTRIC LOADING

For the purposes of the following discussion, it is important to identify the subcategories of AEL, including ballistic, submaximal, maximal, and supramaximal AEL. Specifically, it is necessary to distinguish the differences between each AEL subcategory because the external load and possibly the intent of the exercise may differ significantly. *Ballistic AEL* (AEL-BAL) refers to AEL being used with a ballistic exercise such as a jump (e.g., jump squat) or throw (e.g., Smith machine bench press throw). This method may remove the additional eccentric load by dropping the load (e.g., AEL dumbbell jump) or by using weight releasers. Although previous authors have referred to some AEL jumps as "plyo-accentuated eccentric loading methods" (58), it is important to point out that not all ballistic efforts are plyometric in nature because the duration of the stretch–shortening cycle may be lengthened with loaded exercises. Thus, the term *AEL-BAL* may encompass more exercises. *Submaximal AEL* (AEL-SUB) refers to a nonballistic AEL exercise being performed with a total load that equates to less than 100% of the athlete's 1RM during that same exercise. For example, a bench press may be performed with 30% of the 1RM on weight releasers while 60% of the 1RM is loaded on the barbell, totaling 90% of the athlete's 1RM during a traditional bench press repetition. It should be noted that previous authors have referred to this as "augmented eccentric training" (59); however, this term has also been used to describe banded jumps, dumbbell jumps, and jumps using weight releasers (94, 95), as well as AEL exercises using submaximal (<100% of the 1RM), maximal (100% of the 1RM), and supramaximal (>100% of the 1RM) loading during the eccentric phase of an exercise (60, 93, 94, 95, 151). Therefore, it is important to provide this clarification. Finally, *maximal AEL* (AEL-MAX) and *supramaximal AEL* (AEL-SUPRA) refer to a nonballistic AEL exercise that is performed using a total load equal to or exceeding an athlete's 1RM during the eccentric phase of the movement, respectively.

IMPLEMENTING ACCENTUATED ECCENTRIC LOADING

Strength and conditioning professionals must consider the familiarization, eccentric phase duration, loading, and frequency of implementation when using AEL. The following paragraphs will discuss each point to provide readers with insight into how AEL may be more effectively implemented in training programs.

Familiarization

Unlike VRT where the change in external load is gradual, the change in load during AEL occurs instantaneously. This, in turn, may have a more dramatic impact on an athlete's technique, especially if the difference between eccentric and concentric loads is large. Therefore, it is important that strength and conditioning professionals familiarize their athletes with how to perform the exercises appropriately with the addition of extra equipment (e.g., weight releasers) when applicable. Specifically, it is important that the athletes become familiar with the additional external loads during the eccentric phase as well as the "release" of the loads during an exercise. The most common exercises examined in the AEL literature include various jumps, squatting variations, and the bench press; however, each exercise requires its own familiarization because each movement is unique. To receive the benefits of additional loads during AEL, heavier loads should be applied throughout a large range of motion in which the athlete is actively resisting the load. However, because the loads used may be maximal or supramaximal, it is important that the additional load is released at the proper moment. Therefore, given the range of motion characteristics of each athlete, weight releaser lengths must be adjusted accordingly. Figures 6.4 to 6.6 display the weight releaser setups and release points for the AEL back squat, bench press, and dumbbell jump, respectively.

Regarding the use of weight releasers, strength and conditioning professionals should adjust the weight releaser height so that the load is released just above the end range of motion for each athlete. It should be noted that if the weight releasers are too long, the load will be released too early and prevent the athlete from receiving the same loading stimulus throughout a larger range of motion. The point at which the additional eccentric load will be removed during AEL jump variations may be based on athlete characteristics (e.g., limb length, range of motion), the position of the load (e.g., arm's length vs. upper back), and the concentric load (i.e., unloaded or loaded) the athlete jumps with. If the athlete is releasing the load themselves (e.g., dropping dumbbells during a jump), it is important that they release the load at the bottom of their countermovement rather than before, ideally at the end of the braking phase and before the propulsion phase. Like the load being released too early with weight releasers, a suboptimal stimulus may result if the load is released (dropped) too early. In contrast, if the load is released too late, the propulsion phase of the jump may be negatively impacted.

FIGURE 6.4 *(a-d)* Accentuated eccentric loading back squat.

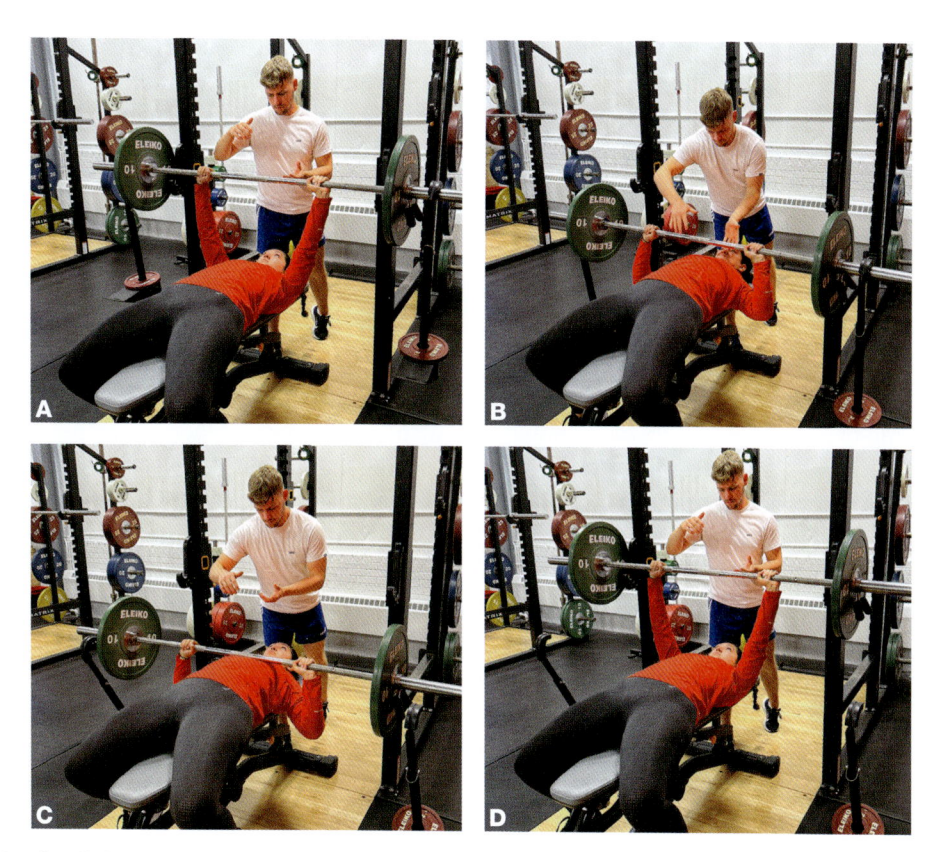

FIGURE 6.5 *(a-d)* Accentuated eccentric loading bench press.

FIGURE 6.6 *(a-c)* Accentuated eccentric loading dumbbell jump.

Eccentric Phase Duration

An important aspect when using AEL within a resistance training program is the way in which the braking phase of the movement is performed. For example, athletes may slowly lower the external load with a selected tempo (e.g., 3 seconds, 5 seconds), perform the movement with the same speed as a typical repetition (i.e., natural movement mechanics), or rapidly drop into the braking phase due to the extra load (i.e., minimally resist the load until late during the braking phase). Researchers have shown that 4 weeks of AEL back squats performed with a slow (3-second) tempo produced greater maximal strength adaptations compared to a subsequent 4-week block in which the AEL squats were performed with a fast tempo (1 second) (36). While this may lead strength and conditioning professionals to prescribe slow AEL movements, these findings should be interpreted with caution. First, it should be noted that the improvements in strength during the slow training phase were small (effect size = 0.38). Furthermore, the order of the training phases in the previous study (slow before fast) was not randomized, so it is unclear whether performing the fast phase first would have produced different effects. Finally, the authors indicated that 8 weeks of AEL training may be inappropriate when training team sport athletes unless sufficient recovery is provided. Furthermore, the authors noted that the fast AEL phase "may have exceeded recovery capabilities compared with fast traditional resistance training" (36). Based on these conclusions, further research on this topic is needed.

Another goal of AEL is to use the braking stimulus to benefit the stretch–shortening cycle and propulsion phase of the movement. From a tempo standpoint, longer durations may negatively affect the contribution of the stretch–shortening cycle when transitioning from the braking phase to the propulsion phase of a task (22, 163). This effect may also be worsened when more repetitions are performed in this manner within an exercise set versus single repetitions. Figure 6.7 shows the differences between a 5-second tempo and the natural tempo mechanics during a back squat performed with weight releasers. Performing AEL movements with a longer tempo increases time under tension during the braking phase of a movement, and potentially the tension produced within the muscle belly itself. This may enhance or provide an effective hypertrophy stimulus; however, it may not contribute to the propulsive force production of the movement itself. Strength and conditioning professionals must consider, however, that eccentric phase duration may be relative to a specific context or each individual. For example, Douglas and colleagues (36) designated fast and slow tempos as 1 and 3 seconds, respectively, whereas other researchers identified 2 and 6 seconds as fast and medium, respectively (48).

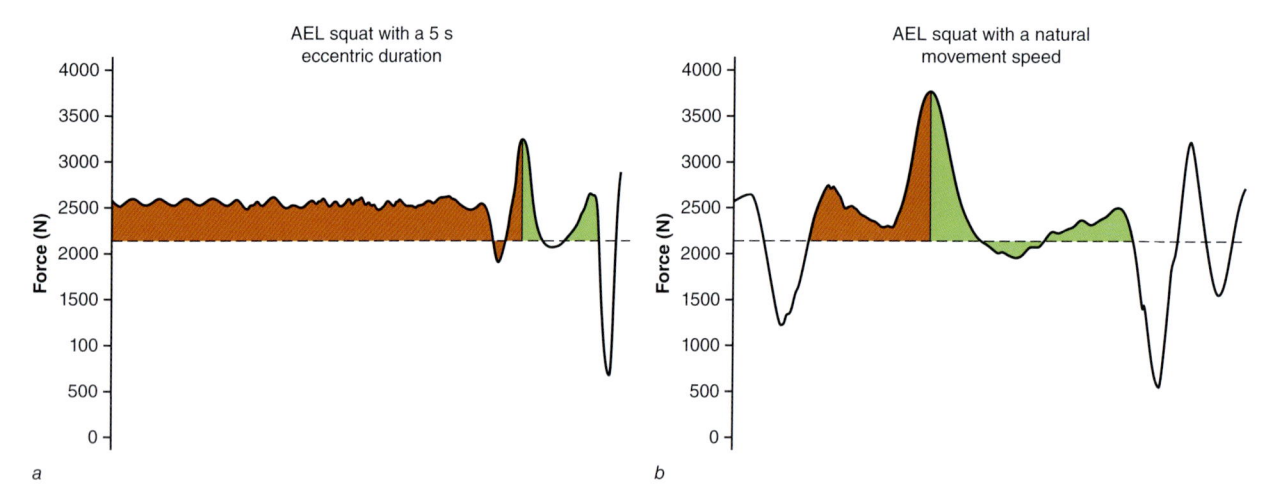

FIGURE 6.7 Force–time comparisons of different AEL squat tempos. The dashed line indicates system weight, red indicates the braking phase, and green indicates the propulsion phase.

Regardless of the duration of the braking phase, it is important that athletes actively resist the additional eccentric load provided by weight releasers or the load they are holding (e.g., dumbbells), because failing to do so may mitigate the training effects of the braking phase itself. As mentioned earlier, some athletes will rapidly unweight and provide minimal resistance during the eccentric phase before rapidly stopping themselves at the end range of motion prior to transitioning to the propulsion phase. However, athletes using this strategy transition from an unweighting phase to a braking phase without applying much force before they have to rapidly produce braking force in a short duration to stop their downward momentum. This, in turn, would force the athlete to withstand high braking forces before transitioning to the propulsive phase of the movement. Although this may force an athlete to adapt to a rapid force production stimulus, it could also negatively affect their propulsion phase if they cannot withstand the higher magnitude of forces, which may potentially increase injury risk during the exercise. While this may not be an issue at lighter loads, heavier concentric loads may place a greater strain on the athlete.

Key Point

Regardless of the duration of the braking phase, it is important that athletes actively resist the additional eccentric load provided by weight releasers or the load they are holding, because failing to do so may mitigate the training effects of the braking phase itself.

Collectively, it appears that the braking phase duration of AEL movements may have its own "Goldilocks zone." For example, if the duration of the braking phase is too slow, the stretch–shortening cycle may not benefit the propulsion phase. In contrast, if the duration is too fast, the athlete may not actively resist the added load during the movement and may thus fail to receive the AEL stimulus. In this light, it is important to consider the natural movement speed of each athlete when choosing an eccentric duration as well as how it will be monitored within training. Furthermore, it is crucial that researchers identify the instructions given to the participants and how the braking phase was performed within their studies.

Loading

Another important aspect of AEL are the loads used for specific training stimuli. An abundance of literature has indicated that using loads greater than 80% of an individual's 1RM may provide a necessary stimulus to enhance strength characteristics (116); however, a benefit of AEL is the ability to use loads in excess of an athlete's 1RM during the braking phase of the movement. In fact, researchers have concluded that eccentric strength may be approximately 40% greater than concentric strength (100), with differences ranging from 20% to 60% based on the exercise (64). Therefore, from a loading perspective, it appears that lowering a greater magnitude of load would contribute to enhancing braking force production characteristics (i.e., eccentric strength). It is important to consider the propulsive or concentric load of the movement as well, given that the stretch–shortening cycle can enhance force production characteristics. Thus, prescribing loads during the concentric phase should also be considered. While greater absolute force production is produced when prescribing loads closer in proximity between eccentric and concentric (e.g., 100%/80% vs. 100%/60%), researchers have shown that prescribing loads farther apart may enhance rapid force production characteristics during both the braking and propulsion phases of a movement (126). This is an important consideration regarding the implementation of AEL during preseason and in-season phases of competition. For example, lighter concentric loads may be prescribed to maintain velocity characteristics during a season, while heavier loads may maintain maximal strength characteristics. In this light, AEL may allow athletes to maintain or improve both such characteristics during a competitive season.

As mentioned earlier, the additional load during the eccentric phase of an AEL exercise may be removed in several ways. Researchers have examined the acute effects of eccentric loads as high as 150% of the participants' 1RM (7, 61, 110) during machine-based AEL, whereas studies involving jumping exercises have examined standardized loads based on specific equipment (143) or promoted the use of 20% of an individual's body mass (17, 95). From a peak force development standpoint, it would appear obvious that the concentric phase of an exercise should be heavily loaded; however, it is also important that strength and conditioning professionals consider the additional load during the eccentric phase. Munger and colleagues (96) used 120% and 90% of the 1RM during the eccentric and concentric phases during front squats, respectively. An obvious consideration of AEL should be whether the weight releasers are used only on the first repetition of a set or during multiple repetitions. Although various studies have used a wide spectrum of loading protocols during AEL, minimal information exists on how the difference between the eccentric and concentric loads affects force production characteristics. However, a larger difference between the eccentric and concentric loads may benefit rapid force production, whereas a smaller difference may benefit peak force production during both the braking and propulsion phases of a back squat in both resistance-trained men (126) and women (21).

Key Point

A larger difference between the eccentric and concentric loads may benefit rapid force production, whereas a smaller difference may benefit peak force production during both the braking and propulsion phases.

Most of the loading information within the AEL literature is based on the 1RM of a traditional eccentric and concentric movement. However, because athletes may be approximately 40% (100) or 20% to 60% stronger eccentrically based on the exercise (64), this may not be the most effective method of loading due to individual capacities of eccentric strength as well as task specificity. This has led researchers to design protocols to assess eccentric maximal strength (45, 49, 60, 121). Interestingly, Harden and colleagues (60) had their participants perform both eccentric and concentric strength assessments on which to base their loads on during AEL leg press training. However, strength and conditioning professionals should note that eccentric "1RMs" within the literature are typically performed in a controlled manner. For example, researchers have required participants to perform 3-second (121) or 5-second (60) eccentric durations during maximal strength assessments. Strength assessments performed in this manner may allow athletes to safely demonstrate that they are able to control a given load; however, it is still unknown whether this is the most effective way of assessing maximal eccentric strength.

Frequency

Beyond the eccentric phase duration and loading considerations, strength and conditioning professionals must decide how often to implement AEL. The frequency of implementation may vary based on the training goals, the physical abilities of an athlete, and the feasibility within training sessions. From a macro-level standpoint, it must be decided which training phases may benefit from the use of AEL. Although AEL training implemented over the course of 2-week, 4-week (36), or 5-week blocks (159) with testing weeks between blocks or over 9 consecutive weeks (16) has improved various measures of maximal strength, it is unknown whether these methods of implementation may provide athletes with an optimal strength stimulus. Moreover, it may be questioned whether implementing AEL during consecutive training blocks or for an extended period of time is ecologically valid if proper recovery is not provided (36). Specific to development of strength characteristics, moderate- to low-volume phases may provide strength and conditioning professionals with the best opportunities to prescribe AEL. This may be due to the potential of within-set potentiation. For example, Wagle and colleagues (156) showed that performing sets of 5 AEL back squat repetitions in which weight releasers were used during each repetition provided the largest benefits in terms of eccentric work as well as concentric force output. However, the same authors also demonstrated that characteristics such as the eccentric rate of force development may remain elevated for the second and third repetitions within a set of 5 despite using weight releasers during only the first repetition of the set (157). Although other researchers have shown that using weight releasers on the first repetition of a bench press set may enhance concentric force production (77), limited repetition-by-repetition information has been provided with this exercise.

Little is known about the effective prescription of AEL during resistance training phases that focus on the development of strength characteristics. Although previous studies have typically implemented AEL at least twice per week for multiple weeks within a training phase (36, 60, 115, 159), it remains unknown whether it is optimal or even necessary to prescribe AEL multiple times per week or multiple weeks within a training phase. Before prescribing AEL during multiple weeks within a training phase, it is first important to consider whether the athlete can physically tolerate such a stimulus. For example, if an athlete has never been exposed to AEL, it will serve as a novel training stimulus regardless of the frequency in which it is implemented; however, overprescribing AEL initially may be detrimental for the athlete if they cannot physically tolerate the prescribed loads. Thus, it is important that strength and conditioning professionals progress each of their athletes in a conservative manner to allow them to adapt to such stimuli.

It is also important to consider the previous training phase that the athlete has completed, given that the training residuals from such training may have left the athlete in either a more prepared or unprepared state. For example, to allow AEL to serve as an overload stimulus during a training phase, it may not be necessary to prescribe AEL every week. Instead, strength and conditioning professionals may prescribe AEL during a *peak week* of a training phase to allow their athletes to develop fatigue resistance to a novel stimulus before providing an additional stimulus (e.g., weight releasers) that may challenge the neuromuscular system further. This would then be followed by a *drop week* that will promote recovery and adaptation. Finally, it is recommended that strength and conditioning professionals consider the accumulated fatigue that may result if AEL is mixed with other training methods or exercises within the same training phase.

Regarding the prescription of AEL multiple times per week, strength and conditioning professionals must consider the ability to provide *heavy days* and *light days* throughout a week. For example, if AEL is prescribed twice per week with the same exercise, a light day may no longer be considered light depending on the duration of the eccentric phase, volume, and both the eccentric and concentric loads prescribed. While the goals of a given resistance training phase may dictate whether AEL is prescribed multiple times per week, strength and conditioning professionals wishing to provide both a heavy and a light AEL stimulus may consider using AEL-MAX or AEL-SUPRA during a strength-based movement (e.g., squat) before prescribing AEL-BAL at the end of the week using a ballistic exercise (e.g., dumbbell jump) (see table 6.2). This may allow athletes to receive both peak force and rapid force production stimuli throughout the week while also receiving a light day. Moreover, this may allow strength and conditioning professionals to effectively manage the accumulated fatigue throughout a training week (34). Like VRT, if strength and conditioning professionals would like to prescribe AEL with multiple exercises during a single training session, they must first determine its feasibility, the ability of the athletes to tolerate such a stimulus, if it is necessary or warranted, and how to redistribute the prescribed volume–load of other exercises if the AEL stimuli are too great and negatively affect subsequent exercise performance.

Key Point

If AEL is prescribed twice per week with the same exercise, a light day may no longer be considered light depending on the duration of the eccentric phase, volume, and both the eccentric and concentric loads prescribed.

Another caveat to AEL is whether the athlete may need to perform each repetition with AEL. As mentioned earlier, the use of weight releasers on the first repetition of a set may benefit the performance of multiple subsequent repetitions (157). However, AEL may also be prescribed during every repetition (77, 156) or even within groups of repetitions

TABLE 6.2 Implementing Accentuated Eccentric Loading (AEL) Into a Strength–Power (3×3) Resistance Training Program Example

Day 1	Day 2	Day 3
Jerk	Hang clean pull	AEL dumbbell jump
AEL back squat	Hex bar jump	Back squat
Bench press	Pull-up	Medicine ball vertical throw

using cluster sets. While more detail on cluster sets is provided in chapter 7, using AEL every repetition or during the first repetition of a smaller set may provide a unique training stimulus. For example, it would appear obvious that a greater number of exposures to AEL within a set may promote greater eccentric maximal force production adaptations. However, using AEL on only the first repetition of a set may promote rapid eccentric force production (157) while still benefiting the concentric phase of the movement (77). Although implementing AEL during every repetition may seem appealing, strength and conditioning professionals must determine whether their athletes can tolerate this type of stimulus because it may require multiple exposures to supramaximal loads, which may result in greater levels of fatigue and affect subsequent performance.

Take-Home Points

- Potentiation complexes, VRT, and AEL may be classified as advanced training methods and may provide a novel training stimulus to athletes with previous training experience if they are not introduced earlier in an athlete's training program. Strength and conditioning professionals must, however, consider the stability of an athlete's technique, their relative strength characteristics, the availability of equipment and spotters, and the practicality of each method before prescribing these methods.

- Potentiation complexes have the potential to acutely improve the maximal and rapid force production characteristics of an athlete; however, it appears the results are highly individualized. Strength and conditioning professionals must consider athlete characteristics (e.g., relative strength, fiber type, sport or event) and the overall design of the complex (e.g., conditioning activity, volume, intensity, rest interval) to effectively implement potentiation complexes in training or competition.

- VRT is typically implemented using chains or elastic bands and may modify the force production characteristics during exercises such as the squat, bench press, and others. When using VRT for peak force production, it is suggested that 20% to 35% of the total weight lifted be accounted for by chains or bands when using heavier training percentages (≥80% of the 1RM); however, further research is needed for optimal rapid force production load prescription.

- AEL may be prescribed in many ways using AEL-BAL, AEL-SUB, AEL-MAX, and AEL-SUPRA protocols; however, strength and conditioning professionals must consider the eccentric duration, loading, and frequency of implementation. Eccentric durations during AEL-BAL are meant to be shorter to benefit the stretch–shortening cycle contribution to performance. In contrast, it is recommended that athletes perform the eccentric phase of AEL-SUB, AEL-MAX, and AEL-SUPRA protocols using their normal movement speed, which is specific to each athlete; this will allow them to use their natural movement mechanics, resist the heavier eccentric load, and still benefit the propulsion phase of the exercise.

- From a loading standpoint, the use of lighter loads (e.g., 20% body weight) may benefit AEL-BAL performance. When using AEL-SUB, AEL-MAX, and AEL-SUPRA protocols, a smaller difference between the training load and the added (i.e., AEL) load may promote maximal strength adaptations, provided the training load is sufficiently heavy. In contrast, a larger difference between the training load and the added load may promote the development of rapid force production characteristics.

CHAPTER 7

Program Design

Designing training programs to improve muscular strength may seem simplistic because lifting heavier loads may benefit maximal strength. However, when using the definition of *strength* identified in chapter 1 (i.e., the ability to produce force against an external resistance within a task-specific context), there are additional program design factors that should be discussed. Whereas the previous chapters highlighted a wide variety of training methods, the purpose of this chapter is to discuss several variables that strength and conditioning professionals should consider when designing training programs to develop both the maximal and rapid force production characteristics of their athletes.

NEEDS ANALYSIS

A strength and conditioning professional should complete a needs analysis before designing training programs for their athletes; these typically include an evaluation of the athlete's sport or event and an assessment of their physical abilities (190). Combining this information may aid in the development of an effective strength and conditioning program aimed at improving the athlete's force production characteristics and overall performance in their sport.

SPORT OR EVENT ANALYSIS

Regarding the athlete's sport or event, strength and conditioning professionals should include a movement, physiological, and injury analysis to determine the characteristics needed to be successful (65, 114). First, a movement analysis consists of determining which movements occur within the sport or event as well as the muscle groups and joints involved, ranges of motion performed, movement velocity, and coordinative patterns of each movement (168).

Using this information and the concept of specificity, strength and conditioning professionals can make decisions regarding exercises that may benefit each athlete. However, it should be noted that although the strength characteristics of sports may differ, similar exercises may be performed by different athletes due to similarities in their movements. For example, the triple extension of the hip, knee, and ankle (plantarflexion) joints may be one of the most common movements performed in sport due to its inclusion in running, jumping, changing direction, and so forth. Thus, exercises such as squatting variations, weightlifting movements and their derivatives, and lunge variations may be incorporated for a variety of athletes.

Second, a physiological analysis requires the strength and conditioning professional to determine the energy systems primarily used during the athlete's sport or event (168). Within this analysis, it is important to verify not only the entire duration of the event but also the actions that determine the outcome of the event. For example, the duration of a collegiate soccer match is 90 minutes broken into two, 45-minute halves. Although this may suggest that the aerobic energy system should be the primary focus of strength and conditioning professionals, researchers have concluded that the most decisive actions that determine the match outcome (e.g., sprinting, jumping, kicking) are anaerobic in nature (265) and that more elite players tax the anaerobic system to a greater extent than non-elite players (200). Within this scenario, programming conditioning activities may include a mixture of both aerobic and anaerobic activities; however, it is important to use a progression to higher-intensity activities as the competitive season gets closer.

Finally, from an injury analysis perspective, the strength and conditioning professional should consult the sports medicine staff (e.g., athletic trainers, physical therapists, team physicians) to gain more

93

insight regarding the injury trends within each sport that they work with. This information should include not only the location of an injury but also the specific tissues involved (e.g., muscle, tendon, ligament, bone) (136). While it is outside the scope of practice for a strength and conditioning professional to assess and treat injuries, it is important that they form a collaborative relationship with sports medicine staff to assist each other when developing an appropriate return-to-play protocol.

PERFORMANCE ANALYSIS

Beyond the analysis of the sport, the strength characteristics of an athlete should be assessed using a battery of tests to determine how they may improve (98). Because strength is task specific, the magnitude and rate of force production characteristics should be determined during eccentric, isometric, concentric, or stretch–shortening cycle tasks based on the movement analysis of an athlete's sport or event. An athlete analysis may also include tests that are specific to the athlete's skill within their sport or event but may be underpinned by their force production characteristics (e.g., sprinting speed, bat speed in baseball or softball, blocking height in volleyball). Testing data should be compared to previous information, if available, to determine the abilities of an athlete and how strength and conditioning professionals may tailor their training program to address their needs. Athlete testing data may also be benchmarked to determine how "good" the performances of an athlete are relative to a wider population (137). The overall assessment of an athlete's strength characteristics and testing considerations is discussed in chapter 9.

PROGRAMMING VARIABLES

While a needs analysis may provide a roadmap as to what an athlete may need to improve, several programming variables should be considered when designing a training program focused on developing the characteristics that underpin muscular strength. Specifically, strength and conditioning professionals should consider exercise selection, the frequency of training sessions, exercise order, volume, set structure, loading and repetition ranges, and rest intervals. Although some may seem obvious, it is important that an evidence-based approach is taken to maximize training adaptations.

EXERCISE SELECTION

Entire chapters have been dedicated to discussing the many facets of exercise selection within resistance training programs (208). Although there may be several reasons for choosing one exercise over another, it is important that strength and conditioning professionals understand the biomechanical and physiological differences between exercises to provide their athletes with the most effective training stimuli. Regarding the use of multi-joint and single-joint exercises, one must consider the movement specificity of each exercise and how it may transfer to the tasks performed in the athlete's sport or event. For example, because single-joint exercises isolate specific muscle groups (12, 201), their ability to transfer to athletic performance tasks may be limited (7, 14, 161, 201). Furthermore, despite the ability of single-joint exercise to improve an athlete's strength characteristics (45), a large emphasis on these exercises in training may be questioned because of a lack of intermuscular coordination patterns (201). Beyond single- or multi-joint exercises, the use of machine-based exercises is still prevalent in the strength and conditioning field. However, researchers have shown that free weight exercises may recruit musculature to a greater extent than machine-based exercises (73, 201), which may ultimately enhance both the magnitude and rate of force development. Combined with greater muscle coordination, free weight exercises may be superior for development of an athlete's force production characteristics. Thus, for the aforementioned reasons, the following content focuses primarily on the use of multi-joint, free weight exercises. However, it should be noted that both single-joint and machine-based exercises exist on a continuum with free weight exercises and may thus serve as exercise progressions or regressions (227).

It is no secret that every strength and conditioning professional has their "bread-and-butter" exercises when it comes to training specific muscle groups. However, it is important to consider the context in which specific exercises are implemented. For example, there are frequent debates among strength and conditioning professionals about whether to implement specific methods or exercises, such as whether athletes should or "need" to perform weightlifting movements or if partial range-of-motion squats should be implemented. Although it may be easier said than done, it is important to set aside any biases and approach each programming decision within a given context. Moreover, it is crucial to understand that no single exercise can address all the training needs of an athlete. Strength and conditioning professionals should first consider what strength characteristic they are trying to improve and use the available evidence to make an informed decision based on movement specificity, the needs of the athlete, the availability of equipment, and the feasibility of teaching the movement correctly.

Maximal Force Production

While the loading of resistance training exercises will be discussed in greater detail next, it should come as no surprise that the use of heavier loads may lead to improvements in maximal force production. It is, however, important that strength and conditioning professionals consider the range of motion in which the progressively heavier loads are lifted. For example, researchers have promoted the use of partial range-of-motion exercises to enhance movement specificity and allow athletes to use heavier loads (149, 174). Meanwhile, additional researchers showed that performing a heavy exercise through a large range of motion may increase the required relative effort compared to a smaller range of motion (17). Thus, it should come as no surprise that full range of motion movements may lead to greater improvements in maximal strength compared to partial ranges of motion in both the squat (162) and bench press (131) exercises. Therefore, it is recommended that strength and conditioning professionals implement full range of motion exercises (e.g., squats, presses, and pulls) to serve as a foundation of maximal force production, although it is important to note that partial range of motion exercises may serve as effective supplemental exercises and be used in combination with the full range of motion movements (11, 149) to further benefit these adaptations. It is important to note that although sprinting (28) and change-of-direction tasks (58, 85, 256) have the potential to produce large magnitudes of force, the durations of these performances are relatively short compared to resistance training exercises, which may hinder greater forces from being produced. Moreover, the loads used during resisted sled pulling (3, 21, 165) and pushing (22) will likely be significantly lower than those used during other resistance training exercises, indicating that sprint and change-of-direction tasks may serve athletes better as a rapid force production stimulus compared to peak force.

Key Point

Full range of motion exercises should serve as the foundation of maximal force production; however, partial range of motion exercises may serve as effective supplemental exercises and be used in combination with the full range of motion movements.

Rapid Force Production

There is a broad spectrum of resistance training exercises that may be used to develop the rapid force production characteristics of an athlete. Typically, the exercises used to develop these characteristics include ballistic intent and may be categorized as either strength–speed (i.e., moving heavy loads quickly) or speed–strength (i.e., moving light loads quickly) (78, 253). Traditional exercises (e.g., squats, presses, and pulls) may be performed with ballistic intent to provide a stimulus for rapid force production (159, 235); during a general or absolute strength phase in which the goals are to develop a greater magnitude and rate of force production (227), strength and conditioning professionals may consider using exercises such as weightlifting movements and their derivatives, especially those that would be classified as more force-dominant or strength–speed exercises such as the mid-thigh pull, countermovement shrug, hang pull/pull from the knee, or pull from the floor (30, 216, 217). These semi-ballistic exercises have the potential to use loads greater than an athlete's 1RM of a catching variation (e.g., power clean, hang power clean) (32, 33, 79, 138, 140), thus providing an effective stimulus for greater rapid force production. Additional weightlifting derivatives such as the jump shrug (111, 112, 212, 233, 238) and hang high pull (213, 233, 238, 240) may provide an effective speed–strength stimulus due to the greatest rapid force production characteristics being produced at lighter loads than the previously mentioned exercises.

Beyond weightlifting derivatives, strength and conditioning professionals should consider that other loaded jumps and plyometric exercises may also serve as effective speed–strength exercises. Compared with the jump shrug, researchers have shown that the jump squat and hexagonal barbell jump possess unique force production characteristics when performed at the same loads (224, 225). The authors noted that although the jump shrug produced greater force at peak power, the jump squat and hexagonal barbell jump produced faster velocities, thus supporting their inclusion as rapid force production stimuli. Similar conclusions were made by other researchers who examined the loads that maximized power development during upper-body exercises (198). As noted in chapter 5, plyometric exercises may serve as effective rapid force production stimuli. There are a variety of plyometric exercises to choose from based on their kinetic characteristics (59, 100, 101, 113, 229); however, these exercises should be implemented based on the abilities of an athlete (237) and in a progressive manner based on the volume and intensity of the chosen exercises (60, 61).

As noted previously, sprinting and change-of-direction tasks may produce large magnitudes of force during short durations. Clark and Weyand (28) showed distinct force peaks in competitive sprinters during the first half of their stance phase. It should, however, be noted that the vertical and horizontal forces produced may vary based on the phase of sprinting (153) and the external load during sled push (22) and pull (21) accelerations. Researchers have also shown peaked impulse shapes during deceleration steps during 135° (156) and 180° (58) change-of-direction tasks. Like sprinting, the rapid force production stimulus during change-of-direction tasks may change based on the angle (40, 88, 186, 192) and entrance velocity (46, 254). In summary, exercises that may be used to develop rapid force production characteristics exist on a spectrum, and it is up to the strength and conditioning professional to determine how best to prescribe these exercises to address the needs of their athletes.

Braking Forces

While much of the previous discussion focused on propulsive force characteristics, it is important not to neglect the braking characteristics of athletes. It should be noted, however, that all the previously discussed exercises and training methods may also provide a braking stimulus due to the inclusion of the stretch–shortening cycle. For example, researchers have shown that weightlifting exercises and their derivatives (34, 150, 218, 219), loaded jumps (121), plyometric exercises (102, 147), the stance phase of sprinting (28), and change-of-direction tasks (58) have all been shown to produce unique braking force–time characteristics. Despite their potential as braking stimuli, the magnitude and rate of force production of each task may be dependent on the exercise, load (if any), and athlete technique. With the exception of plyometric training (236), none of the discussed exercises are typically classified as eccentric training methods, likely due to the lack of focus on the braking phase and the purpose of each movement. However, each of the discussed exercises and the eccentric training methods highlighted in chapters 5 and 6 exist on a continuum, as displayed in figure 7.1. In line with the development of propulsive force production characteristics, strength and conditioning professionals must again understand the sport or event demands and determine the needs of each athlete via testing (chapter 9) to effectively prescribe braking stimuli.

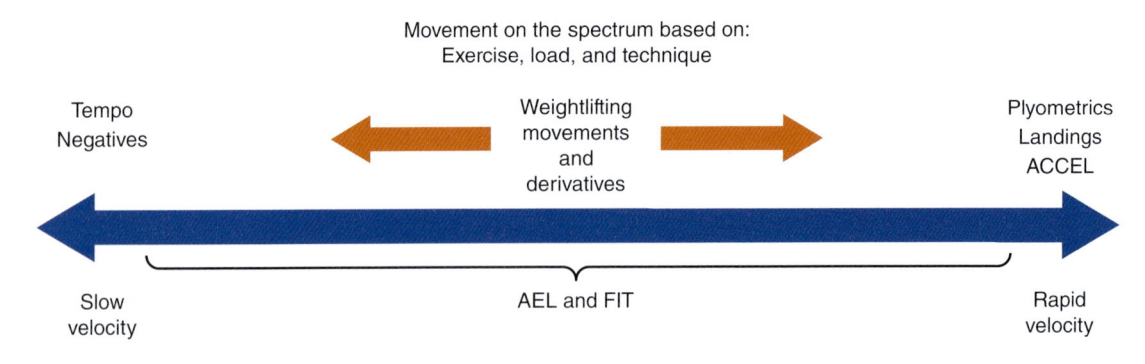

FIGURE 7.1 The eccentric training spectrum.

Combined Training Methods

While specific exercise selection for the development of maximal and rapid force production was discussed earlier, strength and conditioning professionals should also consider the combination of multiple training methods. The implementation of multiple training methods is not a new concept as several reviews have discussed this idea related to power development (78, 117, 158, 253). However, Turner and colleagues (252) emphasized that power development should be targeted by developing its underpinning force production characteristics (i.e., magnitude, rate, and impulse). One of the most common examples of a combined training methods approach is the use of heavy resistance training and plyometric exercises. This combination serves to improve both maximal and rapid force production; alone, these methods may be limited in their ability to address both qualities of strength. Researchers have shown that combined training methods may lead to greater improvements in strength characteristics as well as different markers of performance (2, 35, 86, 108, 127, 244, 245, 246).

Additional research expanded on the previous reviews and discussed the idea that all training methods exist on a "grand" force–velocity curve (figure 7.2) and should consider both concentric, isometric, and eccentric force production qualities (211). As indicated earlier, exercises and training strategies exist on a spectrum based on the magnitude and rates of force production. For example, strength and conditioning professionals implementing weightlifting derivatives should be aware that although the exercises may produce greater velocities compared to traditional resistance training movements, they may not produce the same magnitudes of force. Furthermore, although loaded jumps and plyometric exercises may produce faster velocities compared to weightlifting derivatives, sprinting produces significantly greater velocities and rates of force production due to its task demands. Regarding eccentric force production characteristics, greater forces are developed at faster velocities (63); because many sporting actions require the stretch–shortening cycle, an additional focus on eccentric strength qualities may influence concentric force production as well. While there are a variety of training methods (see chapter 5), using a combination of them may allow strength and conditioning professionals to further develop the force production characteristics of their athletes across a broader spectrum, potentially leading to greater improvements in performance.

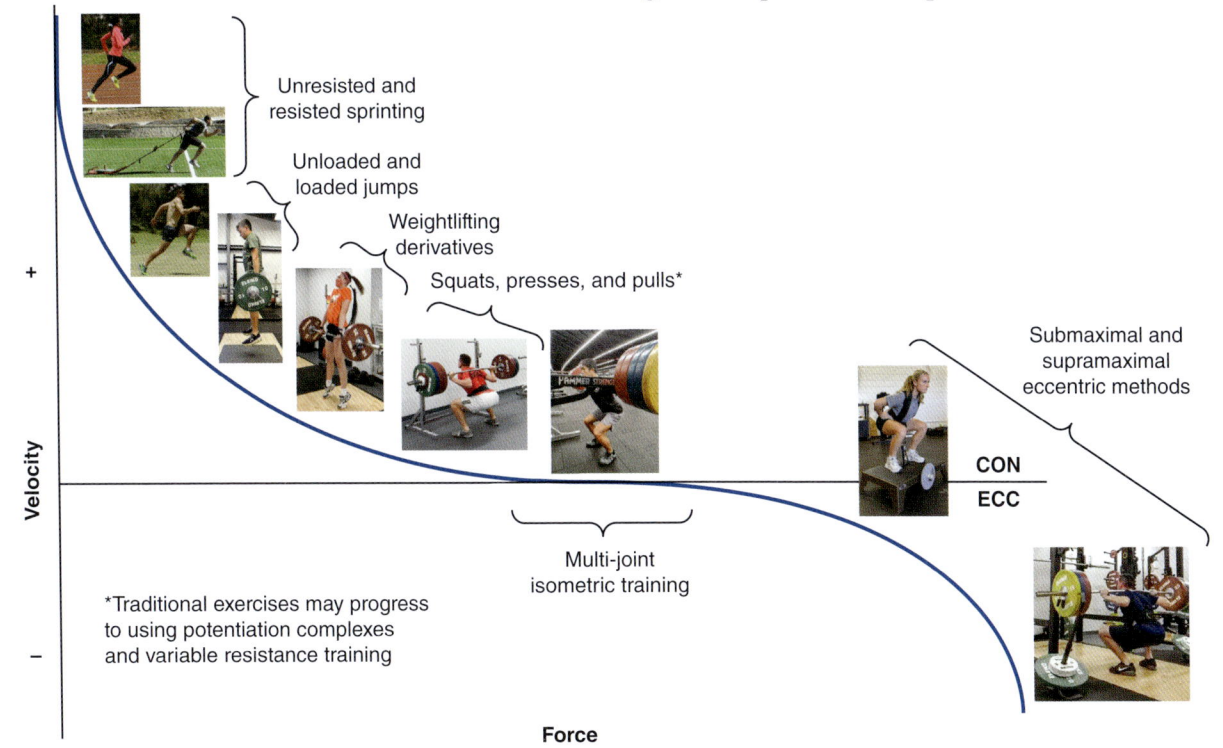

FIGURE 7.2 Training strategies across the force–velocity spectrum.

Photos 1-3 © Human Kinetics.

> **Key Point**
>
> While there are a variety of training methods, using a combination of them may allow strength and conditioning professionals to further develop the force production characteristics of an athlete across a broader spectrum.

FREQUENCY

Training frequency, also known as *training density*, refers to how often an athlete completes training sessions (e.g., number of sessions per week, day). It should be noted that the number of training sessions may vary based on the goals of the training program, the training status of the athlete, and the training season. Training volume and intensity (discussed next) are ultimately influenced by training frequency as well. Interestingly, through various meta-analyses, researchers showed that when volume is matched, there were no differences in maximal strength improvements between a greater or smaller frequency of training sessions (43, 71, 171), with one research group showing greater strength improvements with greater training frequencies and, thus, a greater overall volume of training (71). From a programming standpoint, frequency may be modified during planned, functional overreaching and tapering protocols (6). Altering the number of training sessions in this manner may promote specific adaptations or help strength and conditioning professionals manage the fatigue of their athletes. However, a greater number of training sessions may require each session to have a lower average intensity to prevent nonfunctional overreaching and overtraining. Furthermore, strength and conditioning professionals may split a single training session into multiple sessions to allow for the maintenance of a higher training intensity.

> **Key Point**
>
> Altering the number of training sessions during planned, functional overreaching and taper periods may promote specific adaptations or help strength and conditioning professionals manage the fatigue of their athletes.

Previous recommendations have suggested that novices should train two or three times per week with resistance exercises before progressing to a greater frequency of training sessions (4, 190). It is important that strength and conditioning professionals avoid prescribing a stimulus that may result in negative adaptations, especially because novice athletes may see significant benefits from fewer training sessions. Strength and conditioning professionals should also note that a unique overload stimulus may be provided in the form of additional training sessions; for example, the introductory week of a training phase may require athletes to train three times per week, but the following week may include a fourth training session. In addition, researchers showed that weekly strength training compared to training every other week allowed professional soccer players to maintain improved strength, speed, and jump performances from a preparatory period during a competitive season, whereas the latter group saw decreases in their strength and sprint performances (178). Therefore, when determining the frequency of training sessions per week, it is important to consider the goals of each training phase and the training age of the athletes.

Because of the frequency of games or matches, strength and conditioning professionals have been challenged to continue to implement resistance training sessions throughout a competitive season. This becomes even more of a challenge when athletes may not fully recover from competition muscle soreness despite two resting days between matches (126). Due to the importance of building and maintaining muscular strength characteristics during a competitive season (228), a growing body of literature has focused on the concept of *microdosing*, previously defined as the division of training volume within a microcycle through several short-duration, repeated bouts (43). Kilen and colleagues (110), who termed this frequency of training "micro training," showed that shorter, more frequent training sessions may produce similar knee extensor and finger flexor maximal voluntary isometric force adaptations compared to longer, less frequent training sessions. Additional researchers have found resistance training microdosing to be an effective and efficient strategy for improving sprint performance in field hockey players (41), demonstrating its versatility. Cuthbert and colleagues (44) provided a thorough review that discussed several ways that microdosing could be implemented through strategies such as postactivation performance enhancement, resistance priming, repeated bout effect, training sequencing, and concurrent training. It is recommended that strength and conditioning professionals use the organization of training concepts discussed in chapter 4 to determine the need for microdosing compared to traditional training sessions, its feasibility with athlete training and competition schedules, and the strategy or strategies that could be used when implementing microdosing concepts. Furthermore, it is important to consider the training age of the athletes to determine whether specific microdosing strategies would benefit their strength characteristics as they relate to sport performance.

EXERCISE ORDER

The order of prescribed exercises during a training session may have a significant impact on the strength adaptations realized by the athletes. Researchers have shown that greater maximum strength adaptations may be produced during exercises prescribed earlier in a training session compared to those performed later (57, 196, 199). Additional research indicated that exercises performed earlier in a training session may negatively affect the performance of the exercises performed later in the training session (e.g., decreased force production, number of repetitions performed) (145, 148, 189, 194, 195). However, it should be noted that rating of perceived exertion does not appear to be influenced by exercise order (177, 194, 195). Although it has been previously proposed that pre-exhaustion (i.e., fatiguing) of smaller muscles during isolated exercises may allow athletes to better target the larger muscles that can generate more force (104), researchers who tested this theory found that pre-exhaustion offers no additional strength benefits (64, 197). Collectively, the existing research supports the notion that larger muscle, multi-joint movements should be performed prior to smaller muscle, single-joint movements (193). Beyond large-muscle multi-joint movements, strength and conditioning professionals should prioritize the exercises that place the greatest neuromuscular demand on the athlete and require large magnitudes and rates of force production (e.g., weightlifting movements and derivatives, ballistic exercises). This, in turn, may allow athletes to perform these movements during a less fatigued state and allow for positive strength adaptations to be realized.

Key Point

Exercises performed earlier in a training session may negatively affect the performance of the exercises performed later in the training session.

While the previous paragraph focused on the completion of each exercise in a specific order, several reviews and meta-analyses have focused on the sequencing of exercises when they are paired together (38, 39, 129), often termed *complex training* or *contrast training*. Complex training may be defined as completing all sets of a heavier exercise before performing all sets of a lighter exercise, whereas contrast training refers to alternating heavy and light exercises set for set (129). According to the findings of a 2021 meta-analysis, positive maximal strength adaptations may result from using either complex or contrast training; however, peak force and velocity adaptations favored contrast training (129). It should be noted that these adaptations may be based on the concept of postactivation performance enhancement, which is discussed in greater detail in chapters 5 and 6. Regardless of exercise sequencing, strength and conditioning professionals should be aware that these adaptations may be individualized based on an athlete's characteristics as well as the protocol used (220).

VOLUME

The volume of a resistance training session may be classified as the number of sets and repetitions completed. Briefly, a *set* is a group of repetitions performed to enhance a specific fitness quality (e.g., hypertrophy, strength, power), whereas a *repetition* is the performance of an exercise through a predetermined range of motion one time. It should be noted that sets may be designed with different structures to provide a unique training stimulus; further information on different set structures is discussed next. Specific to resistance training, an abundance of studies support the notion that multiple sets of an exercise may produce greater magnitudes of strength compared to single sets (16, 115, 118, 119, 120, 133, 163, 173, 181, 182, 264); however, strength and conditioning professionals must consider the training status of an athlete as well as the dose–response relationship for developing maximal and rapid force production (164, 170). Regarding athlete training status, smaller doses of resistance training (e.g., 2-3 sets per exercise) may provide a sufficient stimulus to novices and athletes with a lower training age, whereas well-trained athletes may require larger doses (e.g., 4-6 sets per exercise) (227). Regarding the prescription of a greater number of exercise sets, it is important that strength and conditioning professionals do not fall into the mindset of "more is better," because long-term elevations in training volume may promote nonfunctional overreaching, accumulation of undesired fatigue, potential overtraining (207, 226), and an increase in injury potential (62). Moreover, larger training volumes may also negatively affect the athlete's ability to improve their rapid force production characteristics (128, 209); however, these effects may vary based on the training status of the individual.

Strength and conditioning professionals should also note that the addition of more training sets may negatively affect the athlete's ability to train at a sufficient intensity. For example, a greater overall training volume may result in greater fatigue (266), ultimately limiting the athlete from consistently performing training sets at higher intensities and possibly the training stimulus (202, 214). Collectively, the available literature supports the notion that a larger volume of sets may provide a superior stimulus for muscular strength adaptations; however, these effects may be partially dictated by the training status of the athlete, the exercises and loads prescribed, and the goal of a given training phase. Strength and conditioning professionals are therefore encouraged to use methods that may assist them in monitoring the force production capabilities of their athletes to supplement their programming decisions (see chapters 9 and 10 for greater detail) and manage fatigue (81).

Key Point

Regarding athlete training status, smaller doses of resistance training (e.g., 2-3 sets per exercise) may provide a sufficient stimulus to novices and athletes with a lower training age, whereas well-trained athletes may require larger doses (e.g., 4-6 sets per exercise).

SET STRUCTURE

Beyond the number of sets prescribed, it is important for strength and conditioning professionals to consider the structure of the prescribed sets. For example, sets of exercise are most commonly implemented with the athletes performing all the prescribed number of repetitions in succession. While this is a perfectly reasonable method of training prescription to achieve many specific training goals, sets with an increased number of repetitions may cause a decrement in performance characterized by a decline in movement velocity (69, 250) and potentially the technique proficiency of an athlete (83), which may negatively affect the overall training stimulus. Although it could be argued that the length of training sets (i.e., number of repetitions) focused on developing an athlete's strength characteristics is less of a concern given that they typically consist of 6 or fewer repetitions (190), researchers have shown that both maximal and rapid force production capacities can be reduced within 5 to 9 maximal muscle contractions (257). This has led to a growing body of literature that has investigated the idea of manipulating set structure to split a traditional set into multiple sets, often termed *cluster sets*.

Similar to previous definitions (79, 176, 191), cluster sets may be defined as individual or groups of repetitions that are performed with either interrepetition (i.e., between individual repetitions) or intraset (i.e., between clusters) rest intervals, respectively, within a larger set of a predetermined number of repetitions. Another term strength and conditioning professionals may see when examining the cluster set literature is *rest redistribution*, which removes time from interset rest intervals and redistributes it within interrepetition or intraset rest intervals. Although rest intervals are discussed in more detail later in this chapter, it is important to note that the terms cluster sets and rest distribution should not be used synonymously, given that the overall amount of rest provided is not equal (105).

As alluded to earlier, one of the primary benefits of cluster sets is the ability to maintain movement velocity and force production to a greater extent compared to traditional sets (77). From a physiological perspective, athletes using cluster sets may be able to mitigate the reduction of phosphocreatine stores, which has been shown to decrease approximately 27% following sets of 5 leg press repetitions (68), thereby providing more readily available energy sources. The benefits of this include the ability to improve the overall quality of work and increase the potential to use heavier loads (79). Because the number of repetitions can be modified within each cluster, Tufano and colleagues (250) indicated that clusters with fewer repetitions may minimize velocity loss to a greater extent than those that include more repetitions. Furthermore, the use of smaller clusters may allow for greater mean forces to be produced during a set (249). Practically speaking, smaller cluster sets may allow strength and conditioning professionals to prescribe heavier loads, thus providing their athletes with a greater peak force stimulus.

While the acute effects of cluster sets favor the development of strength characteristics (122), additional meta-analyses investigating the chronic effects of cluster sets have shown no differences compared with traditional training sets, due to mixed findings within the literature (48, 107). For example, researchers have shown improved maximal strength in several exercises (e.g., bench press, squat, deadlift) using cluster sets with higher volumes of training (160, 255), whereas other researchers showed no additional benefit to incorporating cluster sets for isometric or dynamic elbow flexion strength (179), bench press or leg press 1RM (20), or bench press 6RM (123). It should, however, be noted that many of the existing longitudinal studies equalized volume–load between conditions, which may reduce participants' ability to fully use the benefits of cluster sets (i.e., use heavier loads). Thus, it appears that implementing cluster sets with higher volumes of training may allow athletes to use heavier training loads than traditional training, which may then lead to a greater capacity to improve maximal strength and peak force. Furthermore, the maintenance of velocity and higher quality of work may promote greater adaptations in rapid force production as well.

When implementing cluster sets, strength and conditioning professionals should be aware that there are several types, including standard, undulating, wave-loaded, ascending, and descending structures (76, 77, 154, 155). Each structure may use intraset or interrepetition rest intervals but possesses unique loading characteristics (figure 7.3). From a practical standpoint, it has been previously recommended that no more than two exercises per training session should be prescribed using cluster sets, with the remaining exercises performed using a traditional format (76, 77). Doing so may allow strength and conditioning professionals to limit the impact on the overall time needed to complete a training session while including cluster sets. Furthermore, both the structure of the cluster set (e.g., standard, undulating, wave-loaded) and the load used during the cluster sets should be considered based on the desired training outcomes. Specific to loading, it is recommended that strength and conditioning professionals define the average load (e.g., %1RM) that they would like the athlete to use when using either undulating, wave-loading, ascending, or descending loading schemes (76). Finally, it is important to prescribe an appropriate duration for either the interrepetition or intraset rest intervals. Further information on cluster set rest intervals is provided later in this chapter. Readers interested in additional insight on the use of cluster sets are referred to several previous reviews (77, 154, 155, 248).

Key Point

No more than two exercises per training session should be prescribed using cluster sets, with the remaining exercises performed using a traditional format to limit the impact on the overall time of the training session.

LOADING AND REPETITION RANGES

A common saying with strength and conditioning is, "If you want to get stronger, you have to lift heavy," and there appears to be considerable support for this statement. Researchers have shown that heavy resistance training may benefit maximal strength to a greater extent than lighter loads (125, 183, 184) and may improve the magnitude (31) and rate of force production (1, 31) as well as impulse characteristics (1). In addition, heavier loads were shown to produce greater magnitudes of muscle activation compared to lighter loads (67, 87). Previous literature has indicated that maximal strength characteristics may be best trained using 6 or fewer repetitions (76, 190) and the loads that coincide with these RMs (i.e., ≥85%

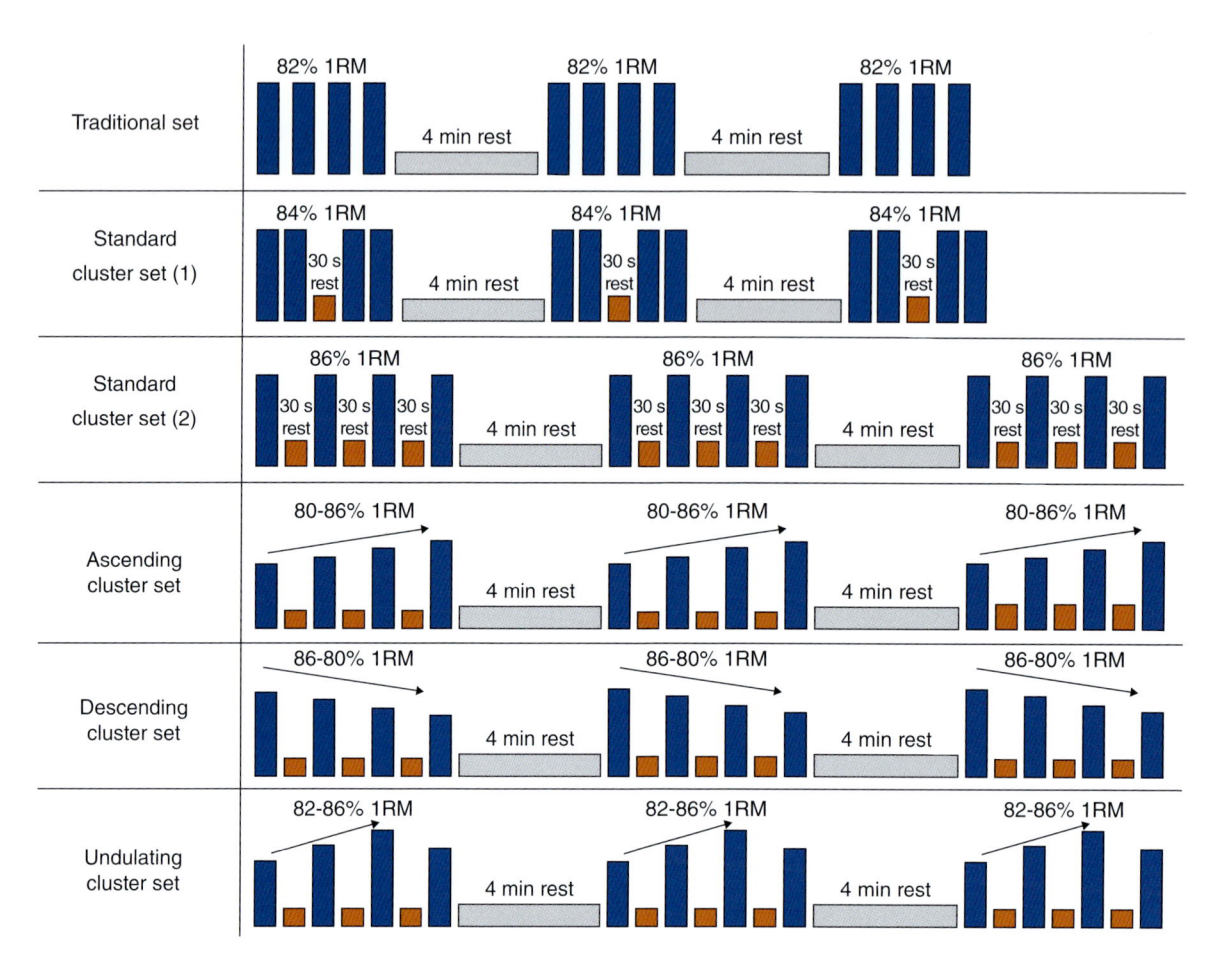

FIGURE 7.3 Cluster set configurations used within strength and conditioning training programs.
Adapted from Nagatani et al. (2022).

1RM) (190). There are several ways that strength and conditioning professionals can prescribe loads, including using percentages of a 1RM or multiple RM, rating of perceived exertion, repetitions in reserve, set–repetition best (i.e., relative intensities based on the best performance of a given set and repetition scheme), and others (226). General loading concepts for the development of maximal and rapid force production are discussed next, and readers are directed to chapter 10 for an in-depth discussion on different methods of loading.

Training to Failure

As noted earlier, consistently lifting heavy loads will lead to improvements in muscular strength; however, a common belief is that it is necessary to train to failure in order to provide a sufficient overload stimulus to achieve optimal strength improvements (18). *Training to failure* has been previously described as the point in which the barbell stops moving during a lift, the duration of a sticking point lasts longer than 1 second, or the athlete can no longer perform full range-of-motion repetitions (94). While the notion of training to failure suggests that training with RM loads may promote greater strength adaptations compared to lighter, submaximal loads, researchers have shown that this may not be the case (47, 164). In fact, the authors of the previous studies suggested that training to failure may be counterproductive and should be used sparingly to mitigate the potential for injury and overtraining. Interestingly, despite the use of relatively heavier loads when training to failure, this strategy does not differ in high-threshold motor unit recruitment compared to training that is not performed to failure (239). It should also be noted that the ability of athletes to train to failure for long training periods may be limited, especially if resistance training is part of a larger, holistic training program used to improve sport performance (47). This notion is supplemented by researchers who showed that consecutive sets of training to failure may significantly reduce the overall volume of repetitions performed at a given load (258, 259, 260), which may ultimately require a reduction in load to maintain the desired training volume for a given phase, possibly reducing the effectiveness of the training stimulus (120, 130, 152). Researchers concluded that the use of relative intensities using set–repetition best for 10 weeks produced greater improvements in squat and countermovement jump and isometric mid-thigh pull rate of force development from 0 to 50 ms and from 0 to 100 ms compared to training to failure using relative maximums (24). The authors also showed that the training-to-failure group displayed significantly greater training strain compared to the relative intensity group. In another study, the same research group showed that training with relative intensities produced favorable improvements in Type I and Type II cross-sectional area as well as myosin heavy chains for Type I, IIa, and IIx fibers compared to training to failure (23). In summary, although it may be necessary to prescribe very heavy loads (e.g., 90%-95% 1RM) to promote peak force development within certain training phases, it does not appear that training to failure is required to further benefit these adaptations.

Key Point

Training to failure may be counterproductive and should be used sparingly to mitigate the potential for injury and overtraining.

Combined Heavy and Light Loading

As noted previously, the consistent prescription of heavy loads may lead to improvements in the maximal force production characteristics of athletes. Beyond peak force characteristics, researchers have shown that heavier loads may also benefit rapid force production characteristics to a greater extent than moderate loads (31). It is important, however, to note that these adaptations may be dependent on the training status of the athlete, exercise selection, and intent during each exercise. Moreover, strength and conditioning professionals must be aware of the physical toll that heavy loads can have on athletes; for example, only prescribing heavier loads may eventually lead to nonfunctional overreaching and potentially overtraining (226) and may negatively affect strength gain if a variation in loading is not provided (242). Despite the potential benefits of heavy strength training, researchers showed that even with improvements in late rate of force development (>200 ms), early rate of force development (<200 ms) decreased following heavy strength training (5). However, further research showed that jump squats performed with loads ranging from 0% to 60% of participants' 1RM back squat resulted in greater impulse and rapid force production characteristics compared to the back squat performed with the same loads (241). The previous findings support the use of lighter loads in combination with heavier loads to benefit both maximal and rapid force production characteristics, a concept previously discussed regarding power development (78, 117, 158, 252, 253).

A combination of heavy and light loads can be implemented in a variety of ways, including through the periodization of training and programming tactics within an individual phase, week, or day of training (211). Regarding the periodization of different training phases, strength and conditioning professionals may plan and organize training to focus on the development of maximal and rapid force production characteristics at specific periods throughout the training year. From a loading standpoint, moderate loads (e.g., 60%-70% 1RM) may be used to first develop strength–endurance characteristics before progressing to heavier loads (e.g., 75%-90% 1RM) during a subsequent maximal strength phase, providing a stimulus for peak force production. Following these phases, strength and conditioning professionals could then prescribe a wide range of loads (0% to 95% 1RM, depending on the exercise and athlete training status) to further expand the loading spectrum and provide effective stimuli for peak and rapid force production characteristics. Apart from exercise selection, sequencing training in this manner may not only provide athletes with a wide loading spectrum, but it also allows for the underpinning mechanisms of strength (see chapter 2) to be developed in a sequential manner that may enhance or potentiate subsequent training goals (144, 205, 227, 267). However, it is important that strength and conditioning professionals invest in the development of maximal and rapid force production characteristics over the long term to maximize such adaptations. The research described earlier provides a snapshot of the development of strength characteristics using periodization, and this concept is discussed in greater detail in chapter 4.

Beyond providing a range of loading stimuli across several phases of training, strength and conditioning professionals can provide athletes with maximal and rapid force production stimuli within a single phase as well. Although training phases vary in length depending on the goal, many last approximately 3 to 6 weeks and use progressive loading across several weeks before including a recovery week. This method of progression allows the use of both heavy and light loads and may provide athletes with maximal and rapid force production stimuli. For example, strength and conditioning professionals may prescribe progressively heavier loads for the first 3 weeks of a 4-week strength phase to focus on the development of peak force production. Beyond peak force production, a recovery week may be used during the fourth week that emphasizes the use of lighter loads (132), which may promote rapid force production adaptations as well as recovery and adaptation from the heavy load training stimulus. It is important to note that different phases of training may also use a combined heavy and light loading approach. For example, the training phase discussed earlier in this paragraph focused on general strength characteristics; researchers have shown that this may be an effective approach during preseason training (97), but a similar approach may be taken in the off-season to build a foundation for maximizing both peak and rapid force production adaptations. An example of weekly variation in loading with a heavy and light emphasis is displayed in figure 7.4.

Heavy and light loading may also be prescribed within a single week of training, in which a heavy day is programmed earlier in the week and the lighter day is later (54, 244). Using this method, the heavy and light days of training provide a greater emphasis on either peak or rapid force production, respectively. Furthermore, it exposes athletes to a greater range of loading stimuli compared to only heavy loading (203), while also promoting fatigue management throughout the week (53, 54). Researchers have shown that a combined heavy and light training emphasis across 9 weeks produced greater strength adaptations compared to either a heavy- or light-only loading program (86). Similarly, Carroll and colleagues (24) showed that favorable improvements in maximal and rapid force production characteristics can be produced using variations in loading compared to RM training. Using this method, strength and conditioning professionals may be able to provide their athletes with a heavy and light emphasis using either the same exercise performed twice (the first day with heavier loads and the second day with lighter loads, biomechanically similar exercises that use lower loads on the light day) or strength–speed and speed–strength exercises on the heavy and light days, respectively (211). A programming example of heavy and light days within a single week of training is displayed in table 7.1. It is important to note that the amount

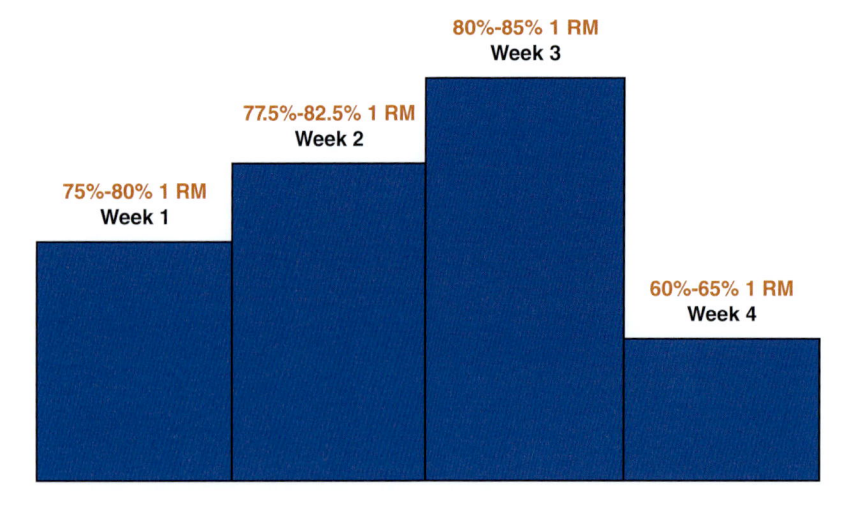

75%-80% 1 RM
Week 1

77.5%-82.5% 1 RM
Week 2

80%-85% 1 RM
Week 3

60%-65% 1 RM
Week 4

FIGURE 7.4 An example of weekly variation in loading. 1RM = one repetition maximum.

TABLE 7.1 Example of Combined Heavy and Light Loading During a Single Week of a General Strength Phase (3×5)

Day 1 (heavy push)		Day 2 (heavy pull)		Day 3 (push-and-pull light mix)	
Push press	70% 1RM	Hang power clean	70% 1RM	Jump shrug	30% of hang power clean 1RM
Back squat	80% 1RM	Stiff-leg deadlift	80% 1RM	Back squat	60% 1RM
Bench press	80% 1RM	Bent-over row	80% 1RM	Dumbbell bench press	60% of bench press 1RM
Barbell lunge	50% of back squat 1RM	Pull-up	Bodyweight	Single-arm dumbbell row	60% of bent-over row 1RM

Note: 1RM = 1 repetition maximum.

of loading variation across a week may vary based on the training emphasis and other necessary skill development (54). For example, a smaller percentage of load may be dropped during a strength–endurance phase in which work capacity is a primary training emphasis, whereas a larger percentage may be dropped during a strength–speed phase to maximize rapid force production.

The programming decisions made during a single day of training may affect the maximal and rapid force production stimuli for an athlete as well. Exercise selection is discussed in greater detail next, but it is important to note that a heavy and light loading stimulus can be provided by prescribing ballistic and strength exercises, strength–speed and speed–strength exercises, potentiation complexes (see chapter 6), or biomechanically similar movements—one with heavy loads and the other with lighter loads (211, 244, 245, 246). Using this strategy, similar muscle groups may be trained; however, the movement mechanics and intent of the exercises may differ. For example, although loaded jumps and squatting variations train the musculature associated with the triple extension of hip, knee, and ankle (plantarflexion) joints, the former may place a greater emphasis on rapid force production due to the use of lighter loads and the ballistic intent, whereas the latter may promote peak force adaptations due to the ability to use heavy loads (36, 241). An example of strength–speed and speed–strength exercises being used within a single session may be to include weightlifting pulling derivatives, which may expand the loading spectrum compared to only using catching variations (210, 216, 217). Researchers have shown that peak and rapid force production characteristics may be improved to a greater extent using a combination of heavy and light loads with weightlifting pulling derivatives compared to submaximal loading with weightlifting catching derivatives (221, 222, 223).

A final programming strategy to implement heavy and light loading may include the use warm-up and drop sets in addition to the working sets during a single exercise. The use of warm-up sets provides athletes with an opportunity to physically prepare themselves for the heavier working sets as well as the ability to emphasize rapid force production with lighter loads. Although this strategy may seem obvious, it is important that athletes are encouraged to perform each warm-up set with maximal intent to receive the greatest adaptations. Following the warm-up sets, the completion of working sets exposes the athlete to heavier loads to emphasize greater force production, especially when performed with near-maximal loads, as mentioned earlier. Finally, *drop sets*, in which a percentage of the working load is removed or "dropped," may provide athletes with another opportunity to enhance their rapid force production characteristics. In theory, the performance of heavier working sets may produce a potentiation effect during a subsequent set performed with a lighter load or ballistic alternative. For example, heavier sets of a squatting (230, 231, 232) or bench press variation (109, 247) may enhance biomechanically similar jumps or bench press throws, respectively. Collectively, the combination of warm-up, working, and drop sets allows strength and conditioning professionals to expose their athletes to both heavy and light loads, which may aid in the development of maximal and rapid force production characteristics. Table 7.2 displays an example of warm-up, working, and drop set programming.

TABLE 7.2 Back Squat Warm-Up, Working, and Drop Set Programming to Emphasize Combined Heavy and Light Loading

Set	Repetitions	Load	Mean barbell velocity (m/s)
Warm-up 1	5	30% 1RM	1.11
Warm-up 2	5	50% 1RM	0.91
Warm-up 3	3	70% 1RM	0.69
Work 1	3	85% 1RM	0.59
Work 2	3	85% 1RM	0.56
Work 3	3	85% 1RM	0.52
Drop	3	50% 1RM	0.91

Note: 1RM = 1 repetition maximum.

Volume–Load

Researchers have shown that a greater amount of overall work completed during a training session may lead to greater energy expenditure and physiological disturbances (19, 80, 135, 141, 166), which may ultimately affect potential training adaptations. From a biomechanics standpoint, work may be calculated as the product of force and displacement (i.e., force × displacement). Thus, researchers have concluded that measuring the force and displacement of each repetition would provide the most valid measurement of mechanical work during training (74, 134). While this may be accomplished in a laboratory setting, few weight rooms are equipped with force plate and linear position transducer technology to be able to measure the force and displacement characteristics of every repetition of each athlete, respectively. Because of this limitation, many strength and conditioning professionals and researchers have calculated volume–load (i.e., sets × repetitions × load) as a measure of training volume. Although researchers have shown that volume–load may provide a reasonable estimate of work during the back squat (134), it fails to account for the differences in athlete height and exercise displacement. As a result, the measurement of volume–load and exercise displacement (i.e., volume–load displacement) has been used as an estimated measure of the work performed within various training studies (9, 10, 11, 24, 91, 92). It should be noted that in a previous study, strong relationships (r = .78-.99) existed between volume–load and volume–load displacement throughout a 20-week training program; however, there were significant differences in the percent change between the measurements during different training phases (91). Because the range of motion of an athlete can change throughout time and exercise variations may result in greater or larger displacements (e.g., full squat vs. partial squat), it is recommended that strength and conditioning professionals measure the displacement of an exercise during different phases throughout the training year. Although it is more time-consuming, the inclusion of volume–load displacement may provide a more valid estimate of the overall work performed during a training session compared to volume–load (134), especially because the measurement of forces during each repetition may be impractical. This can be accomplished by measuring displacements using a linear position transducer, videography, or infrared systems or estimating it using a tape measure at the beginning of different training phases and inputting the information into a spreadsheet with both the volume of repetitions and prescribed load to allow for the calculation of volume–load displacement (91).

From a practical standpoint, the inclusion of volume–load displacement becomes especially important with athletes differing largely in height and when partial movements are used in training. For example, if two athletes are completing the back squat in training, with one athlete being 1.70 m (approximately 5 feet, 6 inches) tall and the other being 2.00 m (approximately 6 feet, 6 inches) tall, strength and conditioning professionals can determine who is performing more overall work during the training session. If the athletes are using the same load throughout training and assuming they squat to the same relative position, it is obvious that the taller individual is performing more overall work. However, if the athletes do not use the same load or squat to the same position, volume–load displacement becomes the differentiating factor. Another example would be the inclusion of both full squats and quarter squats. Full squats require a greater displacement than quarter squats; however, quarter squats allow athletes to use significantly greater loads than full squats. If an athlete is performing the full squat for 3 sets of 3 repetitions using a load of 200 kg but performs 250 kg for the same number of repetitions during the quarter squat, the latter has a significantly higher volume–load (1,800 vs. 2,250 kg); however, if strength and conditioning professionals factor in displacement (e.g., full squat = 0.60 m; quarter squat = 0.25 m), the volume–load displacement for the back squat is much higher than the quarter squat (1,080 kg·m vs. 562.5 kg·m). Although the inclusion of both full squats and partial squats has been shown to be effective for both maximal and rapid force production characteristics (11), the previous example provides a rationale as to why volume–load displacement may be considered as an estimate of workload and should thus be considered during program design.

REST INTERVALS

Other variables to consider when training to improve an athlete's strength characteristics are the interset and intraset rest intervals being prescribed. Briefly, *interset rest interval* refers to the time between sets, whereas *intraset rest interval* may refer the potential time taken between individual or groups of repetitions within a set. Finally, strength and conditioning professionals may have to consider the rest intervals prescribed within potentiation complexes. This subsection discusses each type of rest interval as it relates to the development of strength characteristics.

Interset Rest Intervals

Despite shorter interset rest interval recommendations for muscle hypertrophy (116, 190), researchers have concluded that longer rest intervals may produce greater strength adaptations (72). For example, 1.5- to 3-minute rest intervals produced greater muscle strength adaptations compared to 0.5- to 1-minute intervals (175, 185). Researchers also indicated that longer rest intervals (i.e., 2.5-5 minutes) allowed participants to complete a greater volume of work during training sessions (258, 261), train with heavier loads (259), and improve maximal strength (50, 167, 175) compared to shorter rest intervals (i.e., 0.5-2 minutes). It should, however, be noted that although other researchers found no significant differences in strength improvement between 2- and 4-minute rest intervals (261), the group that trained with longer rest intervals demonstrated a greater magnitude of change (based on effect sizes) than their counterparts. Therefore, it is recommended that strength and conditioning professionals prescribe interset rest intervals ranging from 2 to 5 minutes to maximize strength adaptations, in line with previous recommendations (49, 116, 190). However, the prescribed training loads, an athlete's training age (261), fiber type, and genetics (29) may also factor into rest interval prescription. Simply, longer rest intervals may promote a higher quality of work due to the ability to fully replenish adenosine triphosphate and phosphocreatine stores (68, 69). Although strength and conditioning professionals may be limited in terms of the amount of training time with their athletes, implementing appropriate rest intervals may aid in producing a superior training adaptation. While this also may be difficult in a large group setting, athletes sharing equipment may benefit from the overall rest that they may receive, ultimately producing higher-quality work.

Intraset Rest Intervals

As noted earlier, intraset rest intervals may be implemented within cluster sets between individual repetitions of a set (i.e., interrepetition rest) or between a group of repetitions within a larger set. In a series of studies examining the length of intraset rest intervals during the power clean, Hardee and colleagues showed that longer 40-second interrepetition rest compared to 20 or 0 seconds between power clean repetitions allowed for a greater maintenance of peak force, velocity, and power output characteristics (84), while also allowing participants to maintain their technique to a greater extent (83) and lower their perceived exertion (82). In another study, researchers used a 130-second intraset rest interval between either 2 clusters of 3 repetitions or a cluster of 3 repetitions followed by a cluster performed to failure during the bench press, and they compared these structures to traditional sets of 6 repetitions (51). While there were no differences in eccentric, concentric, or total mean force or impulse between the traditional and basic cluster sets (i.e., 2 clusters of 3 repetitions), the cluster set structure that included sets to failure produced significantly greater mean force and impulse characteristics compared to the other 2 structures. In addition to other findings (93), the previously discussed results support the notion that longer intraset rest intervals may allow athletes to maintain force production characteristics while also allowing for an increased capacity to perform work, ultimately benefiting the strength stimulus they may receive (169). Regarding rapid force production, Zaras and colleagues (268) indicated that 20 seconds of interrepetition rest during 4 sets of 6 repetitions performed twice a week for 7 weeks produced greater increases in isometric leg press rate of force development compared to traditional sets, while improvements in peak force were similar. In 2022, researchers suggested using intraset rest intervals ranging from 20 to 40 seconds when training for maximal or rapid force production characteristics (154); however, strength and conditioning professionals should note that the rest interval length may vary based on the loads and cluster set structure used. In addition, longer intraset rest intervals (e.g., 30-40 seconds) are recommended to maintain the quality of repetitions and promote the development of rapid force production characteristics.

Key Point

Intraset rest intervals of 20 to 40 seconds should be used for the development of maximal and rapid force production characteristics and to maintain the quality of repetitions.

Strength and conditioning professionals must be aware of the amount of training time required to implement cluster set structure training. Simply, using a cluster set strategy takes more time during a training session owing to additional intraset rest and the same interset rest periods as shown with both higher-volume (250) and lower-volume (51) sets. Moreover, smaller cluster sets (i.e., reduced number of repetitions) further increase the overall time needed to complete the prescribed sets as the amount of intraset rest increases.

The use of cluster sets becomes especially cumbersome when larger groups of athletes are training; however, strength and conditioning professionals can use several strategies to effectively implement intraset rest intervals in this scenario. First, athletes may be able to perform antagonistic supersets that use opposing muscle groups (e.g., bench press and bench pulls). Using this method, athletes may perform the opposing exercise during the rest period of the first exercise. Although this strategy may appear logical in context, strength and conditioning professionals should note that the order of exercises may have a significant effect on subsequent exercise performance (38). Moreover, limited research has examined the effectiveness of antagonistic supersets in training.

A second method may be through implementing partner cluster sets, in which one athlete rests during the interrepetition or intraset rest interval while another performs their repetitions (75, 154). It is important to note that athletes should be lifting similar training loads to avoid larger amounts of time needed to change loads between athletes. Moreover, if shorter or longer rest intervals are used, fewer or more athletes may use this strategy together, respectively.

Finally, strength and conditioning professionals may implement rest redistribution to maintain the overall training time relative to traditional sets (248). Simply, *rest redistribution* refers to the strategy of redistributing the overall amount of interset rest intervals to the intraset rest intervals. Researchers have shown that rest redistribution may allow for the maintenance of repetition velocity (15, 42, 106, 143), greater mean force production (26), and similar benefits in explosive strength (99) and may produce a lower perceived exertion (42, 142, 251) compared to traditional sets; however, additional research showed no differences between methods (27, 139). Yet it should be noted that a 2020 meta-analysis concluded that cluster sets may serve as a superior alternative to mitigating fatigue during training (105). Thus, strength and conditioning professionals may have to carefully consider when, how, and with whom to implement intraset rest intervals.

Potentiation Complex Rest Intervals

Beyond interset and intraset rest intervals, strength and conditioning professionals must also consider the rest intervals implemented if they elect to prescribe potentiation complexes in training. Before this topic is discussed further, it is important to recognize that several factors may affect whether potentiation is realized (188, 220, 243); however, the design of a potentiation complex, including the rest interval used, plays a crucial role. It is well established that a state of both fatigue and potentiation exists following a conditioning stimulus within a potentiation complex (66, 90, 172, 180). Based on an acute model of the fitness–fatigue paradigm (269), the interplay of fatigue and potentiation ultimately affects the subsequent performance of the athlete. Researchers have concluded that moderate potentiation benefits may be realized when rest intervals range from 3 to 7 minutes and 7 to 10 minutes (262) or from 8 to 12 minutes (70). Despite these results, it may not be practical to prescribe long durations between a conditioning stimulus and a subsequent exercise within a training session. Thus, strength and conditioning professionals should consider both the potentiation complex design and the characteristics of their athletes to maximize performance during an earlier time. For example, researchers have shown that greater relative strength characteristics may contribute to both a greater magnitude (146, 187, 230, 231) and faster realization (103, 187, 231) of potentiation effects. These findings are likely due to a repeated bout effect in which stronger individuals have been exposed to heavier loads and have developed fatigue resistance to such stimuli (206). From a practical standpoint, developing the strength characteristics of an athlete prior to implementing potentiation complexes may allow strength and conditioning professionals to prescribe shorter rest periods within said complexes (i.e., <3 minutes) and still realize the benefits of the training method. Chapter 6 provides a thorough discussion of potentiation complexes and their use as an advanced training strategy.

TRAINING AGE

Strength and conditioning professionals should note that the emphasis of training may differ between athletes with different training ages to allow each type of athlete to maximize training adaptations. In fact, the training age of an athlete (i.e., training experience) will likely influence programming decisions regarding any of the concepts discussed earlier and possibly the method or methods of training. The following paragraphs discuss differences in training emphasis between athletes with lower and higher training ages and how these athletes may benefit their maximal and rapid force production adaptations.

LOWER TRAINING AGE AND WEAKER ATHLETES

Previous literature supports the notion that maximal muscle strength may serve as the foundation upon which other abilities (e.g., rapid force production, power output, sport performance) are built (37, 144, 228, 234, 267). Therefore, it would appear beneficial for novice athletes with lower training ages to focus on developing a greater capacity to produce force. It is important to note that while any of the training methods discussed in chapters 5 or 6 may allow these individuals to improve their force production characteristics via the neuromuscular adaptations highlighted in chapter 2, almost any structured training stimulus given to inexperienced athletes may lead to improvements in performance.

An issue that arises with the wide variety of training methods is that many strength and conditioning professionals may emphasize high-velocity or ballistic training methods too early. In fact, researchers have shown that youth individuals may benefit more from focusing on developing strength via more traditional methods before an emphasis is placed on more ballistic-type training (13). Additional studies further highlighted this by showing distinct changes in force–velocity characteristics between stronger and weaker participants following 10 weeks of training with weightlifting derivatives, plyometric training, and ballistic exercises (95, 96). Within these studies, weaker participants showed greater improvements in maximal force production; however, stronger individuals showed greater improvements in muscle activation of the hip and knee extensors and velocity metrics. These studies support the idea that although more ballistic-type training methods can be prescribed with athletes with lower training ages, these athletes may not be able to maximize the benefits of such training until they adequately develop their force production capacities. From a motor learning standpoint, basic resistance training strategies may assist youth athletes in optimizing motor control and coordination before shifting to neural and morphological adaptations (227). Simply, athletes need to develop a solid technical foundation with basic movements and learn how to use their newly developed strength during a given motor task (e.g., exercise) (226) before adding exercises or training strategies with increased complexity.

While varied practice is beneficial to skill acquisition (157), prescribing advanced training methods such as variable resistance training or accentuated eccentric loading (in which the load changes during the movement) early in an athlete's training program may introduce unnecessary variability before technique is solidified, which may affect the movement coordination of the athlete. Furthermore, if some training methods are introduced too early, they may not serve as a novel training stimulus later in the athlete's development. Simply, it is not that ballistic-type training cannot be programmed with athletes with lower training ages; rather, these training strategies may not be emphasized until later in their development. For example, basic plyometric training exercises (237) and weightlifting derivatives (151) can be introduced to athletes with lower training ages as foundational exercises; however, an emphasis should be placed on technique before progressing to more demanding exercises.

GREATER TRAINING AGE AND STRONGER ATHLETES

Although the emphasis on maximal muscle strength may lead to significant improvements with weaker athletes and those with a lower training age, the influence on performance from improvements in strength begins to diminish when athletes achieve high relative strength levels (117) or, alternatively named, a *strength reserve* (228). Simply, the rate of improvement is larger when a novel training stimulus is introduced, whereas a slower rate is shown by athletes already familiar with a given stimulus (89). Thus, it may be questionable to continually use the same training methods with athletes of different physical abilities and training ages.

To provide an effective training stimulus to stronger athletes and those with a higher training age, strength and conditioning professionals may begin shifting the emphasis of their training to include exercises and training methods that focus on rapid force production while also maintaining their maximal strength (54, 202, 207, 228). Although some debate surrounds achieving specific strength standards, researchers have shown that individuals able to squat at least twice their body mass produced greater vertical jump power outputs (8, 204), faster sprinting times and higher vertical jump heights (263), and greater magnitudes and earlier potentiation (187, 231) compared to their weaker counterparts. James and colleagues (96) showed that greater improvements in muscle activation and velocity were produced by individuals who squatted twice their body mass following a 10-week training program that featured weightlifting derivatives, plyometric training, and ballistic exercises.

In summary, it appears that individuals who have developed a strength reserve may be able to maximize the benefits of specific training methods (e.g., potentiation complexes, plyometric training, ballistic exercises), indicating the potential to include these methods in training. It is important to note that a shift toward including more advanced training methods does not abandon the development and maintenance of peak force characteristics; rather, strength and conditioning professionals should use a model of emphasis or de-emphasis, depending on the goals of specific training phases (227).

TRAINING YEAR

Training prescription should vary throughout the training year because the goals of each training phase will likely differ. As noted in chapters 2 and 4, an off-season or general preparation phase of training may not focus on the development of peak or rapid force production characteristics. In contrast, a preseason or in-season phase of training may place a greater emphasis on maximizing both of these characteristics in preparation for various competitions (227). Thus, each of the previously discussed training variables may be modified. For example, strength and conditioning professionals will likely select specific exercises to maximize rapid force production while also attempting to maintain the peak force production capabilities of the athlete. In addition, volume and intensity (i.e., loading) will likely decrease and increase, respectively, as athletes move from the off-season to the preseason and, eventually, to the competitive season. Finally, set structure and rest intervals may be modified best on the phases of training to allow the athlete to maximize the training stimulus while promoting recovery and managing fatigue.

OVERALL TRAINING APPROACH

Beyond those who compete in strength sports such as weightlifting, powerlifting, strongman, or highland games, many athletes require various resistance training, sprinting, change of direction, and conditioning stimuli for their overall development. Although this chapter primarily focuses on resistance training considerations, it is important that strength and conditioning professionals do not neglect the development of these other performance capacities. For example, while resistance training practices may aid in the improvement of sprinting (52) and change-of-direction ability (25), each skill has its own constraints and requires a progressive stimulus for the athlete to show demonstrable improvements. However, it is important to note that strength and conditioning professionals can pair exercises that are biomechanically similar and match the physiological demands of other forms of training. DeWeese and colleagues (55, 56) provided a thorough overview and description of the long-term development of both the strength and speed characteristics of bobsled athletes in preparation for the Sochi Olympics. Specifically, the authors highlighted that specific resistance training exercises were selected for their kinetic and kinematic qualities to match the speed development phase that the athlete was in. Additional examples can be found in previous literature that discussed pairing weightlifting movements and their derivatives with speed development tactics (52, 215).

Because each training stimulus places a stress on the athlete to produce specific adaptations, it is important to consider the collective fatigue that the athlete may accumulate, because each individual stress cannot be siloed. This may require strength and conditioning professionals to dedicate various percentages of different stimuli to address the needs of each athlete or plan training sessions in a manner that allows for proper recovery to maximize the desired adaptations. It is also important to consider the added sport practice, academic, and personal stressors to effectively design individualized training programs (124), which highlights the importance of establishing a strong relationship with the sport coach and athlete that includes consistent communication as well as the use of athlete monitoring data.

Take-Home Points

- An effective needs analysis should include the sport or event demands and the assessment of an athlete's force production characteristics during specific tests that relate to the skills that an athlete performs in their sport or event.
- After determining the needs of an athlete, strength and conditioning professionals should use an evidence-based approach when deciding the exercises, frequency of training sessions, order of exercises, number of training sets, set structure, loading, repetition scheme, and rest intervals that may provide the most effective stimuli for individual athletes.
- Strength and conditioning professionals should understand the biomechanical and physiological differences between exercises to provide an effective training stimulus. Exercises may emphasize either maximal or rapid force production; however, the concepts of dynamic correspondence and specificity should lead the decision-making process for exercise selection.
- The frequency of training sessions will be based on the goals of the training program, training status of the athlete, and the training season. Frequency may be modified during planned, functional overreaching and tapering protocols to promote specific adaptations and manage fatigue. Microdosing may provide athletes with busy competition schedules with an effective method of organizing training sessions.
- Exercises that have the highest neuromuscular demand should be prioritized within training sessions before assistance exercises that have less of an impact on the force production characteristics of an athlete. However, combining multiple exercises within a potentiation complex may provide an effective training stimulus, provided that the athlete and protocol characteristics are appropriately accounted for.
- Cluster sets may provide athletes with an effective tool to improve the overall quality of work and increase the potential to use heavier loads during a training session. Although cluster sets are potentially beneficial for maximal and rapid force production adaptations, professionals must consider the amount of time allotted for training because their use may result in a longer training session.
- The use of heavier loads will likely lead to greater adaptations in maximal force production; however, training to failure does not appear necessary to do so. Although a combination of heavy and light loads may allow athletes to maximize rapid force production characteristics, strength and conditioning professionals should consider the needs of their athletes when prescribing different training loads.
- Longer rest intervals promote greater strength adaptations compared to shorter rest intervals. Intraset rest intervals, like those used during cluster sets, may improve the overall quality of training, thus enhancing the potential of the training adaptation.
- The past, current, and future training phases should be considered to determine how best to prescribe training to maximize their performance adaptations at specific time periods. Strength and conditioning professionals must account for the collective stress of different stimuli during training prescription to promote adaptation and manage fatigue.

CHAPTER 8

Concurrent Training

Athlete performance is multifaceted and often requires the development of a variety of technical skills, sport or event tactics, and fitness characteristics. Some sports or events are almost exclusively strength–power oriented (e.g., weightlifting, short sprinting), whereas others may require significant cardiorespiratory endurance (e.g., marathon running, road cycling) and fall on the opposite end of the strength and endurance spectrum (figure 8.1). However, it should be noted that other sports or events may require a combination of strength–power and endurance characteristics (e.g., soccer, lacrosse) to be successful (21). Although some of the previously outlined sports have a singular focus that is underpinned by specific force production characteristics, sports that require a combination of strength–power and endurance characteristics may challenge the strength and conditioning professional when developing both fitness capacities simultaneously, a process called *concurrent training*. The aim of this chapter is to provide a general overview of concurrent training and discuss the different training strategies that can be used within concurrent training and their impact on the development of muscular strength characteristics.

Concurrent training may be defined as a combination of resistance training and training that may benefit cardiorespiratory endurance adaptations; this form of training is typically characterized by heavy resistance training (i.e., >80% of the 1RM) and long-duration, low- to moderate-intensity (approximately 70% maximal oxygen consumption [$\dot{V}O_2max$]) continuous activity (e.g., running, cycling), respectively. Multiple methodologies of implementing concurrent training have been used within the literature, including same-session, same-day, and alternate-day formats. Because each stimulus promotes unique physiological adaptations that may conflict with each other, often termed the *interference effect* (76), it is important that strength and conditioning professionals understand how the two forms of training may be programmed simultaneously to maximize both strength and endurance adaptations as they relate to the needs of an athlete.

THE INTERFERENCE EFFECT

The prospect of an *interference effect* during concurrent training was first proposed in an article published in 1980 by Hickson (52). Since its introduction, a more thorough description of the interference effect has included both acute and longitudinal interference effects. Researchers have shown that performing both strength training and endurance training in proximity to each other may acutely affect maximal strength (24, 52, 57, 84, 107) and endurance performance (19, 22, 30, 32, 33). However, the notion that all concurrent training programming may lead to interference effects may be considered an assumption (4), especially because different programming variables and individual characteristics may impact the effect of a given training stimulus (4, 42). Regarding the training status of an athlete, those with more training experience may be more likely to experience an interference effect due to the need for larger training volume–loads to promote adaptations (110). In contrast, researchers have shown that maximal strength adaptations may not be affected in individuals with less training experience who participate in fewer overall training sessions (23). Although decrements in maximal strength adaptations may be mitigated using a variety of programming tactics, greater interference effects may be shown with rapid force production characteristics. For example, researchers have shown impaired voluntary muscle activation and rate of force development following chronic concurrent training (37, 49, 110).

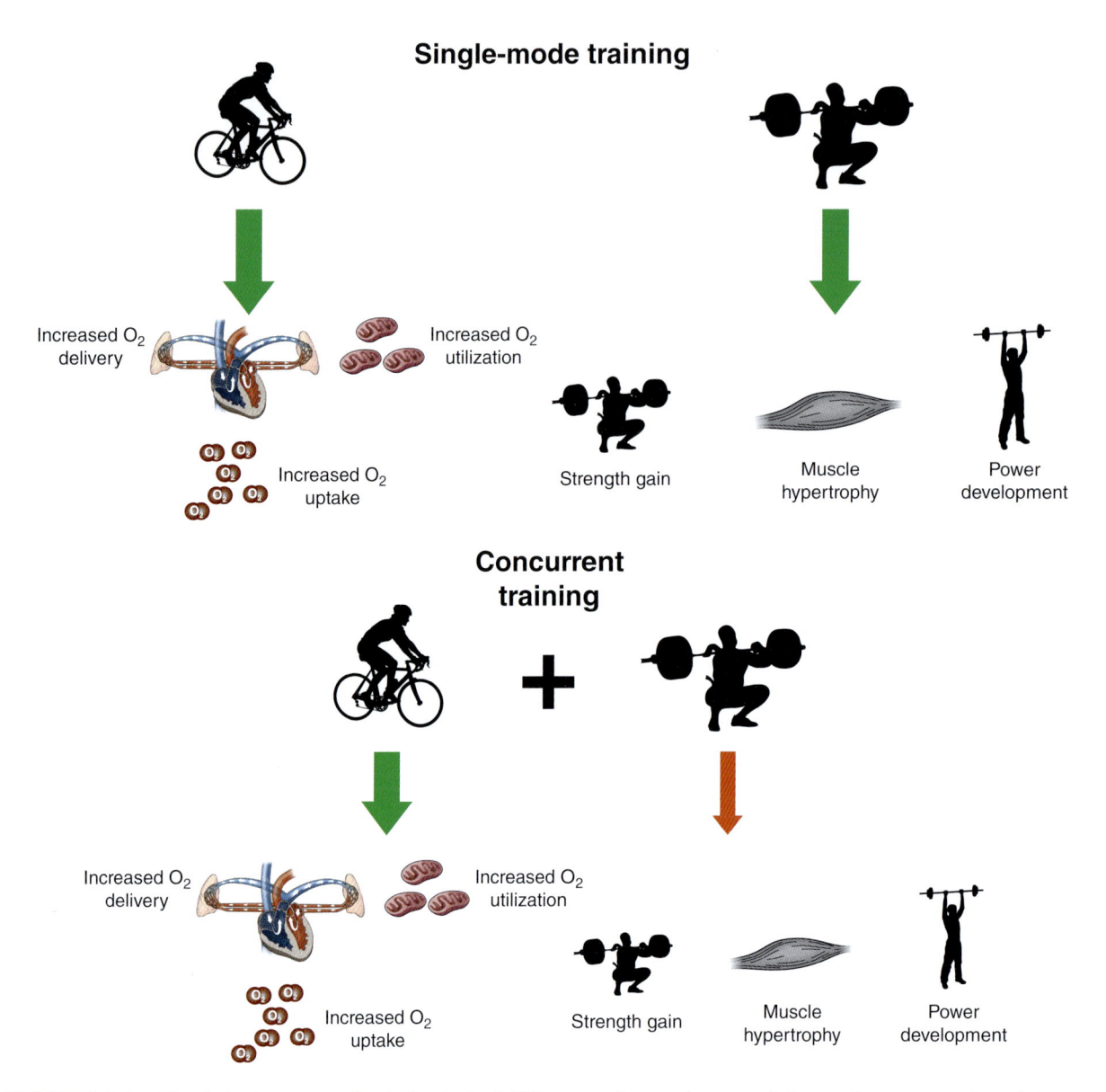

FIGURE 8.1 The interference effect illustrated. When performed separately, endurance and resistance training lead to specific long-term adaptations. When performed concurrently, metabolic conditioning may limit strength and power development; however, strength training is unlikely to reduce (and instead will likely enhance) endurance performance (i.e., average power or speed maintained over prolonged durations or the ability to repeat high-intensity actions over time).

Reprinted by permission from P. Laursen and M. Buchheit, *Science and Application of High-Intensity Interval Training* (Human Kinetics, 2019), 120.

Key Point

The notion that all concurrent training programming may lead to interference effects may be considered an assumption, especially because different programming variables and individual characteristics may impact the effect of a given training stimulus.

Strength and conditioning professionals should note that the acute interference effect of endurance exercise on strength characteristics may be due to the organization and scheduling of concurrent training sessions. For example, many concurrent training programs within the literature have included both endurance and strength training stimuli within the same training session, often with the former preceding the latter. Researchers have shown that strength characteristics may be negatively affected when endurance exercise (e.g., running) is performed immediately prior to strength training due to the residual fatigue from the endurance training (24). Although this may serve as an example of acute fatigue, the chronic effects of such training may lead to a suboptimal stimulus for maximal strength development (46, 111). To compensate for residual fatigue, strength and conditioning professionals may consider implementing a *relief period* between the different training methods because research has shown reduced strength performances for up to 8 hours following endurance exercise (3, 10, 56, 95), whereas periods greater than 8 hours did not show decrements in performance (11, 79, 93, 95). It has been proposed that the full restoration of muscle glycogen that occurs during longer relief periods may be responsible for these findings (46).

Key Point

To compensate for residual fatigue, strength and conditioning professionals may consider implementing a relief period between the different training methods because research has shown reduced strength performances for up to 8 hours following endurance exercise.

As mentioned previously, programming variables may affect the magnitude of the interference effect. For example, it is possible that the mode, duration, intensity, and type of endurance exercise may result in differing degrees of fatigue. Within a meta-analysis, researchers concluded that the mode of exercise (running vs. cycling) within a concurrent training program significantly affected lower-body strength development (110); however, this may not necessarily be the case at the acute level because numerous studies have shown decrements following both running and cycling protocols (57). Interestingly, researchers have shown that running (27), cycling (84, 95), or elliptical (105) endurance stimuli negatively affected lower-body but not upper-body strength, indicating that decrements in performance may be specific to regions of the body. While more research in this area is needed, strength and conditioning professionals may use these results to inform their programming by choosing endurance methods that are upper-body dominant (e.g., upper-body ergometer) when lower-body strength is emphasized during a resistance training session and vice versa.

Regarding the duration of endurance exercise, decrements in acute strength have been shown following activity ranging from 15 minutes (80) up to 150 minutes (3), potentially indicating that the mere act of performing endurance exercise prior to resistance training may lead to negative strength effects. These findings again support the notion that longer relief periods may be needed between the different methods of training if both are completed within a single day.

From an intensity standpoint, very few studies have made direct comparisons between higher- and lower-intensity protocols and their impact on strength characteristics. However, Thomas and colleagues (107) indicated that strength was impaired in male cyclists following both a high-intensity, short-duration time trial (4 kilometers) as well as lower-intensity, longer-duration time trials (20 and 40 kilometers). Interestingly, the authors noted that peripheral and central fatigue mechanisms were dominant following the higher- and lower-intensity protocols, respectively.

Finally, research comparing the type of endurance exercise (i.e., continuous vs. intermittent) and its influence on subsequent strength performance has shown mixed results. Some authors indicated that both types cause decrements in strength (3, 95), whereas others indicated that one or the other may not cause a decrement (26, 27). Like the duration of the endurance stimulus, it appears that the simple inclusion of such an activity may be enough to acutely impair muscular strength.

As noted in chapter 1, an athlete's strength characteristics can be expressed in several ways, and the same can be said for the measurements of strength used within the concurrent training literature. For example, researchers have characterized changes in strength during concurrent training using repetitions until failure (27, 80, 84, 105), 1RM to 5RM or isometric strength assessments (11, 27, 37, 38, 93, 104), and isokinetic testing (3) while also examining lower- and upper-body as well as single- and multi-joint exercises. Despite the amount of literature focused on examining the interference effect that endurance exercise may have on various maximal strength performances, relatively few studies have examined the potential impact on rapid force production characteristics. However, in a pair of studies, Taipale and colleagues (103, 104) indicated that concurrent training had a negative impact on countermovement jump performance and rate of force development during an isometric leg press, which may indicate that rapid force production or ballistic activities may be more susceptible to decrements in performance following endurance activities (24, 49, 90).

Despite the evidence that supports acute interference effects, an abundance of evidence supports the notion that concurrent training may lead to improvements in endurance performance and maximal aerobic capacity in top-level endurance athletes (1, 2, 12, 36), with some evidence showing improved rapid force production characteristics as well (1, 2). These benefits may lead to shorter ground contact times during running (53, 78) and the utilization of elastic energy within the muscle-tendon system (109). However, regarding the chronic effects of concurrent training on an athlete's strength characteristics, the evidence is less clear. Several meta-analyses have indicated that concurrent training may have a negative impact on various strength characteristics, including maximal strength (35, 44, 75, 82, 88) and rapid force production characteristics (90, 110); however, it should be noted that the observed effects may have been based on training status (82), lower- or upper-limb testing (88), or the order of testing (35, 44, 75). Collectively, while the available evidence suggests that concurrent training may not affect an athlete's maximal strength, it may impact their rapid force production characteristics. Thus, strength and conditioning professionals must choose conditioning methods and programming tactics throughout the training year to optimally improve the necessary performance characteristics based on sport specificity.

Key Point

An abundance of evidence supports the notion that concurrent training may lead to improvements in endurance performance and maximal aerobic capacity in top-level endurance athletes, with some evidence showing improved rapid force production characteristics as well.

CONDITIONING METHODS

While there are many different methods of resistance training (see chapter 5), conditioning methods that benefit aerobic adaptations also exist on a spectrum. Strength and conditioning professionals may use long, slow distance, maximal aerobic speed (MAS), and high-intensity interval training (HIIT) to develop the relevant energy systems of their athletes to be successful in their respective sports. Because the method of conditioning combined with resistance training stimuli may have a significant impact on the force production characteristics of an athlete, an understanding of the benefits and limitations of each method is needed. The following paragraphs discuss the various methods used for conditioning and how each stimulus may affect maximal and rapid force production adaptations.

LONG, SLOW DISTANCE

Historically, strength and conditioning professionals and sport coaches have taken the approach of implementing long-duration, steady-state exercise lasting at least 20 to 40 minutes to benefit the aerobic conditioning characteristics of field sport athletes (5). Although this method of training may benefit the cardiorespiratory endurance characteristics of athletes to an extent, researchers have suggested that it may not be as specific or beneficial to their needs as originally thought (20, 102, 111). For example, researchers have indicated that gameplay in soccer has continued to evolve to include more high-intensity efforts and greater physical demands (51, 77). Thus, it would appear beneficial in soccer and similar field-based sports to develop greater aerobic power capabilities. From a practical standpoint, strength and conditioning professionals should consider implementing intensities at or above 100% MAS (discussed next) rather than lower intensities (7, 102). While the development of an "aerobic base" is often discussed among sport coaches, it is important that strength and conditioning professionals mention that higher-intensity conditioning can also develop the desired characteristics and may also be more sport specific. Furthermore, training at higher aerobic intensities may also require athletes to generate greater magnitudes and rates of force development, thus benefiting their strength characteristics within a task-specific manner.

> **Key Point**
>
> While long, slow distance may benefit endurance characteristics, it may not be as specific or beneficial to field sport performance as originally thought.

MAXIMAL AEROBIC SPEED

Maximal aerobic speed (MAS) can be defined as the minimum velocity at which the rate of $\dot{V}O_2$max occurs or, more simply, the velocity at $\dot{V}O_2$max (14). Aerobic training can be applied using a variety of training zones (figure 8.2) based on percentages of MAS (5), and training at or above 100% MAS may provide greater aerobic power benefits than long, slow distance training (102) but also continuous runs at 100% MAS (34). Baker (5, 6) discussed a variety of MAS programming strategies, including long intervals, maximal aerobic grids (figure 8.3), 15/15 Eurofit, Tabata, and unpredictable Tabata, and stressed the importance of implementing these methods in a progressive manner based on the athlete's abilities. However, strength and conditioning professionals should note that a greater emphasis of energy system development via MAS may be included during the general preparation phase of training rather than the specific preparation phase. When combined with resistance training, this sequence may allow for greater strength adaptations to occur as the overall volume of aerobic training decreases and the high-intensity efforts—and need for greater magnitudes and rates of force production—increases.

HIGH-INTENSITY INTERVAL TRAINING

High-intensity interval training (HIIT) is characterized by repeated exercise bouts that are performed at intensities above an athlete's maximal lactate steady-state, anaerobic threshold, or critical power–speed and are separated by rest intervals or low-intensity exercise (17, 18). HIIT consists of a variety of forms including long intervals, short intervals, repeat sprint training, sprint interval training, and small-sided games (table 8.1) (16). Because each HIIT method provides varying degrees of neuromuscular strain, strength and conditioning professionals must be cautious when implementing specific methods so as not to negatively affect the subsequent resistance

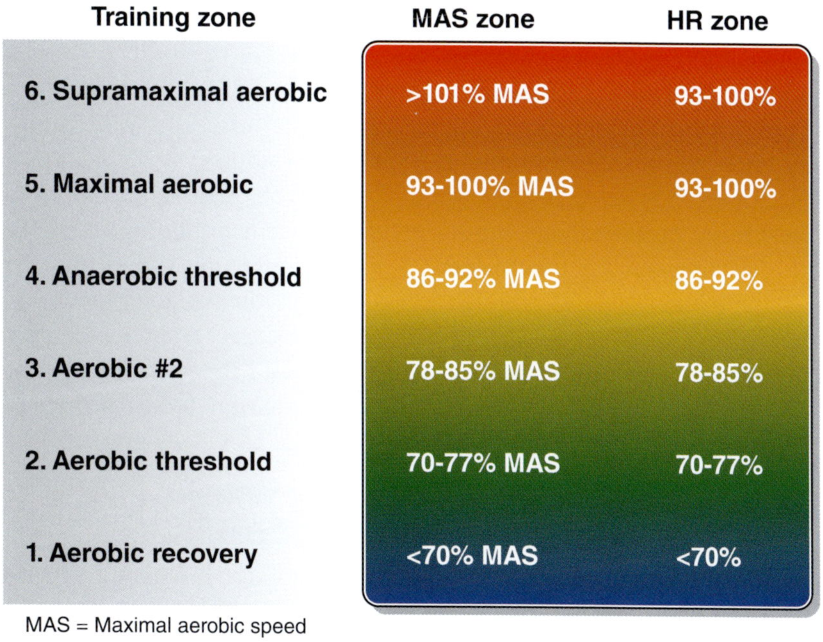

Training zone	MAS zone	HR zone
6. Supramaximal aerobic	>101% MAS	93-100%
5. Maximal aerobic	93-100% MAS	93-100%
4. Anaerobic threshold	86-92% MAS	86-92%
3. Aerobic #2	78-85% MAS	78-85%
2. Aerobic threshold	70-77% MAS	70-77%
1. Aerobic recovery	<70% MAS	<70%

MAS = Maximal aerobic speed

FIGURE 8.2 Aerobic training zones based on maximal aerobic speed (MAS) and heart rate (HR) percentages.

Adapted by permission from D. Baker, "Recent Trends in High-Intensity Aerobic Training for Field Sports," *Professional Strength and Conditioning*, 22 (2011): 4.

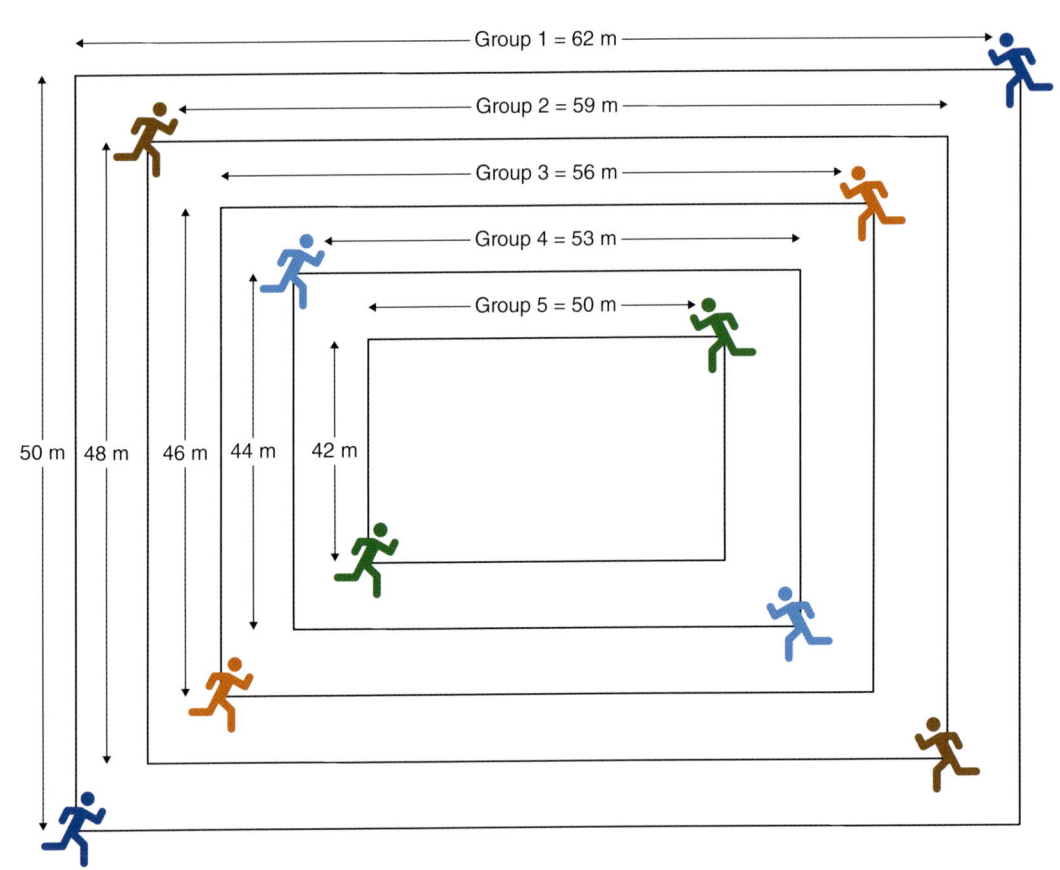

FIGURE 8.3 Maximal aerobic speed (MAS) rectangular grids. The long side is performed at 100% MAS and the short side with 70% MAS for each group. Runners in each group can start in different corners to avoid congestion.

training stimulus (40). Similarly, it is important to be cognizant of the neuromuscular strain placed on an athlete during resistance training sessions because they may affect subsequent HIIT sessions. Researchers indicated that concurrent training involving running and cycling HIIT and resistance training may negatively impact lower-body strength development compared to resistance training alone, while upper-body strength development was unaffected (88); however, the manipulation of HIIT training variables may modify this effect (41).

Strength and conditioning professionals must also consider the mode of HIIT used, because the neuromuscular demand varies based on the involved musculature (74). For example, running and cycling may affect the force production characteristics of an athlete to a greater extent than swimming. Using this knowledge, HIIT-inclusive concurrent training programs may be designed to limit the interference effect and allow for strength improvements in the musculature that is primarily used in sport performance tasks while still providing an effective conditioning stimulus. Researchers have shown that HIIT cycling may have a greater impact on the number of half squat repetitions and volume–load performed compared to treadmill running (80); however, the authors speculated that cycling required a greater anaerobic influence, which may have contributed to their findings.

From an order standpoint, it has been suggested that resistance training should be performed prior to HIIT protocols if they are performed within the same session, while considering the risk for injury based on exercise and loading protocols (40). However, if HIIT is performed before resistance training on the same day, recovery periods of at least 8 hours should be provided between sessions.

Key Point

Because each HIIT method provides varying degrees of neuromuscular strain, strength and conditioning professionals must be cautious when implementing specific methods so as not to negatively affect the subsequent resistance training stimulus.

TABLE 8.1 Overview of High-Intensity Interval Training (HIIT) Methods

Type	Description	Recovery	Training emphases
Long intervals	Intervals lasting >1 minute performed at 95%-105% V-$\dot{V}O_2$max or 80%-90% V-IFT	1-3 minutes of passive recovery or active recovery performed at 60% V-$\dot{V}O_2$max or ≤45% V-IFT	• Aerobic and anaerobic system development • Aerobic and anaerobic system development and neuromuscular stimulus
Short intervals	Intervals lasting <1 minute performed at 100%-120% V-$\dot{V}O_2$max or 90%-105% V-IFT	<1 minute of passive recovery or active recovery at 60% V-$\dot{V}O_2$max or ≤45% V-IFT	• Aerobic system development • Aerobic system development and neuromuscular stimulus • Aerobic and anaerobic system development • Aerobic and anaerobic system development and neuromuscular stimulus
Repeat sprint training	Repeat efforts lasting 3-10 seconds performed with maximum effort	Passive recovery or active recovery performed at 60% V-$\dot{V}O_2$max or ≤45% V-IFT that is variable in duration	• Aerobic and anaerobic system development and neuromuscular stimulus • Peripheral aerobic and anaerobic system development and neuromuscular stimulus
Sprint interval training	Repeat efforts lasting 20-45 seconds performed with maximum effort	Passive recovery lasting 1-4 minutes	• Peripheral aerobic and anaerobic system development and neuromuscular stimulus
Game-based HIIT	Sport-specific intervals lasting 2-4 minutes where an athlete interacts with teammates or opponents during simulated play	Typically passive recovery lasting 90 seconds to 4 minutes	• Aerobic system development and neuromuscular stimulus • Aerobic and anaerobic system development • Aerobic and anaerobic system development and neuromuscular stimulus

Based on Laursen and Buchheit (61).

Note: V-IFT = velocity achieved during the 30-15 intermittent fitness test; V-$\dot{V}O_2$max = velocity achieved during the maximal uptake of oxygen test.

SPRINT TRAINING

Beyond long, slow distance, MAS, and HIIT methods, strength and conditioning professionals may also consider sprint training as a method of conditioning. For example, sprinting may serve as a metabolic stimulus based on the stress placed on both the phosphagen and glycolytic energy systems, which may stimulate an increase in enzyme content and kinetics. Researchers have shown that sprint training may increase myokinase and creatine phosphokinase (55, 81), while both short and long sprint efforts may also increase other glycolytic enzymes (87). Given the number of acceleration efforts that occur within team sport (50) and the high-speed running and sprinting distances covered in field-based sports (85), it would seem logical to include sprint training within a concurrent training program. By doing so, strength and conditioning professionals may improve the force production characteristics of an athlete while also preparing them to accelerate and reach maximal velocities during competition. Previous literature has discussed various speed development progressions when combined with resistance training (29, 99). Like the other methods of conditioning discussed in this chapter, it is important to consider the goals of the training phase to emphasize the development of specific fitness characteristics while temporarily de-emphasizing others. The use of sprinting as a stimulus for improving strength characteristics is discussed in greater detail in chapter 5.

CONCURRENT TRAINING PROGRAMMING CONSIDERATIONS

Beyond the resistance training programming variables discussed in previous chapters, strength and conditioning professionals must also consider endurance training variables when prescribing concurrent training regimens for their athletes. The following paragraphs will provide insight on how the training status of athletes, frequency of implementation, intensity of training, duration and volume of training, session order, and recovery time between sessions affect strength and endurance adaptations during concurrent training.

TRAINING STATUS

As discussed throughout this book, the training age of an athlete can affect not only the programming decisions of a strength and conditioning professional but also the adaptations that may be realized. Concurrent training is no exception; it has been suggested that the training status of an individual may serve as an important factor related to the degree of strength adaptations or interference (43). Researchers have shown that untrained participants may benefit from both endurance and strength adaptations with various concurrent training stimuli (24, 47, 69, 73, 108). While these findings are important for strength and conditioning professionals working with youth and novice athletes who may not have been exposed to structured training stimuli, an understanding of how these adaptations may change as an athlete's training age increases is crucial. For example, additional research has shown that the interference effect may become more apparent as training age increases (23); the authors of a systematic review indicated that concurrent training may negatively affect lower-body maximal strength adaptations in trained individuals but not in those who were moderately trained or untrained (82). It should be noted that the negative effects shown in the latter study were only present when both resistance training and endurance training were included in the same training session, which stresses the importance of separating the stimuli.

FREQUENCY

The frequency of training sessions may be related to the training age of athletes. While more frequent training sessions may provide a larger overall training stimulus, this may present conflicting stimuli in the case of concurrent training, potentially leading to an interference effect. This notion is supported by researchers who showed that 3 or more endurance training sessions per week may negatively affect resistance training adaptations (9, 58, 60). In contrast, additional research has shown that performing up to 2 endurance sessions per week may have less of an effect on strength adaptations (47, 49, 70); however, as previously mentioned, this may be related to the training status of the individuals within the studies. The frequency of training sessions appears to be an important factor in providing a training stimulus to the athlete; however, the intensity of the stimulus may also play a significant role.

> **Key Point**
>
> While more frequent training sessions may provide a larger overall training stimulus, this may present conflicting stimuli in the case of concurrent training, potentially leading to an interference effect.

INTENSITY

Although the endurance stimulus within concurrent training is typically thought of as long, slow distance training, strength and conditioning professionals must also consider the applicability of such a stimulus in sports or events that require frequent high-intensity efforts. While top-flight soccer players may cover over 10 kilometers during a match, approximately 800 meters may involve high-speed running and sprinting (85); however, despite the cumulative distance covered, these high-intensity efforts often provide a tactical advantage that may alter the match outcome (48, 96). Moreover, there is evidence to suggest that soccer is evolving to include more high-intensity efforts and greater physical demands (51, 77). This notion is also supported by data that characterize the number of acceleration and deceleration efforts within elite team sport (50). Thus, while athletes must have the work capacity to tolerate the overall distances that will be covered during athletic events, it should be noted that these capacities may be developed using higher-intensity training methods. For example, concurrent training methods may include endurance stimuli that are provided using MAS (5) or HIIT (40) protocols.

Researchers have indicated that higher-intensity activities may help reduce the interference effect (13, 90), whereas continuous long, slow distance training may be viewed as less specific to the demands of field sports (17, 18). It is, however, important to note that higher endurance training intensities may produce greater residual fatigue compared to lower training intensities (27, 65), which may negatively affect force production characteristics (10, 11) and performance during a subsequent resistance training session (27, 83). Further research showed that HIIT cycling training produced similar detriments in maximal strength but greater detriments to rapid force production characteristics during the countermovement jump compared to work-matched moderate-intensity continuous training (41).

Key Point

Higher-intensity activities may help reduce the strength interference effect, whereas continuous long, slow distance training may be viewed as less specific to the demands of field sports.

DURATION AND VOLUME

The duration or volume of either endurance or resistance training stimuli can be classified as the length of time of the training session or as the overall number of training sets completed, respectively. Like training frequency, greater endurance training durations may provide athletes with a greater overall training stimulus as well as greater residual fatigue (64). Findings from a previous meta-analysis illustrate that there may be a relationship between increases in endurance training duration and the impaired strength adaptations that may follow (110). These findings may be explained by the accumulated fatigue that results over longer training periods and how it may affect other adaptations. For example, additional research has shown that strength adaptations were only impaired during weeks 8 to 10 of a 10-week concurrent training program that included 30 to 40 minutes of running and cycling 6 days per week (52).

Greater volumes of resistance training have been associated with greater muscle damage (8) and inhibited force production characteristics (68, 98, 106). However, progressive increases in resistance training volume have been shown to improve running economy in individuals with various training backgrounds (e.g., recreational, well trained, and highly trained) (28). Given the accumulated fatigue from higher training volumes of current training performed over longer programs, strength and conditioning professionals should interpret these findings with caution because continued increases in volume may eventually lead to nonfunctional overreaching and potentially overtraining (25, 100).

In summary, the existing literature supports the notion that a concurrent training duration can have a significant effect on the subsequent endurance or strength outcomes and should thus be considered when designing training programs. However, it is important that strength and conditioning professionals focus on developing the adaptations that contribute to a positive performance within the athlete's sport or event.

SESSION ORDER

Many concurrent training programs within the literature have included both endurance and resistance training stimuli within the same session. While this may be viewed as an efficient method of programming because it provides multiple training stimuli, the potential for the interference effect on either adaptation must be considered (39, 42, 64). As a result, a large body of research has investigated the order of performing endurance or resistance training first or second within a training session. Although acute studies have shown significant order effects on strength (54, 59) and endurance (31) performance, many of the training studies have shown no differences in strength characteristics (37, 62, 71, 93), endurance (62, 67, 71, 91), or speed and agility (67, 71) between session orders. However, the findings of meta-analyses indicate that the training sequence does not appear to affect maximal aerobic capacity (i.e., $\dot{V}O_2max$), whereas a resistance training session performed before endurance training may negatively impact knee flexion or extension strength (35, 44, 75). Within a subgroup analysis, researchers concluded that female participants favored the resistance training before endurance training sequence for greater training adaptations, with training periods longer than 8 weeks, or when training frequencies were twice per week (44). These results should be interpreted with caution, because the majority of the literature has displayed mixed findings; however, it is recommended that strength and conditioning professionals order the training stimuli in a way that emphasizes the desired fitness characteristic that takes priority during a specific time of the training year (72, 90, 110). It is therefore recommended that the desired fitness characteristic should be programmed first during concurrent training sessions.

RECOVERY TIME BETWEEN SESSIONS

Because strength characteristics may be negatively affected by endurance exercise performed earlier in a concurrent training session and may last up to 8 hours (3, 10, 11, 63, 95), strength and conditioning professionals may need to seek alternative strategies to maximize the force production characteristics of their athletes. Although certainly less common within the literature, researchers have examined the effect of concurrent training sessions performed on alternate days compared to the same session (86, 89) or training twice within the same day (86). The findings of both studies indicated that greater strength adaptations may be achieved by completing concurrent training sessions on alternate days. These findings may be explained by the ability to replenish the necessary energy stores for multiple training efforts, thus promoting higher-quality training sessions that may enhance the force production characteristics of an athlete (15). In an ideal scenario, at least 8 hours are provided between training sessions to maximize strength development (57); however, it is important to note that the mode, duration, and intensity of an endurance stimulus may modify the required recovery time between sessions for optimal adaptation.

PRACTICAL APPLICATION

There exists a wide variety of concurrent training protocols that can be used, which highlights the importance of understanding the available literature when aiming to develop the force production characteristics of an athlete. From a practical standpoint, one of the primary goals while implementing concurrent training should be to mitigate the impact of the interference effect. In this light, it has been recommended to reduce the frequency of aerobic training to less than three times per week and less than 30 minutes per session (110). Further recommendations suggest separating resistance training and endurance stimuli by at least 8 hours; however, the ecological validity of programming in this way is, at best, challenging (57). This is especially true for high school and collegiate athletes who must also adhere to a class schedule throughout the academic year, during which their respective competitive seasons will also occur.

Thus, strength and conditioning professionals should use a periodized approach to determine when larger volumes of endurance exercise can be used compared to when an athlete's strength characteristics need to be maximized. For example, endurance stimuli may be de-emphasized during training periods in which the development of rapid force production is the primary training goal. Researchers who investigated different periodization models in elite kayakers using concurrent training support this notion (45, 46). The authors concluded that a block periodization approach may provide an opportunity to emphasize a specific characteristic while modifying the training intensity in each block and minimizing the interference effect.

Key Point

Strength and conditioning professionals should use a periodized approach to determine when larger volumes of endurance exercise can be used compared to when an athlete's strength characteristics need to be maximized.

Although endurance athletes may require periods of long, slow distance training, these athletes may also benefit from higher-intensity training because it may lead to improvements in both maximal strength (1, 2, 12) and rapid force production (1, 2) as well as running economy (66). Beyond individual sports, concurrent training may also serve as an effective approach to developing the overall performance of team sport athletes (94). While organizing the training for these athletes, it may be possible to use a multitargeted block periodization approach (97,

101); however, strength and conditioning professionals must be cognizant of the number of competitions within individual weeks of the season because they may vary throughout the year. When programming for these athletes, it is important that strength and conditioning professionals consider the specificity of the training stimuli, especially within the endurance training sessions. For example, athletes competing in field- and court-based sports may benefit from higher-intensity training stimuli based on MAS or HIIT and sprint training.

Take-Home Points

- Concurrent training as an overall training stimulus is complex due to the mixture of both resistance and endurance training stimuli. The existing literature suggests that concurrent training has the potential to benefit both maximal strength and endurance adaptations; however, rapid force production characteristics appear to be negatively affected.
- The impact of the interference effect cannot be assumed, because different programming variables and individual characteristics can impact the effect of the training stimulus. Athletes with more training experience may be more likely to experience an interference effect; however, maximal strength adaptations may not be affected in those with less training experience.
- Strength and conditioning professionals must consider the mode, sequence, frequency, volume, intensity, and recovery durations of each stimulus to promote the desired adaptation and minimize the interference effect. If an appropriate relief period is used between resistance training and endurance stimuli, neither appears to be negatively affected to much of an extent.
- The conditioning methods used within concurrent training may have a significant impact on the overall training stimulus given to the athlete. Therefore, strength and conditioning professionals must consider the specificity of using long, slow distance, MAS, HIIT, and sprint training with their athletes and how well it may transfer to their sport.
- It is suggested that a periodized approach is taken with concurrent training to determine when larger volumes of endurance exercise can be used compared to when an athlete's strength characteristics need to be maximized. A block periodization approach may serve as an effective strategy to prioritize a specific characteristic while modifying the training intensity in each block and minimizing the interference effect.
- While this chapter provided a general overview of the different conditioning methods that can be used in concurrent training programs, readers seeking greater detail are referred to Baker (5) and textbooks edited by Laursen and Buchheit (61) and Schumann and Rønnestad (92).

CHAPTER 9

Measuring Strength

Measuring an athlete's maximal and rapid force production characteristics (i.e., strength) can be accomplished using a variety of tests. Because force is expressed in a task-specific context, it is important that strength and conditioning professionals choose tests that relate to the motor tasks performed in an athlete's sport or event. There are benefits and limitations to different types of strength assessment (e.g., isometric, dynamic, reactive); however, each may provide valuable information regarding the force production capabilities of an athlete. The purpose of this chapter is to provide strength and conditioning professionals with an overview of different strength assessments and how to use the information from testing to develop individualized training programs to improve the force production characteristics that affect the tasks performed within an athlete's sport or event.

RATIONALE FOR MEASURING STRENGTH

"If you're not assessing, you're guessing" is a common phrase within the strength and conditioning and sport science community, indicating the importance of athlete testing and monitoring to supplement visual observations of athlete performance. The rationale behind athlete testing is 4-fold. First, regular athlete monitoring may provide strength and conditioning professionals with information regarding the overall training process. Specifically, athlete testing may provide insight into how individual athletes are responding to specific training stimuli (99). Researchers have shown that both maximal and rapid force production characteristics may be monitored effectively during longitudinal training programs (110, 251); however, it should be noted that the latter may be more sensitive to change throughout the program. Moreover, additional research has indicated that it is not appropriate to assume that improvement in one aspect of strength produces improvements in all other aspects as well (116).

Second, coaches are provided with direct feedback regarding the short- and long-term progression or regression of their athletes, while each athlete receives information about their performances. Providing this information in both situations may lead to greater buy-in from both the coaching staff and athletes, provided they see value in the data.

Third, athlete testing and monitoring may help identify the variables that contribute to optimal performances, which in turn may help with talent identification. For example, researchers have shown differences in force production and performance characteristics between athletes at different levels of competition (10, 75, 81, 172). Therefore, if an athlete tests well but does not receive much playing time, this information may be used to highlight the potential of certain athletes; however, it is important to note that just because an athlete tests well does not mean that they will automatically perform well in their respective sport or event.

Finally, regular athlete assessment may help strength and conditioning professionals and sport scientists identify different factors that may be associated with injury, overtraining, and athlete burnout. For example, Sands and colleagues (227, 228) discussed various single-subject design procedures that may be used to monitor the force production characteristics of an athlete. Due to the highlighted importance of muscular strength (268) and the reasons presented earlier, there appears to be a strong rationale for the regular assessment of the force production characteristics of athletes.

STRENGTH TESTING CONSIDERATIONS

There are several aspects of testing that should be accounted for when assessing athlete strength characteristics. Strength and conditioning professionals and sport scientists must consider the reliability, validity, test familiarity, test order, and standardization of methods. The following paragraphs will discuss each concept.

RELIABILITY

Reliability refers to the repeatability of a test or the consistency of a measurement (295). Tests that measure strength characteristics must be repeatable to ensure that the variation produced during the test is accounted for when determining whether real change occurred (17). Equipment setup and calibration, timing of the testing, athlete familiarity with the test, and competency of the tester may contribute to variability in testing (295). Reliability of testing may be expressed in both relative and absolute terms using variables such as intraclass correlation coefficients and coefficients of variation, respectively. To improve the reliability of a performance test, it is recommended that strength and conditioning professionals conduct testing in a consistent manner while controlling as many variables as possible. This notion stresses the importance of testing athletes under very specific circumstances (e.g., time of day, instructions and cueing) to accurately measure the variation produced during each test (109). Although laboratory-based tests may allow for more control of the testing environment, strength and conditioning professionals should attempt to conduct field-based testing under strict circumstances to improve the reliability of testing.

Key Point

Tests that measure strength characteristics must be repeatable to ensure that the variation produced during the test is accounted for when determining whether real change occurred.

VALIDITY

The *validity* of a test refers to its ability to measure what it is meant to measure (i.e., the test accurately measures the desired variables) (111). For example, a skinfold test may be reliable but it does not measure the force production characteristics of an individual, making it an invalid test for measuring strength. Weakley and colleagues (292) summarized the many types of validity as they relate to physical performance characteristics, including those linked to the validity of both the test and the methodology. Perhaps one of the most important forms of validity that strength and conditioning professionals should be concerned with is *construct validity* (i.e., the ability of a test to accurately represent a given construct). In this light, tests should be chosen that accurately represent the ability of an athlete to produce force in different contexts.

Strength and conditioning professionals and sport scientists should also consider the validity of a test as it relates to the performance of an athlete within their respective sport or event, termed *ecological validity*. The more closely a test can represent what an athlete performs in a sport-specific context, the more it may serve as a strong indicator of how an athlete may perform in competition. For example, researchers have measured the force production characteristics of windmill softball pitching (204) and jump shooting in basketball (42) using in-ground force plate technology to mimic performances within each sport.

Key Point

Tests should be chosen that accurately represent the ability of an athlete to produce force in different contexts.

TEST FAMILIARITY

Prior to testing, strength and conditioning professionals and sport scientists should first familiarize themselves and their athletes with the procedures of each test (134). Specific to the athletes, it is important to discuss what is expected of them, the goal of the test, and the cues that will be used. Beyond a verbal description, a demonstration of each test and practice trials should be provided to give the athlete a combination of auditory and visual cues while allowing them to "feel" a trial of the test. Researchers have shown that the reliability of an unfamiliar squat strength test may be reduced when participants are not previously introduced to a modified test (33). Additional studies also showed significant increases in strength between sessions (217, 220, 235) and across trials (240) during the back squat and between sessions for the bench press (217). In contrast, there were no differences in strength performance across multiple testing sessions during the leg press or chest press (25) or during the back squat and power clean (50); however, it should be noted that machine-based protocols may require less of a learning effect compared to free weights, and participants in the study by Comfort and McMahon (50) had more training experience than those in other studies presented earlier.

Readers should note that athletes may also become familiar with specific positions during their training program. For example, researchers indicated that individuals with at least 6 months of experience with weightlifting derivatives showed greater differences in isometric mid-thigh pull (IMTP) performances in different positions, whereas those with less experience showed smaller differences (21). The findings from Comfort and McMahon (50) presented earlier also support the notion that previous training experience with the tested exercises may lead to reliable measurements across multiple testing sessions. Therefore, while allowing individuals to practice the test prior to performing maximal effort trials may not fully alleviate the learning effect, it may reduce its impact to an extent.

Key Point

Beyond a verbal description, a demonstration of each test and practice trials should be provided to give the athlete a combination of auditory and visual cues while allowing them to "feel" a trial of the test.

TEST ORDER

Athletes often perform a battery of tests that are used to assess the various qualities related to their sport performance. However, due to athlete time constraints, it is important to organize the tests in a specific order to establish an efficient testing environment that promotes highly reliable and valid information. If all athlete testing must take place on a single day, anthropometry (e.g., height and body mass) should be performed first, followed by nonfatiguing ballistic (power) (e.g., vertical jump) tests, then change-of-direction or agility, maximal power or strength, speed, muscular endurance, anaerobic capacity, and aerobic capacity tests (161). An advantage to completing all testing on a single day is that each athlete is assessed within a similar physiological state. For example, if three of six tests are assessed on two separate days, it is possible that additional external stressors may affect the athlete's performance on one of the two days, sacrificing the intersession reliability of the given tests. However, the advantage of performing tests on two separate occasions may allow near-maximal or maximal effort tests to be performed without one affecting the other (e.g., maximal back squat test and yo-yo intermittent recovery test).

Regarding the assessment of strength characteristics, it is important to note that the previously listed testing order may allow for force production qualities to be assessed in various contexts provided that the necessary equipment is available. For example, researchers have shown that anthropometric, jump, sprint, change-of-direction, isometric strength, and dynamic strength tests can all be conducted in the same session and provide a wealth of information (260). However, it is important to note that if both maximal isometric and dynamic strength tests are performed during the same session, the former should be performed first because it is less fatiguing and can be conducted in a time-efficient manner.

Key Point

If all athlete testing must take place on a single day, anthropometry (e.g., height and body mass) should be performed first, followed by nonfatiguing ballistic (power) (e.g., vertical jump) tests, then change-of-direction or agility, maximal power or strength, speed, muscular endurance, anaerobic capacity, and aerobic capacity tests.

STANDARDIZATION OF METHODS

As mentioned previously, it is important that strength and conditioning professionals and sport scientists control as many variables as possible when assessing the performance of their athletes. This can be accomplished through the standardization of methods, which may include but is not limited to equipment calibration and setup, warm-up procedures, the instructions given to athletes, and rest periods between trials. Simply, if the equipment used to assess an athlete is not calibrated, it is possible that the data collected may not be accurate. Regarding the setup of equipment, it is important that strength and conditioning professionals and sport scientists are diligent with their placement or positioning of equipment so that the same testing conditions are provided for each athlete and that data can be compared between testing sessions with confidence. Like the familiarization procedures discussed earlier, the same cues should be given to every athlete who performs testing, because cueing may directly affect the way they perform the test and their force production characteristics. For example, cueing an athlete to jump as high as possible or limit the time on the ground during a drop jump may produce significantly different performances (300). It is also important to note that the cues given to the athlete may alter their attentional focus and, subsequently, their force production characteristics. Halperin and colleagues (98) indicated that external compared to internal focus instructions resulted in participants producing greater peak forces during the IMTP test. Further information about the instructions that should be used for specific strength assessments is provided next.

TYPES OF STRENGTH TESTING

As discussed in chapter 1, strength can be expressed in a variety of ways depending on the needs of a given task. As such, it is important that strength and conditioning professionals and sport scientists assess the ability of an athlete to express their strength characteristics using tests that relate to their performance in sport. For example, maximum velocity sprinting requires large magnitudes of force to be expressed within very short periods of time (80-90 ms at the elite level); thus, choosing the appropriate strength assessment and performance variables becomes quite important from an athlete monitoring perspective. The following paragraphs will discuss different methods of strength assessment and provide insight into how the information may be used to inform programming decisions.

ISOMETRIC STRENGTH TESTING

Isometric strength testing may be completed using a variety of devices, including force plates (23, 94), isokinetic dynamometers (192), load cells (114), or strain gauges (176). Depending on the equipment and software available, each device may allow strength and conditioning professionals and sport scientists to generate a force–time curve during the isometric effort and allow for the measurement of many force production variables (e.g., force at various time points, rate of force development, and impulse). While reliable information can be gathered regarding the force production characteristics of an athlete during single-joint isometric tasks, these tests may not relate to tasks performed during athletic competition (31). Two of the most frequently performed multi-joint isometric strength tests in the literature are the IMTP (figure 9.1) and the isometric squat test (ISQT) (figure 9.2). Although both tests assess lower-extremity isometric strength, greater peak forces have been shown during the ISQT, whereas greater magnitudes of rapid force production have been shown during the IMTP (36, 37, 205). However, it should be noted that researchers have also examined force production characteristics during the isometric bench press (195, 298), leg press (155, 302), belt squat (figure 9.3) (138), bench pull (figure 9.4) (151), and knee extension or flexion (230).

Isometric testing results may be related to the 1RM of a similar strength test (20, 63, 205, 248) and even to other lifts (e.g., back squat and bench press) (165, 291). For example, strong relationships have been shown between isometric squat peak force and the 1RM back squat (19, 32) and partial squat (19). Moreover, a strong relationship has been shown between the IMTP and the ISQT (205). Further research showed that a rate of force development from 0 to 250 ms during the isometric leg press was strongly correlated with hang power clean, back squat, and leg press performance (302). Finally, the isometric bench press was strongly correlated with the 1RM bench press and bench press throws with 15%, 30%, and 60% of the 1RM (195).

Researchers have shown that peak force and force produced at specific time points may be reliable,

Chapter 9 ■ Measuring Strength 129

FIGURE 9.1 Isometric mid-thigh pull.

FIGURE 9.2 Isometric squat.

FIGURE 9.3 Isometric belt squat.

FIGURE 9.4 Isometric bench pull.

but the reliability of rate of force development has been inconsistent (93). However, researchers have shown that rate of force development reliability may improve if the time epoch ranges are predetermined (69, 93). Readers should note that because force plate sampling rates ranging from 500 to 2,000 hertz did not significantly affect force–time variables during an IMTP, a minimum sampling rate of 500 hertz may be used to reliably assess these variables (69). Beyond the traditional variables, it may be advantageous to express force at different time intervals as a percentage of maximum force (52).

An important aspect of isometric strength testing is the identification of the onset of the trial (i.e., where the test analysis will start). While the onset may not prevent strength and conditioning professionals and sport scientists from identifying the peak force of a trial, other variables including both rapid force production and impulse may be significantly affected. That said, there have been a variety of methods used to identify the onset of an isometric effort, including manual identification, arbitrary thresholds, percentages of body mass, and relative thresholds that take into account the system (i.e., athlete and force plate) noise and force variation during a quiet standing period (65). The authors of the previous study suggested using 5 times the standard deviation of a 1-second quiet standing phase to identify the start of an IMTP. To understand the ability of an athlete to produce force, it is important that strength and conditioning professionals examine *net force* (i.e., force produced above the athlete's body weight) instead of *absolute force* (i.e., gross force, or force that includes the athlete's body mass) because the latter will benefit heavier athletes. Although absolute net force information is important, especially in scenarios where two athletes similar in size will collide (e.g., American football linemen), isometric forces may also be scaled to allow for comparisons between athletes of different sizes (51, 269) and between different sexes (202, 203).

Isometric Mid-Thigh Pull

As mentioned previously, there are a variety of setups that may be used to assess the isometric force production characteristics of an athlete. When using the IMTP, researchers have demonstrated the importance of the initial starting position; for example, significant differences in force production characteristics were shown in various testing positions (21, 67). Briefly, the IMTP position should resemble the start of the second pull position of the clean exercise, in which the immoveable bar should be positioned on the upper portion of the athlete's thigh while all the slack is removed from their arms. In addition, the athlete's torso should be upright with a hip angle ranging from 140° to 150° and their knee angle ranging from 125° to 145°. After achieving the starting position, athletes should perform warm-up efforts at 50% and 75% of their perceived maximum effort, each for approximately 3 seconds with a minute of rest between trials. To prevent grip from being a limiting factor during the maximum effort trials, the athlete's hands should be strapped and taped to the immoveable bar. Prior to performing a maximal effort trial, the athlete should be instructed to maintain their position during the test and push their feet into the ground "as fast and as hard as possible." IMTP trials typically last around 5 seconds; however, the length of the test may vary based on the ability of an athlete to reach their peak force quickly. A flow chart of the testing procedures for the IMTP is displayed in figure 9.5. For a thorough discussion of IMTP testing best practices, readers are referred to a review by Comfort and colleagues (47).

Isometric Squat

The isometric squat is another common isometric test used to examine the lower-body force production characteristics of an athlete. Like the IMTP, the initial setup of the ISQT is crucial for a reliable and valid test. Researchers have examined a wide variety of positions during the ISQT, with knee angles of 90° and 120° being the most common (19, 32, 200, 299). Briefly, the athlete should stand on the force plate or force plates in a partial squat stance based on the desired knee angle position. Strength and conditioning professionals should encourage their athletes to mimic their squat mechanics to prevent excess strain on the lower back and to ensure maximal vertical force production. Prior to the start of the test, athletes should have their upper back in contact with the bar with minimal pretension. Like the IMTP, athletes should complete warm-up trials prior to maximal effort trials and should follow strict procedures to improve the reliability of each trial. It should be noted that different ISQT testing

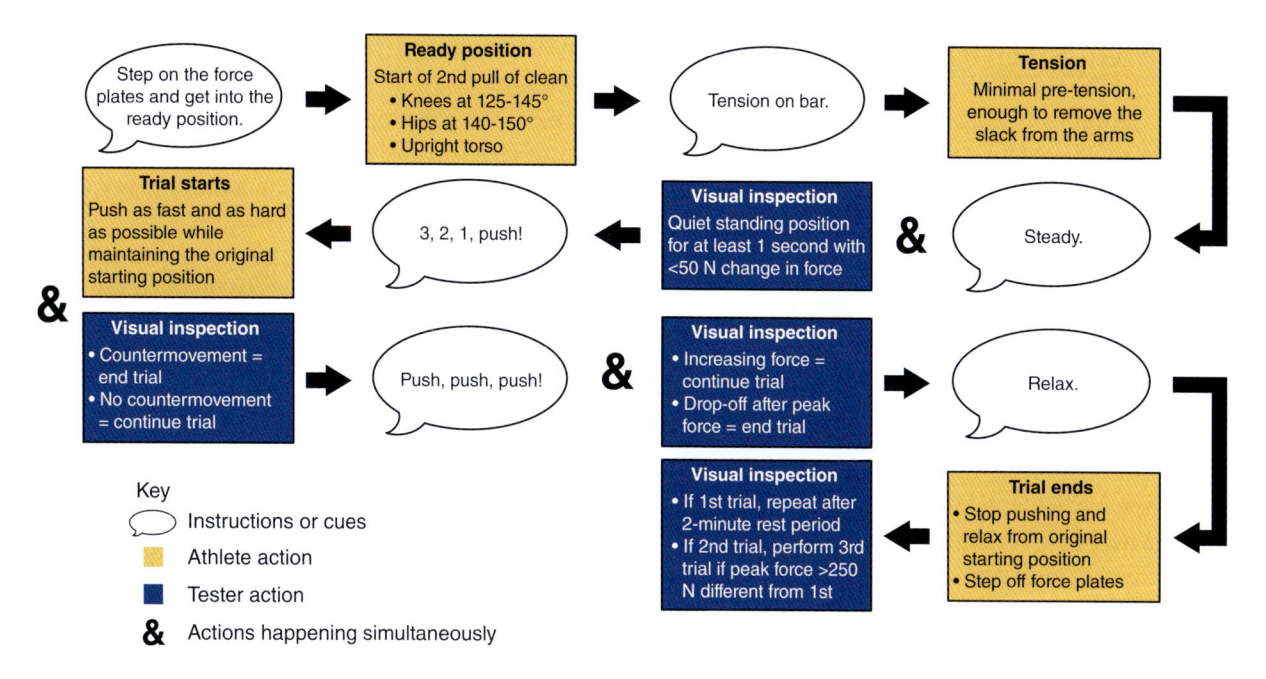

FIGURE 9.5 Flow chart of the isometric mid-thigh pull testing procedures.

protocols may produce different levels of reliability when measuring peak force and rapid force production characteristics (72, 73). While peak force may be measured reliably using similar test durations as the IMTP, it has been recommended that rapid force production should be measured using an *explosive protocol* that lasts 1 second (73). Beyond the duration of the test, the primary difference between the traditional and explosive protocols are the instructions given to the athlete. Specifically, athletes should be cued to "push as *hard* and fast as possible" and "push as *fast* and hard as possible" during the traditional and explosive protocols, respectively.

Isometric Bench Press

Force production assessment during the isometric bench press is typically measured using elbow angles at approximately 90° or 120° (194, 195). However, researchers showed that the greatest peak force was produced at an elbow angle of 150° compared to 60°, 90°, and 120° (298). These authors also noted that while peak force was reliable at each angle, rate of force development did not meet acceptable reliability cut-offs. Strength and conditioning professionals may allow athletes to self-select their "strongest" position; however, they must also standardize the grip width for each athlete so that it may be replicated during future testing sessions (298). When measuring isometric bench press performance, it is important that the athlete's legs and the entire bench are positioned on the force plates because both the upper and lower body will produce force during the movement. Multiple force plates may not always be available for testing, so strength and conditioning professionals and sport scientists should understand the limitations of testing in this manner or using equipment that does not account for lower-body force.

DYNAMIC STRENGTH TESTING

There are a wide variety of dynamic tests that can be used to assess the various strength characteristics of athletes. These tests range from maximal strength to speed tests that can provide a wide range of information about the force production characteristics of athletes. The following paragraphs will discuss maximal strength, relative strength, vertical jump, horizontal jump, loaded jump, sprint, change-of-direction, force production asymmetry, and reactive strength testing.

Maximal Strength

Dynamic strength testing may be viewed as more applicable to an athlete's performance compared to isometric testing, due to the movement similarities completed in various sports or events. The most common form of dynamic strength testing is the completion of a 1RM test, which assesses the maximum amount of weight an athlete can lift for 1 repetition (39). These tests have been classified as reliable despite varying resistance training experience, the number of familiarization sessions, exercise selection, the portion of the body tested (lower or upper), sex, or age of the participants (92). Moreover, researchers have shown that 1RM tests are characterized by measurement error and smallest detectable differences of approximately 5% (50). The most common exercises used for maximal strength testing are the back squat, bench press, and power clean (176). Although these tests may be reliable for both experienced and inexperienced individuals (14, 50, 77), strength and conditioning professionals must consider the technical proficiency of the athlete before deciding to perform a 1RM test. Simply, experienced athletes will be more familiar with the exercise, having previously trained with it; thus, they will likely be able to more accurately predict their 1RM compared to less experienced athletes (198).

Beyond traditional eccentric and concentric movements, 1RM testing may be conducted using isolated muscle actions as well. For example, the concentric-only squat (270, 271, 272, 277), eccentric-only squat (245), concentric-only bench press (285), and eccentric-only leg press (86) have been used to test maximal strength using specific ranges of motion. Regarding concentric-only movements, researchers have completed such testing to investigate the "sticking region" during a bench press (97, 143, 285). Although such testing may be highly applicable to athletes competing in powerlifting, this testing may be less impactful for team sport athletes. It should be noted that 1RM eccentric testing may be possible through the use of a tempo and standardized range of motion (86, 245). Again, while this information may be valuable as it relates to strength within eccentric muscle actions, and potentially for loading purposes during training methods such as accentuated eccentric loading (103), the application of such findings may be limited because many sport-specific tasks involve both eccentric and concentric muscle actions. Because of its taxing nature, frequent 1RM testing may not be practical over the course of a training year. However, because many strength and conditioning professionals use 1RM testing to determine resistance training loads, alternative methods of assessment may be needed.

An athlete's 1RM may also be predicted from using a multiple RM test (e.g., 3RM or 5RM). However, strength and conditioning professionals should be aware that the accuracy of the predicted 1RM decreases as the number of repetitions within an RM test increases (122, 187). Moreover, it has been suggested that the ability to accurately predict a 1RM load may be dependent on the exercise being tested (108, 239). Strength and conditioning professionals should be aware that a wide variety of prediction equations exist and may result in large magnitudes of error (i.e., −24.0% to 22.9%) between the estimated and actual 1RM (158). Several reliable equations with the smallest differences between predicted and actual 1RM and percent error are as follows:

Adams, 1998 (1):
$1RM = RW/(1 - 0.02 \times RTF)$

Mayhew and colleagues, 1992 (157):
$1RM = RW/(0.522 + 0.419\ e^{-0.055RTF})$

O'Conner and colleagues, 1989 (206):
$1RM = 0.025\ (RW \times RTF) + RW$

Reynolds and colleagues, 2006 (216):
$1RM = RW/(0.5551\ e^{-0.0723RTF + 0.4847})$

RW = repetition weight; RTF = repetitions to fatigue

Another potential option to estimate the 1RM of an athlete is through using the set–repetition best method (247, 266). *Set–repetition best* allows strength and conditioning professionals to use the loads that each athlete performs in training to estimate the training loads for other set–repetition schemes, but it also allows for the estimation of a 1RM for a variety of lifts. An obvious benefit of such a method is the ability to use the training facility as the laboratory and remove the need to perform 1RM testing periodically throughout the training year. In other words, the information collected from training sessions—especially those that are heavy—may allow strength and conditioning professionals to estimate and track the 1RM of an athlete as it may change based on the physical, physiological, and psychological demands of a training phase. While this estimation method may be common for traditional resistance training exercises such as squat, press, and pull variations (43, 44, 260, 261, 262), not all exercises have 1RM criteria to allow for comparisons between the estimated and actual value. For example, strength–speed weightlifting pulling derivatives (e.g., mid-thigh pull, countermovement shrug, pull from the knee, hang pull, and pull from the floor) may allow athletes to use loads that are greater than the 1RM of the respective catching variation (e.g., snatch or clean) (256, 257); however, researchers have used relative percentages of a 1RM catching variation to examine various loading effects on force production characteristics during these exercises (49, 55, 95, 177, 178, 179). Thus, the use of set–repetition best with these exercises may allow for the estimation of theoretical 1RM loads for weightlifting pulling derivatives that may be programmed with maximal and supramaximal loads; however, further research is needed on this topic.

Some researchers and strength and conditioning professionals have chosen to indirectly measure 1RM strength by measuring the velocity produced while performing an exercise with submaximal loads and generating a load–velocity profile (14, 89). Researchers have indicated that a range of at least 0.5 meters per second should be used when generating a load–velocity profile (117). Although this method has been shown to be reliable by some researchers (145, 148), many of the existing studies have used a pause (approximately 1.5 seconds) between the eccentric and concentric phases of the assessed exercise (225). It should be noted that while a pause may increase the reliability of the test (208), the ecological validity of the test may suffer, given that most resistance training exercises are not performed in this manner. Strength and conditioning professionals should also consider the reliability and validity of a load–velocity profile before using this method of assessment (219), because researchers have shown large errors in 1RM prediction (14, 78, 231). For example, errors of 5.7% to 17.0%, 8.2% to 20.4%, and 8.6% to 19.9% were present when predicting 1RMs for the squat, bench press, and bent-over row exercises, respectively (78). A 2023 meta-analysis supported these findings and indicated that the use of load–velocity relationships leads to overestimation of 1RMs regardless of the modeling approach used (91). Therefore, the practice of estimating the 1RM of an athlete using load–velocity profiles may be questioned, because the results of such testing may result in strength and conditioning professionals prescribing training loads that fail to maximize the desired training adaptations or expose the athletes to unnecessary stress that may lead to injury.

Key Point

Load–velocity relationships may overestimate 1RMs regardless of the modeling approach used.

Although a 1RM test provides strength and conditioning professionals with information on how much weight an individual can lift, it does not provide a direct indication of how the athlete is producing force. Figure 9.6 displays the force–time curve differences between an estimated 90% of the 1RM squat, a successful 1RM attempt, and an unsuccessful 1RM attempt. Strength and conditioning professionals should note the absence of the second force peak during the unsuccessful 1RM attempt, indicating the failure to effectively tolerate the braking forces or produce the necessary propulsion forces during the transition between lifting phases. Because force plate testing may not always be feasible or practical, a potential alternative for strength and conditioning professionals may be through longitudinal monitoring of body mass and volume–load during resistance training sessions. For example, if an athlete can lift larger loads relative to their body mass or lift the same load for more repetitions, it is possible that they have improved their peak force production capabilities. However, it is important to be aware of the assumptions being made with this method (e.g., physiological state, injury status, time of day).

Key Point

Although a 1RM test provides strength and conditioning professionals with information of how much weight an individual can lift, it does not provide a direct indication of how the athlete is producing force.

Relative Strength

The relative strength of an athlete may influence the programming decisions of strength and conditioning professionals. For example, if an athlete can squat more than twice their body weight, one may be more

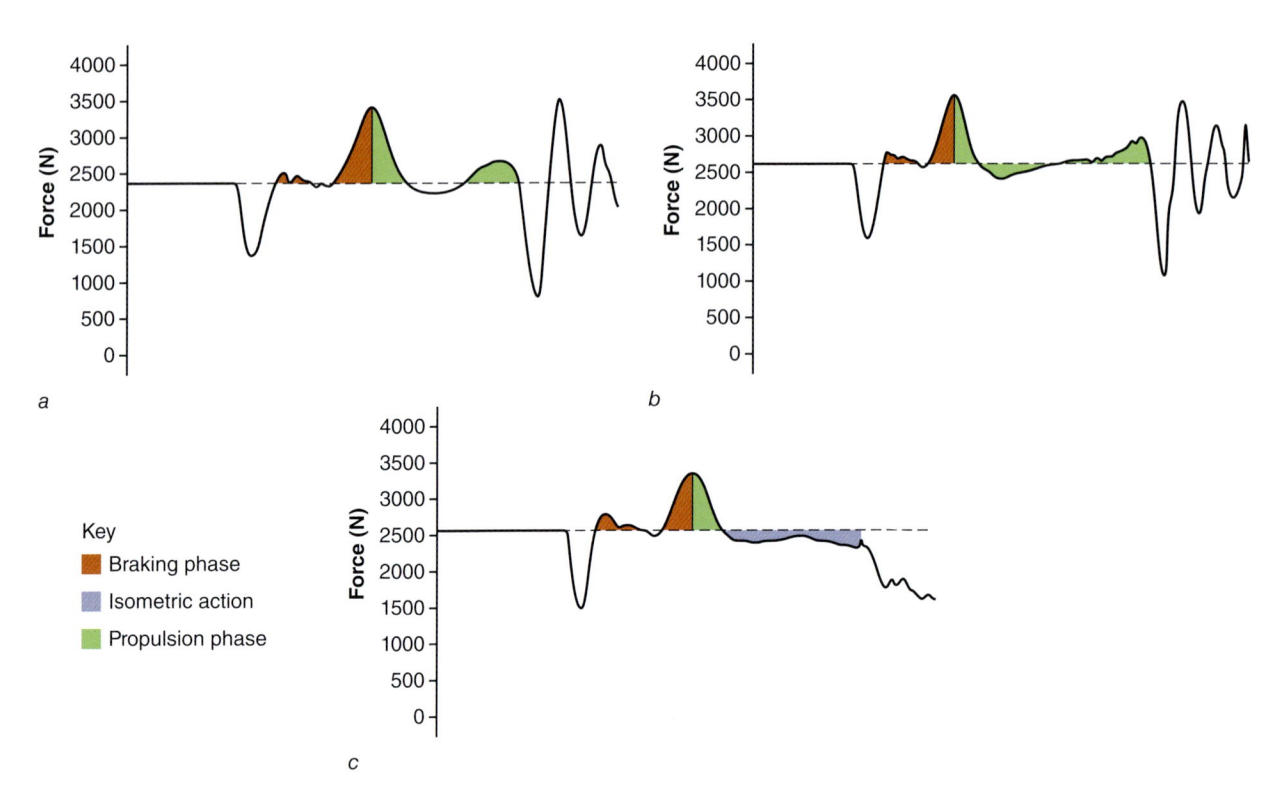

FIGURE 9.6 Force–time curve differences between squatting attempts: *(a)* Estimated 90% 1RM (153kg), *(b)* successful 1RM attempt (175 kg), *(c)* unsuccessful 1RM attempt (181.5 kg).

Key
- Braking phase
- Isometric action
- Propulsion phase

inclined to use advanced training strategies such as potentiation complexes compared to traditional methods of training (see chapter 6 for more detail). While the findings of researchers may support this decision (234, 271), the notion of using potentiation complexes with an athlete who is able to squat double their body weight but not with an athlete who squats 1.9 times their body weight seems farfetched and impractical. This has led many in the strength and conditioning field to challenge the idea that an athlete is only considered "strong" if they can squat twice their body weight. However, it is important that strength and conditioning professionals understand why this "level" of strength has come into focus.

Simply, researchers have grouped individuals based on their strength levels—in the case just described, over or under a double bodyweight back squat—to clearly separate the groups in terms of their force production capabilities. For example, researchers have shown very large effect sizes (Hedges' g >3.00) when comparing groups in this fashion (234, 271). While this may not be an optimal method of comparison, setting specific strength levels as a means of delineating the abilities of athletes may help answer specific research questions. Furthermore, it is important to highlight that although greater levels of strength have been related to a variety of performances (268), strength is task specific when it comes to force application.

From an isometric strength standpoint, strength and conditioning professionals may also scale athlete performance in different ways (269); however, examining the differences in net force per kilogram of body weight may provide the most direct assessment. Isometric relative net peak force may be classified as excellent, good, average, or below average in rugby and soccer players if values are greater than 50, 40.0 to 49.9, 30.0 to 39.9, or less than 30 Newtons/kg of body mass, respectively (182). However, limited normative data exist with other populations; thus, further research is warranted.

Vertical Jump Testing

Vertical jump testing is one of the most common forms of assessment due to its simplicity, familiarity among athletes, and the limited need for familiarization (201). Various types of jumps, including the countermovement jump (CMJ), the squat (static) jump (SJ), and the drop jump, may fall under this category of testing. Due to its ability to assess reactive strength characteristics, drop jump testing will be discussed in greater detail later in this chapter.

Briefly, CMJ testing requires an athlete to perform a countermovement to a self-selected depth before jumping as high as possible (174). In contrast, an SJ requires an athlete to lower themselves down to a predetermined squat depth (typically a 90° knee angle), pause, and then jump as high as possible using concentric-only muscle actions (131). An obvious difference between the jump types is the use of a countermovement, which has been shown to benefit jump heights and force production characteristics due to greater uptake of muscle slack, the buildup of high stimulation, and the storage and use of elastic energy (287).

Both CMJ and SJ trials may be performed with or without the use of an arm swing; however, strength and conditioning professionals should note that the use of an arm swing may significantly modify the jump performance and force production characteristics (101, 102, 139, 140). Thus, strength and conditioning professionals and sport scientists may consider removing arm swings to isolate the legs; this is typically done by having athletes place their hands on their hips (*akimbo*) or holding a wooden dowel or PVC pipe across their upper back, like a back squat. Regardless of the methods used, the same testing protocol should be used when athletes are retested to allow for accurate comparisons between the testing sessions. As noted earlier, vertical jumps may be classified into different types; thus, strength and conditioning professionals need to decide what jump types will reflect the ability of an athlete to perform within the context of their sport or event.

Although vertical jump height may be the most commonly assessed variable and can be determined using several different methods (185), *how* an athlete achieved their performance may provide more valuable insight regarding their neuromuscular status and force production characteristics (84). This has led researchers to examine the phase characteristics of the jump (e.g., unweighting, braking, propulsion, landing), specifically the force, duration, and impulse characteristics of each phase (106, 135, 169, 172, 241). Their findings highlighted the differences in force–time characteristics and showed the differences between athletes performing at a higher or lower level. Figures 9.7 and 9.8 depict the different phases of the CMJ and the SJ, respectively.

It is important to highlight that strength and conditioning professionals must be cautious when making assumptions about an athlete's force production capabilities when assessing vertical jump performance using a device other than a force platform. For example, if an athlete demonstrates the ability to jump higher following a training period, this is reflective of the athlete producing a greater propulsive impulse, which generally would be considered a positive adaptation. However, as noted in chapter 1, the shape of the impulse (i.e., height and width) dictated by the magnitude of force production and the duration of the impulse may determine whether the adaptation may truly be beneficial. Within this context, an athlete could theoretically jump higher with a higher-force, shorter-duration impulse or a lower-force, longer-duration impulse. While the improved jump height may be assumed to be based on the former, a higher jump height that takes a longer duration to achieve may be less applicable in specific sporting scenarios.

Key Point

Although vertical jump height may be the most commonly assessed variable and can be determined using several different methods, *how* an athlete achieved their performance may provide more valuable insight regarding their neuromuscular status and force production characteristics.

Force plate technology may be considered the gold standard when assessing vertical jump performance (83), with researchers demonstrating the reliability of several different devices (4, 136, 147). When assessing vertical jump force–time characteristics, a minimum sampling frequency of 1,000 hertz should be used (207, 250). Furthermore, to accurately assess force–time variables during the vertical jump, strength and conditioning professionals must consider the instructions given to the athlete regarding the use of an arm swing (101, 102, 139, 140) and the cues regarding the performance of the test (149, 300). Like isometric testing, another factor that should be considered is the threshold used to identify the initiation of the jump. For example, researchers have suggested the use of 5 times the standard deviation of a quiet standing phase to identify the start of both the CMJ (207) and the

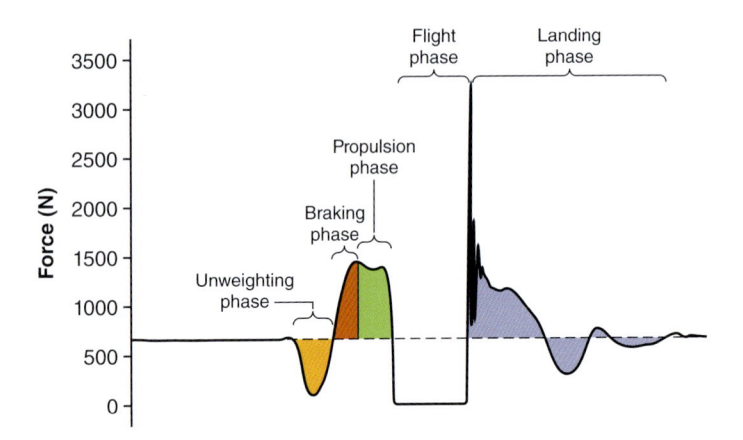

FIGURE 9.7   Countermovement jump phases.

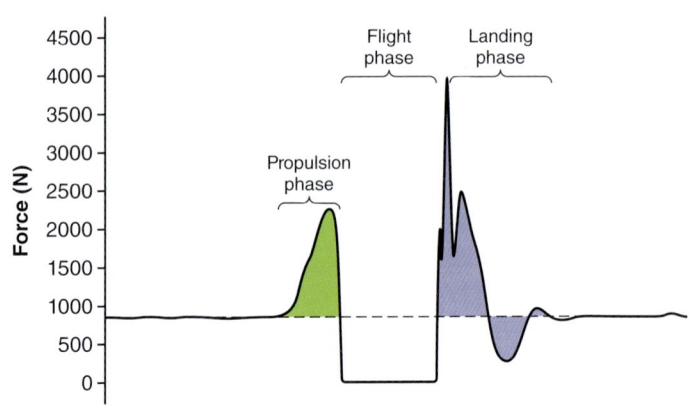

FIGURE 9.8   Squat jump phases.

SJ (64). Unfortunately, force–time characteristics cannot be measured without force plate technology. Although mobile applications may calculate jump height and other variables based on movement and flight times (236), they may be limited in their capacity to accurately assess the force production characteristics of athletes.

Horizontal Jump Testing

Beyond vertical jump testing, the standing long jump is another commonly performed test used within athlete monitoring protocols. Briefly, the standing long jump requires athletes to linearly jump as far as possible from an initial starting line with or without the use of a countermovement or arm swing. Like the vertical jump, researchers have identified specific force–time phase characteristics during the standing long jump and emphasized the importance of increasing propulsive force and decreasing propulsive phase duration (105). Additional research highlighted the usefulness of measuring both vertical and horizontal forces to provide an indication of explosive strength performance during the standing long jump (154). It should be noted that the initial starting position (152), use of an arm swing (2), type of standing long jump (188), and use of external focus (214, 296) may have a significant impact on force production characteristics, jump distance, or both. However, strength and conditioning professionals should be aware that the combined vertical and horizontal force components were not measured in all the previous studies, potentially due to the type of force platforms used. Although considerably more research has been completed on vertical jump testing, further examination of the force production characteristics of the horizontal jump is needed given its specificity to other movements such as sprinting (180, 188).

Loaded Jump Testing

Beyond unloaded jump testing, strength and conditioning professionals can assess the force production characteristics of their athletes using a series of loads. Researchers have examined the force production characteristics of participants using a spectrum of loads during the SJ (131), jump squat (57, 263, 264, 279), hexagonal barbell jump (figure 9.9) (263, 264, 279, 283), and jump shrug (129, 253, 263, 264, 278). Although the purpose of most of the previous research was to identify the load that maximized power output, collecting the force production information may provide strength and conditioning professionals with some insight on how an athlete responds to the addition of an external load as well as on the training needs of an athlete (238, 299). This has often been done by creating a load–power curve and examining the inflection point where power output drops off. However, strength and conditioning professionals may consider examining the force–velocity characteristics that underpin power output at each load to determine the training focus of an athlete (i.e., force or velocity deficit). While force–velocity profiling methods have been discussed previously (189), it is important to put the

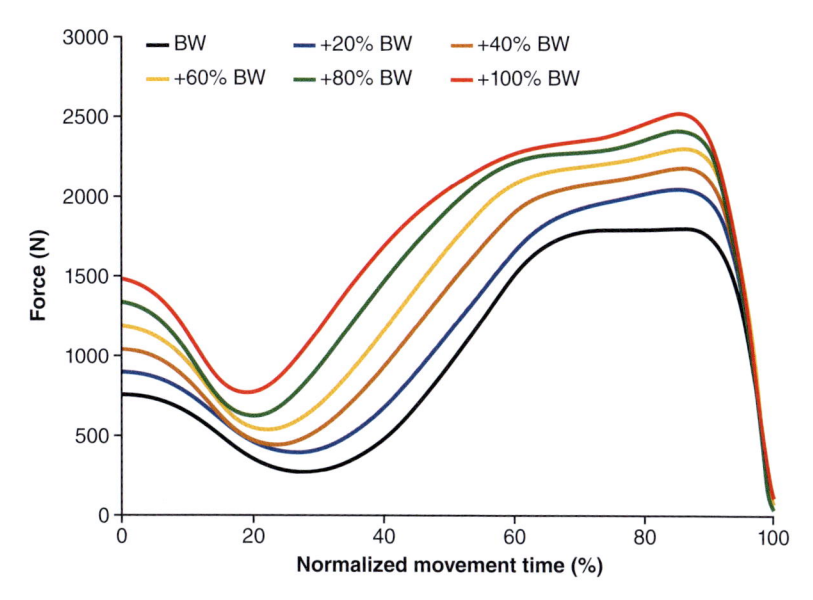

FIGURE 9.9 Load effect on force–time characteristics during time-normalized hexagonal barbell jumps. *Note:* BW = bodyweight.

force–velocity results into context. For example, the existing literature on force–velocity profiling training recommendations may be limited to the task performed but may not translate to other performances (120). Furthermore, it should also be noted that while a mobile application may be used to generate a force–velocity profile (236), the force outputs are estimated but not measured. Force–velocity profiling is discussed in greater detail later in this chapter.

Like the unloaded vertical jump, it is important to consider the impact of a countermovement on force production characteristics when an additional load is added. For example, a jump performed from a static starting position versus one that uses a countermovement may produce vastly different maximal and rapid force production characteristics (35). While some research has examined the effect of a countermovement on jump height across a spectrum of loads (146), limited information exists on the force production characteristics. Because the inclusion or exclusion of a countermovement places unique constraints on an athlete, strength and conditioning professionals must determine which jump type is more specific to the individual to effectively monitor their force production characteristics. Regardless of the decision, it should be noted that very little familiarization is needed to produce reliable force–time information during loaded jumps (184).

Sprint and Change of Direction

The most common method of assessing both sprint and change-of-direction performance is by using a timing system (e.g., timing gates, stopwatch). Although this method may provide information regarding how quickly an athlete is able to cover a specific distance or perform a change-of-direction task, little knowledge is provided on their force production characteristics. Researchers have examined the force production characteristics of both the acceleration and top speed phases during sprinting (196), and they have provided insight as to how force application differs between athletes of varying performance levels (45, 191, 293). From a change-of-direction standpoint, strength and conditioning professionals should consider examining both the deceleration and acceleration portions of the movements. Interestingly, researchers have shown that deceleration forces may be up to 2.7 times greater than acceleration forces (289), and these forces are greatest during the late deceleration phase within the antepenultimate and penultimate steps (104). However, from a testing standpoint, strength and conditioning professionals must also consider the angle and entrance velocities of the change-of-direction tasks being assessed. For example, tasks that include sharper angles (66) and higher entrance velocities (62, 288) require a greater braking demand and, thus, greater magnitudes and rates of force production.

An important consideration of sprint and change-of-direction force testing is its feasibility. First, each type of testing requires equipment (e.g., force plates, high-speed treadmill) and the space necessary for the assessments to take place at full speed. Although some laboratory settings are well equipped with force plate technology (e.g., 60-meters of force plates) (196), many institutions may not be able to afford one force plate, let alone position the force plate in a manner where sprinters would consistently contact it during the examined phase of sprinting.

Another issue is the generalization of forces measured during only several strides compared to an entire trial. In this light, researchers have examined the ability of wireless insoles to measure ground reaction forces, contact time, and impulse (40, 215). The results of the previous research indicates that insole force measurements may be reliable during various walking and running conditions; however, insoles with higher sampling rates (200 hertz) may be needed to improve the reliability of peak force, loading rates, and impulse (215). It should also be noted that other insole devices with a lower sampling rate (50 hertz) significantly underestimated vertical force values (18.9%-48.3% lower, on average) compared to force plate technology during 50-meter sprint accelerations (197), further expressing the need for higher sampling rates and additional research on this technology.

Apart from force plate technology, other researchers have promoted the use of anthropometric, air pressure and temperature, and velocity data (e.g., split times, a radar gun, or a GPS system) to determine force production characteristics of sprint performance (190, 223). One benefit of this method is alleviating the requirement of force plate or force sensor technology to obtain certain kinetic measures, because these force values are indirectly calculated on a step-average basis from the velocity data, rather than measured directly. The calculated forces from this modeling approach have agreed nicely with measured force plate values for horizontal forces (190). This modeling approach assumes that the average vertical forces applied during each step (contact phase + flight phase) are equal to body weight throughout the sprint (223), although it should be noted that the average vertical forces measured only during the contact phase generally increase with each step during the acceleration phase (196). Unfortunately, change-of-direction force assessment appears to be limited to using force plate technology; thus, further research is warranted because it may be difficult to determine the force production capabilities of an athlete without a laboratory-based setup.

It is important to note that changes in both sprint and change-of-direction performance may be based not only on force production characteristics but also on the technique (e.g., body position) used during such tasks. Researchers have shown that athletes who sprint (45) or perform change-of-direction tasks (68) with better mechanics may be able to produce force more effectively. While this may be reflected in terms of faster sprint and change-of-direction times, it is important that strength and conditioning professionals do not automatically assume that athletes have improved their maximal or rapid force production characteristics. Rather, they may have improved their ability to apply their existing strength more effectively via enhanced technique or mechanics. Based on the results of previous research, it would therefore be logical to improve not only technique but also the maximal and rapid force production characteristics of athletes, because both married together may result in optimal improvements in performance.

Force Production Asymmetry

Strength and conditioning professionals may also measure the force production symmetry characteristics of their athletes. Simply, if the athlete produces identical force production on both limbs, they would be considered perfectly symmetrical; however, this is typically not the case, which has led investigators to examine the asymmetric characteristics of athletes. Testing asymmetry can be accomplished using both isometric and dynamic tests performed in a bilateral or unilateral manner. For example, researchers have used a dual force plate system to examine force asymmetry during the IMTP (5, 6, 7), the ISQT (18), and various jumps (6, 76, 137); however, unilateral versions of the IMTP (70), the ISQT (27), and jumps (28, 29, 142) have also been used. Research has shown that the reliability of asymmetry may be biased if reliability is measured before asymmetry is quantified (8). Furthermore, the consistency of the direction of asymmetry may vary based on the chosen variable (137). Thus, if strength and conditioning professionals elect to measure force production asymmetry, it is crucial to establish whether asymmetry is reliable before drawing specific conclusions about the training needs of an athlete.

Key Point

If strength and conditioning professionals elect to measure force production asymmetry, it is crucial to establish whether asymmetry is reliable before drawing specific conclusions about the training needs of an athlete.

Researchers indicated that there may be a negative relationship between strength asymmetries and jumping, kicking, and sprint cycling performances (30). However, an additional review of the literature suggested that the link between symmetry and performance is unclear and attributes asymmetries to be a function of limb dominance and long-term sport participation (153). Bailey and colleagues (5) showed significant relationships between isometric force asymmetries and CMJ and SJ performances, indicating that force production asymmetry may be detrimental to jump performance. Similar results were displayed by Bishop and colleagues (28, 29) who also showed that asymmetries may negatively affect sprint and change-of-direction performances. In contrast, other researchers indicated that there were no significant relationships between asymmetry and sprint performance (76, 181) and that asymmetries less than 15% may not affect multidirectional speed (142).

Interestingly, researchers indicated that weaker athletes showed greater asymmetries compared to those who were stronger (7) and that stronger athletes carry over their force production asymmetries to other tasks to a greater extent (6). Furthermore, researchers displayed that bilateral resistance training may lead to reductions in force asymmetry during the ISQT in weaker individuals (18). While unilateral training may target and reduce existing asymmetries (90, 229), this sort of emphasis may ignore the idea that a limb with a reduced force production capacity has a greater window of adaptation, while the other limb's window is smaller and may experience diminishing returns (153). Thus, strength and conditioning professionals are encouraged to focus on the overall development of maximal and rapid force production characteristics as they relate to performance, rather than reducing force production asymmetries.

Reactive Strength

As discussed in chapter 1, *reactive strength* may be defined as the ability to transition quickly between eccentric and concentric muscle actions (299). An important aspect of assessing reactive strength is the overall duration of the stretch–shortening cycle movement. In this light, the two tests that may allow strength and conditioning professionals to assess the reactive strength characteristics of athletes are the drop jump test or the 10/5 repeated jump test (249). Briefly, a drop jump test requires athletes to step off an elevated surface before contacting the ground and jumping as high as possible. In contrast, the 10/5 repeated jump test requires athletes to perform a maximal CMJ before performing 10 rebound jumps, with the top 5 jumps with ground contact times less than 250 ms being used for analysis (56). Both tests can be used to calculate the reactive strength index (RSI) as the ratio between jump height and ground contact time (i.e., the duration from initial foot contact during landing to when the athlete leaves the ground). Researchers have shown that RSI is a reliable variable (80) that may provide strength and conditioning professionals insight regarding an athlete's neuromuscular fatigue (100) and their response to recent training stimuli (160). In addition, RSI may serve as an indicator of acceleration ability (144) and vertical stiffness (128). Researchers have made it easier to calculate RSI by offering resources to analyze raw force–time data in spreadsheets (210). Furthermore, this information may be gathered using resources other than a force platform, such as a switch mat (79); however, it is important to understand the limitations of these resources (124).

It is important to note that RSI values calculated from both the drop jump and 10/5 repeated jump tests only account for approximately 30% shared variance between the tests (249), indicating that the tests should not be used interchangeably. While this information leaves the decision of which test to use up to strength and conditioning professionals and sport scientists, the ability of the athletes to perform each test and the availability of equipment may support either the drop jump or 10/5 repeated jump test.

The use of a drop jump within an athlete monitoring protocol may not be as common as using a CMJ. This may be due to the need to have athletes step off an elevated surface (e.g., box, riser) before contacting the ground and jumping. Although this equipment is not particularly burdensome to set up, it may be difficult to find equipment that is the same height, especially when the team is traveling. Similarly, choosing the height for the drop jump to be performed may also be a challenge for athletes of different abilities. Researchers have shown that the optimal drop jump height may vary based on the strength of the athletes (156). Simply, higher magnitudes of force must be produced when drop heights are higher (79, 294); while stronger athletes can tolerate and benefit their jump height from more rapid stretch–shortening cycle actions, the jumps of weaker athletes may be negatively affected (171). It should be noted that there does not seem to be a consensus on optimal drop jump height. Some researchers have concluded that the optimal one may be the height that maximizes RSI (41, 79), less than 75% of CMJ height (211), between 50% and 100% of CMJ height (212), or between 75% and 125% of SJ height (16).

Regardless of the drop height, it is also important to note that there may be a discrepancy between the surface height and the actual drop height. In fact, researchers have shown that the surface height may be between 26% and 33% higher than the actual drop height (3, 61). Thus, any within- or between-athlete differences related to drop jump RSI may be, in part, explained by actual drop height. While proposed analysis methods may correct for the difference when using force plates (3), these methods may not be applied during field-based testing.

Another consideration regarding test choice are the instructions used during the test. Researchers have shown that the instructions or cueing used during jumping exercises may have a significant impact on force production characteristics (125, 226, 258, 259). For example, Young and colleagues (300) showed that cueing athletes during drop jumps to maximize jump height, minimize ground contact time, or a combination of both may significantly affect how the jumps are performed. Regarding the 10/5 repeated jump test, it is likely that athletes are more familiar with performing a CMJ and would therefore be more consistent in their movements; however, it is important that strength and conditioning professionals and sport scientists remind athletes to minimize their ground contact time while maximizing their jump height during the rebound jumps that follow.

It should be noted that a CMJ variation of RSI was created in 2010 by Ebben and colleagues (74), termed the *modified reactive strength index* (RSImod). When calculating RSImod, time to take-off (i.e., the duration from the onset of the CMJ to when the athlete leaves the ground) is substituted for ground contact time. Researchers have shown that RSImod is a reliable performance variable (74, 127, 252) that may distinguish the differences in performance within teams (9, 242) and between teams (166, 168, 213, 242, 273) and between male and female athletes (24), provides insight into how an athlete achieves a specific jump height (276), has a large relationship with isometric strength (22), and can be used to monitor performance over the course of a competitive season (275). However, it should be noted that only 22% (175) or 20% to 41% (150) of the variance was explained when RSImod was correlated with RSI, indicating that these variants may assess different performance characteristics. This may be explained by the differences in the speed of the stretch–shortening cycle actions during each jump. For example, the drop jump has been classified as a rapid stretch–shortening cycle action (<250 ms) (79, 232, 299), whereas the CMJ may be classified as a "slow" stretch–shortening cycle action given that the individual jump phases may exceed 250 ms (241). Given these differences, it may be concluded that a CMJ does not assess the reactive strength characteristics of an athlete.

ISOKINETIC TESTING

Another method of muscular strength assessment is isokinetic testing (46, 218). As discussed in chapter 1, isokinetic testing includes the assessment of an isolated joint (e.g., knee, elbow) in which the velocity of the movement is controlled. This method of strength assessment is most often performed in a laboratory setting, and it may provide rehabilitation professionals and strength and conditioning professionals with some insight into how an athlete may produce force during controlled conditions. Specifically, isokinetic testing provides information about the torque produced during opposing movements (e.g., knee flexion and extension). Because force production capacity may be hindered following injury, some rehabilitation professionals may use isokinetic testing results as a measure of how an athlete is responding to various stimuli during treatment, as well as if they are ready to progress to higher-intensity stimuli, and as a potential measure of their physical readiness. In an ideal scenario, rehabilitation professionals would share information about the progress of an athlete as they transition from rehabilitation to sport performance training, because a collaborative environment with strength and conditioning professionals would better serve the athlete.

While isokinetic testing may benefit the progress of athletes within the rehabilitation process, there are several limitations to using such testing. First, joints are rarely used in isolation within sporting or athletic contexts. In fact, most movements involve coordinated multi-joint actions to successfully complete a motor task (e.g., rapid extension of the hip, knee, and ankle joints). Thus, assessing the performance of a joint in isolation yields limited information about how an athlete may or may not need to compensate to achieve a similar level of performance before injury. Second, the controlled speeds of isokinetic testing may not simulate the actual performance of a movement in training or sports. Most isokinetic devices range in velocity from 0 to 300° per second; however, this may have limited applicability to actual performance tasks in sports because the joints may need to perform at a wide variety of speeds depending on the task. Finally, an isokinetic device typically has a large footprint, is limited in its versatility, and can be quite expensive (e.g., $50,000-$80,000 USD). In fact, most commercial isokinetic machines exceed the cost of many force plates.

COMBINED ASSESSMENT METHODS

Combined assessment methods have been defined as those that use either multiple variables produced during a single test or a single variable produced during multiple tests (265). Examples of combined assessment methods may include the use of peak force from the CMJ and the IMTP to determine the dynamic strength index (DSI) or the force and velocity at peak power of several loaded jumps to determine the force–velocity profile of an athlete. The following paragraphs highlight various combined assessment methods as they relate to the force production characteristics of athletes and how they may be used to guide training prescription.

DYNAMIC STRENGTH INDEX

The *dynamic strength index* (DSI), also known as the *dynamic strength deficit*, has previously been defined as the ratio of peak force produced during a ballistic task (e.g., CMJ, SJ, or ballistic bench throw) and an isometric task (e.g., IMTP, ISQT, or isometric bench press) (237). Using DSI in this manner provides strength and conditioning professionals with a percentage of maximal force production during ballistic tasks. Researchers have shown DSI to be a reliable measure in athletes when examined in this manner during both lower-body (53, 167, 237, 280, 281) and upper-body (209, 297, 298) exercises.

When used in practice, strength and conditioning professionals may use the DSI of an athlete to identify their theoretical training emphasis. For example, previous authors have recommended a ballistic training emphasis when DSI is 0.60 or less, whereas a maximal strength emphasis may be recommended with a DSI of 0.80 or greater during lower-body exercises (237). Others have suggested that while DSI results are context specific, a DSI of 50% ± 5% for an ISQT:SJ ratio (299) or of 65% ± 5% for an IMTP:SJ ratio (237) may be ideal. In theory, these ratios suggest that a dual emphasis of ballistic and maximal strength should be prescribed. Because no recommendations for the upper body exist, it is unknown whether a scale similar to the lower-body recommendations may be applicable.

Although DSI may provide a general idea of the physical characteristics that should be focused on within training, it is important to put each ratio into context considering that the distinct abilities of certain athletes are what allow them to perform well within their sport or event (209, 274). Examples of athlete DSI and respective training recommendations are displayed in table 9.1. Beyond serving as a guide to training prescription, researchers

TABLE 9.1 Examples of Dynamic Strength Index (DSI) Ratios and Corresponding Training Recommendations

Athlete	DSI ratio	CMJ height (cm)	Relative squat strength (kg/kg)	Training recommendation
A	0.70	60	2.00	Combined maximal and rapid force emphasis
B	0.80	30	1.40	Maximal force emphasis
C	0.50	40	1.80	Rapid force emphasis
D	0.50	20	1.30	Maximal force emphasis

Note: CMJ = countermovement jump.

have shown differences in DSI between collegiate teams (280) and those that demonstrate unique force–time characteristics during the CMJ (167). Interestingly, the relationships between DSI and other performance variables are generally small (233, 274). While DSI ratios may assist strength and conditioning professionals when prescribing specific training emphases, it should be noted that a higher or lower DSI may not necessarily result in an improved performance. This is likely due to the demands of various sporting events and the interplay between the maximal strength and rapid force production characteristics (183, 246, 268, 277, 301). Although researchers have shown that DSI values can change in response to resistance training stimuli (54), longer longitudinal training studies are needed to determine whether DSI can serve as an effective method of training prescription.

It is important that strength and conditioning professionals and sport scientists consider that DSI is determined using two data points that may occur during 1/1,000th of a second, depending on the sampling rate used to collect the data. Researchers have examined DSI using the impulse (96) or mean force (112) produced across various durations (i.e., epochs). The rationale behind using this method is to account for the duration over which force is produced during an entire movement, rather than using a peak value. However, research focused on this method of DSI is in its infancy, and further information is needed to determine its usefulness as an indicator of an athlete's strength characteristics.

FORCE–VELOCITY PROFILING

Strength and conditioning professionals often design resistance training programs with the end goal of maximizing rapid force production and power output (267). Given that power output is the product of force and velocity, an athlete may perform a given movement with a series of loads to determine their force–velocity profile. While similar, researchers have commonly examined the load–velocity profiles of individual exercises to determine the load that maximizes power output, often termed the *optimal load*. For example, the effect of load on maximal power output has been examined in a variety of upper-body exercises (11, 13, 26, 193, 244, 282), lower-body exercises (12, 58, 159, 243, 282), as well as weightlifting movements and their derivatives (48, 49, 55, 58, 123, 126, 159, 177, 179, 253, 254, 255). Although the identification of an optimal load or load range may be important for training prescription, the measurement of the underpinning force production characteristics may provide strength and conditioning professionals with greater insight into how an athlete produces a given power output.

Researchers have used force–velocity profiling as a method of identifying a training need to enhance both jump and sprint performances (121, 221, 222, 223, 224). Previous authors have indicated that jump and sprint force–velocity profiles may be created using a series of loaded jumps or sprint splits, respectively (189). Using this method, the authors proposed that individualized profiles can be generated and compared to an "optimal" profile to determine whether the individual has a force deficit, velocity deficit, or is well balanced. In a follow-up study, researchers designed individualized training programs based on the jump force–velocity profiles of the participants in effort to train the weaknesses and shift their profiles closer toward the optimal profile (118). While the results of the previous study showed improvements in jumping performance compared to the use of training programs not based on force–velocity profiling, additional literature has suggested that force–velocity profiling does not provide insight into how a specific deficit should be trained (222, 223). For example, if a training program is designed for an individual with a velocity deficit, it is important that strength and conditioning professionals do not ignore the maximal strength characteristics of the individual because they may underpin ballistic force production (183, 301). Moreover, it does not appear that vertical jump and sprinting profiles appear to be related (120), which challenges the design of the training program if multiple aspects of performance are not improved simultaneously.

Strength and conditioning professionals should also note that the type of jump performed may modify the force–velocity profile of an individual (119). Simply, jumps performed with a countermovement may allow for greater force and velocity outputs to be achieved compared to those that start from a static position. Thus, it is important to consider what types of jumps are already being performed within the athlete monitoring protocol if force–velocity profiling is used. Finally, researchers have questioned the use of force–velocity profiles on which to base training prescription altogether, due to unreliable profiles (130, 131, 141) and identification of the force or velocity imbalance (284). Thus, further research is warranted to determine the usefulness of identifying the force–velocity profile of each athlete.

While the previously discussed literature has focused primarily on the concentric or propulsive force–velocity profiles of athletes, strength and conditioning professionals may consider the isometric and eccentric (braking) characteristics as well. For example, beyond using the same exercise with multiple loads, a combined assessment approach may also be used to determine the force–velocity profile of an individual. For example, researchers have used a combination of jump squats and IMTPs or ISQTs to examine the force–velocity curve changes following different training stimuli (59, 262) and the adaptations between stronger and weaker individuals (60, 113). Further research has also examined the eccentric force–velocity profiles of individuals (15, 186). Given that most movements in sports include a stretch–shortening cycle, it would seem logical to consider the eccentric, isometric, and concentric force production capabilities of an athlete when generating force–velocity profiles. By doing so, strength and conditioning professionals may be able to identify the weaknesses of an athlete more accurately; however, further research on this topic is necessary.

ECCENTRIC UTILIZATION RATIO

The ability of an athlete to effectively use the stretch–shortening cycle to benefit their performance has been determined by calculating the *eccentric utilization ratio* (EUR) (164) or the prestretch augmentation percentage (290). Due to the volume of existing research and the perfect relationship (i.e., r = 1.00) between variables (276), the following discussion will focus on EUR. Briefly, EUR is calculated as the ratio between the same CMJ and SJ variable (e.g., CMJ peak force/SJ peak force) (164). Beyond serving as an indicator of how well an athlete uses the stretch–shortening cycle, previous authors have noted that EUR may be used to determine the training focus of an athlete (163). For example, researchers have shown that EUR varies between athletes in different sports but also during the training phase that the athletes are in (164). While the authors showed that EUR increased during a preseason phase in both rugby union and field hockey athletes, these findings may be attributed to the characteristics of the prescribed exercises during this phase (e.g., exercises that emphasize rapid force production). Additional studies also support the notion that EUR may be sensitive to different stimuli prescribed throughout the training year (71, 162).

Despite the previous findings, the use of EUR as a guide for training prescription has been questioned. For example, jump height is ultimately determined by the propulsion net impulse produced by an athlete; however, an increase in propulsion displacement caused from initiating the jump using a deeper squat may modify the shape of the impulse by decreasing propulsion force and increasing duration (171). Thus, any superior CMJ performance may not be accredited to the stretch–shortening cycle if there is a difference in propulsion displacement between a CMJ and an SJ. Moreover, a higher EUR may be attributed to a poor SJ performance (287), potentially related to an athlete's inability to rapidly produce force (34) or greater levels of muscle slack (286). In a study with 770 participants (712 athletes and 58 physical education students), researchers showed that EUR may not provide an accurate representation of overall jumping ability; the highest values were shown by the physical education students, whereas some of the lowest values were shown by track and field athletes (133). Moreover, Kozinc and colleagues (132) showed that EUR calculated several ways (i.e., jump height, peak power, peak force, and average power) was only moderately correlated to approach jump, linear sprint, and change-of-direction tasks in volleyball players.

The aforementioned results should lead strength and conditioning professionals to question the ability of EUR to provide insight regarding an athlete's neuromuscular capabilities and prevent the practices of "chasing" specific EUR values. Researchers have shown that jump performances can increase while there may not be a meaningful change in EUR (85, 107). Thus, if EUR values are used in monitoring athletes, it is recommended that strength and conditioning professionals monitor not only the force production characteristics but also other performances (e.g., sprinting, change of direction, conditioning) to provide a more holistic view of their athletes.

TIMING OF TESTING

Regarding the timing of testing and test selection, it is important that strength and conditioning professionals identify what they hope to gain from testing (82). For example, the purpose of testing could be to establish baseline performance capacities of novice athletes, assess the changes in performance following a specific training phase, or determine the impact of a competitive season. However, it should be noted that the implementation of testing, especially maximal effort testing, may be individual, team, and context dependent (38). In this light, the following paragraphs will highlight how the focus of athlete testing may change at various times throughout the training year.

OFF-SEASON

Testing during the off-season serves as a check and balance to determine whether the athletes are adapting to the prescribed training program or if adjustments need to be made. However, it is important to keep the training age of each athlete in mind, given that younger, less experienced athletes may not show any signs of decreased performance because they have not previously received a sufficient training stimulus. Testing during the off-season can be completed on a weekly basis, assuming that noninvasive tests are used. Given that the off-season is typically characterized by larger training volumes, it is important to recognize that certain characteristics such as rapid force production may not show peak adaptations during this time. However, if large increases are shown, strength and conditioning professionals need to investigate whether it was the adaptation to a new stimulus or if the prescribed training had the proper emphasis. Although improvements in rapid force production are generally viewed as positive, it is important that these adaptations are maximized during the proper phases of the competitive year.

PRESEASON

Preseason testing allows strength and conditioning professionals to gain insight into the preparation of an athlete leading into the competitive season. Namely, it is important to examine the key performance indicators related to an improved performance. It should be noted that key performance indicators may vary based on the demands of an athlete's sport or event or position within the sport; however, preseason testing may provide strength and conditioning professionals with an indication of the effectiveness of the prescribed training program.

IN-SEASON

The purpose of in-season testing is to monitor the ability of athletes to maintain, at a minimum, the force production characteristics demonstrated during preseason testing, but also to ensure that the training program provided to the athletes is producing the desired adaptations. Testing during a competitive season may be questioned by the sport coach, especially if they do not see its value. Therefore, it is the job of the sport scientist or strength and conditioning professional to provide the coach with the *why* and be able to demonstrate the value of such testing using team and individual athlete reports. Furthermore, it is important to choose nonfatiguing tests that yield the most impactful information regarding the neuromuscular status of each athlete. Although it may be unavoidable for all athletes, given the added stress of the competitive season, testing should not feel overly burdensome; thus, strength and conditioning professionals are encouraged to promote testing as a part of the athlete's daily routine. Doing so may increase athlete buy-in and yield more insightful information.

POSTSEASON

The primary purpose of postseason testing is to determine what physical attributes were improved, maintained, or lost over the course of a competitive season. Within a strength-specific context, strength and conditioning professionals may be interested in braking and propulsive force production characteristics and their timing. Because of the impact that competitions may have on athlete testing performance, it is recommended that postseason testing take place between 1 and 2 weeks following the conclusion of the season. This may allow athletes to recover from the end of the season physically and mentally before they begin to set goals for the following season.

INTEGRATION OF SPORT SCIENCE AND EVIDENCE-BASED PRACTICE

Monitoring the strength characteristics of athletes may occur on a daily, weekly, or phasic basis. While information produced from testing can benefit strength and conditioning professionals and their ability to modify and prescribe evidence-based programs to address the needs of athletes, it is important to be able to present this information to sport coaches and athletes in a way that they understand. Although these duties may be one of the primary responsibilities of those hired as sport scientists (87, 88), strength and conditioning professionals may have to communicate this information as well. In this light, McGuigan and colleagues (163) highlighted the ability to display performance data in a radar plot format using Z scores (i.e., [athlete score − average score]/standard deviation) (figure 9.10). However, the authors also indicated that the use of

modified Z scores (i.e., [athlete score − benchmark score]/standard deviation) may provide more useful information, especially when a small number of athletes are being tested. The benchmark score may be determined by the strength and conditioning professional, information within the scientific literature, previous testing data, and feedback from the sport coaching staff (163). It is important to note that while this information may provide insight into a single testing session, testing information can be displayed using information from multiple testing sessions to display trended data as well. While graphing Z scores may provide an effective visualization of the athlete testing data, sport coaches and athletes may have a hard time grasping what it means to be X number of standard deviations above or below the benchmark performance. Thus, researchers have promoted the use of T scores (i.e., [Z score × 10] + 50) to provide a more intuitive value that may be

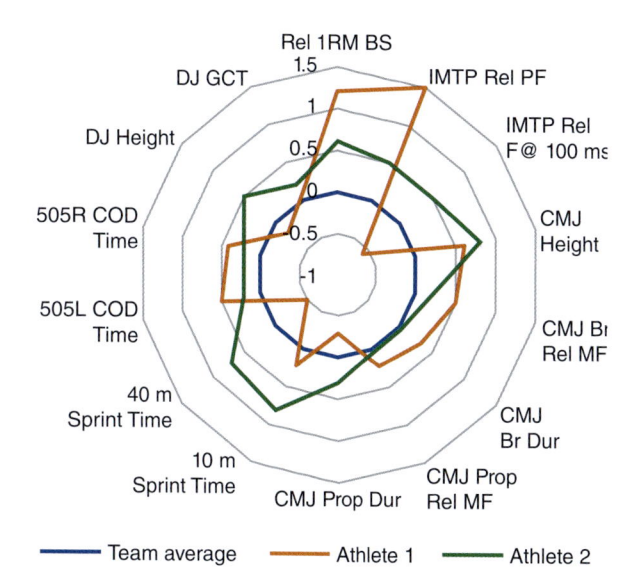

FIGURE 9.10 Strength characteristic visualization. *Notes:* Lines represent different standard deviations from the team average; Rel = relative to body mass; 1RM BS = one repetition maximum back squat; IMTP = isometric mid-thigh pull; PF = peak force; F@100 = force at 100 milliseconds from start of IMTP; CMJ = countermovement jump; Br = braking phase; MF = mean force; Dur = duration; Prop = propulsion phase; 505L COD = 505 change of direction with the left leg; 505R COD = 505 change of direction with the right leg; DJ = drop jump; GCT = ground contact time.

more understandable (173), especially if the values are visualized in a color-coded format, such as a stoplight system.

Beyond visuals, some sport coaches and athletes may prefer to receive raw testing information. If this is the case, it is recommended that strength and conditioning professionals and sport scientists provide percentile rank information (170, 242). Regardless of how data are reported to sport coaches and athletes, data become less valuable if they are not delivered in a timely manner. Previous recommendations suggest that individual and team reports should be delivered within 3 days of testing (163).

As mentioned previously, the ability to visualize information regarding the strength characteristics of athletes may provide insight into their specific needs. It is also important to interpret the data within a greater context by incorporating a variety of different performances, such as the force production characteristics and performances of various 1RMs, vertical jump, sprinting, and change-of-direction tests. This notion has been highlighted within this chapter as well as in previous literature (115, 199). Therefore, it is important for strength and conditioning professionals to identify the specific characteristics needed for athletes within their respective sport or event and to be able to choose, measure, and visualize the chosen variables to determine the current status and individual needs of their athletes.

Take-Home Points

- It is not appropriate to assume that all aspects of strength improve following training; thus, strength and conditioning professionals and sport scientists should choose tests that are specific to an athlete's sport or event.
- When choosing tests for athletes, strength and conditioning professionals and sport scientists should ensure that the tests are reliable and valid while also considering the impact that test familiarity, test order, and standardization of methods may have on the outcomes observed.
- Strength assessments may include isometric, dynamic, reactive, and isokinetic tests; thus, it is important to measure and monitor an athlete's strength characteristics as they relate to the requirements of their respective sport or event. Choosing the testing variables that may reflect a meaningful change in performance should be prioritized.
- Isometric testing may provide a valuable alternative to maximal dynamic strength testing due to its relationships with the latter, its ability to provide insight into the maximal and rapid force production characteristics of an athlete, and its nonfatiguing nature.
- Dynamic testing can include maximal strength, submaximal strength, unloaded and loaded jump, sprint, and change-of-direction tests. Because each athlete requires specific force production capabilities within their sport or event, it is important to limit the number of tests to those that provide the most insight regarding their abilities.
- Sprint and change-of-direction testing may provide limited information on the strength characteristics of an athlete if only a timing system is used. Simply, the quality of the movement is not accounted for.
- Although isokinetic testing may benefit the rehabilitation process, its applicability may be questionable with healthy athletes. Isokinetic testing is limited for healthy athletes because it isolates specific joints, controls the speed of testing, and is quite expensive.
- While use of combined assessment methods such as DSI, force–velocity profiling, and EUR has become common, strength and conditioning professionals should be aware of the shortcomings of each method and put the athlete's testing data into context before making training decisions. Additional context is needed when athlete testing takes place throughout the training year, because training, practice, or competitions may skew certain results.
- Integrating sport science practices may benefit both the strength and conditioning and sport coaching staffs, because the information from athlete testing can help provide an evidence-based rationale for training and practice decisions.

CHAPTER 10

Monitoring and Adjusting Training Loads

Athlete monitoring includes two primary purposes: fatigue management and program efficacy. Managing fatigue requires strength and conditioning professionals and sport scientists to detect acute and accumulative fatigue that may negatively affect the stimulus–recovery–adaptation process, whereas program efficacy provides insight into the extent to which the prescribed training stimuli produced the expected results. Beyond fatigue management and program efficacy, it is also important to consider how training stimuli affect changes in motor capacity (i.e., strength) and motor performance (i.e., learning to use one's newfound strength). Including this information within a monitoring protocol may provide greater insight into the overall development of an athlete regarding their ability to produce force and apply it during specific motor tasks. Although the day-to-day manipulation of volume and intensity may help manage fatigue, it may also allow strength and conditioning professionals to provide an effective training stimulus over time while mitigating performance decrements. Moreover, monitoring the progression or regression of an athlete based on the loads performed in training may provide information regarding the motor skill development of an athlete. The aim of this chapter is to provide an overview of the methods used to monitor and adjust training intensity for strength development.

MONITORING AND LOAD ADJUSTMENT METHODS

There are a wide variety of load adjustment methods that may be used to also monitor athlete performance. The following paragraphs will discuss the benefits and limitations of both objective and subjective methods that strength and conditioning professionals can implement with their athletes in training.

LINEAR LOADING

The concept of *linear loading* focuses on gradually increasing the prescribed load for an exercise beyond what was prescribed during a previous training session (11). From a strength development standpoint, linear loading places a large emphasis on the overload principle (i.e., a training stimulus that may produce an adaptation beyond an athlete's baseline abilities) (25). Although linear loading may benefit the strength characteristics of athletes for a brief period (78, 94), additional variation in loading may be needed to manage fatigue and allow for the recovery–adaptation process to occur (25). Moreover, added variation may assist with an athlete's complex motor skill consolidation where strength and conditioning professionals may observe an improvement in lifting performance between training sessions (54, 81). Practically speaking, a greater emphasis may be placed on recovery–adaptation as a greater emphasis is placed on planned load variation (i.e., increases or decreases in load). However, if the linear loading method is used for an extended period of time (e.g., months to years), the ability of an athlete to recover and adapt to the prescribed training stimuli may be impaired, which in turn could lead to the stagnation of performance, nonfunctional overreaching, and, potentially, overtraining (figure 10.1) (11, 20, 94). Therefore, because linear loading emphasizes a continual increase in loading but provides very little load variation to combat the accumulated fatigue of an athlete, it may not serve as an effective monitoring strategy.

Key Point

If the linear loading method is used for an extended period of time (e.g., months to years), the ability of an athlete to recover and adapt to the prescribed training stimuli may be impaired, which in turn could lead to the stagnation of performance, nonfunctional overreaching, and, potentially, overtraining.

TWO-FOR-TWO RULE

The *two-for-two rule* is based on the ability of an athlete to perform at least 2 repetitions over the prescribed number of repetitions in their final set in 2 consecutive training sessions (5). If this is accomplished, strength and conditioning professionals would then increase the load during the next session that the same exercise is prescribed. Because the load is increased at this point, this method of monitoring and adjustment is entirely based on the completion of the assigned repetition scheme. For example, if an athlete performs 5 repetitions on their final set of a deadlift after being prescribed 3 sets of 3 repetitions (i.e., 3 × 3) in 2 consecutive sessions, the prescribed load should be increased during the next training session where the deadlift is prescribed.

It should be noted that although the two-for-two rule may allow novice athletes to improve their strength characteristics, it may also promote training to failure and does not account for the athlete's technique, training phase goals, or relative intensity. For example, it is important that strength and conditioning professionals first prioritize technique stability with novice athletes before increasing the demand of the task by prescribing heavier loads. In addition, if one of the training phase goals of the athlete is to improve their strength characteristics, the prescribed load may be too light to elicit the desired adaptations. In this case, if an athlete can perform at least 2 more repetitions than what has been prescribed in consecutive training sessions, additional variation in loading during each set may be beneficial. Modifying the load in this manner provides slight alterations in the task demand, which may also benefit the skill acquisition of the athlete (e.g., varied practice) (83).

Finally, it is important to note that the ability of an athlete to perform additional repetitions beyond what has been prescribed may be by design. For example, strength and conditioning professionals may want an athlete to be able to perform additional repetitions beyond what was initially prescribed (i.e., repetitions in reserve) in an effort to continually provide a progressive overload stimulus while simultaneously managing fatigue and avoiding training to failure. Further information on repetitions in reserve is provided later in this chapter.

PERCENTAGE OF 1 REPETITION MAXIMUM

The most common method of prescribing resistance training loads is by expressing the loads as

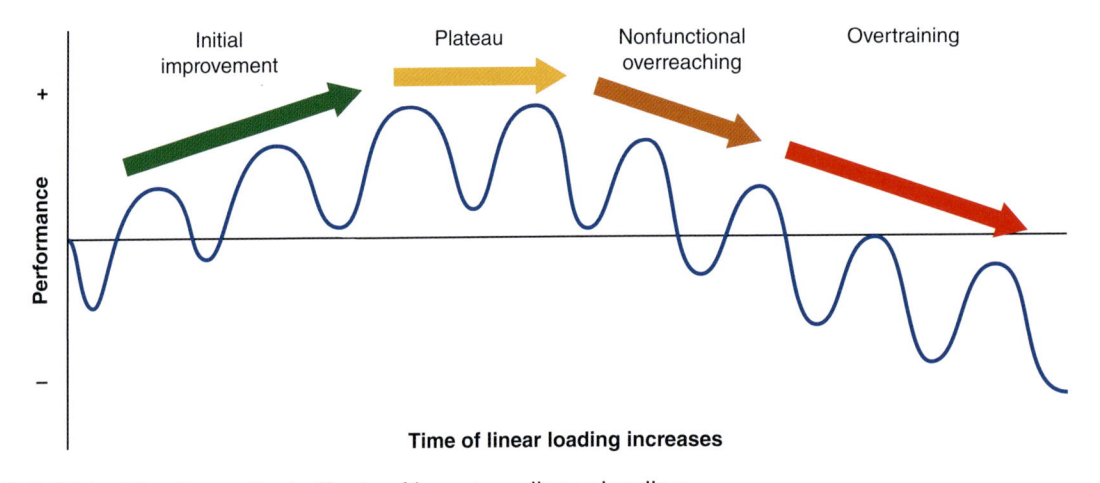

FIGURE 10.1 The theoretical effects of long-term linear loading.

Adapted by permission from T.J. Suchomel, S. Mimphius, C.R. Bellon et al., "Training for Muscular Strength: Methods for Monitoring and Adjusting Training Intensity," *Sports Medicine*, 51 (2021): 2051-2066, Springer Nature.

a percentage of an athlete's 1 repetition maximum (%1RM). Using this method, strength conditioning professionals typically determine the 1RM of select exercises by identifying the heaviest weight an athlete can lift for 1 repetition with proper technique (104). Although performing 1RM testing is logically the most accurate method of determining a 1RM, it should be noted that maximal loads may also be estimated as a function of the load lifted by performing multiple repetition testing (e.g., 93% of the 1RM = 3RM) (12). After the 1RM of an athlete is established, training loads are based on %1RM according to the desired set and repetition scheme and strength characteristics being targeted (12). While 1RM testing may be viewed as strenuous and less safe compared to multiple RM testing (e.g., 3RM or 5RM), strength and conditioning professionals should be aware that 1RM prediction becomes less accurate as higher repetition RMs are performed (64).

While training prescription based on an athlete's %1RM has been shown to be an effective method, it is important that strength and conditioning professionals are aware of its limitations as well. For example, researchers have indicated that an athlete's 1RM may fluctuate based on their physiological or psychological status (10, 79), training-related stressors (e.g., accumulated fatigue) (79), or other life-related stressors (e.g., sleep deprivation, nutrition) (10, 65). A second limitation of this method is that there is considerable variation in the number of repetitions an athlete can perform at a given %1RM (3, 49, 58, 96) or during different exercises performed at the same %1RM (105). In fact, a 2024 meta-regression analysis of multiple studies showed that the standard deviation of the estimated number of repetitions performed at a given load percentage increased as the number of repetitions increased (84). Moreover, the authors concluded that the number of estimated repetitions performed at a given percentage varied between exercises such as the bench press and leg press. Finally, researchers have also reported that the number of repetitions performed at a given %1RM may also be influenced by exercise type, gender or sex, and training status (49). It is possible that the previously discussed limitations may lead to an inconsistent training stimulus and, by extension, less than optimal training adaptations. Taking all of this into account, strength and conditioning professionals who prescribe training using %1RM may benefit from combining it with another load adjustment method to lessen the impact of the highlighted limitations and account for the physiological status of their athletes (102).

> **Key Point**
>
> An athlete's 1RM may fluctuate based on their physiological or psychological status, training-related stressors (e.g., accumulated fatigue), or other life-related stressors (e.g., sleep deprivation, nutrition).

REPETITION MAXIMUM ZONES

Another method of load prescription is the use of repetition maximum (RM) zones. This method typically involves selecting the heaviest loads that may be lifted for a specific repetition range (e.g., 4-6 repetitions), often with the intent of reaching muscular failure on the final prescribed set (15, 17, 87). Advocates of RM zone training suggest that it removes the limitations related to prescription based on %1RM because the prescribed loads for each exercise are adjusted based on the physiological status of the athlete for each exercise (17). Because RM zone prescription allows for loads to be determined without 1RM testing, it is understandable why this method may be appealing during large group training sessions.

Although researchers have shown that training with RM zones may improve maximum strength (15, 50, 94), it is important to highlight the potential limitations of this method as well. The primary limitation to RM zones is that training prescription often requires a relative maximum effort (87). This becomes a potential issue especially when developing rapid force production characteristics, which may be optimized when using a combination of both heavy and light days (45). Simply, if maximum effort is required during each day, the concept of a light day diminishes because performing multiple sets to failure qualifies more as a heavy day. Chronic training in this manner may diminish the strength and conditioning professional's ability to manage fatigue and, in turn, could lead to nonfunctional overreaching and potentially overtraining (56, 92). For example, Carroll and colleagues (17) demonstrated that RM zone training for 10 weeks resulted in significantly greater training strain during the final 7 weeks of training, compared to prescribing training using relative intensities as well as heavy and light days using set–repetition best (SRB) load prescription. The RM zone group also displayed moderate to large reductions in the rate of force development at 50 and 100 ms, whereas the SRB group did not. In contrast, the SRB group displayed moderate to large improvements in their absolute and allometrically scaled strength ($g = 1.05$-1.26). Painter and colleagues (87) indicated that similar improvements in maximal strength and rapid force production were produced following training with either RM zones or SRB; however, RM zone training required participants to perform a significantly larger volume–load ($g = 1.69$, large) and number of repetitions ($g = 3.89$, very large) compared to the SRB group, indicating that latter group was more efficient in their training.

It should be noted that not all RM zone training is prescribed with the intent to train to failure. For example, strength and conditioning professionals may prescribe loads that coincide with the athlete's 4RM to 6RM (i.e., 85%-90% of the 1RM) to be performed for only 3 repetitions. While this variation of RM zone prescription may prevent training to failure, it is important to prescribe zones that coincide with the sought-after adaptations. From a strength development perspective, zones ranging in intensity from 1RM to 6RM may maximize peak force development, whereas the range of RMs may be much wider for rapid force production because the specific nature of exercises may differ significantly (e.g., back squat vs. jump squat or clean pull from the floor vs. hang high pull). Despite these general recommendations, strength and conditioning professionals should still attempt to provide their athletes with a progressive overload stimulus to achieve these adaptations across a range of loads.

Key Point

RM zones ranging in intensity from 1RM to 6RM may maximize peak force development, whereas the range of RMs may be much wider for rapid force production because the specific nature of exercises may differ significantly.

While training prescription based on RM zones may address some of the %1RM limitations, chronic training using maximal or near-maximal loading may result in a performance plateau or even negative training adaptations (126). Although this may be mitigated by prescribing RM zones that prevent training to failure, it is suggested that prescription methods that allow for the use of submaximal loading, as well as heavy and light days, may facilitate greater long-term adaptations in an athlete's force production characteristics.

RATING OF PERCEIVED EXERTION

The original *rating of perceived exertion* (RPE) scale was developed in the 1970s to serve as a subjective complement to other objective measures (e.g., behavioral and physiological) during the completion of physical work and ranged from 6 to 20 (13), likely due to its relationship with heart rate ranges during activity (i.e., 60-200 beats per minute). Although the original scale continues to be used during aerobic activities, a simplified version ranging from 0 to 10 was also developed (14) and has since been used as a subjective measurement of the resistance training intensity of each set of exercise (62, 97) and entire resistance training sessions (i.e., session RPE) (21, 76, 123). Related to the other methods discussed in this chapter, this subsection will focus on RPE following individual training sets rather than session RPE. However, additional content on the differences between set RPE and session RPE as resistance training monitoring tools can be found in a previous review (102).

When using RPE as a load prescription method, strength and conditioning professionals typically prescribe ranges of intensity. For example, athletes may be prescribed 3 sets of 5 repetitions with an RPE of 8 to 9 instead of a range of loads based on %1RM. Interestingly, researchers have shown that prescribing intensities based on RPE produced similar improvements in strength compared to intensities prescribed based on %1RM (46). However, it should be noted that limited research has examined the efficacy of this approach between different populations, including by gender or sex or by training status; thus, caution should be taken when interpreting these findings. Furthermore, due to the existence of multiple RPE scales, additional research is needed to determine which scale may produce the greatest strength benefits.

Researchers have shown inverse relationships between mean barbell velocity and RPE in both experienced and inexperienced lifters (144) and during the deadlift and bench press exercises (46). Therefore, because RPE is based on subjective feedback from the athletes, it may be implemented more effectively if it is combined with methods that include objective measurements such as %1RM or velocity zones when using velocity-based training (VBT) (120). By doing this, strength and conditioning professionals may prescribe specific loads using %1RM and adjust the loads using RPE as a method of autoregulation to adjust the training loads. For example, if an RPE range of 8 to 9 is prescribed with a load of 90% of the 1RM (i.e., 4RM) for a set of 3 repetitions and the athlete reports an RPE of 7, the load may be increased to provide a more effective training stimulus. Using this method, RPE provides information related to the perceived maximal capacity of each athlete on a given day of training. It should be noted, however, that using RPE in the manner described may be difficult with a large number of athletes or a small strength and conditioning staff (102).

The primary limitations of prescribing and monitoring training loads using RPE are the familiarity of the athletes with the scale and, by extension, the underreporting of athletes. For example, researchers have shown that athletes reported submaximal values (i.e., RPE < 10) despite failing during an individual set, which indicates maximal exertion (i.e., RPE = 10) (44, 105). These findings are supported by additional evidence showing that athletic experience affects the ability to accurately gauge exertion (9). Because the inability to accurately gauge exertion may be due to neuromuscular inefficiencies or coordination issues, athlete autonomy may need to be limited when selecting loads until athletes become proficient in accurately reporting RPE following sets of exercise. However, athletes may not display this proficiency until they experience a maximal or near-maximal set of exercise and relate it back to the RPE scale.

Key Point

The primary limitations of prescribing and monitoring training loads using RPE are the familiarity of the athletes with the scale and, by extension, the underreporting of athletes.

REPETITIONS IN RESERVE

Similar to RPE, *repetitions in reserve* (RIR) is a subjective measurement in which an athlete reports the estimated number of repetitions they could have performed beyond those performed in the completed set (table 10.1) (66). The use of RIR has been shown to be a valid and reliable method for prescribing loads during resistance training sessions (4, 40, 66) and is highly correlated with RPE following an exercise set (42, 43, 44, 144). Furthermore, researchers showed strong relationships between mean barbell velocity and RIR in the bench press (86) as well as in the chest press and leg press (43). It should, however, be noted that the strength of the relationship between RIR and RPE may be influenced by both training experience and intensity. For example, researchers have shown strong positive relationships between estimated RIR and the actual RIR during both the bench press and squat within a population of trained bodybuilders (44). Additional findings support the notion that RIR becomes more accurate when exercise sets are performed near failure (42, 73, 143).

Researchers have compared the use of prescribing loads using RIR versus fixed %1RM (40) and RM training (4). Graham and Cleather (40) indicated that although both the RIR and fixed %1RM groups improved back squat and front squat maximal strength, the RIR group showed significantly greater improvements. Similar findings were shown by Arede and colleagues (4), who demonstrated that resistance training based on RIR may produce greater improvements in bench press strength compared to RM training. Collectively, these findings show the effectiveness of implementing RIR as a load prescription method; however, like RPE, additional research is still needed with various populations, because studies using other methods of load prescription vastly outnumber those using RIR.

TABLE 10.1 Methods of Subjective Loading and Correlated Relative Percentages

Load Effort	Rating of Perceived Exertion	Repetitions in Reserve	Relative Percentage
Maximal	10	0 — No additional load	100
	9.5	0 — Could increase load	
Very heavy	9	1	95-99
	8.5	1-2	
Heavy	8	2	90-94
	7.5	2-3	
Moderate-heavy	7	3	85-89
	6-7	3-4	
Moderate	5-6	4	80-84
		4-5	
Light-moderate	5	5	75-79
Light	4-5	5-6	70-74
Very light	3-4	Light effort	<70
Very, very light	1-2	Little or no effort	<60

Adapted by permission from S. McKeever and R. Howard, "Resistance Training," in NSCA's Guide to High School Strength and Conditioning, edited for the National Strength and Conditioning Association by P. McHenry and M.J. Nitka (Human Kinetics, 2022), 240. Adapted from DeWeese, Sams, and Serrano (2014). Adapted by permission from Zourdos, Klemp, Dolan, et al. (2016).

When implementing RIR with athletes, small ranges are typically prescribed (e.g., 1-2, 3-4). This method can theoretically be applied in a progressive manner with any set and repetition scheme (e.g., 3 sets of 5 repetitions; week 1: 3-4 RIR, week 2: 2-3 RIR); however, it should be noted that RIR accuracy may diminish during higher repetition sets (e.g., >12 repetitions) and with loads that correspond to relative low intensities (e.g., >4 RIR) (143). Although this may not be an issue with heavy lifts, the inability to accurately monitor and adjust training loads with lighter, more ballistic exercises (e.g., jump squat) that do not approach muscular failure is concerning, given the ability of these exercises to train rapid force production characteristics. This, in turn, may prevent strength and conditioning professionals from prescribing heavy and light days. Thus, RIR may be implemented more effectively if it is combined with other methods such as %1RM or SRB.

SET–REPETITION BEST

Set–repetition best (SRB) is a method of load prescription that uses relative intensities (percentage ranges) where the maximum weight for a given athlete is estimated based on their performance of a given set and repetition scheme (26, 109). SRB loading prescriptions use 5% ranges to allow strength and conditioning professionals to assess their athletes via observation and athlete feedback to provide an opportunity for autoregulation and confirmation of the prescription of loads (figure 10.2). As discussed previously, researchers have shown that SRB prescription may produce greater skeletal muscle fiber (e.g., Type I and II hypertrophy, myosin heavy chain) and strength (e.g., isometric peak force and rapid force production) adaptations compared to RM zone training (16, 17). Furthermore, the authors justified their findings by suggesting that they were the result of variation in workload distribution through the use of heavy and light days (16) and increase in training strain during RM zone training (17). Due to its autoregulatory characteristics and the abundance of literature supporting its use (16, 17, 53, 87, 110, 116, 117), SRB may serve as an effective load prescription and adjustment strategy during resistance training sessions.

Although prescribing loads based on %1RM may allow strength and conditioning professionals to use a given RM (e.g., 2RM, 4RM) to estimate the 1RM and RM for various repetitions (12), a potential limitation to this method is that the estimations are based on the performance of a single set. Because

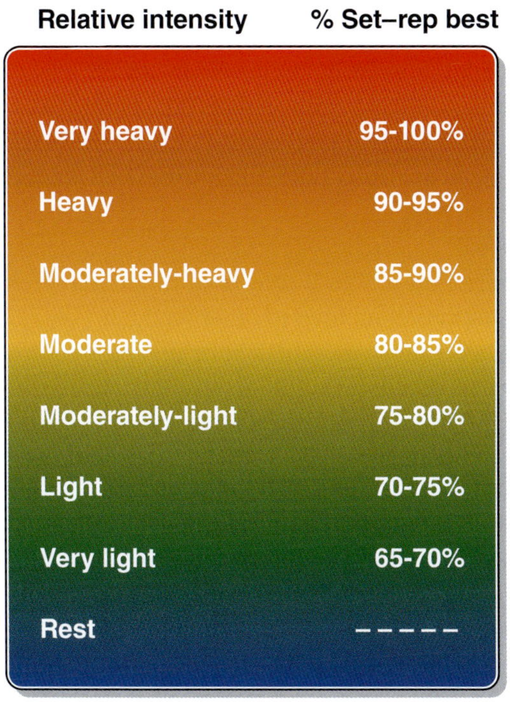

Relative intensity	% Set–rep best
Very heavy	95-100%
Heavy	90-95%
Moderately-heavy	85-90%
Moderate	80-85%
Moderately-light	75-80%
Light	70-75%
Very light	65-70%
Rest	– – – –

FIGURE 10.2 Set–repetition best relative intensities.

Adapted by permission from T.J. Suchomel, S. Mimphius, C.R. Bellon et al., "Training for Muscular Strength: Methods for Monitoring and Adjusting Training Intensity," *Sports Medicine,* 51 (2021): 2051-2066, Springer Nature.

resistance training is typically prescribed using multiple sets, it is important to consider the fatigue accumulated across each set when estimating the 1RM of a particular exercise, especially those with large ranges of motion. SRB allows strength and conditioning professionals to use the loads performed in previous training sessions to adjust the maximal loads of an athlete on a weekly basis (26, 27). In addition, SRB allows for training loads to be estimated when prescribing different set and repetition schemes (table 10.2) (109). It should be noted that estimated maximal loads are based on *ideal conditions*, in which the athlete has become accustomed to training with a specific set and repetition scheme. However, when prescribing loads based on SRB, strength and conditioning professionals should note that 90% for 3 sets of 5 repetitions is not based on %1RM but rather the athlete's previous 3 × 5RM load. Although this method of load prescription may be confusing at first, SRB allows for loads to be prescribed in a more fluid manner based on an athlete's training (relative intensity) rather than on maximal testing (absolute intensity).

Key Point

The SRB method allows strength and conditioning professionals to use the loads performed in previous training sessions to adjust the maximal loads of an athlete on a weekly basis.

Learning to use SRB may be challenging for strength and conditioning professionals at first, especially with novice athletes. Like other load prescription methods, a general recommendation would be to load conservatively before progressing to heavier loads later in a training phase. If resistance training is being prescribed using a 3:1 weekly loading paradigm (i.e., 3 weeks of progressive increases and 1 drop or recovery week) during a strength-focused block of training, the first 3 weeks may serve as an opportunity for observation and input from the athlete before progressing the load. However, strength and conditioning professionals may also use the previous training block as a starting point for a subsequent one. For example, if a general strength block (e.g., 3 × 5) is followed by an absolute strength block (e.g., 3 × 2), the heaviest weights lifted by the athlete may serve as the starting point before progressing during subsequent weeks of training. Another advantage to prescribing loads based on SRB is that there is a built-in goal-setting component in which each athlete may remember their "best" performances during a specific set

TABLE 10.2 Approximate Percentage Changes for Squatting and Pulling Exercises Using Different Set-Repetition Schemes

Set-repetition scheme	Load % change from 3 × 2
3 × 2	—
3 × 3	↓ 5%
3 × 5	↓ 15%
5 × 5	↓ 17.5%
3 × 10	↓ 25%
5 × 10	↓ 27.5%

Modified from Stone and O'Bryant (109).

Strength and conditioning professionals should consider using approximately 10% lower changes in percentages for upper body exercises. A decrease of approximately 10% may exist between an individual's assessed 1 repetition maximum and the 3 × 2 load.

Reprinted by permission from T.J.Suchomel, S. Nimphius, C.R. Bellon et al., "Training for Muscular Strength: Methods of Monitoring and Adjusting Intensity," *Sports Medicine,* 51(2021): 2056, Springer Nature.

and repetition scheme, which provides a target for them to exceed during future training. As a result, SRB may serve as an effective monitoring tool in determining whether an athlete is responding to training in an expected manner or if the strength and conditioning professional needs to modify their training to prevent maladaptation.

AUTOREGULATORY PROGRESSIVE RESISTANCE EXERCISE

Originally termed *progressive resistance exercise*, the origins of *autoregulatory progressive resistance exercise* (APRE) date back to its use in treating orthopedic injuries from World War II (22). The original training prescription required individuals to perform 2 warm-up sets based on 50% and 75% of their 10RM before performing a final set to failure, with the goal of completing at least 10 repetitions (23, 24). The 10RM load for the subsequent training session would then be adjusted based on the performance of the final set. This method was expanded to include a fourth set and an adjustment chart (59) and a protocol that included heavier loads using a 6RM (60). A final addition included a 3RM protocol to shape what is now viewed as APRE (107). As its name suggests, APRE is used to adjust resistance training loads based on an athlete's training readiness during each session (i.e., autoregulation) (72). The loading and adjustments for each APRE protocol are displayed in figures 10.3 and 10.4, respectively.

While researchers have shown that autoregulatory prescription methods may lead to greater strength adaptations compared to other methods (18, 72, 77, 142), it is important to differentiate APRE compared to others that require additional data collection (e.g., RPE, RIR, and VBT). Relatively few studies have examined the effects of APRE compared to other methods of load prescription. However, researchers have shown that load prescription based on APRE may produce greater back squat, bench press, and hang clean strength (70, 72, 138), as well as bench press strength–endurance (72) compared to linear loading. Strength and conditioning professionals should interpret the existing studies with caution, because there may be a plateau effect during linear loading as indicated earlier and by other researchers (48). Furthermore, there are no other comparisons between APRE and other methods of load prescription, which highlights a need for additional research. Collectively, it appears that APRE may aid strength and conditioning professionals in monitoring and

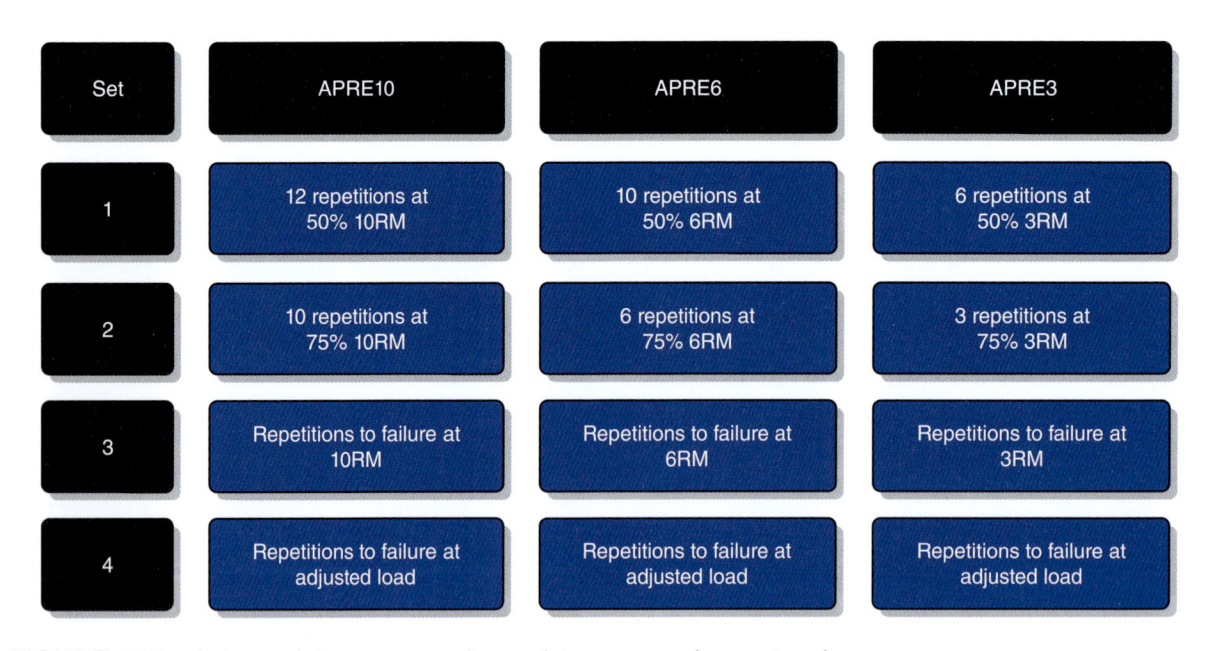

Set	APRE10	APRE6	APRE3
1	12 repetitions at 50% 10RM	10 repetitions at 50% 6RM	6 repetitions at 50% 3RM
2	10 repetitions at 75% 10RM	6 repetitions at 75% 6RM	3 repetitions at 75% 3RM
3	Repetitions to failure at 10RM	Repetitions to failure at 6RM	Repetitions to failure at 3RM
4	Repetitions to failure at adjusted load	Repetitions to failure at adjusted load	Repetitions to failure at adjusted load

FIGURE 10.3 Autoregulatory progressive resistance exercise protocols.

Adapted by permission from T.J. Suchomel, S. Mimphius, C.R. Bellon et al., "Training for Muscular Strength: Methods for Monitoring and Adjusting Training Intensity," *Sports Medicine*, 51 (2021): 2051-2066, Springer Nature.

APRE10

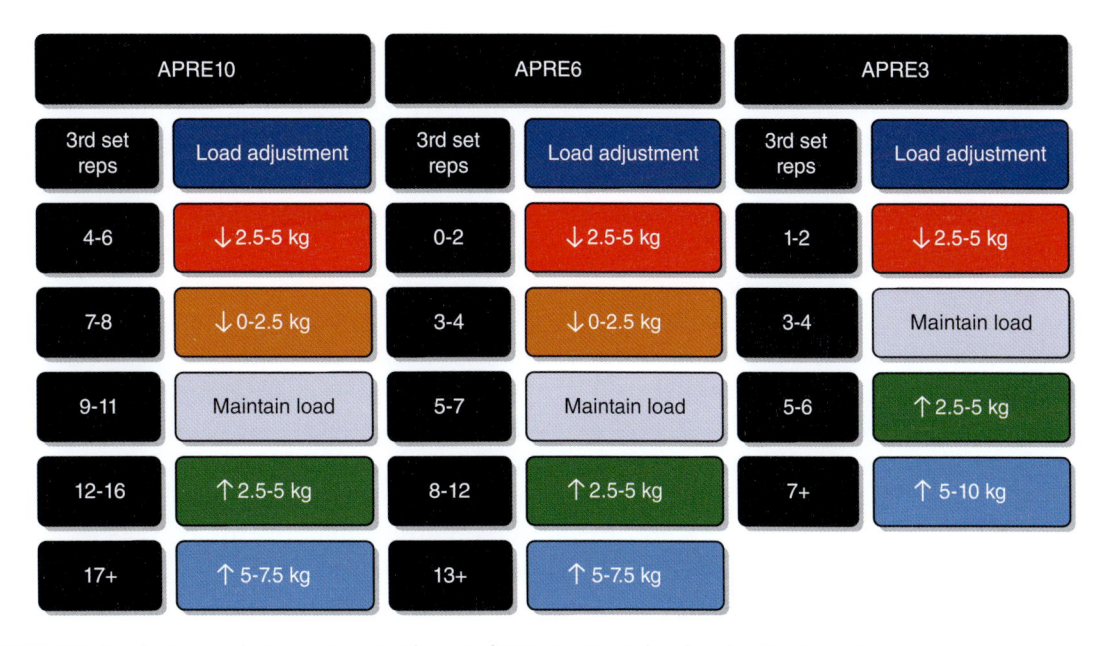

3rd set reps	Load adjustment
4-6	↓ 2.5-5 kg
7-8	↓ 0-2.5 kg
9-11	Maintain load
12-16	↑ 2.5-5 kg
17+	↑ 5-7.5 kg

APRE6

3rd set reps	Load adjustment
0-2	↓ 2.5-5 kg
3-4	↓ 0-2.5 kg
5-7	Maintain load
8-12	↑ 2.5-5 kg
13+	↑ 5-7.5 kg

APRE3

3rd set reps	Load adjustment
1-2	↓ 2.5-5 kg
3-4	Maintain load
5-6	↑ 2.5-5 kg
7+	↑ 5-10 kg

FIGURE 10.4 Autoregulatory progressive resistance exercise load adjustments.

Adapted by permission from T.J. Suchomel, S. Mimphius, C.R. Bellon et al., "Training for Muscular Strength: Methods for Monitoring and Adjusting Training Intensity," *Sports Medicine*, 51 (2021): 2051-2066, Springer Nature.

adjusting loads for athletes due to the integrated 10RM, 6RM, or 3RM lifts within each protocol as well as the individualized load adjustments during each training session. However, as with any method, it is important to understand the limitations and additional considerations of APRE before implementing it as a prescription strategy.

Key Point

APRE may aid strength and conditioning professionals in monitoring and adjusting loads for athletes due to the integrated 10RM, 6RM, or 3RM lifts within each protocol as well as the individualized load adjustments during each training session.

Strength and conditioning professionals interested in using APRE for load prescription and monitoring should consider its unique facets, such as technical failure during multiple sets, psychological momentum, and alterations to the load adjustment chart (69). Regarding technical failure, strength and conditioning professionals must ensure that athletes maintain proper technique during each of the RM sets. It has been recommended that if an athlete sacrifices proper technique to perform additional repetitions, the set should be stopped and loads adjusted accordingly (69). Strength and conditioning professionals must also consider that APRE may be used as a motivational tool for their athletes. Specifically, as athletes become familiar with the load adjustment protocols, APRE can help encourage athletes to perform additional repetitions so that they can increase their training loads during subsequent training sessions. A third consideration of APRE is to ensure that the prescribed loads are put into context relative to the individual athlete. For example, prescribing an additional 5 to 7.5 kg may be a much larger relative percentage for those who cannot lift as much in terms of absolute weight (e.g., 100 kg 6RM = 5%-7.5% increase; 200 kg 6RM = 2.5%-3.75% increase) (120). Therefore, it is recommended that strength and conditioning professionals be aware of the maximal strength characteristics of their athletes and adjust the loads on an individual basis. Finally, those interested in using APRE should consider the autoregulation literature that supports the use of objective measurements (e.g., mean barbell velocity) over subjective methods (e.g., RPE) (103). Specifically, strength and conditioning professionals should consider the use of additional methods that provide objective measurements of performance (e.g., VBT) to supplement APRE and ensure proper load prescription.

VELOCITY-BASED TRAINING

As its name implies, *velocity-based training* (VBT) refers to the measurement and use of velocity to prescribe loads during resistance training sessions. This load prescription and monitoring method may use a variety of devices, such as linear position transducers or inertial measurement units, to measure or calculate the displacement and velocity of exercise repetitions (e.g., mean and peak barbell velocity). Three primary benefits of VBT discussed within the scientific literature include instantaneous feedback, the potential to predict the 1RM loads of athletes during certain exercises, and the ability to use velocity loss thresholds to monitor and adjust the loads used in training (figure 10.5) (131).

Using VBT as a method of feedback has led to improvements in velocity and power output up to approximately 10% (135, 136, 137), which could have been due to intrinsic or extrinsic motivational factors such as competition within or between athletes (135, 136). Shattock and Tee (103) also showed countermovement jump (CMJ) height improvements with VBT feedback; however, these adaptations may have been due to improvements in back squat strength. It should, however, be noted that the authors showed no changes in CMJ force–time characteristics despite increases in jump height (103).

As mentioned previously, a benefit of VBT feedback is that it may increase the competition between athletes to achieve higher velocities. However, it is important to note that strength and conditioning professionals must choose when an athlete receives feedback from VBT devices, which may affect athlete autonomy and potentially skill acquisition (141). Furthermore, the use of velocity feedback may be questioned if the goal is to also improve the motor skill, which is likely a focus during complex exercises such as weightlifting derivatives (e.g., hang power clean, snatch pull from the floor). For example, improvements in velocity have been shown to improve through technical changes rather than velocity-focused instruction (139). These findings

	➕	➖	💭
Real-time feedback	↑ Athlete motivation ↑ Competition in weight room	Loss of focus on exercise technique to achieve faster velocities	Consistent feedback needed during exercise sets Best implemented with heavy, multi-joint movements or ballistic exercises
Daily 1RM prediction	Training percentages based on current state of athlete	↓ Athlete effort during warm-ups may underestimate 1RM Overestimated 1RM using load-velocity profiles ↑ Time needed to determine load-velocity profiles	General equations using velocity 1RM from all athletes simplify load-velocity assessments Two-point method of assessment may be useful for upper but not lower body exercises
Velocity loss thresholds	↑ Ability to monitor fatigue during training sets Use of "flexible" repetition schemes to compensate for fatigue during training	↑ Training time due to frequent load adjustments Individual athlete testing variability ↑ Time needed to determine velocity baselines	High volume (accumulation) phases may warrant the maintenance of velocity despite dropping below threshold "Flexible" repetition schemes may modify the training stimulus of the targeted adaptation

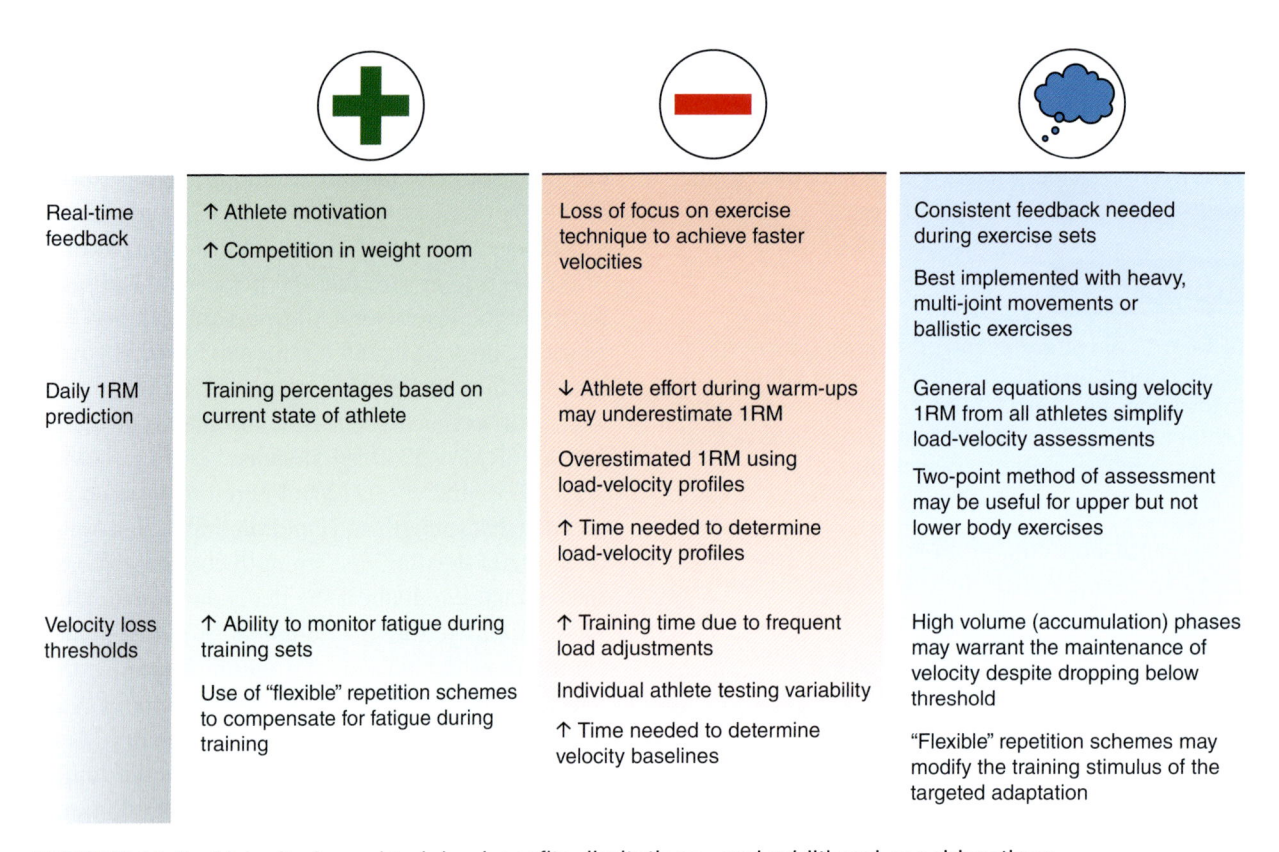

FIGURE 10.5 Velocity-based training benefits, limitations, and additional considerations.

Adapted by permission from T.J. Suchomel, S. Mimphius, C.R. Bellon et al., "Training for Muscular Strength: Methods for Monitoring and Adjusting Training Intensity," *Sports Medicine*, 51 (2021): 2051-2066, Springer Nature.

are likely because weightlifting derivatives have more complex coordination strategies compared to other lifts (68). Therefore, while velocity feedback during traditional resistance training movements (e.g., squats, presses, and pulls) may be effective, further research is needed to determine whether velocity feedback benefits skill acquisition and coordination changes during more complex lifts.

Because lifting velocity decreases as the external load increases (55, 111, 112, 113, 130), it may be possible to predict the 1RM of a lift using either general (38) or individualized load–velocity profile equations (8, 98). Briefly, load–velocity equations may be produced by recording mean barbell velocity during single repetitions (general) of an exercise or repetitions performed with several submaximal loads (individualized). While general and individualized load–velocity equations may allow strength and conditioning professionals to estimate the 1RM of an athlete, both have limitations. A primary limitation of general prediction equations is that exercise type (19, 47, 100, 125), technique (35, 88), gender or sex (6, 127), velocity measurement device (7, 31, 90, 125), and between-athlete variation (91) may affect the relationship between single repetition mean barbell velocity and 1RM percentage. Regarding individual load–velocity 1RM prediction equations, researchers have shown that mean barbell velocities achieved at 1RM may not be reliable (8, 36, 98), which has led others to recommend the use of general prediction equations (131). It should be noted, however, that the use of minimal velocity threshold references values within 1RM prediction equations may produce moderate to high absolute error as well (32).

Additional literature has discussed the use of a two-point method in which only two velocity measurements at different loads are used to predict the 1RM of an athlete (34, 75); however, it is important that the loads used within the equation are on opposing ends of the load–velocity spectrum (e.g., 20% and 70% of the 1RM) (89). Furthermore, in a 2023 study, researchers concluded that load–velocity prediction equations may overestimate the 1RM of an athlete regardless of the selected modeling approach (41). Based on the available literature, it is important that strength and conditioning professionals exercise caution if they choose to predict the 1RM of an exercise using either general or individualized load–velocity prediction equations.

Velocity loss thresholds are another way in which strength and conditioning professionals may use VBT when monitoring resistance training sessions. Because traditional exercise sets require the completion of a given number of repetitions at the same load, it is possible that large reductions in velocity may occur throughout the set (128, 134). Decreases in muscle fiber shortening velocity have been linked to exercise-induced fatigue and additional decrements in movement velocity (39, 99); thus, the loss of velocity throughout an exercise set may serve as an indicator for resistance training volume and intensity prescription (131). For example, Weakley and colleagues (132) showed that 10%, 20%, and 30% velocity loss thresholds displayed linear reductions in neuromuscular function but increases in perceived effort and metabolic responses. This same research group also showed the ability to maintain velocity and power outputs using the same velocity thresholds (133). From a practical standpoint, it is possible to monitor fatigue during single or multiple sets using various velocity loss thresholds, which may allow for the use of *flexible repetition schemes* that use barbell velocity as a measurement of exercise intensity. However, it is important that strength and conditioning professionals understand that using such a method may modify the prescribed training stimulus and alter the focus of the training session or phase (52). For example, if the velocity loss threshold dictates that athletes should perform fewer repetitions than originally prescribed (e.g., 2 repetitions instead of 5), the adjusted stimulus may be less efficient at targeting specific strength characteristics. Similarly, it should be noted that the same velocity threshold may not be appropriate for all individuals, because load, sex, training age, and personality traits may contribute to different responses to the stimulus (57).

While the use of VBT technology may be appealing, strength and conditioning professionals should be aware of the potential disadvantages. First, regarding skill acquisition and coordination, the understanding of frequency of velocity feedback, the type of feedback given to each athlete, and the effect of velocity as a primary focus should be considered. Simply, proper technique should be emphasized over attempting to achieve a higher velocity at the expense of technique. This consideration is why it has been recommended that athletes focus on solidifying their technique before using VBT (71). However, this concept has been largely ignored because the majority of VBT research studies used fixed-axis Smith machines rather than free weights. Furthermore, much of the existing literature has focused on concentric-only movements following a pause versus an eccentric-concentric movement; thus, strength and conditioning professionals should interpret and apply the findings from the existing VBT literature with caution. To be clear, velocity monitoring has its place within load prescription and athlete monitoring; however, it may serve athletes best as a complement to some of the other previously described methods.

MOTOR LEARNING AND SKILL ACQUISITION MONITORING

As mentioned in chapters 2 and 3, the ability to develop an effective and efficient movement may aid in the performance of athletic skills (e.g., sprinting, jumping, throwing) by enhancing an athlete's ability to develop the necessary force characteristics. Thus, monitoring these characteristics, in addition to other strength attributes (chapter 9), may be of utmost importance (120). Although the various measures of strength may exist on the motor capacity or capability spectrum (51), sport coaches must also determine which motor performances are relevant to their athletes and performance in sport. In this light, sport coaches may use their experiential knowledge of the sport to qualitatively assess movement skill, whereas strength and conditioning professionals may also combine quantitative analysis (61), particularly during highly technical exercises that include greater coordination complexity.

Interestingly, researchers have integrated principles and measures from motor behavior literature with strength and conditioning literature to describe the process of skill learning over time (33, 80, 95), which differs from strictly using physiological perspectives within an assessment. This type of research may allow strength and conditioning professionals and sport scientists to quantify and monitor skill acquisition or adaptation to motor performance or motor capabilities during different resistance training phases that aim to develop muscular strength characteristics and determine whether any performance stagnation occurs. Furthermore, this method of monitoring may allow the *process* of motor learning and adaptation or skill acquisition to be combined with monitoring methods that measure the *outcome* of a movement (e.g., VBT), thus providing a more holistic monitoring approach. While the concept of monitoring motor learning and skill acquisition may be appealing, understanding which variable or feature of learning (e.g., adaptation, coordination, transfer) to monitor may prove difficult; however, various measures could be adapted from neurorehabilitation research (106) or sport skill monitoring (93).

The relevance of monitoring the motor learning and skill acquisition process is obvious when the stages of learning (i.e., coordination, control, and skill) of different athletes are considered (83). It is suggested that future athlete monitoring should include a combination of outcome-focused (e.g., RPE, RIR, VBT) and process-focused resistance training methods. Using this combination, measures of bivariate coordination variability (82) or execution variability (140) could serve as the process measures that drive changes in loading during training, instead of outcome measures. Moreover, measures of coordination or variability magnitude or structure may help strength and conditioning professionals when deciding to implement a change in training technique or determine when to modify the exercise task (33). Although there is a lack of research in this area, the qualitative practice of coaches suggests that there is a need to include skill acquisition and motor learning measures within the athlete monitoring process. Therefore, future resistance training monitoring should attempt to combine outcome (e.g., velocity, load, time) and process (e.g., variability magnitude or structure, coordination, efficiency) measures relevant to the stage of learning or to the motor task being performed. This, in turn, may provide a better understanding of the interaction between the motor capability and control of an athlete (101).

ADDITIONAL CONSIDERATIONS FOR MONITORING AND LOAD ADJUSTMENT

This chapter has primarily focused on monitoring and adjusting training loads during traditional resistance training exercises (e.g., squats, presses, and pulls). However, it is important to acknowledge that different types of resistance training, such as weightlifting movements (2, 37, 108, 114, 115), eccentric training methods (74, 121, 122, 129), isometric training (28, 67, 85), plyometric training (1, 29, 30), and loaded jumps (63, 118, 119, 124), may require different methods of monitoring. This is due to unique methods of loading (e.g., eccentric, isometric) or the coordinative complexity of specific exercises (e.g., weightlifting movements and jump variations). Furthermore, because novices may lack consistency during their training due to changes in their strength, technique, and effort, strength and conditioning professionals should consider using less complex monitoring and load adjustment methods such as RPE, RIR, or APRE. In contrast, athletes who are more advanced in their training status may require more frequent load adjustments and information about *how* they are completing an exercise with different loads or the coordinative strategies (behavioral flexibility) they are using. In this case, SRB or VBT in combination with other methods may serve these athletes more effectively.

> **Key Point**
>
> Because novices may lack consistency during their training due to changes in their strength, technique, and effort, strength and conditioning professionals should consider using less complex monitoring and load adjustment methods such as RPE, RIR, or APRE.

Monitoring an athlete's strength characteristics is multifaceted. While the previously discussed methods may help monitor an athlete's progress in the weight room, none of them directly measure how an athlete applies force. Although some aspects of enhanced maximal force production may be assumed by some of these methods (e.g., increased %1RM relative to body mass), indirect measurements are limited because an increase in load lifted does not mean that the athlete has improved their force expression during sport-related tasks. Therefore, it is important that strength and conditioning professionals consider integrating additional measures (e.g., vertical jump, isometric mid-thigh pull, speed, change of direction) when monitoring an athlete to provide a more holistic view of how they are responding to the accumulated stress of different training stimuli, practices, competitions, and psychological stressors (120). This information may then allow for more efficient training prescription that promotes favorable force production characteristics, improves fatigue management, mitigates injury risk, and recalibrates the motor system to help athletes "learn how to use their newfound strength" (120). Irrespective of the chosen methods, strength and conditioning professionals must put the prescribed loads of an athlete into context by using athlete feedback, testing results, and the past, current, and future training goals within the long-term training plan of the athlete. This may promote an evidence-based approach to training prescription and avoid being overly reactive to less monitoring information.

Take-Home Points

- Although there are a wide variety of methods that can be used to monitor and adjust training loads during resistance training sessions, strength and conditioning professionals should be aware of the benefits and limitations of each method.
- Linear loading and the two-for-two methods may be useful for novice athletes because their exercise technique and maximal strength may frequently change; however, they may not provide much variation in training and may be detrimental to an athlete's force production characteristics if they are exclusively used for extended training periods.
- Although %1RM and RM zones may provide more variation and a better training stimulus compared to other methods, they fail to account for external stressors that may affect performance. It is recommended that ranges of loads are provided to account for these stressors.
- Subjective (e.g., RPE, RIR, SRB, and APRE) and objective (e.g., VBT) load adjustment methods may provide greater insight into an athlete's daily readiness to train compared to the other methods discussed.
- The RPE and RIR methods may be limited by athlete familiarity with various training stimuli. Professionals should consider limiting athlete autonomy when selecting loads until they become proficient in accurately reporting RPE or RIR following sets of exercise.
- Strength and conditioning professionals may consider combining subjective methods with an objective one to determine specific loads in training while using the subjective method to provide autoregulation of load adjustments.
- Future athlete monitoring protocols should consider combining output-focused measures, measures of skill acquisition, and motor learning to provide greater insight into an athlete's overall development.

CHAPTER 11

Nutritional Considerations for Muscular Strength

Beyond the design of training programs, strength and conditioning professionals must consider the energy levels of their athletes. Simply, if the athlete does not have the caloric input to be able to complete or withstand the prescribed training, the desired adaptations may suffer. From this standpoint, it is important to consider how athletes fuel themselves and how this may affect their ability to train and recover. Thus, macronutrient, micronutrient, and ergogenic aid consumption should be considered to develop a sound nutrition plan for athletes. The purpose of this chapter is to provide readers with insight on various nutritional factors and how they may affect muscular strength characteristics and recovery from different training stimuli.

ENERGY BALANCE

Before addressing the nutritional needs of an athlete, it is important to determine their energy requirements; this can be accomplished by calculating their total daily energy expenditure, which includes the resting metabolic rate, non-exercise activity thermogenesis, thermic effect of food (i.e., dietary-induced thermogenesis), and thermic effect of exercise (i.e., exercise energy expenditure) (71). An in-depth discussion on the procedures for calculating each energy expenditure factor is beyond the scope of this chapter. Put simply, *energy balance* may be determined as the difference between energy intake and energy expenditure, or "calories in versus calories out." Readers should note that the energy requirements for specific activities (e.g., training, practice, competition) may vary, and the metabolic cost and rate of energy expenditure may be based on training variables such as volume, intensity, duration, and frequency (249, 284). Regarding the frequency of training sessions, it is important to consider the accumulated amount of work performed by the athlete and whether they have consumed enough kilocalories to combat the energy expended to reduce the potential for negative training effects (247) or losses in body mass, lean body mass, or both. In general, more kilocalories are used as more overall work is completed, but this may also depend on an athlete's body size, mass, and composition (250) as well as their position in sport (e.g., American football lineman vs. skill player). Readers should also note that the energy expended during exercise may affect post-exercise consumption and recovery parameters (37); this, in turn, provides a justification for including post-exercise nutrition planning, because the summative effects of energy expenditure from training may relate to various training adaptations, including body mass and composition, cardiovascular function, and sport performance (245).

Key Point

The energy requirements for specific activities (e.g., training, practice, competition) may vary, and the metabolic cost and rate of energy expenditure may be based on training variables such as volume, intensity, duration, and frequency.

Energy availability can be calculated using the following equation (169). Different levels of energy availability based on Cabre and colleagues (38) are displayed in table 11.1.

$$\text{Energy availability} = [\text{Energy intake (kcal)} - \text{Exercise energy expenditure (kcal)}]/\text{Fat free mass (kg)}$$

TABLE 11.1 Energy Availability Level Classification and Description

Level	Description	Energy (kcal\kg of fat free mass)
Optimal	Sufficient energy for weight maintenance and all physiological function	≥45
Low	Acutely sustainable; may be intentional for body composition purposes	30-40
Clinically low	Insufficient energy that may lead to health implications and impaired training adaptations	<30

Adapted by permission from Cabre et al. (2022). Distributed under the terms of the Creative Commons Attribution 4.0 International License (http://creativecommons.org/licenses/by/4.0/).

MACRONUTRIENTS

Three primary macronutrients provide energy to the body while maintaining the body's structure and systems; these include carbohydrates, proteins, and fats. The following sections provide an overview of how each macronutrient contributes to the development of muscular strength.

CARBOHYDRATE

Carbohydrates are chemical compounds that are made up of carbon, hydrogen, and oxygen and serve as the primary fuel source for moderate- to high-intensity exercise (269), making carbohydrate an essential component of an athlete's diet. Based on their structure, they may be classified as either *simple carbohydrates*, where they may take the form of either monosaccharides (i.e., single sugar) or disaccharides (composed of two monosaccharides), or *complex carbohydrates*, where polysaccharides are formed with potentially thousands of monosaccharides. Examples of simple and complex carbohydrates are displayed in table 11.2. While polysaccharides may consist of the storage form of carbohydrate (i.e., glycogen), they generally must be broken down into the monosaccharide glucose to be used as a substrate for the usable form of energy in the human body, adenosine triphosphate (ATP). For additional context, carbohydrate provides an athlete with 4 kcal/g of energy. Although low-carbohydrate diets may be used in certain scenarios, diets with less than 30% of total kilocalories consisting of carbohydrate may be associated with symptoms of fatigue (28) and may contribute to overtraining and a reduction in performance (247).

Carbohydrates may be classified based on their *glycemic index* (GI), which refers to the ability of a given food to raise blood glucose levels. Briefly, white bread serves as the primary reference food for the GI and has a value of 100. Thus, high, moderate, and low GI foods are classified with values of greater than 70, 55 to 70, and less than 55, respectively. Examples of high, moderate, and low GI foods are displayed in table 11.3, and readers are referred to Atkinson and colleagues for more comprehensive information (6, 7). The GI of various foods is important because it relates to how, when, and to what extent glucose levels become elevated and to the use of different substrates for energy (65). For example, low GI foods consumed in the hours prior to training may have a metabolic advantage over high GI foods because they produce a smaller insulin response, which may increase free fatty acid availability and allow a more stable glucose concentration during training, thus sparing muscle glycogen (46). However, because low GI foods may be high in fiber content, it is important that athletes

TABLE 11.2 Examples of Simple and Complex Carbohydrates Commonly Consumed by Athletes

Simple	Complex
Baked goods	Beans
Breakfast cereal	Brown rice
Candy	Fruit
Fruit juice	Oats
Honey	Pasta
Sport drinks	Potatoes
Sugar	Vegetable
Yogurt	Whole-wheat bread

TABLE 11.3 Examples of High, Moderate, and Low Glycemic Index Foods Commonly Consumed by Athletes

High	Moderate	Low
Bagel	Baked beans	Apple
Doughnut	Basmati rice	Black beans
Mashed potatoes	Brown rice	Cashews
Pretzels	Croissant	Kidney beans
Waffles	Chips	Mashed sweet potatoes
Wheat bread	Honey	Orange
White bread	Milk chocolate	Peanuts
White rice	Salted popcorn	Strawberries

allow for the necessary time to digest such foods to prevent gastrointestinal issues during training or competition. Interestingly, two meta-analyses indicated that there may be no difference between high and low GI foods when it comes to glycogen sparing (75, 118); however, high GI foods may be advantageous to consume after exercise because they promote glycogen resynthesis, thus benefiting recovery (32).

While there is little doubt that carbohydrate intake can significantly benefit endurance performance (209, 273), less research supports additional carbohydrate intake beyond a typical diet for resistance training performance. For example, the authors of a 2022 systematic review concluded that carbohydrate intake may not affect an athlete's resistance training performance while in a fed state during workouts that included up to 10 sets per muscle group (113). However, the authors also noted that higher-volume resistance training workouts may benefit from carbohydrate intake, although further research is needed. Additional researchers indicated that carbohydrate ingestion before or during a resistance training session may allow individuals to perform a greater volume of work during sessions that are longer than 45 minutes in duration and include at least 8 to 10 sets (151). The authors also concluded that carbohydrate ingestion after an 8-hour fast may improve resistance training performance. Despite these results, it should be noted that carbohydrate restriction may negatively affect an athlete's ability to consistently perform high-intensity training sets (165), indicating the importance of still having adequate levels of carbohydrates readily available for activity. To counteract carbohydrate restriction, some researchers have investigated the use of *carbohydrate mouth rinsing*, defined as distribution of carbohydrate fluid in the mouth for 5 to 10 seconds before spitting it out (57). Researchers concluded that carbohydrate mouth rinsing did not benefit repeat and all-out running and cycling sprint performance (58); however, others found that mouth rinsing with carbohydrate and caffeine may enhance exercise capacity (145). Collectively, the available literature suggests that additional carbohydrate intake prior to training may not benefit strength performance when activities are shorter in duration. Thus, athletes and strength and conditioning professionals should consider the amount of carbohydrates that are necessary for different types of training and competition.

Key Point

Carbohydrate restriction may negatively affect an athlete's ability to consistently perform high-intensity training sets, indicating the importance of still having adequate levels of carbohydrates readily available for activity.

Daily carbohydrate requirements for athletes should be periodized to ensure adequate availability for training and competition; however, athletes and strength and conditioning professionals should consult sport nutritionists or registered dietitians to improve their knowledge of best practices. While many have suggested that an athlete's diet should consist of a specific percentage of carbohydrate, this method does not seem to be supported because it is poorly correlated with the amount of carbohydrate consumed during activity as well as the fuel requirements of said activities (34). Thus, it is recommended that the amount of carbohydrate needed for each athlete should be based on the activities that they are performing on any given training day. For example, it has been suggested that athletes ingest <2 g/kg, 2-4 g/kg, 4-6 g/kg, 6-8 g/kg, or 8-12 g/kg of carbohydrate if they are participating in little to no activity, low-intensity activities with minimal energy use (e.g., recovery session), moderate- to high-intensity activities that last about an hour per day (e.g., resistance training session), moderate- to high-intensity activities that last between 1 to 3 hours per day (e.g., sport practices or competitions), or moderate- to high-intensity activities that last more than 4 hours per day (e.g., combined sport practice, resistance training session, and recovery session), respectively (46).

Regarding carbohydrate consumption during training or competition, additional recommendations suggest that well-trained athletes competing in events lasting 1 to 2 hours, 2 to 3 hours, or more than 2.5 hours should consume approximately 30, 60, or 90 g of carbohydrate per hour, respectively; less-trained athletes may adjust these values downward (137). Regarding post-training or competition recommendations, researchers have suggested consuming 1.2 g/kg of carbohydrate per hour (136) starting within the first 2 hours of cessation to optimize glycogen resynthesis (133); however, consumption should also be based on how long the training session or event was and when the next one will occur.

The previous recommendations are quite general, but it is important that carbohydrate intake is based on the activities that will be performed. However, flexibility should be exercised, especially when schedules change on short notice. Finally, educating athletes on specific portion sizes and nutrition software programs or apps that remind athletes about specific feeding times may help create an autonomous athlete (71). Readers interested in additional information on how to periodize carbohydrates are directed to a previous review by Impey and colleagues (131).

PROTEIN

Beyond serving as contractile elements for force production (e.g., actin, myosin, titin, nebulin), proteins may be structural (e.g., keratins), part of the immune system (e.g., antibodies), or regulatory (e.g., enzymes) in their functions. This diverse range of actions may ultimately be the result of the large variation in structure among different proteins. Each protein is made up of a wide variety of amino acids, of which 9 are considered essential and 11 are considered non-essential (see table 11.4). Simply, essential amino acids must be consumed within an athlete's diet, whereas non-essential proteins are naturally made within the human body. Because proteins are in a constant state of turnover, being broken down and rebuilt throughout any given day, a diet with adequate protein allows new proteins to be formed to replace damaged ones following a given training stimulus (46). Although protein consumption may not directly affect maximal strength (166), athletes and strength and conditioning professionals should recognize its contribution to the underlying mechanisms and recovery of muscular strength characteristics.

Key Point

Although protein consumption may not directly affect maximal strength, athletes and strength and conditioning professionals should recognize its contribution to the underlying mechanisms and recovery of muscular strength characteristics.

The *nitrogen balance* reflects the ratio of nitrogen intake compared to what is lost and provides an indication of gain or loss of total-body protein (a nitrogen-containing molecule). In 2007, the World Health Organization indicated that 0.8 to 0.9 g/kg

TABLE 11.4 Essential and Non-Essential Amino Acids

Essential	Non-essential
Histidine	Alanine
Isoleucine*	Arginine
Leucine*	Asparagine
Lysine	Aspartic acid
Methionine	Cysteine
Phenylalanine	Glutamine
Threonine	Glutamic acid
Tryptophan	Glycine
Valine*	Proline
	Serine
	Tyrosine

*Branched-chain amino acids.

of protein per day is required to maintain a nitrogen balance in adults (282); however, this conclusion was primarily based on research focused on sedentary individuals. As a result, researchers have concluded that athletes require significantly higher values than the previous recommendations (164, 198), with a minimum intake at around 1.2 to 1.6 g/kg/day (257) to maintain nitrogen balance.

Although the nitrogen balance represents the minimum protein intake for an athlete, this may not be optimal based on specific training goals and demands; moreover, the previous recommendations may not account for different types of athletes. Close and colleagues (46) suggested that endurance athletes should consume 1.2 to 1.4 g/kg of protein day, whereas strength athletes may need to consume 1.8 to 2.0 g/kg/day. It should be noted that these values may be elevated to 2.0 to 2.5 g/kg/day based on the frequency and intensity of training and competitions (3, 182). Additional research has suggested consuming approximately 0.4 g/kg of protein every 3 to 4 hours throughout the day, rather than consuming much larger boluses at individual meals (187). This, in turn, may promote a positive nitrogen balance, leaving the athlete in a more anabolic (building) state rather than a catabolic (breakdown) state throughout the day.

Key Point

Researchers have suggested consuming approximately 0.4 g/kg of protein every 3 to 4 hours, rather than consuming much larger boluses at individual meals; this may promote a positive nitrogen balance, leaving the athlete in a more anabolic (building) state rather than a catabolic (breakdown) state throughout the day.

Beyond the amount of protein that an athlete may consume, the type of protein, including animal based (e.g., whey and casein) and plant based (e.g., soy), should be considered. Researchers have concluded that animal-based protein generally has higher bioavailability (percentage digested and used) and contains a higher percentage of essential amino acids (21, 228), including leucine, the amino acid responsible for activating the mammalian target of rapamycin complex-1 (mTOR) pathway and subsequent upregulation of protein synthesis (199). Consuming 2.5 to 5 g of leucine within a single meal may be required to activate the mTOR pathway and initiate muscle protein synthesis (200). Consumption of leucine alone, however, is not sufficient to synthesize complete muscle proteins, which requires adequate availability of all nine essential amino acids (74). In contrast to animal sources of protein, athletes and strength and conditioning professionals should note that plant-based proteins include incomplete protein, containing insufficient amounts of one or more essential amino acids. This can be overcome by consuming greater overall amounts and varied sources of plant proteins to achieve adequate intake to increase the bioavailability of the essential amino acids required to support muscle protein synthesis. It should, however, be noted that consumption of greater amounts of plant-based protein may lead to greater carbohydrate and caloric consumption as well (e.g., 120 g of chicken breast = 38 g of protein and 0 g of carbohydrates; 520 g mixed beans = 38 g of protein and 84 g of carbohydrate) (71, 203). Beyond the bioavailability of amino acids, consumers may also consider the rate that different protein types increase amino acid concentrations within the blood, a primary driver of muscle protein synthesis (43). For example, researchers showed that whey protein increased protein synthesis to a greater extent after exercise compared to both milk (80% casein and 20% whey) and soy-based protein, while milk was also superior to soy (199). Although these findings support the use of specific forms of protein for protein synthesis, it should be noted that the type of protein may be linked to the timing of ingestion (described later in this subsection).

Another form of protein that should be considered within an athlete's diet is collagen. Briefly, collagen is the most abundant protein in connective tissue and serves as the primary structural component of tendons, ligaments, and intramuscular connective tissue (10), which may then dictate the strength and stiffness of such tissue. Researchers have shown that 15 g of gelatin consumed with vitamin C–rich blackcurrant juice led to a 2-fold increase in collagen synthesis (230). However, additional research showed that ingesting 30 g of collagen protein following 6 sets of a high-volume (8-15 repetitions) barbell squat using 60% of the 1RM resulted in no increase in connective protein synthesis rates (8), indicating that the literature appears to be inconclusive on this topic and that further research is needed (127).

Athletes have the option of consuming protein in a variety of concentrations (e.g., solid, liquid, gels), especially due to food manufacturers adding protein to many of their existing products. It should be noted that liquid forms of protein may produce a faster increase in plasma amino acids compared to other forms (36), especially following training or competition; these findings are likely due to time needed to digest and absorb solid forms of protein. However, as mentioned previously, different forms of protein have their place within the athlete's overall diet plan.

Another aspect of protein consumption that must be discussed is the timing of ingestion relative to activity. As noted previously, it has been suggested that protein should be consumed throughout the day at a rate of 0.4 g/kg every 3 to 4 hours (187), which may place an athlete in a more anabolic state. In fact, researchers have shown that performing resistance training in a fasted state may result in a net loss of protein (201). Thus, based on the time of a given training session, it is important for athletes to consume amino acids either before (261) or immediately after (199) exercise to prevent a net loss of muscle protein. This may be challenging, especially with training sessions that occur early in the morning or late at night, which is why athletes should have snack options containing protein prior to training that do not result in gastrointestinal distress. As a workout commences, athletes begin to experience a breakdown in muscle protein. While this is a normal occurrence, it would seem logical to implement strategies to counteract these activities and set the athlete up for the rebuilding of proteins after exercise. Unfortunately, limited research on intraworkout protein consumption exists; however, Beelen and colleagues (17) indicated that whole-body and muscle protein synthesis rates increased during a workout and early post-workout when a protein and carbohydrate drink was consumed during the training session. Based on these findings, it would appear advantageous to begin consuming protein prior to the end of the resistance training workout to increase protein synthesis and potentially slow protein breakdown.

Following a training session, athletes are generally in a catabolic state that is characterized by greater levels of protein breakdown, and they are therefore in need of protein intake to counteract these effects and promote adaptation. In a classic study, Cribb and Hayes (52) examined the effect of consuming meals before and following resistance training with morning and evening meals that both consisted of 32 g of protein, 34 g of carbohydrate, less than 0.4 g of fat, and 5.6 g of creatine for 10 weeks. The authors showed that pre- and post-training meals produced improvements in lean body mass, body fat percentage, cross-sectional area of Type IIa and IIx fibers, contractile protein content, phosphocreatine (PCr) and total creatine content, muscle glycogen content, and increases in both bench press and squat strength. A previous systematic review and meta-analysis supports the previous findings, and the authors concluded that at least 6 weeks of resistance training and post-training protein supplementation may benefit maximal strength characteristics (41).

Although it has previously been suggested that a "metabolic window" (i.e., anabolic window) exists for approximately 45 minutes after training (134), the literature has indicated that total protein intake may be more important for muscular strength and hypertrophy, to an extent, following a workout (228). In fact, further research showed that muscle protein synthesis remained increased several hours following exercise (210) and that 40 g of protein consumed immediately after exercise and over 3 subsequent hours may produce greater protein synthesis effects compared to 20 g. It should be noted that consuming a combination of carbohydrate and protein following a workout may also promote increased muscle glycogen resynthesis (35). Finally, although discussed less frequently, researchers have suggested that consuming approximately 0.4 to 0.5 g/kg of protein prior to the overnight sleeping period may benefit resistance training adaptations (237, 265). In this light, it is suggested that athletes use casein protein to allow amino acids to be absorbed over a longer duration.

Key Point

It is important for athletes to consume amino acids either before or immediately after exercise to prevent a net loss of muscle protein.

FAT

The third macronutrient that generally comprises the smallest portion of an athlete's diet is fat (i.e., lipids). Although fat within diets is typically viewed in a negative light, fats are an essential nutrient that contributes to a variety of functions within the human body. For example, fats may be used to protect internal organs, absorb vitamins, and facilitate cell membrane, hormone, and prostaglandin production. In addition to carbohydrates, fats are also one of the primary substrates for energy production, especially during long-duration exercise, and may provide 9 kcal/g of energy. Fats are classified based on their structure as either saturated or unsaturated, and the latter may be further subdivided into monounsaturated or polyunsaturated fats. It should, however, be noted that all classifications of fat include a mixture of different fatty acids, and they are classified based on the majority of the fat source they contain (46). Although saturated and unsaturated fats are frequently discussed as bad fats and good fats, respectively, this may be an oversimplification; in some instances, replacing the former with the latter may affect other nutrient levels (e.g., the removal of dairy products may negatively affect calcium intake) (46, 71). However, when possible, it is recommended that athletes avoid consumption of trans fats, because this form of fat may increase low-density lipoprotein (i.e., bad) cholesterol and decrease high-density lipoprotein (i.e., good) cholesterol, which in turn may increase the risk of cardiovascular disease (188). Examples of foods that may contain trans fats include baked goods, margarine, and frozen pizza.

It is important to note that certain fatty acids (i.e., n-3 and n-6 fatty acids, also known as omega-3 and omega-6 fatty acids, respectively) are considered essential because they are not naturally made in the human body. In fact, it has been suggested that a diet that consists of too little dietary fat may negatively affect an individual's overall health (46). Western diets have been reported to include relatively high levels of n-6 fatty acid (e.g., poultry, eggs, nuts, peanut butter, avocado) but much lower levels of n-3 fatty acid (233). This may relate to a reduced consumption of oily fish, such as salmon, tuna, mackerel, herring, and swordfish, all of which are significant sources of n-3 fatty acids. Given that some athletes may not enjoy eating oily fish, n-3 supplements may be beneficial to both their health and performance; however, it is important that athletes consult with a sport nutrition professional to determine whether supplementation is necessary. A 2023 study indicated that 10 weeks of progressive resistance training combined with a fish oil supplement (4 g/day) may lead to greater improvements in absolute and relative bench press and squat strength and reductions in fat mass (112). Additional research concluded that n-3 supplementation may significantly reduce muscle soreness following resistance training (141), which in turn may enhance the recovery process between training sessions (202). Collectively, the previous research findings suggest that greater maximal strength and enhanced recovery may be by-products of fish oil supplementation. It should be noted that the research in this area is evolving, and further research is needed to provide more concrete recommendations.

Some researchers have promoted the idea of high-fat diets for athletes, with the mindset of sparing muscle glycogen and improving the use of fat for energy. For example, Yeo and colleagues (288) suggested that a high-fat, low-carbohydrate diet consumed for up to 2 weeks paired with normal training, followed by 1 to 3 days of a high-carbohydrate diet and taper in training volume, may lead to "fat adaptation," in which greater fat use remains elevated even after carbohydrate availability increases. Although this may be beneficial for endurance-trained athletes, this type of diet may reduce the activity of pyruvate dehydrogenase during rest as well as during submaximal and supramaximal exercise (243), which in turn may negatively affect high-intensity efforts in which carbohydrate is the primary fuel source. Additional research has also shown that high-fat diets may negatively affect protein synthesis (102). Therefore, it is suggested that high-fat diets should be avoided when the goal is to enhance muscular strength.

MICRONUTRIENTS

Micronutrients perform a wide range of functions within the human body. Although some are needed in small quantities and others are required in larger amounts, they may have a significant impact on the development of skeletal muscle and the force production characteristics of an athlete. An overview of common vitamins and minerals that may contribute to strength characteristics is provided in the following paragraphs.

VITAMINS

Vitamins are organic compounds that serve a variety of purposes within the human body; they aid in normal physiological functioning and may play either direct or indirect roles in various muscle actions and metabolic reactions (250). Because vitamins (except for vitamin D) cannot be produced naturally within the body, athletes must consume them as part of their diet. Vitamins are classified as either fat soluble (e.g., vitamins A, D, E, and K) or water soluble (e.g., B vitamins and vitamin C), both of which serve the body in different ways.

Fat-Soluble Vitamins

Fat-soluble vitamins are unique compared to water-soluble vitamins in that they can be stored. Although they are generally stored within the liver and adipose tissue, fat-soluble vitamins may be delivered to other tissues as needed; however, if the vitamins are not used, excessive intake may lead to health issues (46). Vitamins A, D, E, and K have either antioxidant or hormone-like properties that relate to their functions. Table 11.5 outlines the functions of fat-soluble vitamins, potential food sources, and recommended intakes. Vitamin D has received the most attention within the literature, due to its potential benefits to muscular strength and athletic performance, and is discussed further next.

Compared to the other vitamins discussed in this section, vitamin D is naturally produced in the body. Although adequate amounts of vitamin D can be achieved through sufficient sunlight exposure, it may be difficult to achieve optimum intake through diet alone (193). Researchers have shown that athletes may become deficient in vitamin D during the winter months (25, 47), which in turn may negatively affect muscle function and recovery (194). Several systematic reviews and meta-analyses have examined the impact of vitamin D supplementation on maximal strength characteristics (42, 234, 244, 262); however, the findings in the literature have been mixed. While some authors indicated that vitamin D supplementation may improve measurements of maximal strength (42, 262), others indicated that it may not have any effect (234, 244). It should be noted that the mixed findings of the previous literature may be based on the type of vitamin D supplementation (42) or the study inclusion criteria. Despite the existing findings, athletes can prioritize foods that have higher quantities of vitamin D, especially during the winter months. However, it is important to note that athletes can also supplement with too much vitamin D (195); thus, athletes should exercise caution and consult with a sport nutrition professional to determine whether supplementation is necessary.

Water-Soluble Vitamins

In addition to their contribution to muscle actions and energy expenditure reactions, water-soluble vitamins may act as coenzymes that help fully activate enzymes. Table 11.6 outlines the functions of water-soluble vitamins, potential food sources, and recommended intakes. Compared to fat-soluble vitamins, water-soluble vitamins are not stored within body

TABLE 11.5 Fat-Soluble Vitamin Functions, Food Sources, and Recommended Dietary Allowances (RDAs) for Nonpregnant, Healthy Adults

Vitamin	Physiological functions	Food source examples	RDA
A	Vision; white blood cell production, bone remodeling, cell growth and division regulation	Carrots, sweet potatoes, tomatoes, milk, eggs	Males = 900 µg/d RAE Females = 700 µg/d RAE
D	Calcium and phosphorus absorption and retention	Salmon, swordfish, tuna, sardines	15 µg/d
E	Antioxidant; red blood cell production, anti-inflammation	Almonds, peanuts, spinach, sunflower seeds	15 mg/d
K	Blood clotting protein production, bone formation	Spinach, broccoli, cabbage, canola oil	Males = 120 µg/d Females = 90 µg/d

RAE = retinol activity equivalents
Based on the National Institutes of Health Office of Dietary Supplements Dietary Reference Intakes.

tissue, and excess intake is excreted in the urine. Although excess intake of water-soluble vitamins is generally not toxic, vitamin B_6 (pyridoxine) serves as the exception because excess intake may result in peripheral nerve damage (48). Because water-soluble vitamins are not stored, it is important that athletes consume the recommended dietary allowance per day to optimize physiological functioning throughout the day and for training purposes.

B vitamins play an important role in energy production pathways, and the need for proper intake may change with exercise (286). This conclusion is supported by research showing that vitamin requirements may increase with training demands (26, 267). Interestingly, researchers found that 10% to 60% of female athletes were deficient in vitamin B intake, whereas male athletes were deemed to have sufficient levels, potentially due to a much larger (i.e., 2-3 times greater) caloric intake (170, 171). While these findings may be attributable to a low-calorie diet, poor food choices may also be the cause. It should also be noted that food decisions may either be due to naivety or lack of financial resources to pay for higher-quality food. This is an important consideration, especially for athletes who must lose weight for a competition. Regardless, it is recommended that athletes consult with a sport nutrition professional to discuss best practices and potential alternatives to current food choices.

Vitamin C consumption is important to an athlete for several reasons, including boosting the immune system and antioxidant levels (279) and promoting

TABLE 11.6 Water-Soluble Vitamin Functions, Food Sources, and Recommended Dietary Allowances (RDAs) for Nonpregnant, Healthy Adults

Vitamin	Physiological functions	Food source examples	RDA
B_1 (thiamin)	Carbohydrate, fat, and protein metabolism	Whole grains, liver, fish, asparagus, peanuts, green peas, black beans	Males = 1.2 mg/d Females = 1.1 mg/d
B_2 (riboflavin)	Carbohydrate, fat, and protein metabolism	Eggs, lean beef and pork, chicken, dairy products, salmon	Males = 1.3 mg/d Females = 1.1 mg/d
B_3 (niacin)	Carbohydrate, fat, and protein metabolism	Chicken, fish, nuts, bananas, brown rice, eggs, tomatoes	Males = 16 mg/d NE Females = 14 mg/d NE
B_5 (pantothenic acid)	Carbohydrate, fat, and protein metabolism; coenzyme A production	Chicken, beef, eggs, milk, broccoli, potatoes, whole wheat, oats	5 mg/d
B_6 (pyridoxine)	Carbohydrate metabolism, hemoglobin production, nervous system function	Fish, bananas, oranges, chicken, spinach	Males = 1.3 mg/d Females = 1.2 mg/d
B_7 (biotin)	Carbohydrate, fat, and protein metabolism	Sweet potatoes, nuts, fish, eggs, liver, ground beef, spinach	30 µg/d
B_9 (folic acid)	DNA production, cell division, protein production	Green leafy vegetables, beans, peanuts, whole grains, fruits	400 µg/d DFE
B_{12} (cobalamin)	Red blood cell formation, carbohydrate metabolism, nervous system function, DNA production	Chicken, beef, fish, and dairy products	2.4 µg/d
C	Antioxidant; iron absorption; stimulates collagen synthesis (e.g., skin, tendons, ligaments, blood vessels)	Oranges, grapefruits, kiwis, broccoli, cauliflower, bell peppers, tomatoes	Males = 90 mg/d Females = 75 mg/d

Based on the National Institutes of Health Office of Dietary Supplements Dietary Reference Intakes.

DFE = dietary folate equivalents; NE = niacin equivalents.

collagen synthesis (197, 230). It should be noted that researchers have examined the effect of vitamin C supplementation on various aspects of muscular strength and performance. Dutra and colleagues (70) indicated that combined vitamin C and E supplementation with resistance training had no additive effect on maximal strength or muscle growth. In contrast, Lis and colleagues (167) showed that combined collagen and vitamin C supplementation significantly increased rapid force production characteristics during the isometric squat and countermovement jump. Readers should interpret the previous studies with caution because vitamin C was combined with another supplement and thus its individual effects may not be clear. Beyond performance measures, a 2023 review concluded that long-term vitamin C supplementation may not be recommended because the existing literature does not appear to support its use for performance adaptations, muscle damage, or perceived muscle soreness (214).

MINERALS

Minerals are inorganic compounds that play crucial roles in the body's normal physiological processes and in several aspects related to metabolism (176). Minerals can be subdivided into either macrominerals (e.g., sodium, potassium, calcium, and magnesium) or microminerals (e.g., iron, zinc) and are stored in various tissues throughout the body. A discussion of all the minerals that contribute to normal physiological functioning is beyond the scope of this chapter, so the following paragraphs provide a brief overview of the minerals that contribute most to the development of muscular strength (see table 11.7).

Calcium and magnesium play key roles in force production; specifically, each mineral aids muscle contraction and the transmission of nerve impulses (46). In addition, calcium helps support the development of strong bones, which in turn allow them to withstand greater forces, an important concept for injury resilience and force transmission. Because athletes such as male (117) and female (254) distance runners may possess lower bone mineral density, calcium supplementation may be appropriate to build a more robust skeletal structure to withstand training demands (246, 254). Beyond calcium and magnesium, iron serves as an important contributor to oxygen transport to the muscle tissue. Although this may obviously benefit endurance athletes using oxidative energy systems, especially those with iron deficiencies (61), research suggests that iron supplementation may benefit maximal strength performance as well (184). Finally, zinc may contribute to protein digestion, energy pathways, and immune function and may provide antioxidant effects. It should be noted that a systematic review concluded that magnesium and iron may be the only supplemented microminerals that lead to improvements in maximal strength and athletic performance (110); however, further research is still needed. It is recommended that athletes consult a sport nutrition

TABLE 11.7 Mineral Functions, Food Sources, and Recommended Dietary Allowances (RDAs) for Nonpregnant, Healthy Adults

Mineral	Physiological functions	Food source examples	RDA (mg/day)
Calcium	Nerve impulse transmission; facilitates muscle actions, develops strong bones	Dairy products, salmon with bones, almonds, beans	1,000-1,300
Iron	Hemoglobin and myoglobin production to produce oxygen throughout and to muscle cells, respectively.	Beef, chicken, fish, eggs, beans, spinach, nuts, seeds	Males = 8-11 Females = 8-18
Magnesium	Nerve impulse transmission; facilitates muscle actions	Spinach, nuts, bananas, potatoes, bread, dairy products	Males = 400-420 Females = 310-320
Sodium	Nerve impulse transmission; facilitates muscle actions	Breads, soups, sunflower seeds, chicken, meat that is smoked, cured, salted, or canned	1,500
Zinc	Antioxidant; immune system function, cell division and growth, carbohydrate metabolism	Beef, yogurt, chicken, cashews, pork chops, oatmeal, dairy products	Males = 11 Females = 8-9

Based on the National Institutes of Health Office of Dietary Supplements.

professional when considering mineral supplementation, because there may be a relatively small margin between the recommended dietary allowance and toxicity (46).

> **Key Point**
>
> Magnesium and iron may be the only supplemented microminerals that lead to improvements in maximal strength and athletic performance.

While some athletes may be deficient in certain micronutrients, it should be noted that an affordable and easily accessible alternative to additional food intake is to supplement with a daily multivitamin. However, athletes who consume a vegan or vegetarian diet may need to take extra precautions because some vitamins and minerals (e.g., B vitamins, magnesium, iron, and zinc) may only be present in meat sources. Athletes are therefore encouraged to work with a sport nutrition professional when or if they are following specific dietary patterns.

ANTIOXIDANTS

Exercise and training may increase free radical production and other forms of reactive oxygen species, which in turn may place an athlete in a state of oxidative stress. Although these effects are normal and a necessary part of muscle or metabolic adaptation, chronic resting inflammation or free radicals may alter lipids, protein, and DNA and potentially trigger human disease (168). Chronic oxidative stress may also disrupt skeletal muscle contractile properties, resulting in muscle weakness, fatigue, and a reduction in force production capabilities (205). Furthermore, free radicals may lead to muscle damage and post-exercise muscle soreness (45). While the human body possesses its own antioxidants, consumption of vitamins C and E may act as synergists to provide additional protection, due to their interaction in scavenging radicals (268). Because the potential for free radicals may increase with training, it has been suggested that athletes may require additional antioxidants to combat these effects; however, researchers have indicated that training itself may result in increases in antioxidant enzyme content within muscles (82). Readers should note that a well-balanced diet that includes a wide variety of fruits and vegetables typically supplies all the necessary antioxidants that an athlete may require and, thus, additional supplementation may not be necessary. However, supplementation may be necessary if there is a large increase in training volume or the number of competitions within a small timeframe (259).

HYDRATION

The hydration status of an athlete, and the practice of preventing dehydration, is a common concern among athletes, sport coaches, and strength and conditioning professionals; this is especially true in both hot and high-altitude environments (192). Small changes in either intra- or extracellular water have the potential to affect a variety of biological reactions, thermoregulation, and electrolyte balance (250). Researchers have indicated that mild dehydration of 1% to 2% can negatively affect maximal strength (229) as well as intermittent cycling performance (280), cardiovascular function and performance (177, 222), and cognitive function (285). Further research has also shown associations between prolonged mild dehydration and central nervous system damage (285). While the possible effects of dehydration on maximal strength are well documented (142, 224), it may also affect rapid force production characteristics (105) and high force production with high internal temperatures (78), indicating the importance of hydration status during training and competitions.

> **Key Point**
>
> Mild dehydration of 1% to 2% can negatively affect maximal strength.

Strength and conditioning professionals should note that the activity type, temperature, humidity, and clothing type may all affect the sweating rates of athletes. For example, 8% of body mass loss may occur during long-duration exercise (e.g., marathon) or repeat high-intensity exercise (e.g., football) if inadequate hydration practices are used (27, 216). It has been recommended that fluid intake that limits the loss of body mass to less than 2% of pre-exercise values should be used (225); this may consist of approximately 5 to 7 ml/kg of fluid being consumed at least 4 hours ahead of training. However, athletes and strength and conditioning professionals also should consider sports drinks during pre-exercise due to the addition of carbohydrates (46). In addition, athletes may consider incorporating sports drinks that include electrolytes such as sodium, potassium, and chloride, which play a role in regulating membrane potentials within skeletal muscle (114) and may indirectly affect thermoregulation and other metabolic functions (250). Researchers have indicated that hydration practices during exercise should also consist of a cold beverage (e.g., 50°F [10°C] compared to 98.6°F [37°C] or 122°F [50°C]) to attenuate the rise in internal body temperature (162, 163). Furthermore, it has been recommended that sports drinks should consist of approximately 4% to 8% carbohydrate because higher concentrations may delay gastric emptying (275). Following training, it has been recommended that 1.5 liters of fluid be consumed for every kilogram of body mass lost (225); however, athletes should consume fluids (ideally containing electrolytes) over time to benefit fluid retention over a longer duration (152).

It is important to note that the feeling of thirst may lag behind the need for the ingestion of water or other fluids (72); thus, athletes should practice regularly consuming fluids throughout the day whether they are training or not. Fluid ingestion in this manner will instill a mindset to ensure that the athlete is properly hydrated; however, athletes should be educated on their water consumption needs, how to track their intake (e.g., with mobile apps), and the dangers of water intoxication (i.e., hyponatremia) (2). While there are a variety of monitoring methods that may be used to assess the hydration status of athletes (e.g., urine osmolality, urine specific gravity), a simple method may begin with weighing athletes before and after training sessions or competitions. Another strategy that may be used is the scheduled assessment of hydration status during game or match days; this method allows the athlete to know that they will be assessed ahead of time so they may plan their fluid consumption to ensure that they are hydrated.

ERGOGENIC AIDS

The use of dietary supplements as ergogenic aids is prolific among competitive athletes (178, 179). Of all the existing supplements, creatine monohydrate, β-alanine, sodium bicarbonate, and caffeine are among the most prevalent. Thus, the following sections will provide an overview of the existing research related to strength characteristics and the recommended dosages of each supplement. While the information in each section may be general in nature, readers are referred to the International Olympic Committee consensus statement on dietary supplements (178) as well as several position stands from the International Society of Sports Nutrition (95, 101, 156, 264) for further information. Readers are advised to pay attention to third-party-tested supplements to ensure that athletes are not ingesting a banned substance; the most common and recommended certification logos can be found in a review by Smith-Ryan and colleagues (235). Finally, although strength and conditioning professionals should have baseline knowledge of the potential benefits of supplements and should be aware of what supplements their athletes are taking, it is recommended that athletes discuss potential strategies with a sport nutrition professional, especially when combining multiple supplements.

CREATINE MONOHYDRATE

Perhaps the most researched supplement available to athletes is creatine monohydrate. Creatine is an amino acid compound that naturally exists within the human body, of which approximately 95% can be found in skeletal muscle (155). Within the muscle, a primary role of creatine is to bind with a phosphoryl group to form PCr through the creatine kinase reaction. Briefly, PCr serves as a substrate to form and maintain ATP availability within the phosphagen energy system (227, 278), especially during short-duration, maximal efforts. Despite the role of creatine in energy metabolism, the total creatine pool (creatine and PCr) in skeletal muscle averages approximately 120 mmol/kg of dry muscle within a 70-kg individual (130). Provided that researchers have indicated that the upper storage limit of creatine is approximately 160 mmol/kg in most individuals (84, 130), there appears to be a strong rationale for dietary supplementation given the actions of creatine within skeletal muscle.

As just alluded to, creatine supplementation may enhance the ability of PCr to maintain ATP availability during short-duration, high-intensity muscle actions. This, in turn, may lead to a series of training adaptations that may benefit an athlete's strength characteristics. Researchers have indicated that creatine supplementation may increase acute exercise capacity and training adaptations in both adolescents (50, 56, 100, 143, 232) and young adults (19, 44, 148, 154, 248, 276). Specifically, supplementation may lead to increases in single and repetitive sprint performance, work performed during maximal effort sets of exercise, muscle mass and maximal strength, work capacity, and training tolerance (155). Put simply, creatine supplementation may allow athletes to train at higher intensities for longer periods of time, which in turn may produce positive strength adaptations that contribute to improvements in performance. Collectively, the research overwhelmingly supports the use of creatine supplementation for improving maximal strength characteristics (153, 160, 161); however, it should be noted that most of the research has been completed with male participants. Additional research has concluded that although female participants benefit from creatine supplementation, the improvements in maximal strength may not be as large as those compared to men (9, 100, 208, 258, 272).

Key Point

Creatine supplementation may lead to increases in single and repetitive sprint performance, work performed during maximal effort sets of exercise, muscle mass and maximal strength, work capacity, and training tolerance.

Beyond strength improvements, creatine supplementation may also benefit both recovery from training and rehabilitation from an injury. For example, Cooke and colleagues indicated that isokinetic and isometric force recovery was enhanced with creatine supplementation (49). Additional research showed greater isometric force production following eccentric leg extensions after chronic (30 days) versus acute (7 days) creatine supplementation (215). Regarding the response to higher training volumes, an enhanced tolerance of large increases in training volume during an overreaching phase was shown by Volek and colleagues (277), which suggests that recovery from such training may be improved with creatine supplementation. To complement these findings, researchers have also shown that creatine supplementation led to a reduction in inflammatory markers after multiple bouts of the running-based anaerobic sprint test (62) but may also reduce exercise-induced muscle damage (150). It should be noted that additional literature has shown no significant reduction in inflammation markers following high-volume resistance training programs (5 sets of 15-20 repetitions at 50% or 8 sets of 10 target repetitions at 70% of the 1RM) in resistance-trained men (211) or women (109), respectively. From an injury rehabilitation standpoint, researchers showed that a rehabilitation group that supplemented with creatine increased their isokinetic power output and quadriceps cross-sectional area to a greater extent compared to a placebo group, but the two groups displayed no differences in isometric knee extension improvements (115). Additional research supports these findings, in that improvements in strength during ACL reconstruction rehabilitation were no different compared to the placebo group following creatine supplementation (266). Collectively, creatine supplementation may allow for the recovery of force production characteristics, allow athletes to return to higher-intensity training, and decrease exercise-induced muscle damage; however, the existing literature focused on muscle inflammation and rehabilitation from injuries appears to be mixed.

While a typical diet may include 1 to 2 g/day of creatine, this may only saturate creatine stores up to approximately 60% to 80%; thus, creatine supplementation may allow individuals to increase creatine and PCr stores by 20% to 40% (84, 130). It should, however, be noted that vegetarian diets may alter these percentages and the ability to replenish creatine stores (30); thus, athletes who follow a vegan or vegetarian diet may actually benefit more from creatine supplementation. From a diet standpoint, creatine can primarily be found in protein-rich foods such as red meat and seafood (22, 255). Beyond a traditional diet, research findings support the consumption of approximately 0.3 g/kg/day of creatine monohydrate for 5 to 7 days as an initial loading period before ingesting 3 to 5 g/day to maintain elevated creatine stores, although larger athletes may need to ingest greater amounts (e.g., 10 g/day) (153, 156).

Further evidence supports the addition of carbohydrate or a carbohydrate and protein mixture to creatine supplementation to increase the uptake of creatine (86, 156, 239). It should be noted that once elevated, it may take 4 to 6 weeks to return to baseline levels (130, 272); however, there does not appear to be evidence that muscle creatine levels will fall below baseline after the cessation of supplementation (149, 157). Despite anecdotal claims that creatine supplementation increases the risk of musculoskeletal injuries, dehydration, muscle cramping, gastrointestinal distress, kidney dysfunction, or long-term detrimental health effects, the existing evidence does not support these thoughts (85, 87, 88, 135, 149, 157, 213); in fact, the opposite may be true in many cases (156). It should also be noted that creatine supplementation may have ergogenic benefits across a wide age range, including younger individuals, adolescents, and older populations (20, 29, 55, 135, 155, 157), which further promotes the safety of its use.

β-ALANINE

β-alanine is an amino acid that is produced within the liver. While the individual effects of β-alanine may be limited, its primary function is to serve as a rate-limiting precursor to the formation of carnosine (108); specifically, β-alanine combines with L-histidine. Thus, the primary purpose behind β-alanine supplementation is to increase the formation of carnosine within the muscle. Briefly, carnosine is a naturally occurring dipeptide primarily found in skeletal muscle (1) that serves in a buffering capacity to mitigate the decrease in pH during high-intensity exercise (5, 53), although its greatest effects may affect activities ranging from 1 to 4 minutes in duration (123, 219). Interestingly, it has been suggested that the buffering effects of carnosine may precede those of other buffering systems, such as the bicarbonate system (256), perhaps indicating that it may affect shorter efforts that require more rapid force production.

Regarding its effect on strength characteristics, β-alanine may serve athletes by first increasing carnosine concentrations within skeletal muscle, which may increase the capacity to perform high-intensity work. For example, researchers have shown that β-alanine supplementation increased the number of repetitions performed during a back squat (125), mean and peak power output during a cycling sprint after exhaustive exercise (270), repeat sprint performance (185), time to exhaustion during cycling tests (236), and 2,000-meter rowing performance (11). Furthermore, increases in the physical working capacity at fatigue threshold in both men (251) and women (252) were found after 28 days of β-alanine supplementation, indicating a reduction in neuromuscular fatigue. Although the previous results may provide an indication that individuals may increase their capacity to produce and apply force more effectively during high-fatigue conditions, the research focused on various strength characteristics has shown mixed results. Researchers have shown that β-alanine supplementation may improve back squat 1RM (175), isometric knee extension force and countermovement jump performance (132), isometric strength endurance (15, 218), training volume completed (126, 175), and maximal isometric knee extension peak torque (274). However, additional research found no improvements in maximal strength after resistance training following high-intensity interval training (77), in vivo force (103), or isotonic or isokinetic strength–endurance (15); in another study, researchers found no differences in strength improvements with a placebo group following a 10-week training intervention (146).

The available literature supports the notion that β-alanine supplementation may aid in the development of work capacity during high-intensity exercise; however, its effect on maximal strength and other force production characteristics remains inconclusive (53, 123, 207, 219, 264). It should be noted, however, that if β-alanine improves training capacity at higher intensities (124), it is possible that these effects may carry over into force production benefits within a situational context; thus, further research on this topic is warranted.

Additional research on β-alanine has focused on the potential additive effects when combined with other supplements such as creatine or sodium bicarbonate. Hoffman and colleagues (124) indicated that a combination of β-alanine and creatine produced significant improvements in training volume for both the back squat and bench press exercises as well as back squat average training intensity and 1RM in strength–power athletes; however, the latter results did not appear to differ from creatine supplementation alone. Additional research showed no additive benefits of creatine when assessing the physical working capacity at fatigue threshold (251). When combined with sodium bicarbonate, researchers showed improvements in 200-meter swimming time (59), 2,000-meter rowing time (122), and total work performed while cycling 110% of maximum power output (220); however, the additive effects of sodium bicarbonate are unclear, because isolated β-alanine supplementation produced similar effects. The existing literature that has combined β-alanine with other supplements suggests that some positive effects may occur, but further research is needed to draw more concrete conclusions (264).

Chronic dosages of 4 to 6 g of β-alanine ingested per day (multiple dosages of ≤2 g) for at least 2 weeks may result in 20% to 30% increases in muscle carnosine concentrations (12); however, increases of 40% to 60% may be displayed after 4 weeks of supplementation (107, 241). It should be noted that if athletes are ingesting a non-time-release version, daily doses of 6 g may be required to increase carnosine concentrations within the muscle (242). Although the side effects of β-alanine supplementation are limited, athletes may experience a tingling sensation (i.e., paresthesia), especially when larger single doses are ingested (264). Researchers have also shown that consuming β-alanine with a meal may augment the muscle carnosine improvements (240), specifically with foods that contain higher levels of carnosine, such as beef, pork, poultry, and fish (106). In fact, researchers have shown that vegetarianism has been linked to lower carnosine levels (73), which may be an important consideration for athletes considering β-alanine supplementation. Like other supplements, strength and conditioning professionals and athletes should be aware of the individualized nature of the responses with β-alanine supplementation. For example, high responders and low responders may increase their carnosine concentrations following 5 to 6 weeks of supplementation by approximately 55% and 15%, respectively (12); however, it should be noted that high or low responder status may be based on the baseline muscle carnosine content and muscle fiber type of each athlete (106). Similar to the supplementation effects, the washout time (i.e., time to return to baseline values) may also vary, with periods of 6 to 16 weeks noted within the literature (12, 287) and with the timing being related to high-intensity exercise tolerance (287).

SODIUM BICARBONATE

Another common supplement used by athletes is sodium bicarbonate. Compared to creatine supplementation, the primary purpose of sodium bicarbonate supplementation is to buffer hydrogen ions (H^+) that are produced during high-intensity exercise via anaerobic glycolysis to maintain a homeostatic state and prevent a pH reduction (159). This, in turn, may delay the onset of metabolic acidosis and a subsequent reduction in force production, which may benefit athletes in sports that include explosive efforts, repeated bouts, and large muscle mass actions (97). While research has consistently supported the use of sodium bicarbonate supplementation for muscle endurance, the same cannot be said for maximal strength (91, 95, 97). However, some researchers have shown that an increased number of repetitions of resistance training exercises (e.g., squat, leg press) may be performed following supplementation (40, 69). Further research indicated that sodium bicarbonate supplementation may enhance repeated effort performance during high-intensity intervals while running (90, 158, 174), cycling (23, 24, 186), and swimming (80, 231, 290). Because *strength* refers to the ability to produce force in a task-specific context, it could be argued that although maximal strength and peak force may not benefit from sodium bicarbonate supplementation, athletes may display an improved ability to apply forces during repeated high-intensity efforts, suggesting an acute improvement in strength. Although the majority of studies have been completed with male participants, it should be noted that female participants may also benefit from sodium bicarbonate supplementation (95, 223).

Despite the majority of sodium bicarbonate supplementation research focusing on its isolated effects, researchers have also investigated its additive effects when paired with other supplements such as creatine (14, 99, 181), caffeine (39, 119, 206), and β-alanine (54, 59, 122). The collective body of research appears to support the additive effects of sodium bicarbonate combined with creatine and β-alanine; however, less evidence supports its combination with caffeine (95). Given the range of participants in these studies (e.g., physically active to elite-level athletes), it is important to understand that the additive effects may be task, dose, time, and individual dependent; thus, readers should exercise caution when combining different supplements to ensure maximal effectiveness. Moreover, it should be determined when and if an athlete's sport, event, or training calls for this type of supplementation strategy.

Sodium bicarbonate supplementation is typically completed on a single-dose basis in the hours leading up to training or competition, although multiday protocols have also been used (95). For single-dose protocols, the recommended dose is 0.3 g/kg (139, 180), which should be ingested between 60 and 180 minutes before training or competition. Multiday protocols, on the other hand, may range from 3 to 7 days with dosages of 0.4-0.5 g/kg/day, in which 0.1 to 0.2 g/kg may be ingested during multiple meals throughout the day (95). It should be noted that higher doses (i.e., 0.4-0.5 g/kg) may not provide any additional benefits compared to smaller doses (e.g., 0.3 g/kg) (66) and may also increase the potential for side effects such as bloating, vomiting, diarrhea, and abdominal pain (180). However, some researchers have indicated that delayed-release capsules may be used to alleviate potential side effects (121). Beyond the previously discussed dosages, the timing of dosages may have a significant impact on performance adaptations as well as potential side effects, particularly due to individualized responses (111, 140). Thus, it is important that athletes practice various supplementation strategies before using the supplement in training or competition.

CAFFEINE

Typically ingested in the form of beverages such as coffee, soft drinks, tea, or energy drinks, caffeine is the most widely consumed psychoactive substance in the world (101) and its use has risen over the past 2 decades (13). Because of its widespread use, an abundance of research has sought to determine the effect of caffeine on training and sport performance (64, 79, 92, 93, 94, 120). From a physiological standpoint, caffeine serves to block the actions of adenosine (i.e., reductions in neuronal firing and neurotransmitter release) by binding to its receptors (76). This, in turn, allows caffeine to essentially "wake up" the nervous system while also stimulating the secretion of epinephrine (83), which may affect the force production characteristics of an athlete. Additional literature has suggested that caffeine may increase the threshold for pain perception (67, 68, 172), which may increase the ability of an athlete to train at higher intensities for longer durations. However, further research is needed to confirm these findings.

Although there is some conflicting evidence (204, 260), an overwhelming amount of literature supports the notion that caffeine supplementation may benefit maximum dynamic, isometric, and isokinetic strength (16, 96, 98, 281), rapid force production (16, 93), strength–endurance characteristics (204, 281), and movement velocity during resistance training (94, 212). Moreover, researchers have shown that caffeine supplementation may reduce rating of perceived exertion (64) as well as other indicators of neuromuscular fatigue (253), which may explain the potential increases in exercise capacity. Additional research supports the use of caffeine supplementation for ballistic performances such as jumping and throwing (18, 98, 217, 221).

Although the additive effects of caffeine mixed with other supplements have been outlined previously, combined caffeine and creatine supplementation has been commonly studied due to the individual benefits of each (263). Researchers have suggested that both supplements may have counteracting mechanisms on Ca^{+2} clearance and release and muscle relaxation time (116, 271). Thus, further research is needed to identify the optimal dosages to determine whether there are additive benefits to consuming both supplements concurrently. It is important to note that while most of the previous literature used male participants, caffeine supplementation is also effective for women (89, 183, 190). Furthermore, the effectiveness of supplementation may be based on the timing and dosage (191) as well as on the genetic characteristics of the athlete (238).

Key Point

An overwhelming amount of literature supports the notion that caffeine supplementation may benefit maximum dynamic, isometric, and isokinetic strength as well as rapid force production, strength–endurance characteristics, and movement velocity during resistance training.

The general recommendations for caffeine supplementation include ingesting 3 to 6 mg/kg 60 minutes prior to training (101). Readers should note that beyond the beverages highlighted previously, caffeine can also be consumed in other forms, such as chewing gum, energy gels and chews, aerosols, and other caffeinated food products (283). The form of caffeine supplementation may alter the timing of the observed effects; for example, researchers have shown that chewing gum may increase the rate of caffeine delivery compared to caffeine in capsule form (144). Moreover, timing and retention of caffeine concentrations may be based on individual characteristics, as mentioned previously. Products that contain caffeine may range in dose from 1 mg (e.g., chocolate milk) to more than 300 mg (e.g., dietary supplements) per serving (63), making it important for consumers to understand how much caffeine they are ingesting so they may appropriately time its effects to benefit their performance and minimize negative side effects. Consumers should also be aware of the several potential side effects of caffeine supplementation, which may include tachycardia and heart palpitations, anxiety, headaches, insomnia, hindered sleep quality, and increased jitters (60, 196).

PERIODIZATION OF NUTRITION

Nutritional stimuli, like training and recovery stimuli, induce physiological and performance changes and should thus be periodized to address the needs of an athlete while also promoting the desired adaptations at specific periods of time (189). Specific to nutrition, the energy demands of an upcoming practice, training session, or competition may affect the type and timing of macronutrients and micronutrients, hydration practices, and supplementation. Regarding carbohydrate intake, some have promoted the idea of a train-low, compete-high model in which carbohydrates are restricted before, during, and after endurance training sessions but are increased prior to competition. While researchers have shown that this model may improve both whole-body (289) and intramuscular (129) fat metabolism, exercise capacity (104), and performance (173), readers should also be aware that the train-low, compete-high method may reduce the ability to train at higher intensities (289), increases the potential for illness and infection if high-intensity training is consistently performed under these conditions (81), promotes muscle protein breakdown (128), and decreases the ability to use carbohydrates for energy (51). However, Burke and Hawley (33) indicated that a train-high, sleep-low model (figure 11.1) may benefit high-intensity training while also increasing the time period of low carbohydrate availability during recovery and additional aerobic training. This, in turn, may increase the amount of time for transcriptional activation of metabolic genes and their target proteins, thus improving enzyme activation of carbohydrate and fat oxidation and mitochondrial biogenesis. Regardless of the carbohydrate periodization model, athletes are encouraged to "fuel for the work required" (131).

Beyond carbohydrate manipulation, nutrient timing may play a crucial role in whether an athlete has an effective training session, maximizes the training adaptations following training, or is prepared for an upcoming training session (134). As mentioned previously, athletes should prepare for a given training session by ingesting a combination of carbohydrate and protein to promote muscle protein synthesis; this may include low GI carbohydrates and whey protein. Following training, promoting recovery and the adaptations trained during the given session may be produced by ingesting a combination of high GI carbohydrates and whey protein. As previously discussed, post-exercise ingestion should start almost immediately and be sustained for several hours; this, in turn, may promote both glycogen and protein resynthesis. Beyond pre- and post-training recommendations, additional protein synthesis can also be promoted by consuming casein protein prior to sleep.

In summary, athletes need to be properly fueled to be able to produce and apply force during a given session and given the nutritional tools to recover their force production capabilities to prepare for the following session. Put succinctly, athletes need to "repair, refuel, and rehydrate" to properly recover and realize training adaptations (71). Moreover, it should be emphasized that athletes should be fueling throughout the day (e.g., breakfast, mid-morning snack, lunch, afternoon or pre-exercise snack,

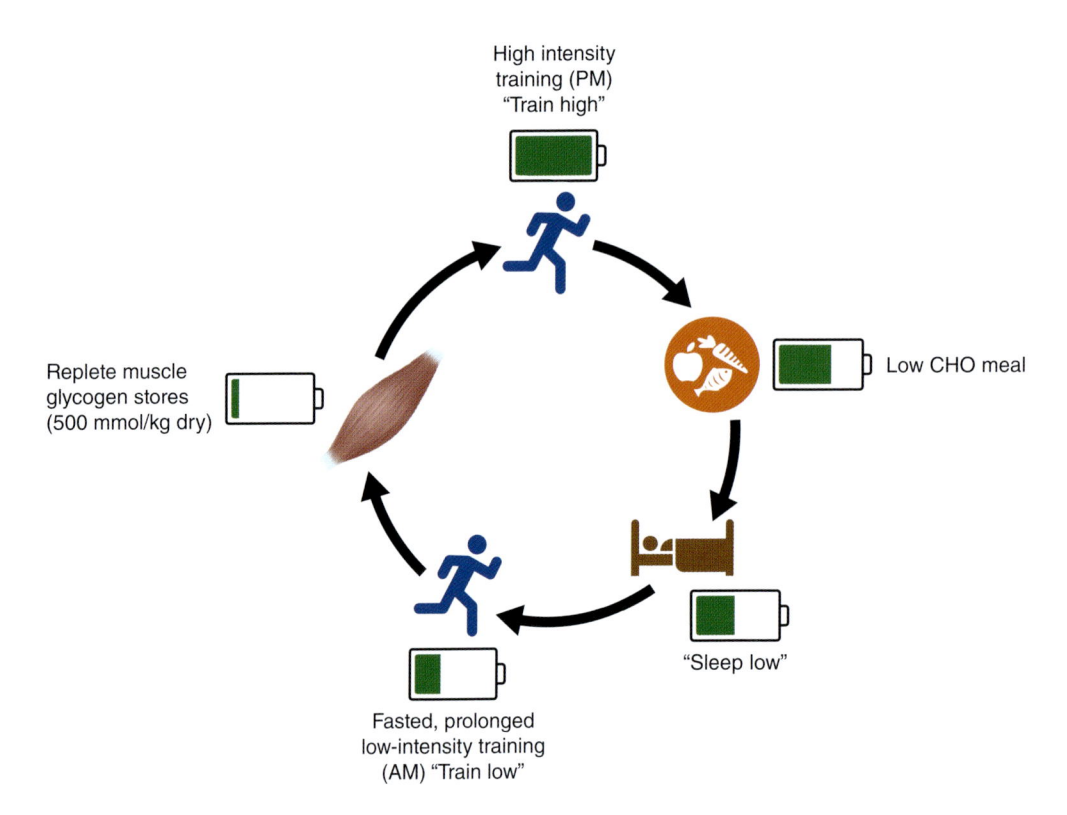

FIGURE 11.1 Periodization of carbohydrate intake using the train-high, sleep-low model.
Adapted by permission from L.M. Burke and J.A. Hawley, "Swifter, Higher, Stronger: What's on the Menu?" *Science*, 362 (2018): 781-787.

post-exercise snack, dinner, and evening snack) to ensure that they are meeting their energy needs (4). Readers interested in more information about periodized nutrition and nutrient timing are referred to reviews by Jeukendrup (138) and Kerksick and colleagues (147).

MONITORING NUTRITION

The ability to track and monitor the dietary intake of athletes may ensure that the desired training goals are being supported by proper nutritional practices. While this may be accomplished by keeping a written journal, a variety of nutrition apps exist as well; however, it is important to understand the benefits and limitations of each (226). There are several obvious benefits to nutrition tracking apps, such as the convenience of being able to enter information on any mobile device, cloud-based information that is available nearly everywhere, and creation of athlete awareness regarding both their food and kilocalorie intake as it relates to their training (31). Regarding the latter, this may create a self-sufficient athlete who takes ownership of their dietary intake and the overall training process (250). A potential limitation to nutrition apps is that some athletes may become overly obsessive about calorie counting, especially if they do not achieve their desired body composition or performance goals. Thus, it is crucial that strength and conditioning professionals are aware of the signs of disordered eating and eating disorder tendencies. Moreover, athletes should be educated on best nutritional practices and the negative impact that disordered eating and associated eating disorders may have on the athlete's health and performance.

Take-Home Points

- The energy requirements of an athlete may vary based on the volume, intensity, duration, and frequency of training; thus, it is important that athletes consume the necessary calories to allow them to consistently train well and perform in competition.
- At the most basic level, athletes should consider their energy balance as calories consumed versus calories expended. Macronutrients (carbohydrates, proteins, and fats) serve as the foundation of nutritional intake and should be planned to supply enough energy for the programmed training sessions and to promote recovery and adaptations following training.
- Adequate vitamin and mineral ingestion is important for the body's normal physiological processes but also the completion of various muscle actions and metabolic reactions related to force production.
- Athletes can supplement their diet using ergogenic aids such as creatine monohydrate, β-alanine, sodium bicarbonate, and caffeine. While creatine supplementation has been shown to have a significant impact on muscular strength characteristics, β-alanine, sodium bicarbonate, and caffeine may relate more to the ability to perform greater amounts of overall work.
- Athletes are encouraged to periodize their nutrition and fuel for the work required within their training and competitions. Consulting with a sport nutrition professional may help athletes develop a plan and monitor their nutritional practices to ensure that the necessary nutrients are consumed to perform and recover throughout their career.

CHAPTER 12

Recovery Considerations for Strength

While the content within the previous chapters has provided readers with insight on underpinning mechanisms of strength and how to effectively design training programs to improve an athlete's strength characteristics, it is important to recognize the reality of competitions in the modern world. Simply, the number of competitions for athletes, even at a young age, has increased, which places additional challenges on how to best prepare athletes to consistently perform at an optimal level. This, in turn, has introduced new challenges to strength and conditioning professionals in how to help athletes recover for specific training sessions or other competitions (129). Combined with a variety of other external stressors, the ability (or inability) of an athlete to perform can be challenged (85).

Although more frequent competition may introduce new challenges, the concept of recovering an athlete's strength characteristics is not new. In fact, the use of recovery methodology has existed in some form since at least ancient Egyptian, Greek, Roman, Chinese, and Japanese civilizations (68). Sands and Murray (121) displayed a wide variety of definitions of recovery and concluded that there is no consensus on the term. To discuss the recovery of strength, it is important to first provide a definition of fatigue. Interestingly, a common theme of most definitions of fatigue refers to a reversible reduction in force that may not be observable but could result in the failure to complete a given task (152). Given the definition of *strength* discussed in chapter 1 (i.e., the ability to produce force against an external resistance within a task-specific context), the term *fatigue* for the purposes of this chapter will refer to the diminished ability to produce the necessary force (magnitude and rate) that results in a performance decrement during a given task. In this light, *recovery* will be defined as the improvement in force production from a fatigued state that allows an athlete to perform a given task with minimal hindrance.

Key Point

Recovery may be defined as the improvement in force production from a fatigued state that allows an athlete to perform a given task with minimal hindrance.

Before discussing different methods of recovery, it is important to acknowledge that training (work) and recovery of strength may run in parallel and can both lead to the return of an athlete's baseline abilities and performance enhancement (117). In this light, a brief discussion of the relationships among training, fatigue, and adaptation is necessary. Han Selye's *general adaptation syndrome* (GAS) is based on the premise that a decrease in performance follows the introduction of a new stimulus (31, 125, 126). This acute performance decrement is likely caused by a given level of fatigue that results from the training stimulus. However, while recovery is meant to return an athlete to a baseline level of performance, the goal of long-term training is to enhance an athlete's performance beyond their baseline. Within the GAS model, Selye (124) noted a substantial improvement in the resistance to stress following a return to baseline levels and termed this "adaptation energy"; this has since led to the term *recovery–adaptation*, with the idea that recovery should replenish adaptation energy (117, 118). Harre (57) further described this within a training context by discussing the interplay of work and recovery, in which the decisive stimulus that produces the sought-after adaptation that occurs during the recovery phase by replenishing the spent energy (*regeneration*), but to a level beyond an athlete's initial baseline (*overcompensation*). Thus, the overall stimulus

181

given to an athlete should combine the effects of both the training and recovery stimuli to produce a desired adaptation (e.g., strength).

Key Point

The overall stimulus given to an athlete should combine the effects of both the training and recovery stimuli to produce a desired adaptation (e.g., strength).

An often overlooked and underappreciated factor is how well (or not well) an individual has recovered from previous activities or competitive events, which may directly affect subsequent training and competition. Thus, it is important to promote recovery practices that will create an environment that allows an athlete to return their force production capabilities (maximal and rapid force production) to a level that will prepare them for these activities. There are a variety of recovery methods to choose from; however, the benefits, limitations, and practicality of these methods should be considered for each athlete. Because specific recovery methods may not be accessible or financially feasible, it is important to understand which methods may provide a solid foundation to promote recovery and adaptation. This chapter will discuss the benefits and limitations of common recovery methods and how to periodize and program various recovery methods to promote an athlete's strength characteristics.

RECOVERY METHODS

The most straight-forward method of recovery is to stop training and rest; however, Sands (117) suggested that because athletes and coaches are "doers," it may serve the best interests of all parties to select recovery methods that involve action. It is important for strength and conditioning professionals to understand the potential benefits and limitations of different recovery methods, so they can promote evidence-based practice rather than anecdotal claims or commercial products that may serve as little more than a gimmick or fad.

The primary goal of recovery is to return the body to its physiological and psychological state of homeostasis, with the strategies of different methods targeting the metabolic, mechanical, or cognitive alterations produced from a training stimulus (e.g., resistance training, practice, or competition) (5, 133). Despite this being the goal, previous researchers have failed to identify which variables should be returned to a homeostatic state (42, 83, 147). Sands (117) suggested that most approaches of recovery aim to enhance blood circulation, with the goal of clearing by-products of heavy exertion, such as edema (indicating an immune response), remnants of damaged muscle tissue, and catabolic markers of stress. In this light, there are a variety of recovery methods that may be used to return an athlete's strength characteristics to their homeostatic state.

Key Point

The primary goal of recovery is to return the body to its physiological and psychological state of homeostasis.

While much of the existing literature shows mixed findings, typically due to different study designs, populations, and protocols, two recovery methods may serve as the recovery foundation for athletes. Stephens and Halson (133) depicted the *recovery pyramid* using both sleep and nutrition as the foundational levels of an athlete's recovery (figure 12.1). This is likely due to the large impact that sleep and nutrition may have on performance but also to the ability to have greater control of these methods (e.g., going to bed at an earlier hour or eating at certain times). Additional literature discussing the preparation of Olympic athletes supports these foundations as well (110). Other methods such as cold-water immersion (CWI), hot-water immersion (HWI), compression, and massage may be less supported by the literature and thus serve as a smaller portion of the recovery pyramid. Finally, the top of the pyramid should be reserved for gimmicks or fads that may be viewed as popular but are solely based on anecdotal evidence. It should be noted that other methods of recovery may be implemented; however, it is important that strength and conditioning professionals and athletes focus on the foundation and control what they have the capability of controlling.

Key Point

Sleep and nutrition serve as the foundation for athlete recovery.

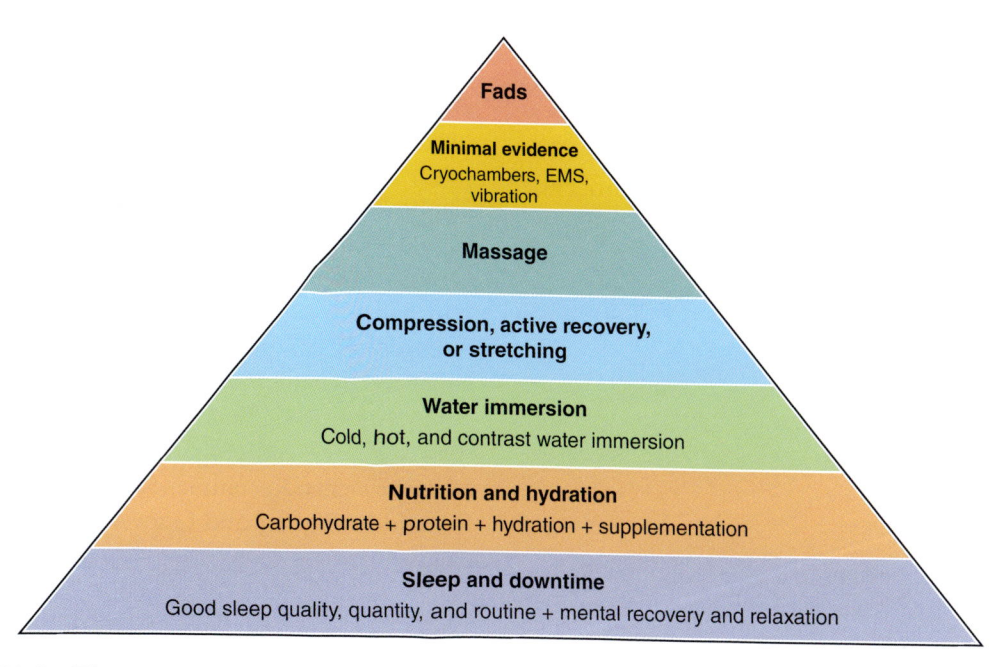

FIGURE 12.1 The recovery pyramid.

Reprinted by permission from J.M. Stephens and S.L. Halson, "Recovery and Sleep," in *NSCA's Essentials of Sport Science*, edited for the National Strength and Conditioning Association by D.N. French and L.T. Ronda (Human Kinetics, 2022), 356.

SLEEP

Sleep is something that all living creatures need; however, it is also something that some athletes may take for granted as it relates to their performance. Researchers have shown that sleep quantity, sleep quality, or both in elite athletes may be less than optimal (14, 77, 81). Additional findings suggest that athletes who report sleep problems display worse sleep hygiene, more health complaints, and mood disturbances (10). While sleep plays an important role in the ability to train, recover, and perform (24), inadequate sleep may negatively affect an athlete's mood, metabolism, immune and cognitive function, and well-being (54). Despite the variety of factors that may influence an athlete's sleep (e.g., muscle soreness, injury, jet lag, climate), it is important to control as many variables as possible and avoid certain behaviors that may be detrimental (e.g., use of smartphones or electronics or caffeine consumption before sleep). Simple changes to some of these behaviors may result in a significant difference. For example, researchers have shown that a minimum of 1 week of improved sleep duration may positively affect an athlete's performance (13).

It is recommended that elite male and female athletes need approximately 8.3 hours of sleep per night to feel rested (122). However, many athletes do not meet this standard or the minimum recommended number of hours (77, 81). The results of a 2022 meta-analysis indicated that sleep loss can have a significant impact on strength (30). Specifically, sleep deprivation (i.e., extended wakefulness in which no sleep is obtained for at least 24 hours) and restricted sleep (i.e., partial sleep deprivation in which the amount of sleep deviates from an individual's normal sleep habit [109]) resulted in a reduction of strength of 3.0% and 2.8%, on average, respectively. It should be noted the sleep restriction findings were heavily weighted on late sleep restriction (i.e., waking up earlier than normal and waking up multiple times during the night [86]), in which an average 4.5% reduction was shown (30). The authors of the meta-analysis also showed that the loss of sleep may significantly affect strength performance in both morning (−1.8%) and afternoon (−4.6%) exercise sessions. Late restriction had the largest negative impact on strength performance during afternoon sessions (−10.5%); however, sleep deprivation also produced significant negative effects on both morning (−2.4%) and afternoon (−3.8%) training sessions. Finally, Craven and colleagues (30) showed that sleep loss significantly affected lower-body strength (−3.4%), while reductions in upper-body strength (−1.6%) were not significant. Specific to the lower body, significant reductions in strength were shown with sleep deprivation (−3.3%), but not for any other type of sleep loss.

> **Key Point**
>
> Sleep deprivation and restricted sleep may result in a reduction of strength of 3.0% and 2.8%, on average, respectively.

Halson (52) suggested that physiological alterations that negatively affect performance can occur when sleep patterns are disturbed. For example, researchers have concluded that sleep loss may delay glycogen repletion and negatively affect subsequent performance (127). Other alterations due to sleep deprivation may include both the immune and endocrine systems. Main and colleagues (87) indicated that sleep loss was associated with an elevated proinflammatory cytokine response. Additional researchers also showed elevated creatine kinase and C-reactive protein levels with sleep loss (128). Combined with negatively impacted testosterone concentrations (32) and sleep-related rises in serum testosterone (86), it is not surprising that reduced amounts of sleep can negatively affect an athlete's strength characteristics.

Educating athletes (19, 37, 98) and improving athlete sleep hygiene (37) may have a positive impact on an athlete's quantity and quality of sleep and, by extension, their performances. Previous literature has noted that sleep can be improved through appropriate training schedules, consistent sleep schedules, proper nutritional habits, a bedtime routine that promotes relaxation, management of psychophysiological stress, and techniques that can help individuals fall asleep and return to sleep (59, 94, 95). Halson (52) also suggested combining positive sleep hygiene with other recovery strategies that initiate sleep onset and decrease inflammation and delayed-onset muscle soreness. Promoting effective sleep hygiene habits becomes especially important because jet lag that results from traveling can have a significant impact on sleep deprivation (44).

NUTRITION

As noted in chapter 11, proper nutritional practices may properly fuel the athlete but also promote a positive environment for recovery and adaptation. The foundations of proper nutrition for an athlete should focus on macronutrient intake, specifically carbohydrates and protein, but should also emphasize hydration. Through proper nutrient timing, individuals can promote positive recovery and enhanced preparedness for subsequent training and competition (69). Finally, athletes may augment their recovery and preparation by using a variety of supplements. Recommended nutritional practices for athlete fueling and recovery are discussed in chapter 11.

WATER IMMERSION

Water immersion therapy typically refers to either full-body or limb-only immersion and can be implemented using CWI, HWI, or their combination, typically referred to as a *contrast bath* or *contrast water therapy* (CWT). Similar to other methods of recovery, the type of water immersion should be chosen based on what the athlete is recovering from or what they may be recovering for (133).

Cold-Water Immersion

The general purposes of CWI (figure 12.2) are to reduce the temperature of the athlete's body tissue and blood flow to specific areas of the body to lessen swelling, inflammation, cardiovascular strain, and perceived pain (134). During CWI, athletes are exposed to water temperatures between 40°F and 68°F (5°C-20°C) either continuously or intermittently for up to 20 minutes (145). In addition to the hydrostatic pressure of the water, CWI promotes vasoconstriction that may reduce muscle perfusion and interstitial fluid diffusion (49) as well as cellular, lymphatic, and capillary permeability (26). These effects may decrease edema, which in turn may lessen pain and allow for force production to be maintained to a greater extent (130).

Compared to other recovery methods, CWI may hold an advantage. Montgomery and colleagues (91) showed greater recovery of 20-meter acceleration, line drill performance, and flexibility with CWI compared to wearing compression garments and combined carbohydrate and stretching routines

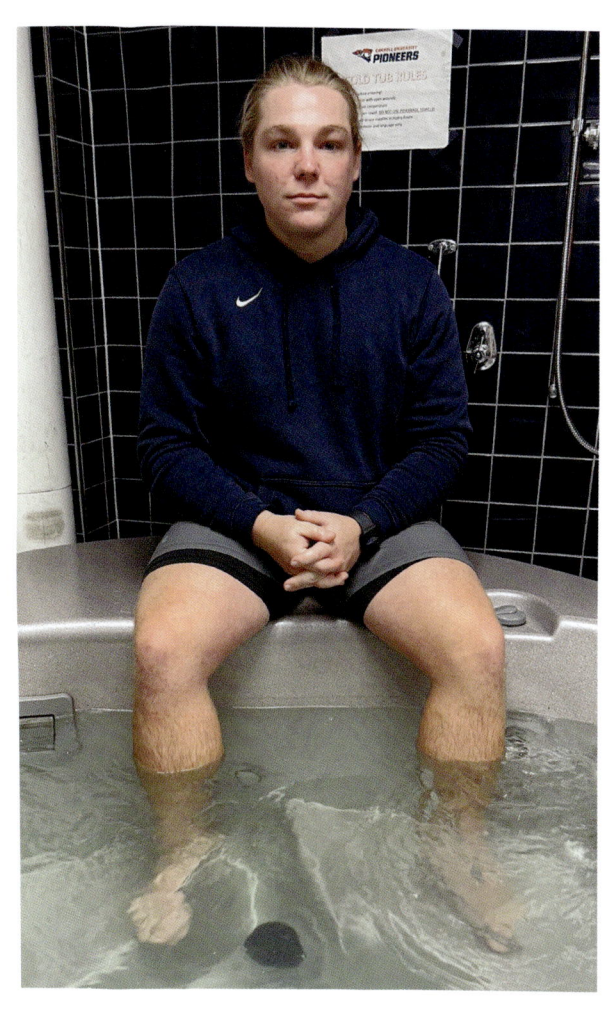

FIGURE 12.2 Cold-water immersion.

during a 3-day basketball tournament. In addition, despite no differences in countermovement or repeat sprint ability, junior soccer players reported decreased leg soreness and general fatigue following CWI compared to a thermoneutral immersion group during a 4-day soccer tournament (112). CWI was also effective at improving the recovery of squat jump and isometric force while reducing creatine kinase concentration 48 hours after exercise compared to a passive control condition (139). Further research has shown enhanced recovery of sprint speed and attenuated release of creatine kinase during a simulated team sport tournament following CWI compared to a controlled condition (82). Researchers have also concluded that CWI may benefit the recovery of neuromuscular performance during vertical jumping and isometric strength testing (63, 139). Another meta-analysis also showed that CWI attenuated strength loss across a variety of time periods (<6, 24, 48, 72, and 96 hours) compared to passive recovery (9).

Despite a larger body of literature compared to other recovery methods, there does not appear to be a gold standard for implementing CWI with regard to water temperature, depth, duration, or mode of immersion (134). Thus, the CWI protocol may be based on the athlete, their characteristics, and what they are recovering from. Stephens and colleagues (132) indicated that athlete body composition, sex, age, and ethnicity may affect the physiological responses to CWI. In fact, customized CWI protocols based on water immersion time and water temperature produced greater recovery of heart rate variability and squat jump mean power compared to active rest and a standardized CWI protocol (155). It has also been suggested that less intense protocols using warmer temperatures or shorter durations should be used with athletes with low body fat or muscle mass, and that female, youth, and masters athletes require less intense protocols compared to the average adult male athlete (132).

Finally, previous literature has suggested that CWI may be effectively periodized over the course of a competitive season (72). For example, the previous authors suggested that CWI may be implemented effectively during periods in which many games or matches are played within a short period of time, but not during periods where the goal is to facilitate positive training adaptations. Taking this into account, CWI may be limited during the off-season training period in order to allow for training adaptations to occur unimpeded (72).

Hot-Water Immersion

The goals of HWI are relaxation and easing muscle tension (133). This method is typically implemented using a hot tub with underwater jets to massage the muscles (145) with water temperatures greater than 96°F (36°C). From a physiological standpoint, it is believed that HWI increases body temperature and blood flow (140), leading to enhanced removal of metabolic waste and increased nutrient delivery to and from the muscle cells (149). Some researchers have stated that these physiological responses may lead to the recovery of neuromuscular performance (140, 149); however, there is minimal evidence to support the use of HWI for strength recovery purposes. Peake and colleagues (101) indicated that regular exposure to HWI over a 10-week training program did not enhance strength adaptation and attenuated improvements in hypertrophy. While isometric force recovery benefited from HWI, weighted squat jump performance, thigh girths, perceived pain, and blood markers of fatigue did not show signs of recovery compared to passive recovery (139). Additional researchers have reported that HWI may diminish potential training adaptations (145), suggesting that HWI should not be combined with additional training stimuli.

Despite the effects of HWI displayed within the literature, the ability to make the athlete comfortable during a recovery session speaks volumes and may lead to positive effects. Thus, HWI may serve as an effective recovery method due to its ability to create a comfortable environment for the athlete to relax in (117). However, it is important that strength and conditioning professionals recognize and understand the potential side effects of HWI if it is not prescribed carefully (e.g., acute burns, loss of vasomotor control, hypotension, faintness, tachycardia) (93, 137).

Contrast Water Therapy

Contrast water therapy (CWT), also known as *contrast baths*, typically requires athletes to transition back and forth between CWI and HWI 3 to 7 times and to spend 1 to 2 minutes in each (145). Like CWI and HWI, the purpose of using CWT for recovery is to enhance blood flow to remove metabolic waste within the muscles. However, unique to CWT, the transitions between cold and hot water temperatures are thought to create a pumping effect in which the blood vessels constrict and dilate, respectively, to enhance blood flow (140, 141).

Meta-analytical results have shown that CWT may provide a small, positive effect on delayed-onset muscle soreness (38); however, additional findings suggest that the effects of CWT on recovery within team sports is minimal (63). While researchers have shown the potential effectiveness of CWT on muscle strength and power recovery (138, 139, 141, 144), other researchers have shown no added benefit of this method (46). Furthermore, there does not appear to be a consensus on the number of CWI and HWI exposures or the duration of each exposure. Crampton and colleagues (29) showed that 1:1 and 1:4 CWI:HWI exposures enhanced recovery of high-intensity cycling performance to a greater extent compared to passive recovery. However, additional researchers also support 1:1 protocols (53, 144), with one of the protocols using 6 minutes of CWI and HWI exposures (144). It should be noted, however, that CWT may not provide any additional recovery benefits compared to CWI, HWI, compression, active recovery, and stretching (9). Furthermore, researchers have shown that CWT did not benefit neuromuscular recovery in team sport at 24, 48, or 72 hours (63). It has been recommended that if CWT is used as a recovery method, it should be prescribed sparingly so as not to blunt its potential benefits (103, 129). The periodization of various recovery methods will be discussed in greater detail later in this chapter.

COMPRESSION

Another method of recovery that may be frequently used by athletes is compression. *Compression* can take multiple forms in terms of providing consistent pressure or varying pressure. The following paragraphs will discuss the different types of compression methods as they relate to the recovery of various force production characteristics.

Compression Garments

The use of compression garments stems from the medical field and they are typically used to enhance circulation, lymphatic flow, and venous return (16), which in turn may reduce swelling (edema). Kraemer and colleagues (75) showed positive effects on bench press throw power output as well as other physiological and psychological markings among resistance-trained men and women who wore a whole-body compression garment after a heavy resistance training workout. It should be noted, however, that no differences were shown between the compression garment and passive recovery on vertical jump power output. Additional findings suggest that custom-fitted compression garments may enhance the recovery of lower-body strength (17). Interestingly, the conclusions from several meta-analyses and systematic reviews were also mixed. In a 2017 study, Brown and colleagues (16) concluded that compression garments had a large effect on strength recovery 2 to 8 hours and more than 24 hours after exercise. In a 2022 meta-analysis, Négyesi and colleagues (96) suggested that compression garments do not provide recovery benefits with regard to strength. Similarly, Weakley and colleagues (148) concluded that compression garments may not have a positive effect on recovery of various strength measures but also do not have a negative impact on them either.

Compression garments may help athletes recover when they are traveling as well. In fact, researchers have shown effective recovery of countermovement jump performance and a reduction of creatine kinase after participants wore compression garments on long flights (15, 76). Like many of the methods presented previously, there does not appear to be a consensus on the optimal protocol for using this method; however, it has been suggested by some that wearing compression garments for longer durations may produce greater benefits (140).

Pneumatic Compression

Similar to the static compression method described earlier, pneumatic compression (figure 12.3) is thought to improve circulation and venous return to help remove metabolic waste products (99). A purported benefit of pneumatic compression devices is that they have the potential to exert greater pressures than traditional compression garments, which may allow for greater lymphatic drainage and possibly greater blood flow velocity, dissipation of swelling, and the removal of cellular debris (20, 60, 73). While previous findings showed that pneumatic compression enhanced blood flow and oxygenation to limbs during recovery (156, 157), Cochrane and colleagues (23) indicated that it did not mitigate muscle force loss during slow (30° per second) or fast (180° per second) angular velocities following strenuous eccentric exercise. This is supported by additional findings showing that pneumatic compression had no impact on the recovery of strength (22). Further research has also shown minimal performance benefits of pneumatic compression (21, 61, 99, 154). Anecdotally, athletes may be drawn to using pneumatic compression as a recovery method because it "feels" like it is working (61). However, researchers have displayed mixed findings regarding subjective measures of recovery with pneumatic compression (21, 36, 61, 65, 99, 120).

FIGURE 12.3 Pneumatic compression.

ACTIVE RECOVERY

Active recovery, or light activity, is typically used to maintain blood flow following another activity to enhance the removal of accumulated waste products (e.g., blood lactate) (143). Despite displaying no significant differences with compression garments or CWT, Gill and colleagues (47) demonstrated that an active recovery in rugby players produced the greatest recovery of creatine kinase (88.2%) 84 hours after the match. It should be noted that active recovery may have a minimal impact on performance or injury reduction if performed greater than 4 hours following the initial activity (143); however, it may be beneficial when an athlete must perform repeat efforts within 30 minutes (59).

STRETCHING

The primary rationale for using stretching for recovery purposes is to reduce muscle soreness and stiffness while relaxing the muscle and improving an athlete's range of motion (119). Through stretching, it is thought that athletes may be able to mitigate muscle soreness to aid recovery by dispersing the accumulated edema. Although it is widely used, there does not appear to be much evidence that supports the use of stretching for recovery of an athlete's strength characteristics (5). In fact, a 2021 meta-analysis concluded that there was no difference between post-exercise stretching and passive recovery regarding the recovery of strength (2).

MASSAGE

The purposes of massage are to increase blood flow, decrease muscle tension and excitability, and promote positive feelings of well-being (59). Interestingly, the findings presented in two meta-analyses stated that the effects of massage on the recovery of athletes are inconclusive (33, 105); however, massage may have the potential to reduce delayed-onset muscle soreness (33). Poppendieck and colleagues (105) indicated that post-exercise massage (manual or automated) had only a trivial effect on the recovery of strength. Interestingly, Sands (117) showed that resident athletes at the United States Olympic Committee Recovery Center preferred massage and cold therapy recovery methods over a wide variety of other methods. This is likely due to the notion that the effects of massage may be more psychological than physiological (150). It should be noted that beyond manual massage, there is some evidence to suggest that self-massage in the form of foam rolling may mitigate decreases in sprint and strength performance, decrease pain perception and delayed-onset muscle soreness, and improve range of motion (151). While there are a variety of tools that may be used for self-massage (e.g., foam rollers, massage balls, roller sticks, percussion devices), additional research is needed before practical recommendations can be made regarding their use. However, strength and conditioning professionals should be aware of the potential side effects of self-massage (e.g., soreness, bruising, and swelling) and communicate this information to their athletes.

CRYOTHERAPY

Another method that uses cold temperatures as the primary mechanism for recovery is cryotherapy. However, in the case of cryotherapy compared to CWI, the temperatures are much colder (<–166°F [–100°C]) (11). This method requires athletes to stand, sit, or lay in either a cryochamber (entire body exposed to cold) or cryosauna (entire body with the exception of the head exposed to cold) (133). The proposed mechanisms for recovery are similar to those for CWI: decreases in muscle tissue temperature and blood flow leading to reductions in swelling and inflammation (4).

Interestingly, Abaidia and colleagues (1) showed that CWI produced greater recovery results for both unilateral and bilateral countermovement jump performances following 5 sets of 15 single-leg hamstring eccentric exercise repetitions compared to whole-body cryotherapy. It should be noted, however, that no differences in isometric strength recovery existed between recovery methods. In contrast, another study found that cryotherapy showed greater increases in isometric peak force and rate of force development 48 hours after exercise in resistance-trained men compared to CWI and placebo methods (153). However, the authors indicated that neither cryotherapy nor CWI appeared to be more effective at accelerating recovery compared to a placebo, thus questioning its implementation.

Another study showed that cryotherapy implemented immediately following plyometric training improved the rate of strength recovery compared to a control condition (41). However, additional findings have shown no positive effects on performance recovery or soreness (50, 113, 146). Compared to single cryotherapy exposures, other researchers have shown positive effects on isometric knee flexion rate of torque development (43) and maximal isometric force (58) after eccentric training and long-distance running after successive treatments, respectively. However, it should be noted that despite their results, one group of authors did not fully endorse the use of cryotherapy as a recovery method (43). It has been noted that certain findings may have been due to the repeated bout effect (41, 43); however, a single bout may not produce a beneficial effect (84).

Selfe and colleagues (123) suggested that cryotherapy may be best implemented for a duration of 30 seconds at −76°F (−60°C) and then 2 minutes at −211°F (−135°C) with 2 minutes between bouts. As mentioned earlier, it may be necessary to implement both immediate and recurrent cryotherapy bouts after exercise to see the best results (43, 58). Strength and conditioning professionals interested in implementing cryotherapy must be aware of the associated risks, such as hypothermia, cold burn (skin damage), hypertension, reduced nerve conduction velocity, and reduced peripheral blood flow (133). Moreover, several safety precautions should be implemented as well; athletes should be supervised throughout their time in cryochambers or cryosaunas, cold-sensitive portions of the body (hands, feet, and mouth) should be protected from cold exposure, the athlete's skin surface should be completely dry and free of sweat and moisture, and athletes should refrain from breathing the liquid nitrogen when using cryosaunas (84, 133). Readers interested in cryotherapy are referred to more thorough reviews of this method (4, 11, 84, 111).

ELECTRIC MUSCLE STIMULATION

Electric muscle stimulation (EMS) requires an athlete to have electrodes placed on the muscles that need to be recovered to receive transcutaneous stimulation (figure 12.4). In general, the stimulation will cause the muscles to twitch, but it is thought that EMS may be an effective recovery method because it may increase blood flow and venous return, which may aid in the removal of metabolites (6, 104). Researchers have indicated that EMS does not appear to produce negative recovery effects (40) but has not proven effective either (104). EMS may serve as an effective method when it comes to providing positive perceptual effects (6, 40), which may justify its use for athletes. These findings were further supported by a systematic review on the use of EMS for recovery purposes (88). There also appears to be a lack of evidence supporting the use of EMS for muscle strength recovery (3). It should be noted that compared to other recovery methods, a limited amount of research has examined the effect of EMS on strength recovery; thus, further research may be warranted.

VIBRATION

Vibration devices such as massage guns (figure 12.5) or whole-body vibration platforms (figure 12.6) have become popular within the strength and conditioning field. Touted benefits of these devices are the ability to decrease muscle stiffness and tension, increase range of motion, and promote circulation (74, 106). Some researchers have found positive effects of vibration massage on the recovery of strength (78), whereas others have not (18, 90, 136). Because a minimal amount of research has been conducted on the effect of vibration massage on the recovery of strength, further research is warranted before drawing concrete conclusions. At present, it appears that most of the effects are based on the positive perceptions of the participants rather than the effectiveness of the devices (74). However, creating a positive environment in which the athlete believes that the recovery method is working can go a long way.

MENTAL FATIGUE AND RECOVERY

Given the potential impact of stressors on an athlete's performance and the potential for burnout (85), research focused on mental fatigue may become significant for the wellness and holistic development of an athlete (115). *Mental fatigue* has been defined as a psychobiological state that is caused by extended periods of demanding cognitive activity and is typi-

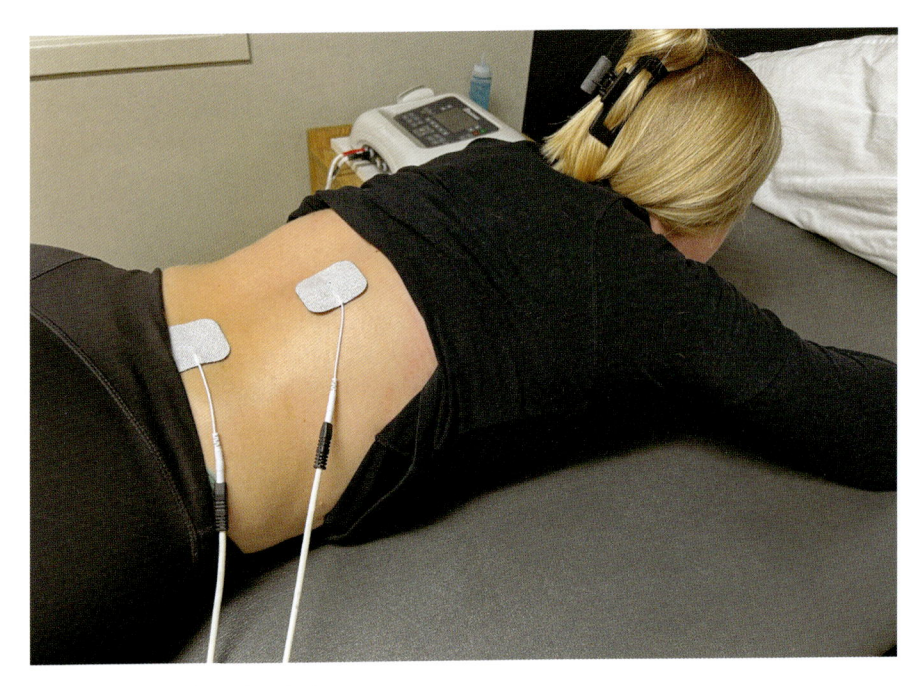

FIGURE 12.4 Electric muscle stimulation.

FIGURE 12.5 Massage gun therapy.

FIGURE 12.6 Whole-body vibration.

cally characterized by feelings of "tiredness" or a "lack of energy" (12). Russell and colleagues (114) identified common themes of perceived causes and the impact of mental fatigue as viewed by professional soccer athletes and staff. Their results showed that mental fatigue may be due to other commitments, the environment the athlete is in, professionalism (e.g., interviews, sponsorships), overanalysis of specific situations, and experience (or lack thereof). In addition, the participants identified that mental fatigue may affect overall performance, decision-making, and response time; participants also said that mental fatigue makes tasks feel harder than they should be, showed greater tendencies to make mistakes, and demonstrated diminished willpower (114). Further findings have shown that in international netball players, mental fatigue may occur during training and preparation camps to a greater extent than during competitive periods (116). These findings emphasize the importance of monitoring mental fatigue throughout an entire training year.

While research examining the effect of mental fatigue on performance is still developing, several reviews have been published on the topic (51, 107, 131, 135, 142). The conclusions from these reviews suggest that mental fatigue can negatively affect endurance performance (100, 142), sport-specific psychomotor performance variables (e.g., decision-making, reaction time, and accuracy outcomes) (51, 100, 131), and skilled performances in soccer, basketball, and table tennis (131, 135). In contrast, mental fatigue does not appear to affect maximal force production, power, or anaerobic work (100, 142). Interestingly, some authors also indicated that mental fatigue may negatively affect submaximal exercises but not exercises performed at maximal or supramaximal intensities (100). The authors suggested that these findings were mediated by the perception of effort. This notion is supported by the anticipatory fight-or-flight response due to a release in catecholamines (45). Although the existing research suggests that mental fatigue may not affect an athlete's strength characteristics, further research is needed before drawing conclusions.

A 2022 systematic review discussed potential strategies to counteract mental fatigue (107). The authors concluded that caffeine ingested before or during the occurring mental fatigue, pleasant odors during the mental fatiguing task, relaxing music, and extrinsic motivation (e.g., rewards) have the most support for counteracting mental fatigue. Although the findings of the previous review provide a general overview of potentially effective methods, there is not enough research to provide specific prescriptions based on individual circumstances. As research focused on mental fatigue progresses, it is hoped that strength and conditioning professionals can effectively implement the aforementioned strategies to avoid decrements in performance.

THE PLACEBO EFFECT

Collectively, the findings of research focused on the effectiveness of different recovery methods appears to be mixed; however, it is important that strength and conditioning professionals do not discount what the athlete perceives as "working." Although the use of various recovery methods may yield positive perceptions based on the *placebo effect*, Halson and Martin (56) noted that there is a legitimate reason to create athlete belief. For example, several studies have demonstrated the positive psychological effects on perceived recovery following massage therapy (33, 34, 35, 39, 62, 64, 89). Because the relationship between the athlete and the practitioner can have a large impact on the athlete's beliefs about a specific recovery method (121), it is important to create buy-in into what is being prescribed. However, it is paramount that practitioners endorse the recovery methods that have supportive evidence, rather than fads or methods based on their own anecdotal evidence.

Key Point

The findings of research focused on the effectiveness of different recovery methods appears to be mixed; however, it is important that strength and conditioning professionals do not discount what the athlete perceives as "working."

PERIODIZATION OF RECOVERY

Planned recovery sessions may enable athletes to use short-term intensive loading, which in turn may produce favorable adaptations (55). Mujika and colleagues (92) discussed the periodization of recovery in a previous review. The authors suggested that the periodization of recovery should be focused on returning the body to a homeostatic state during periods in which athletes must maximize their performance. They discussed the following four primary themes when it comes to periodizing recovery:

- Withholding recovery methods during the general preparation time to maximize training adaptations
- Prescribing recovery methods during the specific preparation period to prepare athletes to complete certain training sessions
- Prescribing recovery methods to decrease acute fatigue during the competitive season in which multiple games or matches may be played within a short period
- Using recovery methods during travel, during return to play from injury, and when managing psychological stress

Based on the themes just mentioned, it is important and potentially necessary in specific situations to alter the prescribed methods of recovery because they still serve as a stimulus provided to an athlete. Thus, if the same recovery method is prescribed in every scenario (e.g., between off-season training sessions, preseason practices, competitive season matches), its effects may become muted and fail to promote the same magnitude of recovery that it may have previously. Furthermore, it has been suggested that certain recovery methods should be either withheld or cycled throughout the training year (92, 117). For example, researchers have suggested withholding CWI during the off-season training period to allow for training adaptations to naturally occur without the influence of another external stimulus (72). Another example would be increasing the use of active recovery or compression garments between preseason matches (92). Therefore, it is suggested that different recovery methods be periodized and programmed throughout the training year based on training sessions, practices, and competitions to appropriately stimulate recovery and adaptation. While strength and conditioning professionals must consider the severity of the fatigue and the available time between training sessions or competitions (8), the prescription of specific recovery methods may not be needed if adequate time is available for the body's natural recovery processes (129). Finally, the implementation of various recovery strategies should be considered on an individual basis, because players who receive less playing time may be better off "training through" different portions of the training year in order to maximize the effects of different training strength stimuli (102, 129).

Key Point

It is important and potentially necessary in specific situations to alter the prescribed methods of recovery because they still serve as a stimulus provided to an athlete.

While related to the periodization of recovery, strength and conditioning professionals must also consider the practicality of implementing specific methods of recovery (133). For example, if an athletic training or recovery room only has a single cold tub for CWI, it may not be practical to get the entire team through in a time-efficient manner. In addition, not all methods of recovery will be available to athletes when they are traveling for competition. For example, it may not be practical to pack pneumatic compression boots and their accessories or baggage check them at the airport, especially if several pairs need to be brought along. Therefore, it is important that practitioners stress the importance of the recovery pyramid and help their athletes focus on managing their sleep and nutrition before placing additional emphasis on any of the remaining recovery methods.

MEASURING RECOVERY

As discussed previously, implementing any or all of the aforementioned recovery methods places a stimulus on an athlete. While this may be done with the intention of returning the athlete to a homeostatic state and increasing their readiness to train or compete, it cannot be assumed that the desired effect occurred unless it is measured. Several studies have identified different markers of recovery (25, 28, 67, 80, 97); however, it should be noted that many of the previous studies used indirect measurements of physiological markers that may not relate the recovery of an athlete's performance. In addition, there is limited information on how to determine whether an athlete is "fully recovered" (27, 108). This issue becomes even more complex when only some, but not all, of the measured variables related to performance are recovered (48, 97).

Whereas some have proposed scales of perceived recovery as a method to monitor recovery (79), others have used the Profile of Mood States (POMS) (66, 70) or the Recovery–Stress Questionnaire (RESTQ) for athletes (70, 71). Although the POMS and the RESTQ may provide strength and conditioning professionals with subjective information from the athlete, direct measurement of the variables that relate to the characteristic being monitored will provide the most accurate information regarding an athlete's recovery status. Regarding the recovery of an athlete's strength characteristics, the variables of interest will include mean and peak force as well as the duration of time that the athlete is producing force during tasks that relate to the sport or event of each athlete. To measure such variables, sport scientists and practitioners would require a force plate and the knowledge to use such a device (7), instead of a device that may estimate force production based on a given load (e.g., velocity measurement tool). Moreover, the stability of each variable must be considered within each phase of training, given that an individual athlete may have unique "normal ranges" within different training phases. However, it is important to consider that practitioners may not always have the luxury of being able to use force plate technology on a regular basis. Thus, alternative methods of measuring strength may be needed. Considerations for measuring and monitoring the strength characteristics of an athlete are discussed in greater detail in chapter 9.

Take-Home Points

▶ Sport performance is multifaceted and despite improvements in magnitude or rate of force development stemming from either training or recovery, this does not mean that an individual or team will necessarily improve their standing in an event or win the contest. This is due to the nature of competition and the factors that may affect it, such as officials, environment and climate, weather, injury, coaching decisions, and the competition's preparedness.

▶ The term *recovery* should not be used in isolation. Although recovery is meant to return an athlete to a baseline level of performance, training stimuli should be appropriately implemented to enhance an athlete's performance beyond their baseline. Thus, the term *recovery–adaptation* should be understood as the training stimulus that produces the desired adaptation that occurs during the recovery phase (regeneration), but to a level beyond an athlete's initial baseline (overcompensation).

▶ Sleep and nutrition serve as the foundation of athlete recovery. Controlling these aspects may provide the greatest benefits when it comes to recovery of athletes and their ability to consistently train and perform at the highest level.

▶ Much of the existing recovery literature on CWI, HWI, CWT, compression, active recovery, stretching, massage, cryotherapy, electric muscle stimulation, vibration, and mental recovery has displayed inconclusive effects. Thus, it is important that professionals consider the preferences of their athletes when selecting different methods. Although athlete preferences may be based on a placebo effect, this effect may be enough to create athlete buy-in and positive adaptations.

▶ Recovery methods should be implemented based on the aim of recovery (e.g., stress to recover from, time frame of recovery, environmental factors affecting recovery, whether the athlete is excessively fatigued, the phase of the season the athlete is in, and what facilities are accessible).

▶ It is important to control the variables that can be controlled to a greater extent (e.g., training prescription, sleep, and nutrition) before implementing one or multiple methods of recovery.

References

Chapter 1

1. Alshewaier, S, Yeowell, G, and Fatoye, F. The effectiveness of pre-operative exercise physiotherapy rehabilitation on the outcomes of treatment following anterior cruciate ligament injury: A systematic review. *Clin Rehabil* 31:34-44, 2017.

2. Austin, D, and Mann, JB. *Powerlifting: The Complete Guide to Technique, Training, and Competition.* Champaign, IL: Human Kinetics, 2021.

3. Barker, M, Wyatt, TJ, Johnson, RL, Stone, MH, O'Bryant, HS, Poe, C, and Kent, M. Performance factors, psychological assessment, physical characteristics, and football playing ability. *J Strength Cond Res* 7:224-233, 1993.

4. Bazyler, CD, Beckham, GK, and Sato, K. The use of the isometric squat as a measure of strength and explosiveness. *J Strength Cond Res* 29:1386-1392, 2015.

5. Beattie, K, Carson, BP, Lyons, M, and Kenny, IC. The relationship between maximal-strength and reactive-strength. *Int J Sports Physiol Perform* 12:548-553, 2016.

6. Behringer, M, Vom Heede, A, Matthews, M, and Mester, J. Effects of strength training on motor performance skills in children and adolescents: A meta-analysis. *Pediatr Exerc Sci* 23:186-206, 2011.

7. Bijker, K, De Groot, G, and Hollander, A. Differences in leg muscle activity during running and cycling in humans. *Eur J Appl Physiol* 87:556-561, 2002.

8. Blazevich, AJ, and Babault, N. Post-activation potentiation (PAP) versus post-activation performance enhancement (PAPE) in humans: Historical perspective, underlying mechanisms, and current issues. *Front Physiol* 10:1-19, 2019.

9. Burger, H, and Weidt, K. *Kraftproben.* Berlin: Sportverlag, 1985.

10. Burt, LA, Greene, DA, Ducher, G, and Naughton, GA. Skeletal adaptations associated with pre-pubertal gymnastics participation as determined by DXA and pQCT: A systematic review and meta-analysis. *J Sci Med Sport* 16:231-239, 2013.

11. Case, MJ, Knudson, DV, and Downey, DL. Barbell squat relative strength as an identifier for lower extremity injury in collegiate athletes. *J Strength Cond Res* 34:1249-1253, 2020.

12. Chang, C-C, and Chiang, C-Y. Using the countermovement jump metrics to assess dynamic eccentric strength: A preliminary study. *Int J Environ Res Public Health* 19:16176, 2022.

13. Comfort, P, and Pearson, SJ. Scaling—Which methods best predict performance? *J Strength Cond Res* 28:1565-1572, 2014.

14. Comfort, P, Suchomel, TJ, and Stone, MH. Normalisation of early isometric force production as a percentage of peak force, during multi-joint isometric assessment. *Int J Sports Physiol Perform* 15:478-482, 2019.

15. Conroy, BP, Kraemer, WJ, Maresh, CM, Fleck, SJ, Stone, MH, Fry, AC, Miller, PD, and Dalsky, GP. Bone mineral density in elite junior Olympic weightlifters. *Med Sci Sports Exerc* 25:1103-1109, 1993.

16. Cronin, JB, and Hansen, KT. Strength and power predictors of sports speed. *J Strength Cond Res* 19:349-357, 2005.

17. Crowther, NB. *Sport in Ancient Times.* Westport, CT: Greenwood Publishing Group, 2010.

18. Doss, WS, and Karpovich, PV. A comparison of concentric, eccentric, and isometric strength of elbow flexors. *J Appl Physiol* 20:351-353, 1965.

19. Ebben, WP, and Petushek, EJ. Using the reactive strength index modified to evaluate plyometric performance. *J Strength Cond Res* 24:1983-1987, 2010.

20. Faigenbaum, AD, MacDonald, JP, and Haff, GG. Are young athletes strong enough for sport? DREAM on. *Curr Sports Med Rep* 18:6-8, 2019.

21. Faigenbaum, AD, and Myer, GD. Resistance training among young athletes: Safety, efficacy and injury prevention effects. *Br J Sports Med* 44:56-63, 2010.

22. Falk, B, and Eliakim, A. Resistance training, skeletal muscle and growth. *Pediatr Endocrinol Rev* 1:120-127, 2003.

23. Garhammer, J, and Gregor, R. Propulsion forces as a function of intensity for weightlifting and vertical jumping. *J Strength Cond Res* 6:129-134, 1992.

24. Gentil, P, Del Vecchio, FB, Paoli, A, Schoenfeld, BJ, and Bottaro, M. Isokinetic dynamometry and 1RM tests produce conflicting results for assessing alterations in muscle strength. *J Hum Kinet* 56:19-27, 2017.

25. Gialourēs, N, and Andronikos, M. *The Olympic Games in Ancient Greece: Ancient Olympia and the Olympic Games.* Athens, Greece: Ekdotike Athenon, 1982.

26. Gorostiaga, EM, Granados, C, Ibanez, J, and Izquierdo, M. Differences in physical fitness and throwing velocity among elite and amateur male handball players. *Int J Sports Med* 26:225-232, 2005.

27. Häkkinen, K, Newton, RU, Walker, S, Häkkinen, A, Krapi, S, Rekola, R, Koponen, P, Kraemer, WJ, Haff, GG, and Blazevich, AJ. Effects of upper body eccentric versus concentric strength training and detraining on maximal force, muscle activation, hypertrophy and serum hormones in women. *J Sports Sci Med* 21:200-213, 2022.

28. Hamill, BP. Relative safety of weightlifting and weight training. *J Strength Cond Res* 8:53-57, 1994.

29. Hollander, DB, Kraemer, RR, Kilpatrick, MW, Ramadan, ZG, Reeves, GV, Francois, M, Hebert, EP, and Tryniecki, JL. Maximal eccentric and concentric strength discrepancies between young men and women for dynamic resistance exercise. *J Strength Cond Res* 21:34-40, 2007.

30. Hori, N, Newton, RU, Andrews, WA, Kawamori, N, McGuigan, MR, and Nosaka, K. Does performance of hang power clean differentiate performance of jumping, sprinting, and changing of direction? *J Strength Cond Res* 22:412-418, 2008.

31. Hunter, JP, Marshall, RN, and McNair, PJ. Relationships between ground reaction force impulse and kinematics of sprint-running acceleration. *J Appl Biomech* 21:31-43, 2005.

32. Izquierdo, M, Häkkinen, K, Gonzalez-Badillo, JJ, Ibanez, J, and Gorostiaga, EM. Effects of long-term training specificity on maximal strength and power of the upper and lower extremities in athletes from different sports. *Eur J Appl Physiol* 87:264-271, 2002.

33. James, LP, Haff, GG, Kelly, VG, Connick, M, Hoffman, B, and Beckman, EM. The impact of strength level on adaptations to combined weightlifting, plyometric and ballistic training. *Scand J Med Sci Sports* 28:1494-1505, 2018.

34. Jaric, S. Muscle strength testing: Use of normalisation for body size. *Sports Med* 32:615-631, 2002.

35. Jaric, S. Role of body size in the relation between muscle strength and movement performance. *Exerc Sport Sci Rev* 31:8-12, 2003.

36. Kawamori, N, Nosaka, K, and Newton, RU. Relationships between ground reaction impulse and sprint acceleration performance in team sport athletes. *J Strength Cond Res* 27:568-573, 2013.

37. Keiner, M, Brauner, T, Kadlubowski, B, Sander, A, and Wirth, K. The influence of maximum squatting strength on jump and sprint performance: A cross-sectional analysis of 492 youth soccer players. *Int J Environ Res Public Health* 19:5835, 2022.

38. Keiner, M, Sander, A, Wirth, K, Caruso, O, Immesberger, P, and Zawieja, M. Strength performance in youth: Trainability of adolescents and children in the back and front squats. *J Strength Cond Res* 27:357-362, 2013.

39. Keiner, M, Sander, A, Wirth, K, and Schmidtbleicher, D. Long-term strength training effects on change-of-direction sprint performance. *J Strength Cond Res* 28:223-231, 2014.

40. Keiner, M, Wirth, K, Fuhrmann, S, Kunz, M, Hartmann, H, and Haff, GG. The influence of upper- and lower-body maximum strength on swim block start, turn, and overall swim performance in sprint swimming. *J Strength Cond Res* 35:2839-2845, 2021.

41. Kirby, TJ, McBride, JM, Haines, TL, and Dayne, AM. Relative net vertical impulse determines jumping performance. *J Appl Biomech* 27:207-214, 2011.

42. Kraska, JM, Ramsey, MW, Haff, GG, Fethke, N, Sands, WA, Stone, ME, and Stone, MH. Relationship between strength characteristics and unweighted and weighted vertical jump height. *Int J Sports Physiol Perform* 4:461-473, 2009.

43. Kugler, F, and Janshen, L. Body position determines propulsive forces in accelerated running. *J Biomech* 43:343-348, 2010.

44. Lauersen, JB, Andersen, TE, and Andersen, LB. Strength training as superior, dose-dependent and safe prevention of acute and overuse sports injuries: A systematic review, qualitative analysis and meta-analysis. *Br J Sports Med* 52:1557-1563, 2018.

45. Lauersen, JB, Bertelsen, DM, and Andersen, LB. The effectiveness of exercise interventions to prevent sports injuries: A systematic review and meta-analysis of randomised controlled trials. *Br J Sports Med* 48:871-877, 2014.

46. Lepley, LK, and Palmieri-Smith, RM. Pre-operative quadriceps activation is related to post-operative activation, not strength, in patients post-ACL reconstruction. *Knee Surg Sports Traumatol Arthrosc* 24:236-246, 2016.

47. Lloyd, RS, Cronin, JB, Faigenbaum, AD, Haff, GG, Howard, R, Kraemer, WJ, Micheli, LJ, Myer, GD, and Oliver, JL. National Strength and Conditioning Association position statement on long-term athletic development. *J Strength Cond Res* 30:1491-1509, 2016.

48. Lloyd, RS, Faigenbaum, AD, Stone, MH, Oliver, JL, Jeffreys, I, Moody, JA, Brewer, C, Pierce, KC, McCambridge, TM, Howard, R, Herrington, L, Hainline, B, Micheli, LJ, Jaques, R, Kraemer, WJ, McBride, MG, Best, TM, Chu, DA, Alvar, BA, and Myer, GD. Position statement on youth resistance training: The 2014 International Consensus. *Br J Sports Med* 48:498-505, 2014.

49. Loturco, I, Nakamura, FY, Artioli, GG, Kobal, R, Kitamura, K, Abad, CCC, Cruz, IF, Romano, F, Pereira, LA, and Franchini, E. Strength and power qualities are highly associated with punching impact in elite amateur boxers. *J Strength Cond Res* 30:109-116, 2016.

50. Louder, T, Thompson, BJ, and Bressel, E. Association and agreement between reactive strength index and reactive strength index-modified scores. *Sports* 9:97, 2021.

51. Lum, D, and Aziz, AR. Relationship between isometric force–time characteristics and sprint kayaking performance. *Int J Sports Physiol Perform* 16:474-479, 2020.

52. Maestroni, L, Read, P, Bishop, C, Papadopoulos, K, Suchomel, TJ, Comfort, P, and Turner, A. The benefits of strength training on musculoskeletal system health: Practical applications for interdisciplinary care. *Sports Med* 50:1431-1450, 2020.

53. Malina, RM. Weight training in youth-growth, maturation, and safety: An evidence-based review. *Clin J Sport Med* 16:478-487, 2006.

54. Malone, S, Hughes, B, Doran, DA, Collins, K, and Gabbett, TJ. Can the workload-injury relationship be moderated by improved strength, speed and repeated-sprint qualities? *J Sci Med Sport* 22:29-34, 2019.

55. Mann, JB, Ivey, PA, Stoner, JD, Mayhew, JL, and Brechue, WF. Efficacy of the National Football League-225 test to track changes in one repetition maximum bench press after training in National Collegiate Athletic Association Division IA football players. *J Strength Cond Res* 29:2997-3005, 2015.

56. Mann, JB, Stoner JD, and Mayhew JL. NFL-225 test to predict 1RM bench press in NCAA Division I football players. *J Strength Cond Res* 26:2623-2631, 2012.

57. Masuda, K, Kikuhara, N, Demura, S, Katsuta, S, and Yamanaka, K. Relationship between muscle strength in various isokinetic movements and kick performance among soccer players. *J Sports Med Phys Fitness* 45:44-52, 2005.

58. McGuigan, MR, Newton, MJ, Winchester, JB, and Nelson, AG. Relationship between isometric and dynamic strength in recreationally trained men. *J Strength Cond Res* 24:2570-2573, 2010.

59. McGuigan, MR, and Winchester, JB. The relationship between isometric and dynamic strength in college football players. *J Sports Sci Med* 7:101-105, 2008.

60. McMahon, JJ, Jones, PA, Suchomel, TJ, Lake, JP, and Comfort, P. Influence of reactive strength index modified on force- and power-time curves. *Int J Sports Physiol Perform* 13:220-227, 2018.

61. McMahon, JJ, Murphy, S, Rej, SJE, and Comfort, P. Countermovement-jump-phase characteristics of senior and academy rugby league players. *Int J Sports Physiol Perform* 12:803-811, 2017.
62. McMahon, JJ, Suchomel, TJ, Lake, JP, and Comfort, P. Understanding the key phases of the countermovement jump force–time curve. *Strength Cond J* 40:96-106, 2018.
63. McMahon, JJ, Suchomel, TJ, Lake, JP, and Comfort, P. Relationship between reactive strength index variants in rugby league players. *J Strength Cond Res* 35:280-285, 2021.
64. Meckel, Y, Atterbom, H, Grodjinovsky, A, Ben-Sira, D, and Rotstein, A. Physiological characteristics of female 100 metre sprinters of different performance levels. *J Sports Med Phys Fitness* 35:169-175, 1995.
65. Mercuriale, G. *De Arte Gymnastica Libri Sex*. Venice: Apub Luntas [Lucantonio Giunta], 1569.
66. Miyamoto, N, Wakahara, T, Ema, R, and Kawakami, Y. Further potentiation of dynamic muscle strength after resistance training. *Med Sci Sports Exerc* 45:1323-1330, 2013.
67. Morin, J-B, Slawinski, J, Dorel, S, Couturier, A, Samozino, P, Brughelli, M, and Rabita, G. Acceleration capability in elite sprinters and ground impulse: Push more, brake less? *J Biomech* 48:3149-3154, 2015.
68. Murphy, AJ, Wilson, GJ, Pryor, JF, and Newton, RU. Isometric assessment of muscular function: The effect of joint angle. *J Appl Biomech* 11:205-215, 1995.
69. Myer, GD, Quatman, CE, Khoury, J, Wall, EJ, and Hewett, TE. Youth versus adult "weightlifting" injuries presenting to United States emergency rooms: Accidental versus nonaccidental injury mechanisms. *J Strength Cond Res* 23:2054-2060, 2009.
70. Nevill, AM, Ramsbottom, R, and Williams, C. Scaling physiological measurements for individuals of different body size. *Eur J Appl Physiol Occup Physiol* 65:110-117, 1992.
71. Nimphius, S, McGuigan, MR, and Newton, RU. Relationship between strength, power, speed, and change of direction performance of female softball players. *J Strength Cond Res* 24:885-895, 2010.
72. Nuzzo, JL, Pinto, MD, Nosaka, K, and Steele, J. The eccentric:concentric strength ratio of human skeletal muscle in vivo: Meta-analysis of the influences of sex, age, joint action, and velocity. *Sports Med* 53:1125-1136, 2023.
73. Paasuke, M, Ereline, J, and Gapeyeva, H. Knee extension strength and vertical jumping performance in Nordic combined athletes. *J Sports Med Phys Fitness* 41:354-361, 2001.
74. Pallarés, JG, Hernández-Belmonte, A, Martínez-Cava, A, Vetrovsky, T, Steffl, M, and Courel-Ibáñez, J. Effects of range of motion on resistance training adaptations: A systematic review and meta-analysis. *Scand J Med Sci Sports* 31:1866-1881, 2021.
75. Pichardo, AW, Oliver, JL, Harrison, CB, Maulder, PS, Lloyd, RS, and Kandoi, R. Effects of combined resistance training and weightlifting on injury risk factors and resistance training skill of adolescent males. *J Strength Cond Res* 35:3370-3377, 2021.
76. Qian, S, and Dawson, R. *Historical Records*. Oxford: Oxford University Press, 1994.
77. Requena, B, González-Badillo, JJ, de Villareal, ESS, Ereline, J, García, I, Gapeyeva, H, and Pääsuke, M. Functional performance, maximal strength, and power characteristics in isometric and dynamic actions of lower extremities in soccer players. *J Strength Cond Res* 23:1391-1401, 2009.
78. Schödl, G. *The Lost Past*. Budapest: International Weightlifting Federation, 1992.
79. Seitz, LB, de Villarreal, ESS, and Haff, GG. The temporal profile of postactivation potentiation is related to strength level. *J Strength Cond Res* 28:706-715, 2014.
80. Seitz, LB, and Haff, GG. Factors modulating post-activation potentiation of jump, sprint, throw, and upper-body ballistic performances: A systematic review with meta-analysis. *Sports Med* 46:231-240, 2016.
81. Sheppard, JM, Cronin, JB, Gabbett, TJ, McGuigan, MR, Etxebarria, N, and Newton, RU. Relative importance of strength, power, and anthropometric measures to jump performance of elite volleyball players. *J Strength Cond Res* 22:758-765, 2008.
82. Sommerfield, LM, Harrison, CB, Whatman, CS, and Maulder, PS. Relationship between strength, athletic performance, and movement skill in adolescent girls. *J Strength Cond Res* 36:674-679, 2022.
83. Speranza, MJA, Gabbett, TJ, Johnston, RD, and Sheppard, JM. Effect of strength and power training on tackling ability in semiprofessional rugby league players. *J Strength Cond Res* 30:336-343, 2016.
84. Spiteri, T, Nimphius, S, Hart, NH, Specos, C, Sheppard, JM, and Newton, RU. Contribution of strength characteristics to change of direction and agility performance in female basketball athletes. *J Strength Cond Res* 28:2415-2423, 2014.
85. Stone, MH, Moir, G, Glaister, M, and Sanders, R. How much strength is necessary? *Phys Ther Sport* 3:88-96, 2002.
86. Stone, MH, and O'Bryant, HS. *Weight Training: A Scientific Approach*. Minneapolis, MN: Burgess International, 1987.
87. Stone, MH, Sands, WA, Carlock, J, Callan, S, Dickie, D, Daigle, K, Cotton, J, Smith, SL, and Hartman, M. The importance of isometric maximum strength and peak rate-of-force development in sprint cycling. *J Strength Cond Res* 18:878-884, 2004.
88. Suchomel, TJ, Bailey, CA, Sole, CJ, Grazer, JL, and Beckham, GK. Using reactive strength index-modified as an explosive performance measurement tool in Division I athletes. *J Strength Cond Res* 29:899-904, 2015.
89. Suchomel, TJ, Comfort, P, and Lake, JP. Enhancing the force–velocity profile of athletes using weightlifting derivatives. *Strength Cond J* 39:10-20, 2017.
90. Suchomel, TJ, Lamont, HS, and Moir, GL. Understanding vertical jump potentiation: A deterministic model. *Sports Med* 46:809-828, 2016.
91. Suchomel, TJ, McKeever, SM, Sijuwade, O, and Carpenter, L. Propulsion phase characteristics of loaded jump variations in resistance-trained women. *Sports* 11:44, 2023.
92. Suchomel, TJ, McKeever, SM, Sijuwade, O, Carpenter, L, McMahon, JJ, Loturco, I, and Comfort, P. The effect of load placement on the power production characteristics of three lower extremity jumping exercises. *J Hum Kinet* 68:109-122, 2019.

93. Suchomel, TJ, Nimphius, S, Bellon, CR, and Stone, MH. The importance of muscular strength: Training considerations. *Sports Med* 48:765-785, 2018.

94. Suchomel, TJ, Nimphius, S, and Stone, MH. The importance of muscular strength in athletic performance. *Sports Med* 46:1419-1449, 2016.

95. Suchomel, TJ, Nimphius, S, and Stone, MH. Scaling isometric mid-thigh pull maximum strength in Division I athletes: Are we meeting the assumptions? *Sports Biomech* 19:532-546, 2020.

96. Suchomel, TJ, Sands, WA, and McNeal, JR. Comparison of static, countermovement, and drop jumps of the upper and lower extremities in U.S. junior national team male gymnasts. *Sci Gymnastics J* 8:15-30, 2016.

97. Suchomel, TJ, Sato, K, DeWeese, BH, Ebben, WP, and Stone, MH. Potentiation effects of half-squats performed in a ballistic or non-ballistic manner. *J Strength Cond Res* 30:1652-1660, 2016.

98. Suchomel, TJ, Sato, K, DeWeese, BH, Ebben, WP, and Stone, MH. Potentiation following ballistic and non-ballistic complexes: The effect of strength level. *J Strength Cond Res* 30:1825-1833, 2016.

99. Suchomel, TJ, Sato, K, DeWeese, BH, Ebben, WP, and Stone, MH. Relationships between potentiation effects following ballistic half-squats and bilateral symmetry. *Int J Sports Physiol Perform* 11:448-454, 2016.

100. Tillin, NA, and Bishop, D. Factors modulating post-activation potentiation and its effect on performance of subsequent explosive activities. *Sports Med* 39:147-166, 2009.

101. Tsiokanos, A, Kellis, E, Jamurtas, A, and Kellis, S. The relationship between jumping performance and isokinetic strength of hip and knee extensors and ankle plantar flexors. *Isokinet Exerc Sci* 10:107-115, 2002.

102. Undheim, MB, Cosgrave, C, King, E, Strike, S, Marshall, B, Falvey, É, and Franklyn-Miller, A. Isokinetic muscle strength and readiness to return to sport following anterior cruciate ligament reconstruction: Is there an association? A systematic review and a protocol recommendation. *Br J Sports Med* 49:1305-1310, 2015.

103. van den Tillaar, R, and Ettema, G. A comparison of muscle activity in concentric and counter movement maximum bench press. *J Hum Kinet* 38:63-71, 2013.

104. Virvidakis, K, Georgiou, E, Korkotsidis, A, Ntalles, K, and Proukakis, C. Bone mineral content of junior competitive weightlifters. *Int J Sports Med* 11:244-246, 1990.

105. Warneke, K, Wagner, CM, Konrad, A, Kadlubowski, B, Sander, A, Wirth, K, and Keiner, M. The influence of age and sex on speed–strength performance in children between 10 and 14 years of age. *Front Physiol* 14:179, 2023.

106. Webster, DP. *The Iron Game: An Illustrated History of Weight-Lifting.* Irvine, Scotland: John Geddes (Printers), 1976.

107. Webster, DP. *World History of Highland Games.* Edinburgh: Luath Press Ltd, 2011.

108. Wisløff, U, Castagna, C, Helgerud, J, Jones, R, and Hoff, J. Strong correlation of maximal squat strength with sprint performance and vertical jump height in elite soccer players. *Br J Sports Med* 38:285-288, 2004.

109. Yang, S. History of weightlifting. *Sport Sci* 6:25-27, 1987.

110. Yang, S. A brief history of weightlifting in ancient China. *Asian Weightlifting* 4, 1996.

111. Young, WB. Laboratory strength assessment of athletes. *New Stud Athl* 10:89-96, 1995.

112. Young, WB, Miller, IR, and Talpey, SW. Physical qualities predict change-of-direction speed but not defensive agility in Australian rules football. *J Strength Cond Res* 29:206-212, 2015.

113. Zaras, N, Stasinaki, A-N, Spiliopoulou, P, Arnaoutis, G, Hadjicharalambous, M, and Terzis, G. Rate of force development, muscle architecture, and performance in elite weightlifters. *Int J Sports Physiol Perform* 16:216-223, 2020.

Chapter 2

1. Aagaard, P, Andersen, JL, Dyhre-Poulsen, P, Leffers, AM, Wagner, A, Magnusson, SP, Halkjær-Kristensen, J, and Simonsen, EB. A mechanism for increased contractile strength of human pennate muscle in response to strength training: Changes in muscle architecture. *J Physiol* 534:613-623, 2001.

2. Aagaard, P, Simonsen, EB, Andersen, JL, Magnusson, P, and Dyhre-Poulsen, P. Increased rate of force development and neural drive of human skeletal muscle following resistance training. *J Appl Physiol* 93:1318-1326, 2002.

3. Aagaard, P, Simonsen, EB, Andersen, JL, Magnusson, P, and Dyhre-Poulsen, P. Neural adaptation to resistance training: Changes in evoked V-wave and H-reflex responses. *J Appl Physiol* 92:2309-2318, 2002.

4. Aagaard, P, Simonsen, EB, Andersen, JL, Magnusson, SP, Halkjaer-Kristensen, J, and Dyhre-Poulsen, P. Neural inhibition during maximal eccentric and concentric quadriceps contraction: Effects of resistance training. *J Appl Physiol* 89:2249-2257, 2000.

5. Ahtiainen, JP, Pakarinen, A, Kraemer, WJ, and Häkkinen, K. Acute hormonal responses to heavy resistance exercise in strength athletes versus nonathletes. *Can J Appl Physiol* 29:527-543, 2004.

6. Alegre, LM, Jiménez, F, Gonzalo-Orden, JM, Martín-Acero, R, and Aguado, X. Effects of dynamic resistance training on fascicle length and isometric strength. *J Sports Sci* 24:501-508, 2006.

7. Andersen, LL, and Aagaard, P. Influence of maximal muscle strength and intrinsic muscle contractile properties on contractile rate of force development. *Eur J Appl Physiol* 96:46-52, 2006.

8. Augustsson, J, Esko, A, Thomeé, R, and Svantesson, U. Weight training of the thigh muscles using closed versus open kinetic chain exercises: A comparison of performance enhancement. *J Orthop Sports Phys Ther* 27:3-8, 1998.

9. Balshaw, TG, Massey, GJ, Maden-Wilkinson, TM, Morales-Artacho, AJ, McKeown, A, Appleby, CL, and Folland, JP. Changes in agonist neural drive, hypertrophy and pre-training strength all contribute to the individual strength gains after resistance training. *Eur J Appl Physiol* 117:631-640, 2017.

10. Bang, M-L, Caremani, M, Brunello, E, Littlefield, R, Lieber, RL, Chen, J, Lombardi, V, and Linari, M. Nebulin plays a direct role in promoting strong actin-myosin interactions. *FASEB J* 23:4117-4125, 2009.

11. Banister, EW. Relationship of effort and performance. Presented at Pre-Commonwealth Games Sports Science Symposium, Edmonton, Alberta, Canada, 1978.
12. Banister, EW. The perception of effort: An inductive approach. *Eur J Appl Physiol Occup Physiol* 41:141-150, 1979.
13. Banister, EW. Modeling elite athletic performance. In *Physiological Testing of the High-Performance Athlete*. MacDougall, JD, Wenger, HA, and Green, HJ, eds. Champaign, IL: Human Kinetics, 403-424, 1991.
14. Banister, EW, Calvert, TW, Savage, MV, and Bach, T. A systems model of training for athletic performance. *Aust J Sports Med* 7:57-61, 1975.
15. Barnard, RJ, Edgerton, VR, Furukawa, T, and Peter, J. Histochemical, biochemical, and contractile properties of red, white, and intermediate fibers. *Am J Physiol* 220:410-414, 1971.
16. Behm, DG, and Sale, DG. Intended rather than actual movement velocity determines velocity-specific training response. *J Appl Physiol* 74:359-368, 1993.
17. Billeter, R, Jostarndt-Fögen, K, Günthör, W, and Hoppeler, H. Fiber type characteristics and myosin light chain expression in a world champion shot putter. *Int J Sports Med* 24:203-207, 2003.
18. Blackburn, JR, and Morrissey, MC. The relationship between open and closed kinetic chain strength of the lower limb and jumping performance. *J Ortho Sports Phys Ther* 27:430-435, 1998.
19. Blazevich, AJ, Gill, ND, Bronks, R, and Newton, RU. Training-specific muscle architecture adaptation after 5-wk training in athletes. *Med Sci Sports Exerc* 35:2013-2022, 2003.
20. Bojsen-Møller, J, Magnusson, SP, Rasmussen, LR, Kjaer, M, and Aagaard, P. Muscle performance during maximal isometric and dynamic contractions is influenced by the stiffness of the tendinous structures. *J Appl Physiol* 99:986-994, 2005.
21. Bonifazi, M, Sardella, F, and Lupo, C. Preparatory versus main competitions: Differences in performances, lactate responses and pre-competition plasma cortisol concentrations in elite male swimmers. *Eur J Appl Physiol* 82:368-373, 2000.
22. Bosquet, L, Berryman, N, Dupuy, O, Mekary, S, Arvisais, D, Bherer, L, and Mujika, I. Effect of training cessation on muscular performance: A meta-analysis. *Scand J Med Sci Sports* 23:e140-149, 2013.
23. Bosquet, L, Montpetit, J, Arvisais, D, and Mujika, I. Effects of tapering on performance: A meta-analysis. *Med Sci Sports Exerc* 39:1358-1365, 2007.
24. Bottinelli, R, Canepari, M, Pellegrino, MA, and Reggiani, C. Force-velocity properties of human skeletal muscle fibres: Myosin heavy chain isoform and temperature dependence. *J Physiol* 495:573-586, 1996.
25. Boullosa, DA, Abreu, L, Beltrame, LG, and Behm, DG. The acute effect of different half squat set configurations on jump potentiation. *J Strength Cond Res* 27:2059-2066, 2013.
26. Brooks, GA, Fahey, TD, and Baldwin, KM. *Exercise Physiology: Human Bioenergetics and Its Applications*. New York: McGraw Hill, 2005.
27. Bullock, N, and Comfort, P. An investigation into the acute effects of depth jumps on maximal strength performance. *J Strength Cond Res* 25:3137-3141, 2011.
28. Burke, RE, Levine, DN, Zajac, FE, Tsairis, P, and Engel, WK. Mammalian motor units: Physiological-histochemical correlation in three types in cat gastrocnemius. *Science* 174:709-712, 1971.
29. Campos, GE, Luecke, TJ, Wendeln, HK, Toma, K, Hagerman, FC, Murray, TF, Ragg, KE, Ratamess, NA, Kraemer, WJ, and Staron, RS. Muscular adaptations in response to three different resistance-training regimens: Specificity of repetition maximum training zones. *Eur J Appl Physiol* 88:50-60, 2002.
30. Ceaser, T, and Hunter, G. Black and white race differences in aerobic capacity, muscle fiber type, and their influence on metabolic processes. *Sports Med* 45:615-623, 2015.
31. Cermak, NM, Snijders, T, McKay, BR, Parise, G, Verdijk, LB, Tarnopolsky, MA, Gibala, MJ, and Van Loon, LJ. Eccentric exercise increases satellite cell content in Type II muscle fibers. *Med Sci Sports Exerc* 45:230-237, 2013.
32. Costill, DL, Daniels, J, Evans, W, Fink, W, Krahenbuhl, G, and Saltin, B. Skeletal muscle enzymes and fiber composition in male and female track athletes. *J Appl Physiol* 40:149-154, 1976.
33. Costill, DL, Fink, WJ, and Pollock, ML. Muscle fiber composition and enzyme activities of elite distance runners. *Med Sci Sports Exerc* 8:96-100, 1976.
34. Crewther, BT, Kilduff, LP, Cook, CJ, Middleton, MK, Bunce, PJ, and Yang, GZ. The acute potentiating effects of back squats on athlete performance. *J Strength Cond Res* 25:3319-3325, 2011.
35. Cunanan, AJ, DeWeese, BH, Wagle, JP, Carroll, KM, Sausaman, R, Hornsby, WG, Haff, GG, Triplett, NT, Pierce, KC, and Stone, MH. The general adaptation syndrome: A foundation for the concept of periodization. *Sports Med* 48:787-797, 2018.
36. Cuthbert, M, Haff, GG, Arent, SM, Ripley, N, McMahon, JJ, Evans, M, and Comfort, P. Effects of variations in resistance training frequency on strength development in well-trained populations and implications for in-season athlete training: A systematic review and meta-analysis. *Sports Med* 51:1967-1982, 2021.
37. Dai, B, Garrett, WE, Gross, MT, Padua, DA, Queen, RM, and Yu, B. The effect of performance demands on lower extremity biomechanics during landing and cutting tasks. *J Sport Health Sci* 8:228-234, 2019.
38. Davids, K, Button, C, and Bennett, S. *Dynamics of Skills Acquisition: A Constraints-Led Approach*. Champaign, IL: Human Kinetics, 2008.
39. Davies, KJA. Cardiovascular adaptive homeostasis in exercise. *Front Physiol* 9:369, 2018.
40. Davies, T, Orr, R, Halaki, M, and Hackett, D. Effect of training leading to repetition failure on muscular strength: A systematic review and meta-analysis. *Sports Med* 46:487-502, 2016.
41. Davis, GW. Homeostatic control of neural activity: From phenomenology to molecular design. *Annu Rev Neurosci* 29:307-323, 2006.
42. Delignières, D, Nourrit, D, Sioud, R, Leroyer, P, Zattara, M, and Micaleff, J-P. Preferred coordination modes in the first steps of the learning of a complex gymnastics skill. *Hum Mov Sci* 17:221-241, 1998.

43. Dias, I, de Salles, BF, Novaes, J, Costa, PB, and Simão, R. Influence of exercise order on maximum strength in untrained young men. *J Sci Med Sport* 13:65-69, 2010.

44. Erskine, RM, Fletcher, G, and Folland, JP. The contribution of muscle hypertrophy to strength changes following resistance training. *Eur J Appl Physiol* 114:1239-1249, 2014.

45. Fahey, TD, Rolph, R, Moungmee, P, Nagel, J, and Mortara, S. Serum testosterone, body composition, and strength of young adults. *Med Sci Sports* 8:31-34, 1976.

46. Feher, J. Physiology of excitable cells. In *Quantitative Human Physiology: An Introduction*. London: Academic Press, 253-362, 2017.

47. French, DN. The endocrine responses to training. In *Strength and Conditioning for Sports Performance*. Jeffreys, I, and Moody, J, eds. New York: Routledge, 132-152, 2021.

48. French, DN, Kraemer, WJ, Volek, JS, Spiering, BA, Judelson, DA, Hoffman, JR, and Maresh, CM. Anticipatory responses of catecholamines on muscle force production. *J Appl Physiol* 102:94-102, 2007.

49. Fry, AC, and Kraemer, WJ. Resistance exercise overtraining and overreaching. *Sports Med* 23:106-129, 1997.

50. Fry, AC, Kraemer, WJ, Lynch, JM, Triplett, NT, and Koziris, LP. Does short-term near-maximal intensity machine resistance training induce overtraining? *J Strength Cond Res* 8:188-191, 1994.

51. Fry, AC, Kraemer, WJ, Stone, MH, Warren, BJ, Fleck, SJ, Kearney, JT, and Gordon, SE. Endocrine responses to overreaching before and after 1 year of weightlifting. *Can J Appl Physiol* 19:400-410, 1994.

52. Fry, AC, Kraemer, WJ, van Borselen, F, Lynch, JM, Marsit, JL, Roy, EP, Triplett, NT, and Knuttgen, HG. Performance decrements with high-intensity resistance exercise overtraining. *Med Sci Sports Exerc* 26:1165-1173, 1994.

53. Fry, AC, Kraemer, WJ, Van Borselen, F, Lynch, JM, Triplett, NT, Koziris, LP, and Fleck, SJ. Catecholamine responses to short-term high-intensity resistance exercise overtraining. *J Appl Physiol* 77:941-946, 1994.

54. Fry, AC, Schilling, BK, Staron, RS, Hagerman, FC, Hikida, RS, and Thrush, JT. Muscle fiber characteristics and performance correlates of male Olympic-style weightlifters. *J Strength Cond Res* 17:746-754, 2003.

55. Gage, PW, and Eisenberg, RS. Action potentials, afterpotentials, and excitation-contraction coupling in frog sartorius fibers without transverse tubules. *J Gen Physiol* 53:298-310, 1969.

56. Gans, C, and Gaunt, AS. Muscle architecture in relation to function. *J Biomech* 24:53-65, 1991.

57. Gardiner, P, Dai, Y, and Heckman, CJ. Effects of exercise training on α-motoneurons. *J Appl Physiol* 101:1228-1236, 2006.

58. Gorostiaga, EM, Navarro-Amézqueta, I, Cusso, R, Hellsten, Y, Calbet, JAL, Guerrero, M, Granados, C, González-Izal, M, Ibáñez, J, and Izquierdo, M. Anaerobic energy expenditure and mechanical efficiency during exhaustive leg press exercise. *PLoS One* 5:e13486, 2010.

59. Gowitzke, BA, and Milner, M. *Scientific Basis of Human Movement*. Baltimore, MD: Williams & Wilkins, 1988.

60. Grgic, J, Schoenfeld, BJ, Davies, TB, Lazinica, B, Krieger, JW, and Pedisic, Z. Effect of resistance training frequency on gains in muscular strength: A systematic review and meta-analysis. *Sports Med* 48:1207-1220, 2018.

61. Hackney, AC, Szczepanowska, E, and Viru, AM. Basal testicular testosterone production in endurance-trained men is suppressed. *Eur J Appl Physiol* 89:198-201, 2003.

62. Haff, GG, and Nimphius, S. Training principles for power. *Strength Cond J* 34:2-12, 2012.

63. Haff, GG, Whitley, A, McCoy, LB, O'Bryant, HS, Kilgore, JL, Haff, EE, Pierce, K, and Stone, MH. Effects of different set configurations on barbell velocity and displacement during a clean pull. *J Strength Cond Res* 17:95-103, 2003.

64. Haizlip, KM, Harrison, BC, and Leinwand, LA. Sex-based differences in skeletal muscle kinetics and fiber-type composition. *Physiology* 30:30-39, 2015.

65. Häkkinen, K. Neuromuscular and hormonal adaptations during strength and power training: A review. *J Sports Med Phys Fitness* 29:9, 1989.

66. Häkkinen, K, and Keskinen, KL. Muscle cross-sectional area and voluntary force production characteristics in elite strength- and endurance-trained athletes and sprinters. *Eur J Appl Physiol Occup Physiol* 59:215-220, 1989.

67. Häkkinen, K, and Komi, PV. Electromyographic changes during strength training and detraining. *Med Sci Sports Exerc* 15:455-460, 1983.

68. Häkkinen, K, Komi, PV, Alen, M, and Kauhanen, H. EMG, muscle fibre and force production characteristics during a 1 year training period in elite weight-lifters. *Eur J Appl Physiol Occup Physiol* 56:419-427, 1987.

69. Häkkinen, K, Newton, RU, Gordon, SE, McCormick, M, Volek, JS, Nindl, BC, Gotshalk, LA, Campbell, WW, Evans, WJ, and Häkkinen, A. Changes in muscle morphology, electromyographic activity, and force production characteristics during progressive strength training in young and older men. *J Gerontol A Biol Sci Med Sci* 53:B415-B423, 1998.

70. Häkkinen, K, Pakarinen, A, Kyrolainen, H, Cheng, S, Kim, DH, and Komi, PV. Neuromuscular adaptations and serum hormones in females during prolonged power training. *Int J Sports Med* 11:91-98, 1990.

71. Hardee, JP, Lawrence, MM, Utter, AC, Triplett, NT, Zwetsloot, KA, and McBride, JM. Effect of inter-repetition rest on ratings of perceived exertion during multiple sets of the power clean. *Eur J Appl Physiol* 112:3141-3147, 2012.

72. Hardee, JP, Lawrence, MM, Zwetsloot, KA, Triplett, NT, Utter, AC, and McBride, JM. Effect of cluster set configurations on power clean technique. *J Sports Sci* 31:488-496, 2013.

73. Hardee, JP, Triplett, NT, Utter, AC, Zwetsloot, KA, and McBride, JM. Effect of interrepetition rest on power output in the power clean. *J Strength Cond Res* 26:883-889, 2012.

74. Havens, KL, and Sigward, SM. Whole body mechanics differ among running and cutting maneuvers in skilled athletes. *Gait Posture* 42:240-245, 2015.

75. Henneman, E. Recruitment of motor units: The size principle. In *Motor Unit Types, Recruitment and Plasticity in Health and Disease*. Desmedt, JR, ed. New York: Karger, 1982.

76. Henneman, E, Somjen, G, and Carpenter, DO. Excitability and inhibitability of motoneurons of different sizes. *J Neurophysiol* 28:599-620, 1965.

77. Herzog, W. The multiple roles of titin in muscle contraction and force production. *Biophys Rev* 10:1187-1199, 2018.

78. Herzog, W, Duvall, M, and Leonard, TR. Molecular mechanisms of muscle force regulation: A role for titin? *Exerc Sport Sci Rev* 40:50-57, 2012.

79. Herzog, W, Leonard, TR, Joumaa, V, and Mehta, A. Mysteries of muscle contraction. *J Appl Biomech* 24:1-13, 2008.

80. Hessel, AL, Lindstedt, SL, and Nishikawa, KC. Physiological mechanisms of eccentric contraction and its applications: A role for the giant titin protein. *Front Physiol* 8:70, 2017.

81. Higuchi, H, Yoshioka, T, and Maruyama, K. Positioning of actin filaments and tension generation in skinned muscle fibres released after stretch beyond overlap of the actin and myosin filaments. *J Muscle Res Cell Motil* 9:491-498, 1988.

82. Hoffman, JR, Cooper, J, Wendell, M, and Kang, J. Comparison of Olympic vs. traditional power lifting training programs in football players. *J Strength Cond Res* 18:129-135, 2004.

83. Hornsby, WG, Haff, GG, Suarez, DG, Ramsey, MW, Triplett, NT, Hardee, JP, Stone, ME, and Stone, MH. Alterations in adiponectin, leptin, resistin, testosterone, and cortisol across eleven weeks of training among Division One collegiate throwers: A preliminary study. *J Funct Morphol Kinesiol* 5:44, 2020.

84. Horowits, R, Kempner, ES, Bisher, ME, and Podolsky, RJ. A physiological role for titin and nebulin in skeletal muscle. *Nature* 323:160-164, 1986.

85. Huxley, AF, and Niedergerke, R. Structural changes in muscle during contraction. Interference microscopy of living muscle fibers. *Nature* 173:971-973, 1954.

86. Huxley, HE. The contraction of muscle. *Sci Am* 199:66-86, 1958.

87. Huxley, HE, and Hanson, J. Changes in the cross-striations of muscle during contraction and stretch and their structural interpretation. *Nature* 173:973-976, 1954.

88. Imbach, F, Sutton-Charani, N, Montmain, J, Candau, R, and Perrey, S. The use of fitness–fatigue models for sport performance modelling: Conceptual issues and contributions from machine-learning. *Sports Med Open* 8:1-6, 2022.

89. Izquierdo, M, Ibáñez, J, Häkkinen, K, Kraemer, WJ, Ruesta, M, and Gorostiaga, EM. Maximal strength and power, muscle mass, endurance and serum hormones in weightlifters and road cyclists. *J Sports Sci* 22:465-478, 2004.

90. Jensen, J, Oftebro, H, Breigan, B, Johnsson, A, Ohlin, K, Meen, HD, Strømme, SB, and Dahl, HA. Comparison of changes in testosterone concentrations after strength and endurance exercise in well trained men. *Eur J Appl Physiol Occup Physiol* 63:467-471, 1991.

91. Jezová, D, Vigas, M, Tatár, P, Kvetnanský, R, Nazar, K, Kaciuba-Uścilko, H, and Kozlowski, S. Plasma testosterone and catecholamine responses to physical exercise of different intensities in men. *Eur J Appl Physiol Occup Physiol* 54:62-66, 1985.

92. Jovy, D, Bruner, H, Klein, KE, and Am, W. Adaptive responses of adrenal cortex to some environmental stressors, exercise, and acceleration. In *Hormonal Steroids: Biochemistry, Pharmacology, and Therapeutics*. New York: Academic Press, 545-553, 1965.

93. Kadlec, D, Miller-Dicks, M, and Nimphius, S. Training for "worst-case" scenarios in sidestepping: Unifying strength and conditioning and perception-action approaches. *Sports Med Open* 9:22, 2023.

94. Kandell, ER, Schwartz, JH, Jessell, TM, Sieglebaum, SA, and Hudspeth, AJ. *Principles of Neural Science*. New York: McGraw Hill, 2013.

95. Kawakami, Y, Abe, T, and Fukunaga, T. Muscle-fiber pennation angles are greater in hypertrophied than in normal muscles. *J Appl Physiol* 74:2740-2744, 1993.

96. Kindermann, W, Schnabel, A, Schmitt, WM, Biro, G, Cassens, J, and Weber, F. Catecholamines, growth hormone, cortisol, insulin, and sex hormones in anaerobic and aerobic exercise. *Eur J Appl Physiol Occup Physiol* 49:389-399, 1982.

97. Kiss, B, Lee, EJ, Ma, W, Li, FW, Tonino, P, Mijailovich, SM, Irving, TC, and Granzier, HL. Nebulin stiffens the thin filament and augments cross-bridge interaction in skeletal muscle. *Proc Natl Acad Sci U S A* 115:10369-10374, 2018.

98. Kraemer, WJ, Fry, AC, Frykman, PN, Conroy, B, and Hoffman, J. Resistance training and youth. *Pediatr Exerc Sci* 1:336-350, 1989.

99. Kraemer, WJ, Fry, AC, Warren, BJ, Stone, MH, Fleck, SJ, Kearney, JT, Conroy, BP, Maresh, CM, Weseman, CA, and Triplett, NT. Acute hormonal responses in elite junior weightlifters. *Int J Sports Med* 13:103-109, 1992.

100. Kraemer, WJ, Häkkinen, K, Newton, RU, Nindl, BC, Volek, JS, McCormick, M, Gotshalk, LA, Gordon, SE, Fleck, SJ, and Campbell, WW. Effects of heavy-resistance training on hormonal response patterns in younger vs. older men. *J Appl Physiol* 87:982-992, 1999.

101. Kraemer, WJ, Marchitelli, L, Gordon, SE, Harman, E, Dziados, JE, Mello, R, Frykman, P, McCurry, D, and Fleck, SJ. Hormonal and growth factor responses to heavy resistance exercise protocols. *J Appl Physiol* 69:1442-1450, 1990.

102. Kraemer, WJ, Noble, BJ, Clark, MJ, and Culver, BW. Physiologic responses to heavy-resistance exercise with very short rest periods. *Int J Sports Med* 8:247-252, 1987.

103. Krieger, JW. Single versus multiple sets of resistance exercise: A meta-regression. *J Strength Cond Res* 23:1890-1901, 2009.

104. Kubo, K, Ikebukuro, T, Yata, H, Tsunoda, N, and Kanehisa, H. Time course of changes in muscle and tendon properties during strength training and detraining. *J Strength Cond Res* 24:322-331, 2010.

105. Kubo, K, Ishigaki, T, and Ikebukuro, T. Effects of plyometric and isometric training on muscle and tendon stiffness in vivo. *Physiol Rep* 5:e13374, 2017.

106. Kubo, K, Yata, H, Kanehisa, H, and Fukunaga, T. Effects of isometric squat training on the tendon stiffness and jump performance. *Eur J Appl Physiol* 96:305-314, 2006.

107. Lauersen, JB, Andersen, TE, and Andersen, LB. Strength training as superior, dose-dependent and safe prevention of acute and overuse sports injuries: A systematic review, qualitative analysis and meta-analysis. *Br J Sports Med* 52:1557-1563, 2018.

108. Lauersen, JB, Bertelsen, DM, and Andersen, LB. The effectiveness of exercise interventions to prevent sports injuries: A systematic review and meta-analysis of randomised controlled trials. *Br J Sports Med* 48:871-877, 2014.

109. Lehmann, M, Schmid, P, and Keul, J. Age- and exercise-related sympathetic activity in untrained volunteers, trained athletes and patients with impaired left-ventricular contractility. *Eur Heart J* 5:1-7, 1984.

110. Leong, B, Kamen, G, Patten, C, and Burke, JR. Maximal motor unit discharge rates in the quadriceps muscles of older weight lifters. *Med Sci Sports Exerc* 31:1638-1644, 1999.

111. Lopes dos Santos, M, Uftring, M, Stahl, CA, Lockie, RG, Alvar, B, Mann, JB, and Dawes, JJ. Stress in academic and athletic performance in collegiate athletes: A narrative review of sources and monitoring strategies. *Front Sports Act Living* 2:42, 2020.

112. MacIntosh, BR, Gradiner, PF, and McComas, AJ. *Muscle Architecture and Muscle Fiber Anatomy*. Champaign, IL: Human Kinetics, 2006.

113. Maffiuletti, NA, Aagaard, P, Blazevich, AJ, Folland, J, Tillin, N, and Duchateau, J. Rate of force development: Physiological and methodological considerations. *Eur J Appl Physiol* 116:1091-1116, 2016.

114. Mangine, GT, Hoffman, JR, Gonzalez, AM, Townsend, JR, Wells, AJ, Jajtner, AR, Beyer, KS, Boone, CH, Miramonti, AA, Wang, R, LaMonica, MB, Fukuda, DH, Ratamess, NA, and Stout, JR. The effect of training volume and intensity on improvements in muscular strength and size in resistance-trained men. *Physiol Rep* 3:e12472, 2015.

115. Mangine, GT, Hoffman, JR, Wang, R, Gonzalez, AM, Townsend, JR, Wells, AJ, Jajtner, AR, Beyer, KS, Boone, CH, and Miramonti, AA. Resistance training intensity and volume affect changes in rate of force development in resistance-trained men. *Eur J Appl Physiol* 116:2367-2374, 2016.

116. Mann, JB, Bryant, KR, Johnstone, B, Ivey, PA, and Sayers, SP. Effect of physical and academic stress on illness and injury in Division I college football players. *J Strength Cond Res* 30:20-25, 2016.

117. Marshall, J, Bishop, C, Turner, A, and Haff, GG. Optimal training sequences to develop lower body force, velocity, power, and jump height: A systematic review with meta-analysis. *Sports Med* 51:1245-1271, 2021.

118. McArdle, WD, Katch, FI, and Katch, VI. *Essentials of Exercise Physiology*. Wolters Kluwer, 2015.

119. McComas, AJ. *Skeletal Muscle*. Champaign, IL: Human Kinetics, 1996.

120. McCurdy, KW, O'Kelley, E, Kutz, M, Langford, G, Ernest, J, and Torres, M. Comparison of lower extremity EMG between the 2-leg squat and modified single-leg squat in female athletes. *J Sport Rehabil* 19:57-70, 2010.

121. McMillan, JL, Stone, MH, Sartin, J, Keith, R, Marples, D, Brown, C, and Lewis, RD. 20-hour physiological responses to a single weight-training session. *J Strength Cond Res* 7:9-21, 1993.

122. McMurray, RG, Eubank, TK, and Hackney, AC. Nocturnal hormonal responses to resistance exercise. *Eur J Appl Physiol Occup Physiol* 72:121-126, 1995.

123. American College of Sports Medicine. American College of Sports Medicine position stand. Progression models in resistance training for healthy adults. *Med Sci Sports Exerc* 41:687-708, 2009. =Medlar, S. Mixing it up: The biological significance of hybrid skeletal muscle fibers. *J Exp Biol* 222:jeb200832, 2019.

124. Miller, MS, Bedrin, NG, Ades, PA, Palmer, BM, and Toth, MJ. Molecular determinants of force production in human skeletal muscle fibers: Effects of myosin isoform expression and cross-sectional area. *J Physiol Cell Physiol* 308:C473-C484, 2015.

125. Moir, GL. Skill acquisition. In *Strength and Conditioning: A Biomechanical Approach*. Burlington, MA: Jones & Bartlett Learning, 387-430, 2016.

126. Monroy, JA, Powers, KL, Gilmore, LA, Uyeno, TA, Lindstedt, SL, and Nishikawa, KC. What is the role of titin in active muscle? *Exerc Sport Sci Rev* 40:73-78, 2012.

127. Moore, CA, and Fry, AC. Nonfunctional overreaching during off-season training for skill position players in collegiate American football. *J Strength Cond Res* 21:793-800, 2007.

128. Moritani, T. Neural factors versus hypertrophy in the time course of muscle strength gain. *Am J Phys Med* 58:115-130, 1979.

129. Morrison, S, and Newell, KM. Strength training as a dynamical model of motor learning. *J Sports Sci* 41:408-423, 2023.

130. Mujika, I. *Tapering and Peaking*. Champaign, IL: Human Kinetics, 2009.

131. Nagahara, R, Mizutani, M, Matsuo, A, Kanehisa, H, and Fukunaga, T. Association of sprint performance with ground reaction forces during acceleration and maximal speed phases in a single sprint. *J Appl Biomech* 34:104-110, 2018.

132. Nagatani, T, Haff, GG, Guppy, SN, and Kendall, KL. Practical application of traditional and cluster set configurations within a resistance training program. *Strength Cond J* 44:87-101, 2022.

133. Narici, MV, Roi, GS, Landoni, L, Minetti, AE, and Cerretelli, P. Changes in force, cross-sectional area and neural activation during strength training and detraining of the human quadriceps. *Eur J Appl Physiol Occup Physiol* 59:310-319, 1989.

134. Newell, KM. Coordination, control, and skill. In *Differing Perspectives in Motor Learning, Memory, and Control*. Goodman, D, Wilberg, RB, and Franks, IM, eds. Amsterdam, Netherlands: North-Holland, 295-317, 1985.

135. Newton, RU, Häkkinen, K, Häkkinen, A, McCormick, M, Volek, J, and Kraemer, WJ. Mixed-methods resistance training increases power and strength of young and older men. *Med Sci Sports Exerc* 34:1367-1375, 2002.

136. Newton, RU, and Kraemer, WJ. Developing explosive muscular power: Implications for a mixed methods training strategy. *Strength Cond J* 16:20-31, 1994.

137. Nindl, BC, Kraemer, WJ, Gotshalk, LA, Marx, JO, Volek, JS, Bush, FA, Häkkinen, K, Newton, RU, and Fleck, SJ. Testosterone responses after resistance exercise in women: Influence of regional fat distribution. *Int J Sport Nutr Exerc Metab* 11:451-465, 2001.

138. Nourrit, D, Deschamps, T, Lauriot, B, Caillou, N, and Delignieres, D. The effects of required amplitude and practice on frequency stability and efficiency in a cyclical task. *J Sports Sci* 18:201-212, 2000.

139. Nuzzo, JL. Sex differences in skeletal muscle fiber types: A meta-analysis. *Clin Anat* 37:81-91, 2024.

140. Östenberg, A, Roos, E, Ekdah, C, and Roos, H. Isokinetic knee extensor strength and functional performance in healthy female soccer players. *Scand J Med Sci Sports* 8:257-264, 1998.

141. Ottenheijm, CAC, and Granzier, H. New insights into the structural roles of nebulin in skeletal muscle. *J Biomed Biotechnol* 2010:968139, 2010.

142. Ottenheijm, CAC, Granzier, H, and Labeit, S. The sarcomeric protein nebulin: Another multifunctional giant in charge of muscle strength optimization. *Front Physiol* 3:37, 2012.

143. Painter, KB, Haff, GG, Triplett, NT, Stuart, C, Hornsby, G, Ramsey, MW, Bazyler, CD, and Stone, MH. Resting hormone alterations and injuries: Block vs. DUP weight-training among D-1 track and field athletes. *Sports* 6:3, 2018.

144. Pellegrino, MA, Canepari, M, Rossi, R, D'Antona, G, Reggiani, C, and Bottinelli, R. Orthologous myosin isoforms and scaling of shortening velocity with body size in mouse, rat, rabbit and human muscles. *J Physiol* 546:677-689, 2003.

145. Peterson, MD, Rhea, MR, and Alvar, BA. Applications of the dose–response for muscular strength development: A review of meta-analytic efficacy and reliability for designing training prescription. *J Strength Cond Res* 19:950-958, 2005.

146. Pette, D, and Staron, RS. Cellular and molecular diversities of mammalian skeletal muscle fibers. *Rev Physiol Biochem Pharmacol* 116:1-76, 1990.

147. Pette, D, and Staron, RS. Myosin isoforms, muscle fiber types and transitions. *Microsc Res Tech* 50:500-509, 2000.

148. Pette, D, and Staron, RS. Transitions of muscle fiber phenotypic profiles. *Histochem Cell Biol* 115:359-372, 2001.

149. Plotkin, DL, Roberts, MD, Haun, CT, and Schoenfeld, BJ. Muscle fiber type transitions with exercise training: Shifting perspectives. *Sports* 9:127, 2021.

150. Powers, K, Nishikawa, K, Joumaa, V, and Herzog, W. Decreased force enhancement in skeletal muscle sarcomeres with a deletion in titin. *J Exp Biol* 219:1311-1316, 2016.

151. Pullinen, T, Huttunen, P, and Komi, PV. Plasma catecholamine responses and neural adaptation during short-term resistance training. *Eur J Appl Physiol* 82:68-75, 2000.

152. Ralston, GW, Kilgore, L, Wyatt, FB, and Baker, JS. The effect of weekly set volume on strength gain: A meta-analysis. *Sports Med* 47:2585-2601, 2017.

153. Ralston, GW, Kilgore, L, Wyatt, FB, Buchan, D, and Baker, JS. Weekly training frequency effects on strength gain: A meta-analysis. *Sports Med Open* 4:1-24, 2018.

154. Roberts, TJ. Contribution of elastic tissues to the mechanics and energetics of muscle function during movement. *J Exp Biol* 219:266-275, 2016.

155. Robinson, JM, Stone, MH, Johnson, RL, Penland, CM, Warren, BJ, and Lewis, RD. Effects of different weight training exercise/rest intervals on strength, power, and high intensity exercise endurance. *J Strength Cond Res* 9:216-221, 1995.

156. Ross, A, Leveritt, M, and Riek, S. Neural influences on sprint running. *Sports Med* 31:409-425, 2001.

157. Roy, RR, and Edgerton, VR. Skeletal muscle architecture and performance. In *Strength and Power in Sport*. Komi, PV, ed. Oxford: Blackwell Scientific, 115-129, 1992.

158. Rutherford, OM. Muscular coordination and strength training: Implications for injury rehabilitation. *Sports Med* 5:196-202, 1988.

159. Rutherford, OM, and Jones, DA. The role of learning and coordination in strength training. *Eur J Appl Physiol Occup Physiol* 55:100-105, 1986.

160. Sale, DG. Postactivation potentiation: Role in human performance. *Exerc Sport Sci Rev* 30:138-143, 2002.

161. Sale, DG. Neural adaptations to strength training. In *Strength and Power in Sport*. Komi, PV, ed. Oxford: Blackwell Scientific, 281-313, 2003.

162. Sant'Ana Pereira, JA, Sargeant, AJ, Rademaker, AC, de Haan, A, and van Mechelen, M. Myosin heavy chain isoform expression and high energy phosphate content in muscle fibres at rest and post-exercise. *J Physiol* 496:583-588, 1996.

163. Saplinskas, JS, Chobotas, MA, and Yashchaninas, II. The time of completed motor acts and impulse activity of single motor units according to the training level and sport specialization of tested persons. *Electromyogr Clin Neurophysiol* 20:529-539, 1980.

164. Schoenfeld, BJ, Contreras, B, Vigotsky, AD, and Peterson, M. Differential effects of heavy versus moderate loads on measures of strength and hypertrophy in resistance-trained men. *J Sports Sci Med* 15:715-722, 2016.

165. Schoenfeld, BJ, Grgic, J, Ogborn, D, and Krieger, JW. Strength and hypertrophy adaptations between low-versus high-load resistance training: A systematic review and meta-analysis. *J Strength Cond Res* 31:3508-3523, 2017.

166. Schoenfeld, BJ, Grgic, J, Van Every, DW, and Plotkin, DL. Loading recommendations for muscle strength, hypertrophy, and local endurance: A re-examination of the repetition continuum. *Sports* 9:32, 2021.

167. Schoenfeld, BJ, Peterson, MD, Ogborn, D, Contreras, B, and Sonmez, GT. Effects of low- vs. high-load resistance training on muscle strength and hypertrophy in well-trained men. *J Strength Cond Res* 29:2954-2963, 2015.

168. Schoenfeld, BJ, Pope, ZK, Benik, FM, Hester, GM, Sellers, J, Nooner, JL, Schnaiter, JA, Bond-Williams, KE, Carter, AS, and Ross, CL. Longer inter-set rest periods enhance muscle strength and hypertrophy in resistance-trained men. *J Strength Cond Res* 30:1805-1812, 2016.

169. Schwab, R, Johnson, GO, Housh, TJ, Kinder, JE, and Weir, JP. Acute effects of different intensities of weight lifting on serum testosterone. *Med Sci Sports Exerc* 25:1381-1385, 1993.

170. Selye, H. A syndrome produced by diverse nocuous agents. *Nature* 138:32, 1936.

171. Selye, H. Experimental evidence supporting the conception of "adaptation energy." *Am J Physiol* 123:758-765, 1938.

172. Selye, H. The general adaptation syndrome and the diseases of adaptation. *J Clin Endocrinol* 6:117-230, 1946.

173. Selye, H. Stress and the general adaptation syndrome. *Br Med J* 1:1383-1392, 1950.

174. Selye, H. The general-adaptation-syndrome. *Annu Rev Med* 2:327-342, 1951.

175. Selye, H. Forty years of stress research: Principal remaining problems and misconceptions. *Can Med Assoc J* 115:53-56, 1976.

176. Semmler, JG, Kutscher, DV, Zhou, S, and Enoka, RM. Motor unit synchronization is enhanced during slow shortening and lengthening contractions of the first dorsal interosseus muscle. *Soc Neurosci Abstr* 26:463, 2000.

177. Serrano, N, Colenso-Semple, LM, Lazauskus, KK, Siu, JW, Bagley, JR, Lockie, RG, Costa, PB, and Galpin, AJ. Extraordinary fast-twitch fiber abundance in elite weightlifters. *PLoS One* 14:e0207975, 2019.

178. Shepard, R. Hormonal control systems. In *Physiology and Biochemistry of Exercise*. New York: Praeger, 1982.

179. Sheppard, JM, and Triplett, NT. Program design for resistance training. In *Essentials of Strength Training and Conditioning*. 4th ed. Haff, GG, and Triplett, NT, eds. Champaign, IL: Human Kinetics, 439-468, 2016.

180. Sherrington, CS. Some functional problems attaching to convergence. *Proc Royal Soc B* 105:332-362, 1929.

181. Simão, R, de Salles, BF, Figueiredo, T, Dias, I, and Willardson, JM. Exercise order in resistance training. *Sports Med* 42:251-265, 2012.

182. Simão, R, Spineti, J, de Salles, BF, Oliveira, LF, Matta, T, Miranda, F, Miranda, H, and Costa, PB. Influence of exercise order on maximum strength and muscle thickness in untrained men. *J Sports Sci Med* 9:1-7, 2010.

183. Spineti, J, de Salles, BF, Rhea, MR, Lavigne, D, Matta, T, Miranda, F, Fernandes, L, and Simão, R. Influence of exercise order on maximum strength and muscle volume in nonlinear periodized resistance training. *J Strength Cond Res* 24:2962-2969, 2010.

184. St. Clair Gibson, A, Goedecke, JH, Harley, YX, Myers, LJ, Lambert, MI, Noakes, TD, and Lambert, EV. Metabolic setpoint control mechanisms in different physiological systems at rest and during exercise. *J Theor Biol* 236:60-72, 2005.

185. Staron, RS, Hagerman, FC, Hikida, RS, Murray, TF, Hostler, DP, Crill, MT, Ragg, KE, and Toma, K. Fiber type composition of the vastus lateralis muscle of young men and women. *J Histochem Cytochem* 48:623-629, 2000.

186. Staron, RS, and Hikida, RS. Histochemical, biochemical and ultrastructural analyses of single human muscle fibers with special reference to the C-fiber population. *J Histochem Cytochem* 40:563-568, 1992.

187. Staron, RS, Karapondo, DL, Kraemer, WJ, Fry, AC, Gordon, SE, Falkel, JE, Hagerman, FC, and Hikida RS. Skeletal muscle adaptations during early phase of heavy-resistance training in men and women. *J Appl Physiol* 76:1247-1255, 1994.

188. Stone, MH, Collins, D, Plisk, S, Haff, GG, and Stone, ME. Training principles: Evaluation of modes and methods of resistance training. *Strength Cond J* 22:65-76, 2000.

189. Stone, MH, Cormie, P, Lamont, H, and Stone, ME. Developing strength and power. In *Strength and Conditioning for Sports Performance*. Jeffreys, I, and Moody, J, eds. New York: Routledge, 230-260, 2016.

190. Stone, MH, and Fry, AC. Increased training volume in strength/power athletes. In *Overtraining in Sport*. Kreider, R, Fry, AC, and O'Toole, M, eds. Champaign, IL: Human Kinetics, 87-106, 1997.

191. Stone, MH, Hornsby, WG, Haff, GG, Fry, AC, Suarez, DG, Liu, J, Gonzalez-Rave, JM, and Pierce, KC. Periodization and block periodization in sports: Emphasis on strength–power training—A provocative and challenging narrative. *J Strength Cond Res* 35:2351-2371, 2021.

192. Stone, MH, Potteiger, JA, Pierce, KC, Proulx, CM, O'Bryant, HS, Johnson, RL, and Stone, ME. Comparison of the effects of three different weight-training programs on the one repetition maximum squat. *J Strength Cond Res* 14:332-337, 2000.

193. Stone, MH, Sanborn, K, O'Bryant, HS, Hartman, M, Stone, ME, Proulx, C, Ward, B, and Hruby, J. Maximum strength–power–performance relationships in collegiate throwers. *J Strength Cond Res* 17:739-745, 2003.

194. Stone, MH, Suchomel, TJ, Hornsby, WG, Wagle, JP, and Cunanan, AJ. General concepts and training principles for athlete development. In *Strength and Conditioning in Sports: From Science to Practice*. New York: Routledge, 221-251, 2022.

195. Stone, MH, Suchomel, TJ, Hornsby, WG, Wagle, JP, and Cunanan, AJ. Neuromuscular physiology. In *Strength and Conditioning in Sports: From Science to Practice*. New York: Routledge, 3-56, 2022.

196. Stuart, CA, Stone, WL, Howell, MEA, Brannon, MF, Hall, HK, Gibson, AL, and Stone, MH. Myosin content of individual human muscle fibers isolated by laser capture microdissection. *Am J Physiol Cell Physiol* 310:C381-C389, 2016.

197. Suarez, DG, Mizuguchi, S, Hornsby, WG, Cunanan, AJ, Marsh, DJ, and Stone, MH. Phase-specific changes in rate of force development and muscle morphology throughout a block periodized training cycle in weightlifters. *Sports* 7:129, 2019.

198. Suchomel, TJ. Resistance training strategies to train the force–velocity characteristics of athletes. In *Central Virginia Sport Performance: The Manual*, Vol. 7. DeMayo, J, ed. Amazon Publishing, 95-118, 2022.

199. Suchomel, TJ, and Comfort, P. Developing muscular strength and power. In *Advanced Strength and Conditioning: An Evidence-Based Approach*. Turner, A, and Comfort, P, eds. New York: Routledge, 13-39, 2022.

200. Suchomel, TJ, Lamont, HS, and Moir, GL. Understanding vertical jump potentiation: A deterministic model. *Sports Med* 46:809-828, 2016.

201. Suchomel, TJ, Nimphius, S, Bellon, CR, Hornsby, WG, and Stone, MH. Training for muscular strength: Methods for monitoring and adjusting training intensity. *Sports Med* 51:2051-2066, 2021.

202. Suchomel, TJ, Nimphius, S, Bellon, CR, and Stone, MH. The importance of muscular strength: Training considerations. *Sports Med* 48:765-785, 2018.

203. Suchomel, TJ, Nimphius, S, and Stone, MH. The importance of muscular strength in athletic performance. *Sports Med* 46:1419-1449, 2016.

204. Suchomel, TJ, Sato, K, DeWeese, BH, Ebben, WP, and Stone, MH. Potentiation effects of half-squats performed in a ballistic or non-ballistic manner. *J Strength Cond Res* 30:1652-1660, 2016.

205. Suchomel, TJ, Sato, K, DeWeese, BH, Ebben, WP, and Stone, MH. Potentiation following ballistic and non-ballistic complexes: The effect of strength level. *J Strength Cond Res* 30:1825-1833, 2016.

206. Suchomel, TJ, Taber, CB, Sole, CJ, and Stone, MH. Force–time differences between ballistic and non-ballistic half-squats. *Sports* 6:79, 2018.

207. Tharp, GD. The role of glucocorticoids in exercise. *Med Sci Sports* 7:6-11, 1975.

208. Tillin, NA, and Bishop, D. Factors modulating post-activation potentiation and its effect on performance of subsequent explosive activities. *Sports Med* 39:147-166, 2009.

209. Trappe, S, Luden, N, Minchev, K, Raue, U, Jemiolo, B, and Trappe, T. Skeletal muscle signature of a champion sprint runner. *J Appl Physiol* 118:1460-1466, 2015.

210. Travis, SK, Ishida, A, Taber, CB, Fry, AC, and Stone, MH. Emphasizing task-specific hypertrophy to enhance sequential strength and power performance. *J Funct Morphol Kinesiol* 5:7, 2020.

211. Travis, SK, Zwetsloot, KA, Mujika, I, Stone, MH, and Bazyler, CD. Skeletal muscle adaptations and performance outcomes following a step and exponential taper in strength athletes. *Front Physiol* 12:735932, 2021.

212. Tremblay, MS, Copeland, JL, and Van Helder, W. Effect of training status and exercise mode on endogenous steroid hormones in men. *J Appl Physiol* 96:531-539, 2004.

213. Trinick, J. Elastic filaments and giant proteins in muscle. *Curr Opin Cell Biol* 3:112-119, 1991.

214. Tufano, JJ, Conlon, JA, Nimphius, S, Brown, LE, Seitz, LB, Williamson, BD, and Haff, GG. Maintenance of velocity and power with cluster sets during high-volume back squats. *Int J Sports Physiol Perform* 11:885-892, 2016.

215. van Cutsem, M, Duchateau, J, and Hainaut, K. Changes in single motor unit behaviour contribute to the increase in contraction speed after dynamic training in humans. *J Physiol* 513:295-305, 1998.

216. Vermeire, K, Ghijs, M, Bourgois, JG, and Boone, J. The fitness–fatigue model: What's in the numbers? *Int J Sports Physiol Perform* 17:810-813, 2022.

217. Vingren, JL, Kraemer, WJ, Hatfield, DL, Volek, JS, Ratamess, NA, Anderson, JM, Häkkinen, K, Ahtiainen, J, Fragala, MS, and Thomas, GA. Effect of resistance exercise on muscle steroid receptor protein content in strength-trained men and women. *Steroids* 74:1033-1039, 2009.

218. Walker, PM, Brunotte, F, Rouhier-Marcer, I, Cottin, Y, Casillas, J-M, Gras, P, and Didier, J-P. Nuclear magnetic resonance evidence of different muscular adaptations after resistance training. *Arch Phys Med Rehabil* 79:1391-1398, 1998.

219. Weiss, LW, Cureton, KJ, and Thompson, FN. Comparison of serum testosterone and androstenedione responses to weight lifting in men and women. *Eur J Appl Physiol Occup Physiol* 50:413-419, 1983.

220. West, DW, and Phillips, SM. Associations of exercise-induced hormone profiles and gains in strength and hypertrophy in a large cohort after weight training. *Eur J Appl Physiol* 112:2693-2702, 2012.

221. Wickiewicz, TL, Roy, RR, Powell, PL, and Edgerton, VR. Muscle architecture of the human lower limb. *Clin Orthop Relat Res* 179:275-283, 1983.

222. Wickiewicz, TL, Roy, RR, Powell, PL, Perrine, JJ, and Edgerton, VR. Muscle architecture and force–velocity relationships in humans. *J Appl Physiol* 57:435-443, 1984.

223. Wilkerson, JE, Horvath, SM, and Gutin, B. Plasma testosterone during treadmill exercise. *J Appl Physiol Respir Environ Exerc Physiol* 49:249-253, 1980.

224. Yakovlev, NN. Biochemical foundations of muscle training. *Uspekhi Sovr Biol* 27:257-271, 1949.

225. Yakovlev, NN. Biochemistry of sport in the Soviet Union: Beginning, development, and present status. *Med Sci Sports* 7:237-247, 1975.

226. Young, WB, and Bilby, GE. The effect of voluntary effort to influence speed of contraction on strength, muscular power, and hypertrophy development. *J Strength Cond Res* 7:172-178, 1993.

227. Yuen, M, and Ottenheijm, CAC. Nebulin: Big protein with big responsibilities. *J Muscle Res Cell Motil* 41:103-124, 2020.

228. Zając, A, Chalimoniuk M, Maszczyk A, Gołaś A, and Lngfort J. Central and peripheral fatigue during resistance exercise—A critical review. *J Hum Kinet* 49:159, 2015.

Chapter 3

1. Aagaard, P, Andersen, JL, Dyhre-Poulsen, P, Leffers, AM, Wagner, A, Magnusson, SP, Halkjær-Kristensen, J, and Simonsen, EB. A mechanism for increased contractile strength of human pennate muscle in response to strength training: Changes in muscle architecture. *J Physiol* 534:613-623, 2001.

2. Aagaard, P, Simonsen, EB, Andersen, JL, Magnusson, P, and Dyhre-Poulsen, P. Increased rate of force development and neural drive of human skeletal muscle following resistance training. *J Appl Physiol* 93:1318-1326, 2002.

3. Aagaard, P, Simonsen, EB, Andersen, JL, Magnusson, P, and Dyhre-Poulsen, P. Neural adaptation to resistance training: Changes in evoked V-wave and H-reflex responses. *J Appl Physiol* 92:2309-2318, 2002.

4. Aagaard, P, Simonsen, EB, Andersen, JL, Magnusson, SP, Halkjaer-Kristensen, J, and Dyhre-Poulsen, P. Neural inhibition during maximal eccentric and concentric quadriceps contraction: Effects of resistance training. *J Appl Physiol* 89:2249-2257, 2000.

5. Ahtiainen, JP, Pakarinen, A, Kraemer, WJ, and Häkkinen, K. Acute hormonal responses to heavy resistance exercise in strength athletes versus nonathletes. *Can J Appl Physiol* 29:527-543, 2004.

6. Alegre, LM, Jiménez, F, Gonzalo-Orden, JM, Martín-Acero, R, and Aguado, X. Effects of dynamic resistance training on fascicle length and isometric strength. *J Sports Sci* 24:501-508, 2006.

7. Alen, M, Pakarinen, A, Häkkinen, K, and Komi, PV. Responses of serum androgenic-anabolic and catabolic hormones to prolonged strength training. *Int J Sports Med* 9:229-233, 1988.

8. Allen, SV, and Hopkins, WG. Age of peak competitive performance of elite athletes: A systematic review. *Sports Med* 45:1431-1441, 2015.

9. Andersen, LL, and Aagaard, P. Influence of maximal muscle strength and intrinsic muscle contractile properties on contractile rate of force development. *Eur J Appl Physiol* 96:46-52, 2006.

10. Ash, GI, Scott, RA, Deason, M, Dawson, TA, Wolde, B, Bekele, Z, Teka, S, and Pitsiladis, YP. No association between ACE gene variation and endurance athlete status in Ethiopians. *Med Sci Sports Exerc* 43:590-597, 2011.

11. Azizi, E, Brainerd, EL, and Roberts, TJ. Variable gearing in pennate muscles. *Proc Natl Acad Sci USA* 105:1745-1750, 2008.

12. Baddeley, AD, and Longman, DJA. The influence of length and frequency of training session on the rate of learning to type. *Ergonomics* 21:627-635, 1978.

13. Balagué, N, Torrents, C, Hristovski, R, Davids, K, and Araújo, D. Overview of complex systems in sport. *J Syst Sci Complex* 26:4-13, 2013.

14. Balshaw, TG, Massey, GJ, Maden-Wilkinson, TM, Morales-Artacho, AJ, McKeown, A, Appleby, CL, and Folland, JP. Changes in agonist neural drive, hypertrophy and pre-training strength all contribute to the individual strength gains after resistance training. *Eur J Appl Physiol* 117:631-640, 2017.

15. Bang, M-L, Caremani, M, Brunello, E, Littlefield, R, Lieber, RL, Chen, J, Lombardi, V, and Linari, M. Nebulin plays a direct role in promoting strong actin-myosin interactions. *FASEB J* 23:4117-4125, 2009.

16. Barnard, RJ, Edgerton, VR, and Peter, JB. Effect of exercise on skeletal muscle. I. Biochemical and histochemical properties. *J Appl Physiol* 28:762-766, 1970.

17. Barr, MJ, Sheppard, JM, Gabbett, TJ, and Newton, RU. The effect of ball carrying on the sprinting speed of international rugby union players. *Int J Sports Sci Coach* 10:1-9, 2015.

18. Beiter, T, Nieß, AM, and Moser, D. Transcriptional memory in skeletal muscle. Don't forget (to) exercise. *J Cell Physiol* 235:5476-5489, 2020.

19. Bhasin, S, Woodhouse, L, Casaburi, R, Singh, AB, Bhasin, D, Berman, N, Chen, X, Yarasheski, KE, Magliano, L, and Dzekov, C. Testosterone dose-response relationships in healthy young men. *Am J Physiol Endocrinol Metab* 281:E1172-E1181, 2001.

20. Billeter, R, Jostarndt-Fögen, K, Günthör, W, and Hoppeler, H. Fiber type characteristics and myosin light chain expression in a world champion shot putter. *Int J Sports Med* 24:203-207, 2003.

21. Blazevich, AJ, Cannavan, D, Coleman, DR, and Horne, S. Influence of concentric and eccentric resistance training on architectural adaptation in human quadriceps muscles. *J Appl Physiol* 103:1565-1575, 2007.

22. Blazevich, AJ, Gill, ND, Bronks, R, and Newton, RU. Training-specific muscle architecture adaptation after 5-wk training in athletes. *Med Sci Sports Exerc* 35:2013-2022, 2003.

23. Bleisch, W, Luine, VN, and Nottebohm, F. Modification of synapses in androgen-sensitive muscle. I. Hormonal regulation of acetylcholine receptor number in the songbird syrinx. *J Neurosci* 4:786-792, 1984.

24. Bleisch, WV, and Harrelson, A. Androgens modulate endplate size and ACh receptor density at synapses in rat levator ani muscle. *J Neurobiol* 20:189-202, 1989.

25. Bobbert, MF, Gerritsen, KGM, Litjens, MCA, and Van Soest, AJ. Why is countermovement jump height greater than squat jump height? *Med Sci Sports Exerc* 28:1402-1412, 1996.

26. Bobbert, MF, and van Soest, AJ. Why do people jump the way they do? *Exerc Sport Sci Rev* 29:95-102, 2001.

27. Bojsen-Møller, J, Magnusson, SP, Rasmussen, LR, Kjaer, M, and Aagaard, P. Muscle performance during maximal isometric and dynamic contractions is influenced by the stiffness of the tendinous structures. *J Appl Physiol* 99:986-994, 2005.

28. Bonifazi, M, Sardella, F, and Lupo, C. Preparatory versus main competitions: Differences in performances, lactate responses and pre-competition plasma cortisol concentrations in elite male swimmers. *Eur J Appl Physiol* 82:368-373, 2000.

29. Bosco, C, Tihanyi, J, and Viru, A. Relationships between field fitness test and basal serum testosterone and cortisol levels in soccer players. *Clin Physiol* 16:317-322, 1996.

30. Bottinelli, R, Canepari, M, Pellegrino, MA, and Reggiani, C. Force-velocity properties of human skeletal muscle fibres: Myosin heavy chain isoform and temperature dependence. *J Physiol* 495:573-586, 1996.

31. Brainerd, EL, and Azizi, E. Muscle fiber angle, segment bulging and architectural gear ratio in segmented musculature. *J Exp Biol* 208:3249-3261, 2005.

32. Brechue, WF, Mayhew, JL, and Piper, FC. Equipment and running surface alter sprint performance of college football players. *J Strength Cond Res* 19:821-825, 2005.

33. Brooke, MH, and Kaiser, KK. Three "myosin adenosine triphosphatase" systems: The nature of their pH lability and sulfhydryl dependence. *J Histochem Cytochem* 18:670-672, 1970.

34. Buckner, SL, Mouser, JG, Jessee, MB, Dankel, SJ, Mattocks, KT, and Loenneke, JP. What does individual strength say about resistance training status? *Muscle Nerve* 55:455-457, 2017.

35. Burke, RE. Motor units: Anatomy, physiology, and functional organization. In *Handbook of Physiology, Section I, The Nervous System II*. Brooks, VB, ed. Washington, DC: American Physiological Society, 345-422, 1981.

36. Burke, RE, Levine, DN, Zajac, FE, Tsairis, P, and Engel, WK. Mammalian motor units: Physiological-histochemical correlation in three types in cat gastrocnemius. *Science* 174:709-712, 1971.

37. Busso, T, Häkkinen, K, Pakarinen, A, Kauhanen, H, Komi, PV, and Lacour, JR. Hormonal adaptations and modelled responses in elite weightlifters during 6 weeks of training. *Eur J Appl Physiol Occup Physiol* 64:381-386, 1992.

38. Campos, GE, Luecke, TJ, Wendeln, HK, Toma, K, Hagerman, FC, Murray, TF, Ragg, KE, Ratamess, NA, Kraemer, WJ, and Staron, RS. Muscular adaptations in response to three different resistance-training regimens: Specificity of repetition maximum training zones. *Eur J Appl Physiol* 88:50-60, 2002.

39. Cassidy, J, Young, W, Gorman, A, and Kelly, V. Merging athletic development with skill acquisition: Developing agility using an ecological dynamics approach. *Strength Cond J* 46:202-213, 2024.

40. Castillo, A, Nowak, R, Littlefield, KP, Fowler, VM, and Littlefield, RS. A nebulin ruler does not dictate thin filament lengths. *Biophys J* 96:1856-1865, 2009.

41. Cermak, NM, Snijders, T, McKay, BR, Parise, G, Verdijk, LB, Tarnopolsky, MA, Gibala, MJ, and Van Loon, LJ. Eccentric exercise increases satellite cell content in Type II muscle fibers. *Med Sci Sports Exerc* 45:230-237, 2013.

42. Cholewa, JM, Atalag, O, Zinchenko A, Johnson K, and Henselmans M. Anthropometrical determinants of deadlift variant performance. *J Sports Sci Med* 18:448-453, 2019.

43. Clarkson, PM, Devaney, JM, Gordish-Dressman, H, Thompson, PD, Hubal, MJ, Urso, M, Price, TB, Angelopoulos, TJ, Gordon, PM, and Moyna, NM. ACTN3 genotype is associated with increases in muscle strength in response to resistance training in women. *J Appl Physiol* 99:154-163, 2005.

44. Collins, M, Xenophontos, SL, Cariolou, MA, Mokone, GG, Hudson, DE, Anastasiades, L, and Noakes, TD. The ACE gene and endurance performance during the South African Ironman Triathlons. *Med Sci Sports Exerc* 36:1314-1320, 2004.

45. Coons, AH, Leduc, EH, and Connolly, JM. Studies on antibody production: I. A method for the histochemical demonstration of specific antibody and its application to a study of the hyperimmune rabbit. *J Exp Med* 102:49-60, 1955.

46. Cormie, P, McGuigan, MR, and Newton, RU. Developing maximal neuromuscular power: Part 2—Training considerations for improving maximal power production. *Sports Med* 41:125-146, 2011.

47. Costill, DL, Daniels, J, Evans, W, Fink, W, Krahenbuhl, G, and Saltin, B. Skeletal muscle enzymes and fiber composition in male and female track athletes. *J Appl Physiol* 40:149-154, 1976.

48. Crewther, B, Cronin, J, Keogh, J, and Cook, C. The salivary testosterone and cortisol response to three loading schemes. *J Strength Cond Res* 22:250-255, 2008.

49. Crewther, B, Keogh, J, Cronin, J, and Cook, C. Possible stimuli for strength and power adaptation: Acute hormonal responses. *Sports Med* 36:215-238, 2006.

50. Crewther, BT, Cook, C, Cardinale, M, Weatherby, RP, and Lowe, T. Two emerging concepts for elite athletes: The short-term effects of testosterone and cortisol on the neuromuscular system and the dose-response training role of these endogenous hormones. *Sports Med* 41:103-123, 2011.

51. Cronin, JB, McNair, PJ, and Marshall, RN. Magnitude and decay of stretch-induced enhancement of power output. *Eur J Appl Physiol* 84:575-581, 2001.

52. Crowell, HP, and Davis, IS. Gait retraining to reduce lower extremity loading in runners. *Clin Biomech* 26:78-83, 2011.

53. Cumming, DC, Wall, SR, Galbraith, MA, and Belcastro, AN. Reproductive hormone responses to resistance exercise. *Med Sci Sports Exerc* 19:234-238, 1987.

54. Davids, K, Button, C, and Bennett, S. *Dynamics of Skills Acquisition: A Constraints-Led Approach*. Champaign, IL: Human Kinetics, 2008.

55. Davids, K, Glazier, P, Araújo, D, and Bartlett, R. Movement systems as dynamical systems: The functional role of variability and its implications for sports medicine. *Sports Med* 33:245-260, 2003.

56. Delignières, D, Nourrit, D, Sioud, R, Leroyer, P, Zattara, M, and Micaleff, J-P. Preferred coordination modes in the first steps of the learning of a complex gymnastics skill. *Hum Mov Sci* 17:221-241, 1998.

57. DeLong, TH. The effects of the trunk, arm, thigh, and shank lengths on the initial lift-off position of the deadlift movement. In *Kinesiology*. Long Beach: California State University, 105, 2005.

58. Desmedt, JE, and Godaux, E. Ballistic contractions in man: Characteristic recruitment pattern of single motor units of the tibialis anterior muscle. *J Physiol* 264:673-693, 1977.

59. Desmedt, JE, and Godaux, E. Ballistic contractions in fast or slow human muscles: Discharge patterns of single motor units. *J Physiol* 285:185-196, 1978.

60. Doerr, P, and Pirke, KM. Cortisol-induced suppression of plasma testosterone in normal adult males. *J Clin Endocrinol Metab* 43:622-629, 1976.

61. Douglas, J, Pearson, S, Ross, A, and McGuigan, MR. Eccentric exercise: Physiological characteristics and acute responses. *Sports Med* 47:663-675, 2017.

62. Duchateau, J and Hainaut, K. Mechanisms of muscle and motor unit adaptation to explosive power training. In *Strength and Power in Sport*. Paavov, VK, ed. Oxford: Blackwell Scientific, 315-330, 2003.

63. Ehlert, T, Simon, P, and Moser, DA. Epigenetics in sports. *Sports Med* 43:93-110, 2013.

64. Ema, R, Wakahara, T, Miyamoto, N, Kanehisa, H, and Kawakami, Y. Inhomogeneous architectural changes of the quadriceps femoris induced by resistance training. *Eur J Appl Physiol* 113:2691-2703, 2013.

65. Enoka, RM. Morphological features and activation patterns of motor units. *J Clin Neurophysiol* 12:538-559, 1995.

66. Erskine, RM, Fletcher, G, and Folland, JP. The contribution of muscle hypertrophy to strength changes following resistance training. *Eur J Appl Physiol* 114:1239-1249, 2014.

67. Eynon, N, Hanson, ED, Lucia, A, Houweling, PJ, Garton, F, North, KN, and Bishop, DJ. Genes for elite power and sprint performance: ACTN3 leads the way. *Sports Med* 43:803-817, 2013.

68. Eynon, N, Ruiz, JR, Oliveira, J, Duarte, JA, Birk, R, and Lucia, A. Genes and elite athletes: A roadmap for future research. *J Physiol* 589:3063-3070, 2011.

69. Fahey, TD, Rolph, R, Moungmee, P, Nagel, J, and Mortara, S. Serum testosterone, body composition, and strength of young adults. *Med Sci Sports* 8:31-34, 1976.

70. Florini, JR. Hormonal control of muscle growth. *Muscle Nerve* 10:577-598, 1987.

71. Floyd, RT. *Manual of Structural Kinesiology*. New York: McGraw Hill, 2012.

72. Fragala, MS, Kraemer, WJ, Denegar, CR, Maresh, CM, Mastro, AM, and Volek, JS. Neuroendocrine-immune interactions and responses to exercise. *Sports Med* 41:621-639, 2011.

73. Franchi, MV, Atherton, PJ, Reeves, ND, Flück, M, Williams, J, Mitchell, WK, Selby, A, Beltran Valls, RM, and Narici, MV. Architectural, functional and molecular responses to concentric and eccentric loading in human skeletal muscle. *Acta Physiol* 210:642-654, 2014.

74. French, DN. The endocrine responses to training. In *Strength and Conditioning for Sports Performance.* Jeffreys, I, and Moody, J, eds. New York: Routledge, 132-152, 2021.

75. French, DN, Kraemer, WJ, Volek, JS, Spiering, BA, Judelson, DA, Hoffman, JR, and Maresh, CM. Anticipatory responses of catecholamines on muscle force production. *J Appl Physiol* 102:94-102, 2007.

76. Fry, AC, and Kraemer, WJ. Resistance exercise overtraining and overreaching. *Sports Med* 23:106-129, 1997.

77. Fry, AC, Kraemer, WJ, Stone, MH, Koziris, LP, Thrush, JT, and Fleck, SJ. Relationships between serum testosterone, cortisol, and weightlifting performance. *J Strength Cond Res* 14:338-343, 2000.

78. Fry, AC, Kraemer, WJ, Stone, MH, Warren, BJ, Fleck, SJ, Kearney, JT, and Gordon, SE. Endocrine responses to overreaching before and after 1 year of weightlifting. *Can J Appl Physiol* 19:400-410, 1994.

79. Fry, AC, Kraemer, WJ, Van Borselen, F, Lynch, JM, Triplett, NT, Koziris, LP, and Fleck, SJ. Catecholamine responses to short-term high-intensity resistance exercise overtraining. *J Appl Physiol* 77:941-946, 1994.

80. Fry, AC, and Schilling, BK. Weightlifting training and hormonal responses in adolescent males: Implications for program design. *Strength Cond J* 24:7-12, 2002.

81. Fry, AC, Schilling, BK, Fleck, SJ, and Kraemer, WJ. Relationships between competitive wrestling success and neuroendocrine responses. *J Strength Cond Res* 25:40-45, 2011.

82. Fry, AC, Schilling, BK, Staron, RS, Hagerman, FC, Hikida, RS, and Thrush, JT. Muscle fiber characteristics and performance correlates of male Olympic-style weightlifters. *J Strength Cond Res* 17:746-754, 2003.

83. Fry, AC, Webber, JM, Weiss, LW, Arber, MPH, Vaczi, M, and Pattison, NA. Muscle fiber characteristics of competitive power lifters. *J Strength Cond Res* 17:402-410, 2003.

84. Fukunaga, T, Ichinose, Y, Ito, M, Kawakami, Y, and Fukashiro, S. Determination of fascicle length and pennation in a contracting human muscle in vivo. *J Appl Physiol* 82:354-358, 1997.

85. Gabriel, DA, Kamen, G, and Frost, G. Neural adaptations to resistive exercise. *Sports Med* 36:133-149, 2006.

86. Galbo, H, and Gollnick, PD. Hormonal changes during and after exercise. In *Medicine and Sport Science.* Marconnet, P, Poortmans, J, and Hermansen, L, eds. Basel: Karger, 97-110, 1984.

87. Gans, C, and Gaunt, AS. Muscle architecture in relation to function. *J Biomech* 24:53-65, 1991.

88. Georgiades, E, Klissouras, V, Baulch, J, Wang, G, and Pitsiladis, Y. Why nature prevails over nurture in the making of the elite athlete. *BMC Genomics* 18:835, 2017.

89. Gilbert, C. Optimal physical performance in athletes: Key roles of dopamine in a specific neurotransmitter/hormonal mechanism. *Mech Ageing Dev* 84:83-102, 1995.

90. Gleason, BH, Hornsby, WG, Suarez, DG, Nein, MA, and Stone, MH. Troubleshooting a nonresponder: Guidance for the strength and conditioning coach. *Sports (Basel)* 9:83, 2021.

91. Goodwin, J. Maximum velocity is when we can no longer accelerate: Using biomechanics to inform speed development. *Prof Strength Cond* 21:3-9, 2011.

92. Gotshalk, LA, Loebel, CC, Nindl, BC, Putukian, M, Sebastianelli, WJ, Newton, RU, Häkkinen, K, and Kraemer, WJ. Hormonal responses of multiset versus single-set heavy-resistance exercise protocols. *Can J Appl Physiol* 22:244-255, 1997.

93. Granzier, HL, Akster, HA, and Ter Keurs, HE. Effect of thin filament length on the force-sarcomere length relation of skeletal muscle. *Am J Physiol* 260:C1060-C1070, 1991.

94. Hackney, AC, Szczepanowska, E, and Viru, AM. Basal testicular testosterone production in endurance-trained men is suppressed. *Eur J Appl Physiol* 89:198-201, 2003.

95. Haizlip, KM, Harrison, BC, and Leinwand, LA. Sex-based differences in skeletal muscle kinetics and fiber-type composition. *Physiology* 30:30-39, 2015.

96. Häkkinen, K. Neuromuscular and hormonal adaptations during strength and power training: A review. *J Sports Med Phys Fitness* 29:9, 1989.

97. Häkkinen, K, and Keskinen, KL. Muscle cross-sectional area and voluntary force production characteristics in elite strength- and endurance-trained athletes and sprinters. *Eur J Appl Physiol Occup Physiol* 59:215-220, 1989.

98. Häkkinen, K, Komi, PV, Alen, M, and Kauhanen, H. EMG, muscle fibre and force production characteristics during a 1 year training period in elite weight-lifters. *Eur J Appl Physiol Occup Physiol* 56:419-427, 1987.

99. Häkkinen, K, Komi, PV, and Tesch, PA. Effect of combined concentric and eccentric strength training and detraining on force-time, muscle fiber and metabolic characteristics of leg extensor muscles. *Scand J Med Sci Sports* 3:50-58, 1981.

100. Häkkinen, K, and Pakarinen, A. Serum hormones in male strength athletes during intensive short term strength training. *Eur J Appl Physiol Occup Physiol* 63:194-199, 1991.

101. Häkkinen, K, and Pakarinen, A. Muscle strength and serum testosterone, cortisol and SHBG concentrations in middle-aged and elderly men and women. *Acta Physiol Scand* 148:199-207, 1993.

102. Häkkinen, K, Pakarinen, A, Alen, M, Kauhanen, H, and Komi, PV. Relationships between training volume, physical performance capacity, and serum hormone concentrations during prolonged training in elite weight lifters. *Int J Sports Med* 8:S61-S65, 1987.

103. Häkkinen, K, Pakarinen, A, Alen, M, and Komi, PV. Serum hormones during prolonged training of neuromuscular performance. *Eur J Appl Physiol Occup Physiol* 53:287-293, 1985.

104. Häkkinen, K, Pakarinen, A, Kyrolainen, H, Cheng, S, Kim, DH, and Komi, PV. Neuromuscular adaptations and serum hormones in females during prolonged power training. *Int J Sports Med* 11:91-98, 1990.

105. Handelsman, DJ, Hirschberg, AL, and Bermon, S. Circulating testosterone as the hormonal basis of sex differences in athletic performance. *Endocr Rev* 39:803-829, 2018.

106. Hansen, S, Kvorning, T, Kjaer, M, and Sjøgaard, G. The effect of short-term strength training on human skeletal muscle: The importance of physiologically elevated hormone levels. *Scand J Med Sci Sports* 11:347-354, 2001.

107. Harris, GR, Stone, MH, O'Bryant, HS, Proulx, CM, and Johnson, RL. Short-term performance effects of high power, high force, or combined weight-training methods. *J Strength Cond Res* 14:14-20, 2000.

108. Hart, CL, Ward, TE, and Mayhew, JL. Anthropometric correlates of bench press performance following resistance training. *Res Sports Med* 2:89-95, 1991.

109. Haug, WB, Drinkwater, EJ, and Chapman, DW. Learning the hang power clean: Kinetic, kinematic, and technical changes in four weightlifting naive athletes. *J Strength Cond Res* 29:1766-1779, 2015.

110. Haun, CT, Vann, CG, Osburn, SC, Mumford, PW, Roberson, PA, Romero, MA, Fox, CD, Johnson, CA, Parry, HA, and Kavazis, AN. Muscle fiber hypertrophy in response to 6 weeks of high-volume resistance training in trained young men is largely attributed to sarcoplasmic hypertrophy. *PLoS One* 14:e0215267, 2019.

111. Henneman, E, Clamann, HP, Gillies, JD, and Skinner, RD. Rank order of motoneurons within a pool: Law of combination. *J Neurophys* 37:1338-1349, 1974.

112. Henneman, E, Somjen, G, and Carpenter, DO. Excitability and inhibitability of motoneurons of different sizes. *J Neurophysiol* 28:599-620, 1965.

113. Hernández, D, de la Rosa, A, Barragán, A, Barrios, Y, Salido, E, Torres, A, Martín, B, Laynez, I, Duque, A, De Vera, A, Lorenzo, V, and González, A. The ACE/DD genotype is associated with the extent of exercise-induced left ventricular growth in endurance athletes. *J Am Coll Cardiol* 42:527-532, 2003.

114. Herzog, W. The multiple roles of titin in muscle contraction and force production. *Biophys Rev* 10:1187-1199, 2018.

115. Herzog, W, Duvall, M, and Leonard, TR. Molecular mechanisms of muscle force regulation: A role for titin? *Exerc Sport Sci Rev* 40:50-57, 2012.

116. Herzog, W, Powers, K, Johnston, K, and Duvall, M. A new paradigm for muscle contraction. *Front Physiol* 6:174, 2015.

117. Higuchi, H, Yoshioka, T, and Maruyama, K. Positioning of actin filaments and tension generation in skinned muscle fibres released after stretch beyond overlap of the actin and myosin filaments. *J Muscle Res Cell Motil* 9:491-498, 1988.

118. Hilton, EN, and Lundberg, TR. Transgender women in the female category of sport: Perspectives on testosterone suppression and performance advantage. *Sports Med* 51:199-214, 2021.

119. Holsbeeke, L, Ketelaar, M, Schoemaker, MM, and Gorter, JW. Capacity, capability, and performance: Different constructs or three of a kind? *Arch Phys Med Rehab* 90:849-855, 2009.

120. Hornsby, WG, Gentles, J, Comfort, P, Suchomel, TJ, Mizuguchi, S, and Stone, MH. Resistance training volume load with and without exercise displacement. *Sports* 6:137, 2018.

121. Hornsby, WG, Haff, GG, Suarez, DG, Ramsey, MW, Triplett, NT, Hardee, JP, Stone, ME, and Stone, MH. Alterations in adiponectin, leptin, resistin, testosterone, and cortisol across eleven weeks of training among Division One collegiate throwers: A preliminary study. *J Funct Morphol Kinesiol* 5:44, 2020.

122. Huxley, AF. Muscular contraction. *J Physiol* 243:1-43, 1974.

123. Issurin, VB. Generalized training effects induced by athletic preparation. A review. *J Sports Med Phys Fit* 49:333-345, 2009.

124. Izquierdo, M, Ibáñez, J, Häkkinen, K, Kraemer, WJ, Ruesta, M, and Gorostiaga, EM. Maximal strength and power, muscle mass, endurance and serum hormones in weightlifters and road cyclists. *J Sports Sci* 22:465-478, 2004.

125. Jensen, J, Oftebro, H, Breigan, B, Johnsson, A, Ohlin, K, Meen, HD, Strømme, SB, and Dahl, HA. Comparison of changes in testosterone concentrations after strength and endurance exercise in well trained men. *Eur J Appl Physiol Occup Physiol* 63:467-471, 1991.

126. Jezová, D, Vigas, M, Tatár, P, Kvetnanský, R, Nazar, K, Kaciuba-Uścilko, H, and Kozlowski, S. Plasma testosterone and catecholamine responses to physical exercise of different intensities in men. *Eur J Appl Physiol Occup Physiol* 54:62-66, 1985.

127. Johnson, CC, Stone, MH, Byrd, RJ, and Lopez, SA. The response of serum lipids and plasma androgens to weight training exercise in sedentary males. *J Sports Med Phys Fitness* 23:39-44, 1983.

128. Jovy, D, Bruner, H, Klein, KE, and Am, W. Adaptive responses of adrenal cortex to some environmental stressors, exercise, and acceleration. In *Hormonal Steroids: Biochemistry, Pharmacology, and Therapeutics*. New York: Academic Press, 545-553, 1965.

129. Kadlec, D, Miller-Dicks, M, and Nimphius, S. Training for "worst-case" scenarios in sidestepping: Unifying strength and conditioning and perception-action approaches. *Sports Med Open* 9:22, 2023.

130. Kawakami, Y, Abe, T, and Fukunaga, T. Muscle-fiber pennation angles are greater in hypertrophied than in normal muscles. *J Appl Physiol* 74:2740-2744, 1993.

131. Kelly, A, Lyons, G, Gambki, B, and Rubinstein, N. Influences of testosterone on contractile proteins of the guinea pig temporalis muscle. *Adv Exp Med Biol* 182:155-168, 1985.

132. Kindermann, W, Schnabel, A, Schmitt, WM, Biro, G, Cassens, J, and Weber, F. Catecholamines, growth hormone, cortisol, insulin, and sex hormones in anaerobic and aerobic exercise. *Eur J Appl Physiol Occup Physiol* 49:389-399, 1982.

133. Kiss, B, Lee, EJ, Ma, W, Li, FW, Tonino, P, Mijailovich, SM, Irving, TC, and Granzier, HL. Nebulin stiffens the thin filament and augments cross-bridge interaction in skeletal muscle. *Proc Natl Acad Sci U S A* 115:10369-10374, 2018.

134. Kraemer, WJ. Endocrine responses and adaptations to strength training. In *Strength and Power in Sport*. Komi, PV, ed. Oxford: Blackwell Scientific, 291-304, 1992.

135. Kraemer, WJ. Hormonal mechanisms related to the expression of muscular strength and power. In *Strength and Power in Sport*. Komi, PV, ed. Oxford: Blackwell Scientific, 64-76, 1992.

136. Kraemer, WJ, Fleck, SJ, and Deschenes, MR. *Exercise Physiology: Integrating Theory and Application*. Baltimore, MD: Lippincott Williams & Wilkins, 2012.

137. Kraemer, WJ, Fleck, SJ, Dziados, JE, Harman, EA, Marchitelli, LJ, Gordon, SE, Mello, R, Frykman, PN, Koziris, LP, and Triplett, NT. Changes in hormonal concentrations after different heavy-resistance exercise protocols in women. *J Appl Physiol* 75:594-604, 1993.

138. Kraemer, WJ, Fleck, SJ, and Evans, WJ. Strength and power training: Physiological mechanisms of adaptation. *Exerc Sport Sci Rev* 24:363-397, 1996.

139. Kraemer, WJ, Fry, AC, Frykman, PN, Conroy, B, and Hoffman, J. Resistance training and youth. *Pediatr Exerc Sci* 1:336-350, 1989.

140. Kraemer, WJ, Fry, AC, Warren, BJ, Stone, MH, Fleck, SJ, Kearney, JT, Conroy, BP, Maresh, CM, Weseman, CA, and Triplett, NT. Acute hormonal responses in elite junior weightlifters. *Int J Sports Med* 13:103-109, 1992.

141. Kraemer, WJ, Gordon, SE, Fleck, SJ, Marchitelli, LJ, Mello, R, Dziados, JE, Friedl, K, Harman, E, Maresh, C, and Fry, AC. Endogenous anabolic hormonal and growth factor responses to heavy resistance exercise in males and females. *Int J Sports Med* 12:228-235, 1991.

142. Kraemer, WJ, Häkkinen, K, Newton, RU, Nindl, BC, Volek, JS, McCormick, M, Gotshalk, LA, Gordon, SE, Fleck, SJ, and Campbell, WW. Effects of heavy-resistance training on hormonal response patterns in younger vs. older men. *J Appl Physiol* 87:982-992, 1999.

143. Kraemer, WJ, Marchitelli, L, Gordon, SE, Harman, E, Dziados, JE, Mello, R, Frykman, P, McCurry, D, and Fleck, SJ. Hormonal and growth factor responses to heavy resistance exercise protocols. *J Appl Physiol* 69:1442-1450, 1990.

144. Kraemer, WJ, Noble, BJ, Clark, MJ, and Culver, BW. Physiologic responses to heavy-resistance exercise with very short rest periods. *Int J Sports Med* 8:247-252, 1987.

145. Kraemer, WJ, Patton, JF, Gordon, SE, Harman, EA, Deschenes, MR, Reynolds, K, Newton, RU, Triplett, NT, and Dziados, JE. Compatibility of high-intensity strength and endurance training on hormonal and skeletal muscle adaptations. *J Appl Physiol* 78:976-989, 1995.

146. Kraemer, WJ, Patton, JF, Knuttgen, HG, Hannan, CJ, Kettler, T, Gordon, SE, Dziados, JE, Fry, AC, Frykman, PN, and Harman, EA. Effects of high-intensity cycle exercise on sympathoadrenal-medullary response patterns. *J Appl Physiol* 70:8-14, 1991.

147. Kraemer, WJ, and Ratamess, NA. Hormonal responses and adaptations to resistance exercise and training. *Sports Med* 35:339-361, 2005.

148. Kraemer, WJ, Vingren, JL, and Spiering, B. Endocrine responses to resistance training. In *Essentials of Strength Training and Conditioning*. 4th ed. Haff, GG, and Triplett, NT, eds. Champaign, IL: Human Kinetics, 65-86, 2016.

149. Kubo, K, Yata, H, Kanehisa, H, and Fukunaga, T. Effects of isometric squat training on the tendon stiffness and jump performance. *Eur J Appl Physiol* 96:305-314, 2006.

150. Kuijper, EA, Lambalk, CB, Boomsma, DI, van der Sluis, S, Blankenstein, MA, de Geus, EJ, and Posthuma, D. Heritability of reproductive hormones in adult male twins. *Hum Reprod* 22:2153-2159, 2007.

151. Kuoppasalmi, K, and Adlercreutz, H. Interaction between catabolic and anabolic steroid hormones in muscular exercise. In *Exercise Endocrinology*. Fotherby, K, and Pal, SB, eds. Berlin: De Gruyter, 65-98, 1985.

152. Kvorning, T, Andersen, M, Brixen, K, and Madsen, K. Suppression of endogenous testosterone production attenuates the response to strength training: A randomized, placebo-controlled, and blinded intervention study. *Am J Physiol Endocrinol Metab* 291:E1325-E1332, 2006.

153. Kvorning, T, Andersen, M, Brixen, K, Schjerling, P, Suetta, C, and Madsen, K. Suppression of testosterone does not blunt mRNA expression of myoD, myogenin, IGF, myostatin or androgen receptor post strength training in humans. *J Physiol* 578:579-593, 2007.

154. Labeit, S, Ottenheijm, CAC, and Granzier, H. Nebulin, a major player in muscle health and disease. *FASEB J* 25:822-829, 2011.

155. Lehmann, M, Schmid, P, and Keul, J. Age- and exercise-related sympathetic activity in untrained volunteers, trained athletes and patients with impaired left-ventricular contractility. *Eur Heart J* 5:1-7, 1984.

156. Leong, B, Kamen, G, Patten, C, and Burke, JR. Maximal motor unit discharge rates in the quadriceps muscles of older weight lifters. *Med Sci Sports Exerc* 31:1638-1644, 1999.

157. Linnamo, V, Pakarinen, A, Komi, PV, Kraemer, WJ, and Häkkinen, K. Acute hormonal responses to submaximal and maximal heavy resistance and explosive exercises in men and women. *J Strength Cond Res* 19:566-571, 2005.

158. Linthorne, NP. The effect of wind on 100-m sprint times. *J Appl Biomech* 10:110-131, 1994.

159. Lloyd, RS, Cronin, JB, Faigenbaum, AD, Haff, GG, Howard, R, Kraemer, WJ, Micheli, LJ, Myer, GD, and Oliver, JL. National Strength and Conditioning Association position statement on long-term athletic development. *J Strength Cond Res* 30:1491-1509, 2016.

160. Lockie, RG, Moreno, MR, Orjalo, AJ, Lazar, A, Liu, TM, Stage, AA, Birmingham-Babauta, SA, Stokes, JJ, Giuliano, DV, and Risso, FG. Relationships between height, arm length, and leg length on the mechanics of the conventional and high-handle hexagonal bar deadlift. *J Strength Cond Res* 32:3011-3019, 2018.

161. Lopes dos Santos, M, Uftring, M, Stahl, CA, Lockie, RG, Alvar, B, Mann, JB, and Dawes, JJ. Stress in academic and athletic performance in collegiate athletes: A narrative review of sources and monitoring strategies. *Front Sports Act Living* 2:42, 2020.

162. Ma, F, Yang, Y, Li, X, Zhou, F, Gao, C, Li, M, and Gao, L. The association of sport performance with ACE and ACTN3 genetic polymorphisms: A systematic review and meta-analysis. *PLoS One* 8:e54685, 2013.

163. MacArthur, DG, and North, KN. ACTN3: A genetic influence on muscle function and athletic performance. *Exerc Sport Sci Rev* 35:30-34, 2007.

164. Maffiuletti, NA, Aagaard, P, Blazevich, AJ, Folland, J, Tillin, N, and Duchateau, J. Rate of force development: Physiological and methodological considerations. *Eur J Appl Physiol* 116:1091-1116, 2016.

165. Magill, RA. *Motor Learning and Control: Concepts and Applications*. New York: McGraw Hill, 2011.

166. Mann, JB, Mayhew, JL, Dos Santos, ML, Dawes, JJ, and Signorile, JF. Momentum, rather than velocity, is a more effective measure of improvements in Division IA football player performance. *J Strength Cond Res* 36:551-557, 2022.

167. Marx, JO, Ratamess, NA, Nindl, BC, Gotshalk, LA, Volek, JS, Dohi, K, Bush, JA, Gomez, AL, Mazzetti, SA, and Fleck, SJ. Low-volume circuit versus high-volume periodized resistance training in women. *Med Sci Sports Exerc* 33:635-643, 2001.

168. Mazzeo, RS. Catecholamine responses to acute and chronic exercise. *Med Sci Sports Exerc* 23:839-845, 1991.

169. McBride, JM, Triplett-McBride, T, Davie, AJ, Abernethy, PJ, and Newton, RU. Characteristics of titin in strength and power athletes. *Eur J Appl Physiol* 88:553-557, 2003.

170. McElhinny, AS, Kolmerer, B, Fowler, VM, Labeit, S, and Gregorio, CC. The N-terminal end of nebulin interacts with tropomodulin at the pointed ends of the thin filaments. *J Biol Chem* 276:583-592, 2001.

171. McKeever, S, and Howard, R. Resistance training. In *NSCA's Guide to High School Strength and Conditioning.* McHenry, P, and Nitka, MJ, eds. Champaign, IL: Human Kinetics, 223-258, 2022.

172. McMillan, JL, Stone, MH, Sartin, J, Keith, R, Marples, D, Brown, C, and Lewis, RD. 20-hour physiological responses to a single weight-training session. *J Strength Cond Res* 7:9-21, 1993.

173. McMurray, RG, Eubank, TK, and Hackney, AC. Nocturnal hormonal responses to resistance exercise. *Eur J Appl Physiol Occup Physiol* 72:121-126, 1995.

174. McMurray, RG, Forsythe, WA, Mar, MH, and Hardy, CJ. Exercise intensity-related responses of beta-endorphin and catecholamines. *Med Sci Sports Exerc* 19:570-574, 1987.

175. Medlar, S. Mixing it up: The biological significance of hybrid skeletal muscle fibers. *J Exp Biol* 222:jeb200832, 2019.

176. Meechan, D, McMahon, JJ, Suchomel, TJ, and Comfort, P. A comparison of kinetic and kinematic variables during the pull from the knee and hang pull, across loads. *J Strength Cond Res* 34:1819-1829, 2020.

177. Meechan, D, Suchomel, TJ, McMahon, JJ, and Comfort, P. A comparison of kinetic and kinematic variables during the mid-thigh pull and countermovement shrug, across loads. *J Strength Cond Res* 34:1830-1841, 2020.

178. Miller, MS, Bedrin, NG, Ades, PA, Palmer, BM, and Toth, MJ. Molecular determinants of force production in human skeletal muscle fibers: Effects of myosin isoform expression and cross-sectional area. *J Physiol Cell Physiol* 308:C473-C484, 2015.

179. Milner-Brown, HS, and Lee, RG. Synchronization of human motor units: Possible roles of exercise and supraspinal reflexes. *Electroencephalogr Clin Neurophysiol* 38:245-254, 1975.

180. Milner-Brown, HS, and Stein, RB. The relation between the surface electromyogram and muscular force. *J Physiol* 246:549-569, 1975.

181. Minetti, AE. On the mechanical power of joint extensions as affected by the change in muscle force (or cross-sectional area), ceteris paribus. *Eur J Appl Physiol* 86:363-369, 2002.

182. Moir, GL. Skill acquisition. In *Strength and Conditioning: A Biomechanical Approach.* Burlington, MA: Jones & Bartlett Learning, 387-430, 2016.

183. Monroy, JA, Powers, KL, Gilmore, LA, Uyeno, TA, Lindstedt, SL, and Nishikawa, KC. What is the role of titin in active muscle? *Exerc Sport Sci Rev* 40:73-78, 2012.

184. Morrison, S, and Newell, KM. Strength training as a dynamical model of motor learning. *J Sports Sci* 41:408-423, 2023.

185. Nagaya, N, and Herrera, AA. Effects of testosterone on synaptic efficacy at neuromuscular junctions in a sexually dimorphic muscle of male frogs. *J Physiol* 483:141-153, 1995.

186. Narici, MV, Roi, GS, Landoni, L, Minetti, AE, and Cerretelli, P. Changes in force, cross-sectional area and neural activation during strength training and detraining of the human quadriceps. *Eur J Appl Physiol Occup Physiol* 59:310-319, 1989.

187. Newell, KM. Coordination, control, and skill. In *Differing Perspectives in Motor Learning, Memory, and Control.* Goodman, D, Wilberg, RB, and Franks, IM, eds. Amsterdam, Netherlands: North-Holland, 295-317, 1985.

188. Nicklas, BJ, Ryan, AJ, Treuth, MM, Harman, SM, Blackman, MR, Hurley, BF, and Rogers, MA. Testosterone, growth hormone and IGF-I responses to acute and chronic resistive exercise in men aged 55-70 years. *Int J Sports Med* 16:445-450, 1995.

189. Nimphius, S. Lag time: The effect of a two week cessation from resistance training on force, velocity and power in elite softball players. *J Strength Cond Res* 24:1, 2010.

190. Nindl, BC, Kraemer, WJ, Gotshalk, LA, Marx, JO, Volek, JS, Bush, FA, Häkkinen, K, Newton, RU, and Fleck, SJ. Testosterone responses after resistance exercise in women: Influence of regional fat distribution. *Int J Sport Nutr Exerc Metab* 11:451-465, 2001.

191. Norman, B, Esbjornsson, M, Rundqvist, H, Osterlund, T, Von Walden, F, and Tesch, PA. Strength, power, fiber types, and mRNA expression in trained men and women with different ACTN3 R577X genotypes. *J Appl Physiol* 106:959-965, 2009.

192. Nourrit, D, Deschamps, T, Lauriot, B, Caillou, N, and Delignieres, D. The effects of required amplitude and practice on frequency stability and efficiency in a cyclical task. *J Sports Sci* 18:201-212, 2000.

193. Nuzzo, JL. Sex differences in skeletal muscle fiber types: A meta-analysis. *Clin Anat* 37:81-91, 2024.

194. Orysiak, J, Zmijewski, P, Klusiewicz, A, Kaliszewski, P, Malczewska-Lenczowska, J, Gajewski, J, and Pokrywka, A. The association between ACE gene variation and aerobic capacity in winter endurance disciplines. *Biol Sport* 30:249-253, 2013.

195. Ostrander, EA, Huson, HJ, and Ostrander, GK. Genetics of athletic performance. *Annu Rev Genomics Hum Genet* 10:407-429, 2009.

196. Ostrowski, KJ, Wilson, GJ, Weatherby, R, Murphy, PW, and Lyttle, AD. The effect of weight training volume on hormonal output and muscular size and function. *J Strength Cond Res* 11:148-154, 1997.

197. Otten, E. Concepts and models of functional architecture in skeletal muscle. *Exerc Sports Sci Rev* 26:89-137, 1988.

198. Ottenheijm, CAC, Fong, C, Vangheluwe, P, Wuytack, F, Babu, GJ, Periasamy, M, Witt, CC, Labeit, S, and Granzier, H. Sarcoplasmic reticulum calcium uptake and speed of relaxation are depressed in nebulin-free skeletal muscle. *FASEB J* 22:2912-2919, 2008.

199. Ottenheijm, CAC, and Granzier, H. New insights into the structural roles of nebulin in skeletal muscle. *J Biomed Biotechnol* 2010:968139, 2010.

200. Ottenheijm, CAC, Granzier, H, and Labeit, S. The sarcomeric protein nebulin: Another multifunctional giant in charge of muscle strength optimization. *Front Physiol* 3:37, 2012.

201. Painter, KB, Haff, GG, Triplett, NT, Stuart, C, Hornsby, G, Ramsey, MW, Bazyler, CD, and Stone, MH. Resting hormone alterations and injuries: Block vs. DUP weight-training among D-1 track and field athletes. *Sports* 6:3, 2018.

202. Papadimitriou, ID, Lucia, A, Pitsiladis, YP, Pushkarev, VP, Dyatlov, DA, Orekhov, EF, Artioli, GG, Guiherme, JPLF, Lancha, AH, Jr, Ginevičienė, V, Cieszczyk, P, Maciejewska-Karlowska, A, Sawczuk, M, Muniesa, CA, Kouvatsi, A, Massidda, M, Maria Calò, C, Garton, F, Houweling, PJ, Wang, G, Austin, K, Druzhevskaya, AM, Astratenkova, IV, Ahmetov, II, Bishop, DJ, North, KN, and Eynon, N. ACTN3 R577X and ACE I/D gene variants influence performance in elite sprinters: A multi-cohort study. *BMC Genom* 17:285, 2016.

203. Pappas, CT, Krieg, PA, and Gregorio, CC. Nebulin regulates actin filament lengths by a stabilization mechanism. *J Cell Biol* 189:859-870, 2010.

204. Paradisis, GP, and Cooke, CB. Kinematic and postural characteristics of sprint running on sloping surfaces. *J Sports Sci* 19:149-159, 2001.

205. Peeters, MW, Thomis, MA, Loos, RJ, Derom, CA, Fagard, R, Claessens, AL, Vlietinck, RF, and Beunen, GP. Heritability of somatotype components: A multivariate analysis. *Int J Obes* 31:1295-1301, 2007.

206. Pellegrino, MA, Canepari, M, Rossi, R, D'Antona, G, Reggiani, C, and Bottinelli, R. Orthologous myosin isoforms and scaling of shortening velocity with body size in mouse, rat, rabbit and human muscles. *J Physiol* 546:677-689, 2003.

207. Peter, JB, Barnard, RJ, Edgerton, VR, Gillespie, CA, and Stempel, KE. Metabolic profiles of three fiber types of skeletal muscle in guinea pigs and rabbits. *Biochem* 11:2627-2633, 1972.

208. Pette, D, and Staron, RS. Cellular and molecular diversities of mammalian skeletal muscle fibers. *Rev Physiol Biochem Pharmacol* 116:1-76, 1990.

209. Pette, D, and Staron, RS. Myosin isoforms, muscle fiber types and transitions. *Microsc Res Tech* 50:500-509, 2000.

210. Pette, D, and Staron, RS. Transitions of muscle fiber phenotypic profiles. *Histochem Cell Biol* 115:359-372, 2001.

211. Pimenta, EM, Coelho, DB, Veneroso, CE, Barros Coelho, EJ, Cruz, IR, Morandi, RF, De A Pussieldi, G, Carvalho, MRS, Garcia, ES, and De Paz Fernández, JA. Effect of ACTN3 gene on strength and endurance in soccer players. *J Strength Cond Res* 27:3286-3292, 2013.

212. Pitsiladis, YP, Tanaka, M, Eynon, N, Bouchard, C, North, KN, Williams, AG, Collins, M, Moran, CN, Britton, SL, Fuku, N, Ashley, EA, Klissouras, V, Lucia, A, Ahmetov, II, de Geus, E, and Alsayrafi, M. Athlome Project Consortium: A concerted effort to discover genomic and other "omic" markers of athletic performance. *Physiol Genomics* 48:183-190, 2016.

213. Plotkin, DL, Roberts, MD, Haun, CT, and Schoenfeld, BJ. Muscle fiber type transitions with exercise training: Shifting perspectives. *Sports* 9:127, 2021.

214. Powers, K, Nishikawa, K, Joumaa, V, and Herzog, W. Decreased force enhancement in skeletal muscle sarcomeres with a deletion in titin. *J Exp Biol* 219:1311-1316, 2016.

215. Pullinen, T, Huttunen, P, and Komi, PV. Plasma catecholamine responses and neural adaptation during short-term resistance training. *Eur J Appl Physiol* 82:68-75, 2000.

216. Roberts, MD, Haun, CT, Vann, CG, Osburn, SC, and Young, KC. Sarcoplasmic hypertrophy in skeletal muscle: A scientific "unicorn" or resistance training adaptation? *Front Physiol* 11:1-16, 2020.

217. Roberts, TJ. Contribution of elastic tissues to the mechanics and energetics of muscle function during movement. *J Exp Biol* 219:266-275, 2016.

218. Rodríguez-Romo, G, Yvert, T, de Diego, A, Santiago, C, Díaz de Durana, AL, Carratalá, V, Garatachea, N, and Lucia, A. No association between ACTN3 R577X polymorphism and elite judo athletic status. *Int J Sports Physiol Perform* 8:579-581, 2013.

219. Rønnestad, BR, Nygaard, H, and Raastad, T. Physiological elevation of endogenous hormones results in superior strength training adaptation. *Eur J Appl Physiol* 111:2249-2259, 2011.

220. Roy, RR, and Edgerton, VR. Skeletal muscle architecture and performance. In *Strength and Power in Sport*. Komi, PV, ed. Oxford: Blackwell Scientific, 115-129, 1992.

221. Ruiz, JR, Arteta, D, Buxens, A, Artieda, M, Gómez-Gallego, F, Santiago, C, Yvert, T, Morán, M, and Lucia, A. Can we identify a power-oriented polygenic profile? *J Appl Physiol* 108:561-566, 2010.

222. Ruiz, JR, Fernández del Valle, M, Verde, Z, Díez-Vega, I, Santiago, C, Yvert, T, Rodríguez-Romo, G, Gómez-Gallego, F, Molina, JJ, and Lucia, A. ACTN3 R577X polymorphism does not influence explosive leg muscle power in elite volleyball players. *Scand J Med Sci Sports* 21:e34-e41, 2011.

223. Russell, B. The collected papers of Bertrand Russell. In *Détente or Destruction 1955-1957*. Bone, AG, ed. New York: Routledge, 2012.

224. Sands, WA, and McNeal, JR. Predicting athlete preparation and performance: A theoretical perspective. *J Sport Behav* 23:289-310, 2000.

225. Sant'Ana Pereira, JA, Sargeant, AJ, Rademaker, AC, de Haan, A, and van Mechelen, M. Myosin heavy chain isoform expression and high energy phosphate content in muscle fibres at rest and post-exercise. *J Physiol* 496:583-588, 1996.

226. Saplinskas, JS, Chobotas, MA, and Yashchaninas, II. The time of completed motor acts and impulse activity of single motor units according to the training level and sport specialization of tested persons. *Electromyogr Clin Neurophysiol* 20:529-539, 1980.

227. Schakman, O, Kalista, S, Barbé, C, Loumaye, A, and Thissen, J-P. Glucocorticoid-induced skeletal muscle atrophy. *Int J Biochem Cell Biol* 45:2163-2172, 2013.

228. Schoenfeld, B. *Science and Development of Muscle Hypertrophy*. Champaign, IL: Human Kinetics, 2021.

229. Schoenfeld, BJ, Grgic, J, Ogborn, D, and Krieger, JW. Strength and hypertrophy adaptations between low-versus high-load resistance training: A systematic review and meta-analysis. *J Strength Cond Res* 31:3508-3523, 2017.

230. Schöner, G, and Kelso, JAS. Dynamic pattern generation in behavioral and neural systems. *Science* 239:1513-1520, 1988.

231. Schwab, R, Johnson, GO, Housh, TJ, Kinder, JE, and Weir, JP. Acute effects of different intensities of weight lifting on serum testosterone. *Med Sci Sports Exerc* 25:1381-1385, 1993.

232. Sedliak, M, Finni, T, Cheng, S, Kraemer, WJ, and Häkkinen, K. Effect of time-of-day-specific strength training on serum hormone concentrations and isometric strength in men. *Chronobiol Int* 24:1159-1177, 2007.

233. Semmler, JG. Motor unit synchronization and neuromuscular performance. *Exerc Sport Sci Rev* 30:8-14, 2002.

234. Semmler, JG, and Enoka, RM. Neural contributions to changes in muscle strength. In *Biomechanics in Sport: Performance Enhancement and Injury Prevention*. Zatsiorsky, V, ed. Oxford: Blackwell Scientific, 2-20, 2000.

235. Semmler, JG, Kutscher, DV, Zhou, S, and Enoka, RM. Motor unit synchronization is enhanced during slow shortening and lengthening contractions of the first dorsal interosseus muscle. *Soc Neurosci Abstr* 26:463, 2000.

236. Semmler, JG, and Nordstrom, MA. Motor unit discharge and force tremor in skill- and strength-trained individuals. *Exp Brain Res* 119:27-38, 1998.

237. Serrano, N, Colenso-Semple, LM, Lazauskus, KK, Siu, JW, Bagley, JR, Lockie, RG, Costa, PB, and Galpin, AJ. Extraordinary fast-twitch fiber abundance in elite weightlifters. *PLoS One* 14:e0207975, 2019.

238. Shepard, R. Hormonal control systems. In *Physiology and Biochemistry of Exercise*. New York: Praeger, 1982.

239. Silventoinen, K, Maia, J, Jelenkovic, A, Pereira, S, Gouveia, É, Antunes, A, Thomis, M, Lefevre, J, Kaprio, J, and Freitas, D. Genetics of somatotype and physical fitness in children and adolescents. *Am J Hum Biol* 33:e23470, 2021.

240. Smilios, I, Pilianidis, T, Karamouzis, M, and Tokmakidis, SP. Hormonal responses after various resistance exercise protocols. *Med Sci Sports Exerc* 35:644-654, 2003.

241. Spiteri, T, McIntyre, F, Specos, C, and Myszka, S. Cognitive training for agility: The integration between perception and action. *Strength Cond J* 40:39-46, 2018.

242. Staron RS, Hagerman FC, Hikida RS, Murray TF, Hostler DP, Crill MT, Ragg KE, and Toma K. Fiber type composition of the vastus lateralis muscle of young men and women. *J Histochem Cytochem* 48:623-629, 2000.

243. Staron, RS, and Hikida, RS. Histochemical, biochemical and ultrastructural analyses of single human muscle fibers with special reference to the C-fiber population. *J Histochem Cytochem* 40:563-568, 1992.

244. Staron, RS, Karapondo, DL, Kraemer, WJ, Fry, AC, Gordon, SE, Falkel, JE, Hagerman, FC, and Hikida, RS. Skeletal muscle adaptations during early phase of heavy-resistance training in men and women. *J Appl Physiol* 76:1247-1255, 1994.

245. Stone, MH, Cormie, P, Lamont, H, and Stone, ME. Developing strength and power. In *Strength and Conditioning for Sports Performance*. Jeffreys, I, and Moody, J, eds. New York: Routledge, 230-260, 2016.

246. Stone, MH, and Fry, AC. Increased training volume in strength/power athletes. In *Overtraining in Sport*. Kreider, R, Fry, AC, and O'Toole, M, eds. Champaign, IL: Human Kinetics, 87-106, 1997.

247. Stone, MH, O'Bryant, H, and Garhammer, J. A hypothetical model for strength training. *J Sports Med Phys Fitness* 21:342-351, 1981.

248. Stone, MH, O'Bryant, H, Garhammer, J, McMillan, J, and Rozenek, R. A theoretical model of strength training. *Strength Cond J* 4:36-39, 1982.

249. Stone, MH, O'Bryant, HS, McCoy, L, Coglianese, R, Lehmkuhl, M, and Schilling, B. Power and maximum strength relationships during performance of dynamic and static weighted jumps. *J Strength Cond Res* 17:140-147, 2003.

250. Stone, MH, Stone, M, and Sands, WA. *Principles and Practice of Resistance Training*. Champaign, IL: Human Kinetics, 2007.

251. Stone, MH, Suchomel, TJ, Hornsby, WG, Wagle, JP, and Cunanan, AJ. Exercise selection. In *Strength and Conditioning in Sports: From Science to Practice*. New York: Routledge, 252-272, 2022.

252. Stone, MH, Suchomel, TJ, Hornsby, WG, Wagle, JP, and Cunanan, AJ. General concepts and training principles for athlete development. In *Strength and Conditioning in Sports: From Science to Practice*. New York: Routledge, 221-251, 2022.

253. Stone, MH, Suchomel, TJ, Hornsby, WG, Wagle, JP, and Cunanan, AJ. Neuroendocrine factors. In *Strength and Conditioning in Sports: From Science to Practice*. New York: Routledge, 108-144, 2022.

254. Stone, MH, Suchomel, TJ, Hornsby, WG, Wagle, JP, and Cunanan, AJ. *Strength and Conditioning in Sports: From Science to Practice*. New York: Routledge, 2022.

255. Storer, TW, Magliano, L, Woodhouse, L, Lee, ML, Dzekov, C, Dzekov, J, Casaburi, R, and Bhasin, S. Testosterone dose-dependently increases maximal voluntary strength and leg power, but does not affect fatigability or specific tension. *J Clin Endocrinol Metab* 88:1478-1485, 2003.

256. Stromme, SB, Meen, HD, and Aakvaag, A. Effects of an androgenic-anabolic steroid on strength development and plasma testosterone levels in normal males. *Med Sci Sports* 6:203-208, 1974.

257. Stuart, CA, Stone, WL, Howell, MEA, Brannon, MF, Hall, HK, Gibson, AL, and Stone, MH. Myosin content of individual human muscle fibers isolated by laser capture microdissection. *Am J Physiol Cell Physiol* 310:C381-C389, 2016.

258. Suchomel, TJ, and Comfort, P. Developing muscular strength and power. In *Advanced Strength and Conditioning: An Evidence-Based Approach*. Turner, A, and Comfort, P, eds. New York: Routledge, 13-39, 2022.

259. Suchomel, TJ, Nimphius, S, Bellon, CR, and Stone, MH. The importance of muscular strength: Training considerations. *Sports Med* 48:765-785, 2018.

260. Suchomel, TJ, Wagle, JP, Douglas, J, Taber, CB, Harden, M, Haff, GG, and Stone, MH. Implementing eccentric resistance training—Part 1: A brief review of existing methods. *J Funct Morphol Kinesiol* 4:38, 2019.

261. Taber, CB, Vigotsky, A, Nuckols, G, and Haun, CT. Exercise-induced myofibrillar hypertrophy is a contributory cause of gains in muscle strength. *Sports Med* 49:993-997, 2019.

262. Tharp, GD. The role of glucocorticoids in exercise. *Med Sci Sports* 7:6-11, 1975.

263. Trappe, S, Luden, N, Minchev, K, Raue, U, Jemiolo, B, and Trappe, T. Skeletal muscle signature of a champion sprint runner. *J Appl Physiol*, 118:1460-1466, 2015.

264. Travis, SK, Ishida, A, Taber, CB, Fry, AC, and Stone, MH. Emphasizing task-specific hypertrophy to enhance sequential strength and power performance. *J Funct Morphol Kinesiol* 5:7, 2020.

265. Travis, SK, Zwetsloot, KA, Mujika, I, Stone, MH, and Bazyler, CD. Skeletal muscle adaptations and performance outcomes following a step and exponential taper in strength athletes. *Front Physiol* 12:735932, 2021.

266. Tremblay, MS, Copeland, JL, and Van Helder, W. Effect of training status and exercise mode on endogenous steroid hormones in men. *J Appl Physiol* 96:531-539, 2004.

267. Trinick, J. Elastic filaments and giant proteins in muscle. *Curr Opin Cell Biol* 3:112-119, 1991.

268. Trombitás, K, and Pollack, GH. Elastic properties of the titin filament in the Z-line region of vertebrate striated muscle. *J Muscle Res Cell Motil* 14:416-422, 1993.

269. Tsianos, G, Sanders, J, Dhamrait, S, Humphries, S, Grant, S, and Montgomery, H. The ACE gene insertion/deletion polymorphism and elite endurance swimming. *Eur J Appl Physiol* 92:360-362, 2004.

270. Tsolakis, C, Messinis D, Stergioulas A, and Dessypris A. Hormonal responses after strength training and detraining in prepubertal and pubertal boys. *J Strength Cond Res* 14:399-404, 2000.

271. Turvey, MT, Fitch, HL, and Tuler, B. The Berstein perspective I: The problem of degrees of freedom and context conditioned variability. In *Human Motor Behavior: An Introduction*. Kelso, JAS, ed. Hillsdale, NJ: Erlbaum, 239-252, 1982.

272. van Cutsem, M, Duchateau, J, and Hainaut, K. Changes in single motor unit behaviour contribute to the increase in contraction speed after dynamic training in humans. *J Physiol* 513:295-305, 1998.

273. Vann, CG, Roberson, PA, Osburn, SC, Mumford, PW, Romero, MA, Fox, CD, Moore, JH, Haun, CT, Beck, DT, and Moon, JR. Skeletal muscle myofibrillar protein abundance is higher in resistance-trained men, and aging in the absence of training may have an opposite effect. *Sports* 8:7, 2020.

274. Varillas-Delgado, D, Del Coso, J, Gutiérrez-Hellín, J, Aguilar-Navarro, M, Muñoz, A, Maestro, A, and Morencos, E. Genetics and sports performance: The present and future in the identification of talent for sports based on DNA testing. *Eur J Appl Physiol* 122:1811-1830, 2022.

275. Verhoeven, FM, and Newell, KM. Unifying practice schedules in the timescales of motor learning and performance. *Hum Mov Sci* 59:153-169, 2018.

276. Vincent, B, De Bock, K, Ramaekers, M, Van den Eede, E, Van Leemputte, M, Hespel, P, and Thomis, MA. ACTN3 (R577X) genotype is associated with fiber type distribution. *Physiol Genomics* 32:58-63, 2007.

277. Vingren, JL, Kraemer, WJ, Hatfield, DL, Volek, JS, Ratamess, NA, Anderson, JM, Häkkinen, K, Ahtiainen, J, Fragala, MS, and Thomas, GA. Effect of resistance exercise on muscle steroid receptor protein content in strength-trained men and women. *Steroids* 74:1033-1039, 2009.

278. Vingren, JL, Kraemer, WJ, Ratamess, NA, Anderson, JM, Volek, JS, and Maresh, CM. Testosterone physiology in resistance exercise and training: The up-stream regulatory elements. *Sports Med* 40:1037-1053, 2010.

279. Viru, A, and Viru, M. *Biochemical Monitoring of Sport*. Champaign, IL: Human Kinetics, 2001.

280. Wagle, JP, Carroll, KM, Cunanan, AJ, Wetmore, A, Taber, CB, DeWeese, BH, Sato, K, Stuart, CA, and Stone, MH. Preliminary investigation into the effect of ACTN3 and ACE polymorphisms on muscle and performance characteristics. *J Strength Cond Res* 35:688-694, 2021.

281. Wakahara, T, Miyamoto, N, Sugisaki, N, Murata, K, Kanehisa, H, Kawakami, Y, Fukunaga, T, and Yanai, T. Association between regional differences in muscle activation in one session of resistance exercise and in muscle hypertrophy after resistance training. *Eur J Appl Physiol* 112:1569-1576, 2012.

282. Walker, PM, Brunotte, F, Rouhier-Marcer, I, Cottin, Y, Casillas, J-M, Gras, P, and Didier, J-P. Nuclear magnetic resonance evidence of different muscular adaptations after resistance training. *Arch Phys Med Rehabil* 79:1391-1398, 1998.

283. Wdowski, MM, and Gittoes, MJR. Kinematic adaptations in sprint acceleration performances without and with the constraint of holding a field hockey stick. *Sports Biomech* 12:143-153, 2013.

284. Webborn, N, Williams, A, McNamee, M, Bouchard, C, Pitsiladis, Y, Ahmetov, I, Ashley, E, Byrne, N, Camporesi, S, and Collins, M. Direct-to-consumer genetic testing for predicting sports performance and talent identification: Consensus statement. *Br J Sports Med* 49:1486-1491, 2015.

285. Weiss, LW, Cureton, KJ, and Thompson, FN. Comparison of serum testosterone and androstenedione responses to weight lifting in men and women. *Eur J Appl Physiol Occup Physiol* 50:413-419, 1983.

286. Wells, AJ, Fukuda, DH, Hoffman, JR, Gonzalez, AM, Jajtner, AR, Townsend, JR, Mangine, GT, Fragala, MS, and Stout, JR. Vastus lateralis exhibits non-homogenous adaptation to resistance training. *Muscle Nerve* 50:785-793, 2014.

287. Wickiewicz, TL, Roy, RR, Powell, PL, and Edgerton, VR. Muscle architecture of the human lower limb. *Clin Orthop Relat Res* 179:275-283, 1983.

288. Wickiewicz, TL, Roy, RR, Powell, PL, Perrine, JJ, and Edgerton, VR. Muscle architecture and force-velocity relationships in humans. *J Appl Physiol* 57:435-443, 1984.

289. Widmann, M, Nieß, AM, and Munz, B. Physical exercise and epigenetic modifications in skeletal muscle. *Sports Med* 49:509-523, 2019.

290. Wilkerson, JE, Horvath, SM, and Gutin, B. Plasma testosterone during treadmill exercise. *J Appl Physiol Respir Environ Exerc Physiol* 49:249-253, 1980.

291. Wortsman, J. Role of epinephrine in acute stress. *Endocrinol Metab Clin* 31:79-106, 2002.

292. Wulf, G. Attentional focus and motor learning: A review of 15 years. *Int Rev Sport Exerc Psychol* 6:77-104, 2013.

293. Yang, N, MacArthur, DG, Gulbin, JP, Hahn, AG, Beggs, AH, Easteal, S, and North, K. ACTN3 genotype is associated with human elite athletic performance. *Am J Hum Genet* 73:627-631, 2003.

294. Yao, W, Fuglevand, RJ, and Enoka, RM. Motor-unit synchronization increases EMG amplitude and decreases force steadiness of simulated contractions. *J Neurophysiol* 83:441-452, 2000.
295. Young, WB, Pryor, JF, and Wilson, GJ. Effect of instructions on characteristics of countermovement and drop jump performance. *J Strength Cond Res* 9:232-236, 1995.
296. Yuen, M, and Ottenheijm, CAC. Nebulin: Big protein with big responsibilities. *J Muscle Res Cell Motil* 41:103-124, 2020.
297. Zamparo, P, Minetti, A, and di Prampero, P. Interplay among the changes of muscle strength, cross-sectional area and maximal explosive power: Theory and facts. *Eur J Appl Physiol* 88:193-202, 2002.
298. Zaras, N, Methenitis, S, Stasinaki, A-N, Spiliopoulou, P, Anousaki, E, Karampatsos, G, Hadjicharalambous, M, and Terzis, G. Differences in rate of force development, muscle morphology and maximum strength between weightlifters and track and field throwers. *Appl Sci* 12:8031, 2022.

Chapter 4

1. Androulakis-Korakakis, P, Fisher, JP, and Steele, J. The minimum effective training dose required to increase 1RM strength in resistance-trained men: A systematic review and meta-analysis. *Sports Med* 50:751-765, 2020.
2. Bazyler, CD, Abbott, HA, Bellon, CR, Taber, CB, and Stone, MH. Strength training for endurance athletes: Theory to practice. *Strength Cond J* 37:1-12, 2015.
3. Bazyler, CD, Mizuguchi, S, Harrison, AP, Sato, K, Kavanaugh, AA, DeWeese, BH, and Stone, MH. Changes in muscle architecture, explosive ability, and track and field throwing performance throughout a competitive season and following a taper. *J Strength Cond Res* 31:2785-2793, 2017.
4. Bazyler, CD, Mizuguchi, S, Sole, CJ, Suchomel, TJ, Sato, K, Kavanaugh, AA, DeWeese, BH, and Stone, MH. Jumping performance is preserved, but not muscle thickness in collegiate volleyball players after a taper. *J Strength Cond Res* 32:1029-1035, 2018.
5. Bazyler, CD, Sato, K, Wassinger, CA, Lamont, HS, and Stone, MH. The efficacy of incorporating partial squats in maximal strength training. *J Strength Cond Res* 28:3024-3032, 2014.
6. Bompa, TO, and Haff, G. *Periodization: Theory and Methodology of Training.* Champaign, IL: Human Kinetics, 2009.
7. Bondarchuk, A. Periodization of sports training. *Legkaya Atletika* 12:8-9, 1986.
8. Bosquet, L, Berryman, N, Dupuy, O, Mekary, S, Arvisais, D, Bherer, L, and Mujika, I. Effect of training cessation on muscular performance: A meta-analysis. *Scand J Med Sci Sports* 23:e140-149, 2013.
9. Branscheidt, M, Kassavetis, P, Anaya, M, Rogers, D, Huang, HD, Lindquist, MA, and Celnik, P. Fatigue induces long-lasting detrimental changes in motor-skill learning. *eLife* 8:40578, 2019.
10. Bryanton, MA, Kennedy, MD, Carey, JP, and Chiu, LZF. Effect of squat depth and barbell load on relative muscular effort in squatting. *J Strength Cond Res* 26:2820-2828, 2012.
11. Burgomaster, KA, Howarth, KR, Phillips, SM, Rakobowchuk, M, MacDonald, MJ, McGee, SL, and Gibala, MJ. Similar metabolic adaptations during exercise after low volume sprint interval and traditional endurance training in humans. *J Physiol* 586:151-160, 2008.
12. Burke, RE. The control of muscle force: Motor unit recruitment and firing patterns. In *Human Muscle Power*. Jones, NL, McCartney, M, and McComas, AJ. eds. Champaign, IL: Human Kinetics, 97-106, 1986.
13. Bushnell, T, and Hunter, I. Differences in technique between sprinters and distance runners at equal and maximal speeds. *Sports Biomech* 6:261-268, 2007.
14. Busso, T, Candau, R, and Lacour, JR. Fatigue and fitness modelled from the effects of training on performance. *Eur J Appl Physiol Occup Physiol* 69:50-54, 1994.
15. Campeiz, JM, and de Oliveira, PR. Effects of concentrated charges of strength training on anaerobic variables and body composition of professional soccer players. *J Sports Sci Med* 10:172, 2007.
16. Carroll, KM, Bazyler, CD, Bernards, JR, Taber, CB, Stuart, CA, DeWeese, BH, Sato, K, and Stone, MH. Skeletal muscle fiber adaptations following resistance training using repetition maximums or relative intensity. *Sports* 7:169, 2019.
17. Carroll, KM, Bernards, JR, Bazyler, CD, Taber, CB, Stuart, CA, DeWeese, BH, Sato, K, and Stone, MH. Divergent performance outcomes following resistance training using repetition maximums or relative intensity. *Int J Sports Physiol Perform* 14:46-54, 2019.
18. Carvalho, L, Morrigi, R, Jr, Truffi, G, Serra, A, Sander, R, De Souza, EO, and Barroso, R. Is stronger better? Influence of a strength phase followed by a hypertrophy phase on muscular adaptations in resistance-trained men. *Res Sports Med* 29:536-546, 2021.
19. Comfort, P, Haff, GG, Suchomel, TJ, Soriano, MA, Pierce, KC, Hornsby, WG, Haff, EE, Sommerfield, LM, Chavda, S, Morris, SJ, Fry, AC, and Stone, MH. National Strength and Conditioning Association position statement on weightlifting for sports performance. *J Strength Cond Res* 37:1163-1190, 2023.
20. Comfort, P, McMahon, JJ, and Suchomel, TJ. Optimizing squat technique—Revisited. *Strength Cond J* 40:68-74, 2018.
21. Contreras, B, Vigotsky, AD, Schoenfeld, BJ, Beardsley, C, McMaster, DT, Reyneke, J, and Cronin, J. Effects of a six-week hip thrust versus front squat resistance training program on performance in adolescent males: A randomized-controlled trial. *J Strength Cond Res* 31:999-1008, 2017.
22. Coratella, G, Tornatore, G, Caccavale, F, Longo, S, Esposito, F, and Cè, E. The activation of gluteal, thigh, and lower back muscles in different squat variations performed by competitive bodybuilders: Implications for resistance training. *Int J Environ Res Public Health* 18:772, 2021.
23. Cormie, P, McGuigan, MR, and Newton, RU. Adaptations in athletic performance after ballistic power versus strength training. *Med Sci Sports Exerc* 42:1582-1598, 2010.
24. Cormie, P, McGuigan, MR, and Newton, RU. Changes in the eccentric phase contribute to improved stretch-shorten cycle performance after training. *Med Sci Sports Exerc* 42:1731-1744, 2010.
25. Cormie, P, McGuigan, MR, and Newton, RU. Influence of strength on magnitude and mechanisms of adaptation to power training. *Med Sci Sports Exerc* 42:1566-1581, 2010.

26. Cormie, P, McGuigan, MR, and Newton, RU. Developing maximal neuromuscular power: Part 2—training considerations for improving maximal power production. *Sports Med* 41:125-146, 2011.
27. Cunanan, AJ, DeWeese, BH, Wagle, JP, Carroll, KM, Sausaman, R, Hornsby, WG, Haff, GG, Triplett, NT, Pierce, KC, and Stone, MH. The general adaptation syndrome: A foundation for the concept of periodization. *Sports Med* 48:787-797, 2018.
28. Dellal, A, Lago-Peñas, C, Rey, E, Chamari, K, and Orhant, E. The effects of a congested fixture period on physical performance, technical activity and injury rate during matches in a professional soccer team. *Br J Sports Med* 49:390-394, 2015.
29. DeWeese, BH, Bellon, CR, Magrum, E, Taber, CB, and Suchomel, TJ. Strengthening the springs: Improving sprint performance via strength training. *Techniques* 9:8-20, 2016.
30. DeWeese, BH, Hornsby, G, Stone, M, and Stone, MH. The training process: Planning for strength–power training in track and field. Part 1: Theoretical aspects. *J Sport Health Sci* 4:308-317, 2015.
31. DeWeese, BH, Hornsby, G, Stone, M, and Stone, MH. The training process: Planning for strength–power training in track and field. Part 2: Practical and applied aspects. *J Sport Health Sci* 4:318-324, 2015.
32. DeWeese, BH, and Scruggs, SK. The countermovement shrug. *Strength Cond J* 34:20-23, 2012.
33. DeWeese, BH, Serrano, AJ, Scruggs, SK, and Burton, JD. The midthigh pull: Proper application and progressions of a weightlifting movement derivative. *Strength Cond J* 35:54-58, 2013.
34. DeWeese, BH, Serrano, AJ, Scruggs, SK, and Sams, ML. The clean pull and snatch pull: Proper technique for weightlifting movement derivatives. *Strength Cond J* 34:82-86, 2012.
35. DeWeese, BH, Serrano, AJ, Scruggs, SK, and Sams, ML. The pull to knee—Proper biomechanics for a weightlifting movement derivative. *Strength Cond J* 34:73-75, 2012.
36. DeWeese, BH, Suchomel, TJ, Serrano, AJ, Burton, JD, Scruggs, SK, and Taber, CB. The pull from the knee: Proper technique and application. *Strength Cond J* 38:79-85, 2016.
37. Elliott, MCCW, Wagner, PP, and Chiu, L. Power athletes and distance training: Physiological and biomechanical rationale for change. *Sports Med* 37:47-57, 2007.
38. Ema, R, Sakaguchi, M, and Kawakami, Y. Thigh and psoas major muscularity and its relation to running mechanics in sprinters. *Med Sci Sports Exerc* 50:2085-2091, 2018.
39. Ema, R, Wakahara, T, Yanaka, T, Kanehisa, H, and Kawakami, Y. Unique muscularity in cyclists' thigh and trunk: A cross-sectional and longitudinal study. *Scand J Med Sci Sports* 26:782-793, 2016.
40. Fitzpatrick, DA, Cimadoro, G, and Cleather, DJ. The magical horizontal force muscle? A preliminary study examining the "force-vector" theory. *Sports* 7:30, 2019.
41. Folland, JP, Buckthorpe, MW, and Hannah, R. Human capacity for explosive force production: Neural and contractile determinants. *Scand J Med Sci Sports* 24:894-906, 2014.
42. Foster, C. Monitoring training in athletes with reference to overtraining syndrome. *Med Sci Sports Exerc* 30:1164-1168, 1998.
43. Franco-Márquez, F, Rodríguez-Rosell, D, González-Suárez, JM, Pareja-Blanco, F, Mora-Custodio, R, Yañez-García, JM, and González-Badillo, JJ. Effects of combined resistance training and plyometrics on physical performance in young soccer players. *Int J Sports Med* 36:906-914, 2015.
44. Fry, AC, Kraemer, WJ, Lynch, JM, Triplett, NT, and Koziris, LP. Does short-term near-maximal intensity machine resistance training induce overtraining? *J Strength Cond Res* 8:188-191, 1994.
45. Fry, AC, Kraemer, WJ, van Borselen, F, Lynch, JM, Marsit, JL, Roy, EP, Triplett, NT, and Knuttgen, HG. Performance decrements with high-intensity resistance exercise overtraining. *Med Sci Sports Exerc* 26:1165-1173, 1994.
46. Fyfe, JJ, Hamilton, DL, and Daly, RM. Minimal-dose resistance training for improving muscle mass, strength, and function: A narrative review of current evidence and practical considerations. *Sports Med* 52:463-479, 2022.
47. Gehri, DJ, Ricard, MD, Kleiner, DM, and Kirkendall, DT. A comparison of plyometric training techniques for improving vertical jump ability and energy production. *J Strength Cond Res* 12:85-89, 1998.
48. Gergley, JC. Comparison of two lower-body modes of endurance training on lower-body strength development while concurrently training. *J Strength Cond Res* 23:979-987, 2009.
49. Green, CM, and Comfort, P. The affect of grip width on bench press performance and risk of injury. *Strength Cond J* 29:10-14, 2007.
50. Haff, GG, and Nimphius, S. Training principles for power. *Strength Cond J* 34:2-12, 2012.
51. Harris, GR, Stone, MH, O'Bryant, HS, Proulx, CM, and Johnson, RL. Short-term performance effects of high power, high force, or combined weight-training methods. *J Strength Cond Res* 14:14-20, 2000.
52. Haug, WB, Drinkwater, EJ, and Chapman, DW. Learning the hang power clean: Kinetic, kinematic, and technical changes in four weightlifting naive athletes. *J Strength Cond Res* 29:1766-1779, 2015.
53. Hawkins, SB, Doyle, TLA, and McGuigan, MR. The effect of different training programs on eccentric energy utilization in college-aged males. *J Strength Cond Res* 23:1996-2002, 2009.
54. Hawley, JA. Specificity of training adaptation: Time for a rethink? *J Physiol* 586:1-2, 2008.
55. Hodges, NJ, Hayes, S, Horn, RR, and Williams, AM. Changes in coordination, control and outcome as a result of extended practice on a novel motor skill. *Ergonomics* 48:1672-1685, 2005.
56. Hoshikawa, Y, Muramatsu, M, Iida, T, Uchiyama, A, Nakajima, Y, Kanehisa, H, and Fukunaga, T. Influence of the psoas major and thigh muscularity on 100-m times in junior sprinters. *Med Sci Sports Exerc* 38:2138-2143, 2006.
57. Hunter, SK. Performance fatigability: Mechanisms and task specificity. *Cold Spring Harb Perspect Med* 8:a029728, 2018.
58. Ishida, A, Rochau, K, Findlay, KP, Devero, B, Duca, M, and Stone, MH. Effects of an initial muscle strength level on sports performance changes in collegiate soccer players. *Sports* 8:127, 2020.

59. Issurin, VB. Block periodization versus traditional training theory: A review. *J Sports Med Phys Fitness* 48:65-75, 2008.
60. Issurin, VB. Generalized training effects induced by athletic preparation. A review. *J Sports Med Phys Fit* 49:333-345, 2009.
61. Issurin, VB. New horizons for the methodology and physiology of training periodization. *Sports Med* 40:189-206, 2010.
62. Issurin, VB. Benefits and limitations of block periodized training approaches to athletes' preparation: A review. *Sports Med* 46:329-338, 2016.
63. James, LP, Haff, GG, Kelly, VG, Connick, M, Hoffman, B, and Beckman, EM. The impact of strength level on adaptations to combined weightlifting, plyometric and ballistic training. *Scand J Med Sci Sports* 28:1494-1505, 2018.
64. James, LP, Haycraft, J, Pierobon, A, Suchomel, TJ, and Connick, M. Mixed versus focused resistance training during an Australian football pre-season. *J Funct Morphol Kinesiol* 5:99, 2020.
65. Koceja, DM, Davison, E, and Robertson, CT. Neuromuscular characteristics of endurance- and power-trained athletes. *Res Quart Exerc Sport* 75:23-30, 2004.
66. Kotani, Y, Lake, J, Guppy, SN, Poon, W, Nosaka, K, and Haff, GG. Agreement in squat jump force-time characteristics between Smith machine and free-weight squat jump force-time characteristics. *J Strength Cond Res* 37:1955-1962, 2023.
67. Kotani, Y, Lake, JP, Guppy, SN, Poon, W, Nosaka, K, Hori, N, and Haff, GG. Reliability of the squat jump force-velocity and load-velocity profiles. *J Strength Cond Res* 36:3000-3007, 2022.
68. Kotzamanidis, C, Chatzopoulos, D, Michailidis C, Papaiakovou G, and Patikas D. The effect of a combined high-intensity strength and speed training program on the running and jumping ability of soccer players. *J Strength Cond Res* 19:369-375, 2005.
69. Kozinc, Ž, Pleša, J, and Šarabon, N. Questionable utility of the eccentric utilization ratio in relation to the performance of volleyball players. *Int J Environ Res Public Health* 18:11754, 2021.
70. Kozinc, Ž, Žitnik, J, Smajla, D, and Šarabon, N. The difference between squat jump and countermovement jump in 770 male and female participants from different sports. *Eur J Sport Sci* 22:985-993, 2022.
71. Krieger, JW. Single versus multiple sets of resistance exercise: A meta-regression. *J Strength Cond Res* 23:1890-1901, 2009.
72. Kruger, A. From Russia with love? Sixty years of proliferation of L.P. Matveyev's concept of periodisation? *Staps* 114:51-59, 2016.
73. Lago-Peñas, C, Rey, E, Lago-Ballesteros, J, Casáis, L, and Domínguez, E. The influence of a congested calendar on physical performance in elite soccer. *J Strength Cond Res* 25:2111-2117, 2011.
74. Leslie, KLM, and Comfort, P. The effect of grip width and hand orientation on muscle activity during pull-ups and the lat pull-down. *Strength Cond J* 35:75-78, 2013.
75. Lindberg, K, Solberg, P, Bjørnsen, T, Helland, C, Rønnestad, B, Thorsen Frank, M, Haugen, T, Østerås, S, Kristoffersen, M, and Midttun, M. Force-velocity profiling in athletes: Reliability and agreement across methods. *PLoS One* 16:e0245791, 2021.
76. Loeb, GE. Hard lessons in motor control from the mammalian spinal cord. *Trends Neurosci* 10:108-113, 1987.
77. Loturco, I, Contreras, B, Kobal, R, Fernandes, V, Moura, N, Siqueira, F, Winckler, C, Suchomel, TJ, and Pereira, LA. Vertically and horizontally directed muscle power exercises: Relationships with top-level sprint performance. *PLoS One* 13:e0201475, 2018.
78. Mangine, GT, Hoffman, JR, Wang, R, Gonzalez, AM, Townsend, JR, Wells, AJ, Jajtner, AR, Beyer, KS, Boone, CH, and Miramonti, AA. Resistance training intensity and volume affect changes in rate of force development in resistance-trained men. *Eur J Appl Physiol* 116:2367-2374, 2016.
79. Matveyev, LP. *Periodization of Sports Training.* Moskow: Fizkultura i Sport, 1965.
80. Matveyev, LP. *Fundamentals of Sport Training.* Moscow: FIS Publisher, 1977.
81. Matveyev, LP, and Zdornyj, AP. *Fundamentals of Sports Training.* Moscow: Progress Publishers, 1981.
82. McGee, D, Jessee, TC, Stone, MH, and Blessing, D. Leg and hip endurance adaptations to three weight-training programs. *J Appl Sport Sci Res* 6:92-95, 1992.
83. McGuigan, MR, La Doyle, T, Newton, M, Edwards, DJ, Nimphius, S, and Newton, RU. Eccentric utilization ratio: Effect of sport and phase of training. *J Strength Cond Res* 20:992-995, 2006.
84. Minetti, AE. On the mechanical power of joint extensions as affected by the change in muscle force (or cross-sectional area), ceteris paribus. *Eur J Appl Physiol* 86:363-369, 2002.
85. Morais, JE, Silva, AJ, Garrido, ND, Marinho, DA, and Barbosa, TM. The transfer of strength and power into the stroke biomechanics of young swimmers over a 34-week period. *Eur J Sport Sci* 18:787-795, 2018.
86. Morris, SJ, Oliver, JL, Pedley, JS, Haff, GG, and Lloyd, RS. Taking a long-term approach to the development of weightlifting ability in young athletes. *Strength Cond J* 42:71-90, 2020.
87. Mujika, I. *Tapering and Peaking.* Champaign, IL: Human Kinetics, 2009.
88. Mujika, I, and Padilla, S. Detraining: Loss of training-induced physiological and performance adaptations. Part I: Short term insufficient training stimulus. *Sports Med* 30:79-87, 2000.
89. Mujika, I, and Padilla, S. Detraining: Loss of training-induced physiological and performance adaptations. Part II: Long term insufficient training stimulus. *Sports Med* 30:145-154, 2000.
90. Nagahara, R, Mizutani M, Matsuo A, Kanehisa H, and Fukunaga T. Association of sprint performance with ground reaction forces during acceleration and maximal speed phases in a single sprint. *J Appl Biomech* 34:104-110, 2018.
91. Newton, RU, Häkkinen, K, Häkkinen, A, McCormick, M, Volek, J, and Kraemer, WJ. Mixed-methods resistance training increases power and strength of young and older men. *Med Sci Sports Exerc* 34:1367-1375, 2002.
92. Newton, RU, and Kraemer, WJ. Developing explosive muscular power: Implications for a mixed methods training strategy. *Strength Cond J* 16:20-31, 1994.

93. O'Bryant, HS, Byrd, R, and Stone, MH. Cycle ergometer performance and maximum leg and hip strength adaptations to two different methods of weight-training. *J Strength Cond Res* 2:27-30, 1988.

94. Painter, KB, Haff, GG, Ramsey, MW, McBride, J, Triplett, T, Sands, WA, Lamont, HS, Stone, ME, and Stone, MH. Strength gains: Block versus daily undulating periodization weight training among track and field athletes. *Int J Sports Physiol Perform* 7:161-169, 2012.

95. Piper, TJ, and Waller, MA. Variations of the deadlift. *Strength Cond J* 23:66-73, 2001.

96. Plisk, SS, and Stone, MH. Periodization strategies. *Strength Cond J* 25:19-37, 2003.

97. Pyne, DB, Mujika, I, and Reilly, T. Peaking for optimal performance: Research limitations and future directions. *J Sports Sci* 27:195-202, 2009.

98. Rhea, MR, and Alderman, BL. A meta-analysis of periodized versus nonperiodized strength and power training programs. *Res Q Exerc Sport* 75:413-422, 2004.

99. Rhea, MR, Alvar, BA, and Burkett, LN. Single versus multiple sets for strength: A meta-analysis to address the controversy. *Res Quart Exerc Sport* 73:485-488, 2002.

100. Rhea, MR, Kenn, JG, Peterson, MD, Massey, D, Simão, R, Marin, PJ, Favero, M, Cardozo, D, and Krein, D. Joint-angle specific strength adaptations influence improvements in power in highly trained athletes. *Hum Mov* 17:43-49, 2016.

101. Rodríguez-Rosell, D, Torres-Torrelo, J, Franco-Márquez, F, González-Suárez, JM, and González-Badillo, JJ. Effects of light-load maximal lifting velocity weight training vs. combined weight training and plyometrics on sprint, vertical jump and strength performance in adult soccer players. *J Sci Med Sport* 20:695-699, 2017.

102. Rønnestad, BR, Hansen, EA, and Raastad, T. High volume of endurance training impairs adaptations to 12 weeks of strength training in well-trained endurance athletes. *Eur J Appl Physiol* 112:1457-1466, 2012.

103. Sale, DG. Neural adaptations to strength training. In *Strength and Power in Sport*. Komi, PV, ed. Oxford: Blackwell Scientific, 281-313, 2003.

104. Samozino, P, Rabita, G, Dorel, S, Slawinski, J, Peyrot, N, de Villarreal, ESS, and Morin, JB. A simple method for measuring power, force, velocity properties, and mechanical effectiveness in sprint running. *Scand J Med Sci Sports* 26:648-658, 2016.

105. Samozino, P, Rejc, E, Di Prampero, PE, Belli, A, and Morin, J-B. Optimal force–velocity profile in ballistic movements—Altius: Citius or Fortius? *Med Sci Sports Exerc* 44:313-322, 2012.

106. Schoenfeld, BJ, Grgic, J, Ogborn, D, and Krieger, JW. Strength and hypertrophy adaptations between low-versus high-load resistance training: A systematic review and meta-analysis. *J Strength Cond Res* 31:3508-3523, 2017.

107. Schot, P, Dart, J, and Schuh, M. Biomechanical analysis of two change-of-direction maneuvers while running. *J Orthop Sports Phys Ther* 22:254-258, 1995.

108. Sheppard, JM, Chapman, D, and Taylor, K-L. An evaluation of a strength qualities assessment method for the lower body. *J Aust Strength Cond* 19:4-10, 2011.

109. Sigward, SM, Cesar, GM, and Havens, KL. Predictors of frontal plane knee moments during side-step cutting to 45 and 110 men and women: Implications for ACL injury. *Clin J Sport Med* 25:529-534, 2015.

110. Stone, MH. Muscle conditioning and muscle injuries. *Med Sci Sports Exerc* 22:457-462, 1990.

111. Stone, MH, Hornsby, WG, Haff, GG, Fry, AC, Suarez, DG, Liu, J, Gonzalez-Rave, JM, and Pierce, KC. Periodization and block periodization in sports: Emphasis on strength-power training—A provocative and challenging narrative. *J Strength Cond Res* 35:2351-2371, 2021.

112. Stone, MH, Hornsby, WG, Suarez, DG, Duca, M, and Pierce, KC. Training specificity for athletes: Emphasis on strength-power training: A narrative review. *J Funct Morphol Kinesiol* 7:102, 2022.

113. Stone, MH, Moir, G, Glaister, M, and Sanders, R. How much strength is necessary? *Phys Ther Sport* 3:88-96, 2002.

114. Stone, MH, O'Bryant, H, and Garhammer, J. A hypothetical model for strength training. *J Sports Med Phys Fitness* 21:342-351, 1981.

115. Stone, MH, O'Bryant, H, Garhammer, J, McMillan, J, and Rozenek, R. A theoretical model of strength training. *Strength Cond J* 4:36-39, 1982.

116. Stone, MH, Plisk, S, and Collins, D. Training principles: Evaluation of modes and methods of resistance training—A coaching perspective. *Sports Biomech* 1:79-103, 2002.

117. Stone, MH, Stone, M, and Sands, WA. *Principles and Practice of Resistance Training*. Champaign, IL: Human Kinetics, 2007.

118. Stone, MH, Suchomel, TJ, Hornsby, WG, Wagle, JP, and Cunanan, AJ. Exercise selection. In *Strength and Conditioning in Sports: From Science to Practice*. New York: Routledge, 252-272, 2022.

119. Stone, MH, Suchomel, TJ, Hornsby, WG, Wagle, JP, and Cunanan, AJ. General concepts and training principles for athlete development. In *Strength and Conditioning in Sports: From Science to Practice*. New York: Routledge, 221-251, 2022.

120. Stowers, T, McMillan, J, Scala, D, Davis, V, Wilson, D, and Stone, M. The short-term effects of three different strength-power training methods. *Strength Cond J* 5:24-27, 1983.

121. Suarez, DG, Mizuguchi, S, Hornsby, WG, Cunanan, AJ, Marsh, DJ, and Stone, MH. Phase-specific changes in rate of force development and muscle morphology throughout a block periodized training cycle in weightlifters. *Sports* 7:129, 2019.

122. Suarez, DG, Wagle, JP, Cunanan, AJ, Sausaman, RW, and Stone, MH. Dynamic correspondence of resistance training to sport: A brief review. *Strength Cond J* 41:80-88, 2019.

123. Suchomel, TJ. The gray area of programming weightlifting exercises. *Natl Strength Cond Assoc Coach* 7:6-14, 2020.

124. Suchomel, TJ. Resistance training strategies to train the force-velocity characteristics of athletes. In *Central Virginia Sport Performance: The Manual*, Vol 7. DeMayo, J, ed. Amazon Publishing, 95-118, 2022.

125. Suchomel, TJ, and Comfort, P. Weightlifting for sports performance. In *Advanced Strength and Conditioning: An Evidence-Based Approach*. Turner, A, and Comfort, P, eds. New York: Routledge, 283-306, 2022.

126. Suchomel, TJ, DeWeese, BH, Beckham, GK, Serrano, AJ, and French, SM. The hang high pull: A progressive exercise into weightlifting derivatives. *Strength Cond J* 36:79-83, 2014.

127. Suchomel, TJ, DeWeese, BH, Beckham, GK, Serrano, AJ, and Sole, CJ. The jump shrug: A progressive exercise into weightlifting derivatives. *Strength Cond J* 36:43-47, 2014.

128. Suchomel, TJ, DeWeese, BH, and Serrano, AJ. The power clean and power snatch from the knee. *Strength Cond J* 38:98-105, 2016.

129. Suchomel, TJ, McKeever, SM, and Comfort, P. Training with weightlifting derivatives: The effects of force and velocity overload stimuli. *J Strength Cond Res* 34:1808-1818, 2020.

130. Suchomel, TJ, McKeever, SM, McMahon, JJ, and Comfort, P. The effect of training with weightlifting catching or pulling derivatives on squat jump and countermovement jump force-time adaptations. *J Funct Morphol Kinesiol* 5:28, 2020.

131. Suchomel, TJ, McKeever, SM, Nolen, JD, and Comfort, P. Muscle architectural and force-velocity curve adaptations following 10 weeks of training with weightlifting catching and pulling derivatives. *J Sports Sci Med* 21:504-516, 2022.

132. Suchomel, TJ, Nimphius, S, Bellon, CR, Hornsby, WG, and Stone, MH. Training for muscular strength: Methods for monitoring and adjusting training intensity. *Sports Med* 51:2051-2066, 2021.

133. Suchomel, TJ, Nimphius, S, Bellon, CR, and Stone, MH. The importance of muscular strength: Training considerations. *Sports Med* 48:765-785, 2018.

134. Suchomel, TJ, Nimphius, S, and Stone, MH. The importance of muscular strength in athletic performance. *Sports Med* 46:1419-1449, 2016.

135. Suchomel, TJ, Sole, CJ, Bellon, CR, and Stone, MH. Dynamic strength index: Relationships with common performance variables and contextualization of training recommendations. *J Hum Kinet* 74:59-70, 2020.

136. Suchomel, TJ, Taber, CB, Sole, CJ, and Stone, MH. Force-time differences between ballistic and non-ballistic half-squats. *Sports* 6:79, 2018.

137. Taber, CB, Bellon, CR, Abbott, H, and Bingham, GE. Roles of maximal strength and rate of force development in maximizing muscular power. *Strength Cond J* 38:71-78, 2016.

138. Thompson, SW, Rogerson, D, Ruddock, A, and Barnes, A. The effectiveness of two methods of prescribing load on maximal strength development: A systematic review. *Sports Med* 50:919-938, 2020.

139. Travis, SK, Mujika, I, Gentles, JA, Stone, MH, and Bazyler, CD. Tapering and peaking maximal strength for powerlifting performance: A review. *Sports (Basel)* 8:125, 2020.

140. Valenzuela, PL, Sánchez-Martínez, G, Torrontegi, E, Vázquez-Carrión, J, Montalvo, Z, and Haff, GG. Should we base training prescription on the force–velocity profile? Exploratory study of its between-day reliability and differences between methods. *Int J Sports Physiol Perform* 16:1001-1007, 2021.

141. Verkhoshansky, Y, and Tatyan, V. Speed-strength preparation for future champions. *Sov Sports Rev* 18:166-170, 1983.

142. Verkhoshansky, YV. *Programming and organization of training*. Livonia, MI: Sportivny Press, 1985.

143. Verkhoshansky, YV, and Siff, MC. *Supertraining*. Verkhoshansky, 2009.

144. Wetmore, AB, Moquin, PA, Carroll, KM, Fry, AC, Hornsby, WG, and Stone, MH. The effect of training status on adaptations to 11 weeks of block periodization training. *Sports* 8:145, 2020.

145. Williams, TD, Tolusso, DV, Fedewa, MV, and Esco, MR. Comparison of periodized and non-periodized resistance training on maximal strength: A meta-analysis. *Sports Med* 47:2083-2100, 2017.

146. Willoughby, DS. A comparison of three selected weight training programs on the upper and lower body strength of trained males. *Ann J Appl Res Coach Athl* 7:124-146, 1992.

147. Willoughby, DS. The effects of mesocycle-length weight training programs involving periodization and partially equated volumes on upper and lower body strength. *J Strength Cond Res* 7:2-8, 1993.

148. Yavuz, HU, Erdağ, D, Amca, AM, and Aritan, S. Kinematic and EMG activities during front and back squat variations in maximum loads. *J Sports Sci* 33: 1058-1066, 2015.

149. Young, WB. Transfer of strength and power training to sports performance. *Int J Sports Physiol Perform* 1:74-83, 2006.

150. Zamparo, P, Minetti, A, and di Prampero, P. Interplay among the changes of muscle strength, cross-sectional area and maximal explosive power: Theory and facts. *Eur J Appl Physiol* 88:193-202, 2002.

151. Zatsiorsky, V. *Science and Practice of Strength Training*. Champaign, IL: Human Kinetics, 1995.

Chapter 5

1. Aboodarda, SJ, Byrne, JM, Samson, M, Wilson, BD, Mokhtar, AH, and Behm, DG. Does performing drop jumps with additional eccentric loading improve jump performance? *J Strength Cond Res* 28:2314-2323, 2014.

2. Aboodarda, SJ, Page, PA, and Behm, DG. Eccentric and concentric jumping performance during augmented jumps with elastic resistance: A meta-analysis. *Int J Sports Phys Ther* 10:839-849, 2015.

3. Aboodarda, SJ, Yusof, A, Osman, NAA, Thompson, MW, and Mokhtar, AH. Enhanced performance with elastic resistance during the eccentric phase of a countermovement jump. *Int J Sports Physiol Perform* 8:181-187, 2013.

4. Aerts, I, Cumps, E, Verhagen, E, Verschueren, J, and Meeusen, R. A systematic review of different jump-landing variables in relation to injuries. *J Sports Med Phys Fitness* 53:509-519, 2013.

5. Alcaraz, PE, Carlos-Vivas, J, Oponjuru, BO, and Martinez-Rodriguez, A. The effectiveness of resisted sled training (RST) for sprint performance: A systematic review and meta-analysis. *Sports Med* 48:2143-2165, 2018.

6. Allen, WJC, de Keijzer, KL, Raya-González, J, Castillo, D, Coratella, G, and Beato, M. Chronic effects of flywheel training on physical capacities in soccer players: A systematic review. *Res Sports Med* 31:228-248, 2023.

7. Andersen, LL, and Aagaard, P. Influence of maximal muscle strength and intrinsic muscle contractile properties on contractile rate of force development. *Eur J Appl Physiol* 96:46-52, 2006.

8. Anderson, K, and Behm, DG. Trunk muscle activity increases with unstable squat movements. *Can J Appl Physiol* 30:33-45, 2005.

9. Andrade, DC, Manzo, O, Beltrán, AR, Alvares, C, Del Rio, R, Toledo, C, Moran, J, and Ramirez-Campillo, R. Kinematic and neuromuscular measures of intensity during plyometric jumps. *J Strength Cond Res* 34:3395-3402, 2020.

10. Appleby, BB, Cormack, SJ, and Newton, RU. Specificity and transfer of lower-body strength: Influence of bilateral or unilateral lower-body resistance training. *J Strength Cond Res* 33:318-326, 2019.

11. Appleby, BB, Cormack, SJ, and Newton, RU. Unilateral and bilateral lower-body resistance training does not transfer equally to sprint and change of direction performance. *J Strength Cond Res* 34:54-64, 2020.

12. Arabatzi, F, and Kellis, E. Olympic weightlifting training causes different knee muscle-coactivation adaptations compared with traditional weight training. *J Strength Cond Res* 26:2192-2201, 2012.

13. Asadi, A, Arazi, H, Young, WB, and de Villarreal, ESS. The effects of plyometric training on change-of-direction ability: A meta-analysis. *Int J Sports Physiol Perform* 11:563-573, 2016.

14. Augustsson, J, Esko, A, Thomeé, R, and Svantesson, U. Weight training of the thigh muscles using closed versus open kinetic chain exercises: A comparison of performance enhancement. *J Orthop Sports Phys Ther* 27:3-8, 1998.

15. Azevedo, P, Oliveira, MGD, and Schoenfeld, BJ. Effect of different eccentric tempos on hypertrophy and strength of the lower limbs. *Biol Sport* 39:443-449, 2022.

16. Bakrac, ND. Dynamics of muscle strength improvement during isokinetic rehabilitation of athletes with ACL rupture and chondromalacia patellae. *J Sports Med Phys Fitness* 43:69-74, 2003.

17. Beato, M, and Dello Iacono, A. Implementing flywheel (isoinertial) exercise in strength training: Current evidence, practical recommendations, and future directions. *Front Physiol* 11:569, 2020.

18. Beato, M, Maroto-Izquierdo, S, Hernández-Davó, JL, and Raya-González, J. Flywheel training periodization in team sports. *Front Physiol* 12:732802, 2021.

19. Behm, DG, and Anderson, KG. The role of instability with resistance training. *J Strength Cond Res* 20:716-722, 2006.

20. Berg, HE, and Tesch, A. A gravity-independent ergometer to be used for resistance training in space. *Aviat Space Environ Med* 65:752-756, 1994.

21. Berning, JM, Adams, KJ, DeBeliso, M, Sevene-Adams, PG, Harris, C, and Stamford, BA. Effect of functional isometric squats on vertical jump in trained and untrained men. *J Strength Cond Res* 24:2285-2289, 2010.

22. Blackburn, JR, and Morrissey, MC. The relationship between open and closed kinetic chain strength of the lower limb and jumping performance. *J Ortho Sports Phys Ther* 27:430-435, 1998.

23. Bobbert, MF, and Van Soest, AJ. Effects of muscle strengthening on vertical jump height: A simulation study. *Med Sci Sports Exerc* 26:1012-1020, 1994.

24. Brown, LE, and Whitehurst, M. The effect of short-term isokinetic training on force and rate of velocity development. *J Strength Cond Res* 17:88-94, 2003.

25. Brumitt, J, Gilpin, HE, Brunette, M, and Meira, EP. Incorporating kettlebells into a lower extremity sports rehabilitation program. *N Am J Sports Phys Ther* 5:257-265, 2010.

26. Buchheit, M, Samozino, P, Glynn, JA, Michael, BS, Al Haddad, H, Mendez-Villanueva, A, and Morin, JB. Mechanical determinants of acceleration and maximal sprinting speed in highly trained young soccer players. *J Sports Sci* 32:1906-1913, 2014.

27. Cabri, JMH, and Clarys, JP. Isokinetic exercise in rehabilitation. *Appl Ergon* 22:295-298, 1991.

28. Cahill, MJ, Oliver, JL, Cronin, JB, Clark, K, Cross, MR, Lloyd, RS, and Lee, JE. Influence of resisted sled-pull training on the sprint force-velocity profile of male high-school athletes. *J Strength Cond Res* 34:2751-2759, 2020.

29. Cahill, MJ, Oliver, JL, Cronin, JB, Clark, KP, Cross, MR, and Lloyd, RS. Influence of resisted sled-push training on the sprint force-velocity profile of male high school athletes. *Scand J Med Sci Sports* 30:442-449, 2020.

30. Campbell, BI, and Otto, WH, III. Should kettlebells be used in strength and conditioning? *Strength Cond J* 35:27-29, 2013.

31. Campney, HK, and Wehr, RW. Effects of calisthenics on selected components of physical fitness. *Res Q Am Assoc Health Phys Educ* 36:393-402, 1965.

32. Carlson, L, Jonker, B, Westcott, WL, Steele, J, and Fisher, JP. Neither repetition duration nor number of muscle actions affect strength increases, body composition, muscle size, or fasted blood glucose in trained males and females. *Appl Physiol Nutr Metab* 44:200-207, 2019.

33. Carroll, KM, Wagle, JP, Sato, K, Taber, CB, Yoshida, N, Bingham, GE, and Stone, MH. Characterising overload in inertial flywheel devices for use in exercise training. *Sports Biomech* 18:390-401, 2019.

34. Carzoli, JP, Sousa, CA, Belcher, DJ, Helms, ER, Khamoui, AV, Whitehurst, M, and Zourdos, MC. The effects of eccentric phase duration on concentric outcomes in the back squat and bench press in well-trained males. *J Sports Sci* 37:2676-2684, 2019.

35. Channell, BT, and Barfield, JP. Effect of Olympic and traditional resistance training on vertical jump improvement in high school boys. *J Strength Cond Res* 22:1522-1527, 2008.

36. Chaouachi, A, Hammami, R, Kaabi, S, Chamari, K, Drinkwater, EJ, and Behm, DG. Olympic weightlifting and plyometric training with children provides similar or greater performance improvements than traditional resistance training. *J Strength Cond Res* 28:1483-1496, 2014.

37. Chiu, LZF, Fry, AC, Weiss, LW, Schilling, BK, Brown, LE, and Smith, SL. Postactivation potentiation response in athletic and recreationally trained individuals. *J Strength Cond Res* 17:671-677, 2003.

38. Clark, BC, and Manini, TM. Can KAATSU exercise cause rhabdomyolysis? *Clin J Sport Med* 27:e1-e2, 2017.

39. Clark, KP. Determinants of top speed sprinting: Minimum requirements for maximum velocity. *Appl Sci* 12:8289, 2022.

40. Clark, KP, and Weyand, PG. Are running speeds maximized with simple-spring stance mechanics? *J Appl Physiol* 117:604-615, 2014.
41. Clos, P, Laroche, D, Stapley, PJ, and Lepers, R. Neuromuscular and perceptual responses to sub-maximal eccentric cycling: A mini-review. *Front Physiol* 10:1-10, 2019.
42. Comfort, P, Dos'Santos, T, Thomas, C, McMahon, JJ, and Suchomel, TJ. An investigation into the effects of excluding the catch phase of the power clean on force-time characteristics during isometric and dynamic tasks: An intervention study. *J Strength Cond Res* 32:2116-2129, 2018.
43. Comfort, P, Fletcher, C, and McMahon, JJ. Determination of optimal loading during the power clean, in collegiate athletes. *J Strength Cond Res* 26:2970-2974, 2012.
44. Comfort, P, Haff, GG, Suchomel, TJ, Soriano, MA, Pierce, KC, Hornsby, WG, Haff, EE, Sommerfield, LM, Chavda, S, Morris, SJ, Fry, AC, and Stone, MH. National Strength and Conditioning Association position statement on weightlifting for sports performance. *J Strength Cond Res* 37:1163-1190, 2023.
45. Comfort, P, Jones, PA, and Udall, R. The effect of load and sex on kinematic and kinetic variables during the mid-thigh clean pull. *Sports Biomech* 14:139-156, 2015.
46. Comfort, P, Udall, R, and Jones, PA. The effect of loading on kinematic and kinetic variables during the midthigh clean pull. *J Strength Cond Res* 26:1208-1214, 2012.
47. Comfort, P, Williams, R, Suchomel, TJ, and Lake, JP. A comparison of catch phase force-time characteristics during clean derivatives from the knee. *J Strength Cond Res* 31:1911-1918, 2017.
48. Cormie, P, McGuigan, MR, and Newton, RU. Influence of strength on magnitude and mechanisms of adaptation to power training. *Med Sci Sports Exerc* 42:1566-1581, 2010.
49. Cortes, N, Onate, J, and Van Lunen, B. Pivot task increases knee frontal plane loading compared with sidestep and drop-jump. *J Sports Sci* 29:83-92, 2011.
50. Cotter, S. *Kettlebell Training*. Champaign, IL: Human Kinetics, 2022.
51. Coudeyre, E, Jegu, AG, Giustanini, M, Marrel, JP, Edouard, P, and Pereira, B. Isokinetic muscle strengthening for knee osteoarthritis: A systematic review of randomized controlled trials with meta-analysis. *Ann Phys Rehabil Med* 59:207-215, 2016.
52. Cowley, ES, Olenick, AA, McNulty, KL, and Ross, EZ. "Invisible sportswomen": The sex data gap in sport and exercise science research. *Women Sport Phys Act J* 29:146-151, 2021.
53. Coyle, EF, Feiring, DC, Rotkis, TC, Cote, RW, Roby, FB, Lee, W, and Wilmore, JH. Specificity of power improvements through slow and fast isokinetic training. *J Appl Phys* 51:1437-1442, 1981.
54. Crewther, BT, Kilduff, LP, Cook, CJ, Middleton, MK, Bunce, PJ, and Yang, GZ. The acute potentiating effects of back squats on athlete performance. *J Strength Cond Res* 25:3319-3325, 2011.
55. Cross, MR, Brughelli, M, Samozino, P, Brown, SR, and Morin, J-B. Optimal loading for maximizing power during sled-resisted sprinting. *Int J Sports Physiol Perform* 12:1069-1077, 2017.
56. Cuenca-Fernández, F, Smith, IC, Jordan, MJ, MacIntosh, BR, López-Contreras, G, Arellano, R, and Herzog, W. Nonlocalized postactivation performance enhancement (PAPE) effects in trained athletes: A pilot study. *Appl Physiol Nutr Metab* 42:1122-1125, 2017.
57. Dai, B, Garrett, WE, Gross, MT, Padua, DA, Queen, RM, and Yu, B. The effect of performance demands on lower extremity biomechanics during landing and cutting tasks. *J Sport Health Sci* 8:228-234, 2019.
58. Dalen, T, Welde, B, van den Tillaar, R, and Aune, TK. Effect of single vs. multi joint ballistic resistance training upon vertical jump performance. *Acta Kinesiol Univ Tartu* 19:86-97, 2013.
59. Davies, J, Parker, DF, Rutherford, OM, and Jones, DA. Changes in strength and cross sectional area of the elbow flexors as a result of isometric strength training. *Eur J Appl Physiol Occup Physiol* 57:667-670, 1988.
60. De Bleecker, C, Vermeulen, S, De Blaiser, C, Willems, T, De Ridder, R, and Roosen, P. Relationship between jump-landing kinematics and lower extremity overuse injuries in physically active populations: A systematic review and meta-analysis. *Sports Med* 50:1515-1532, 2020.
61. de Keijzer, KL, Gonzalez, JR, and Beato, M. The effect of flywheel training on strength and physical capacities in sporting and healthy populations: An umbrella review. *PLoS One* 17:e0264375, 2022.
62. de Keijzer, KL, McErlain-Naylor, SA, Brownlee, TE, Raya-González, J, and Beato, M. Perception and application of flywheel training by professional soccer practitioners. *Biol Sport* 39:809-817, 2022.
63. de Keijzer, KL, Raya-González J, López Samanés, Á, Moreno Perez, V, and Beato, M. Perception and use of flywheel resistance training amongst therapists in sport. *Front Sports Act Living* 5:1141431, 2023.
64. de Villarreal, ESS, Requena, B, and Cronin, JB. The effects of plyometric training on sprint performance: A meta-analysis. *J Strength Cond Res* 26:575-584, 2012.
65. de Villarreal, ESS, Requena, B, Izquierdo, M, and Gonzalez-Badillo, JJ. Enhancing sprint and strength performance: Combined versus maximal power, traditional heavy-resistance and plyometric training *J Sci Med Sport* 16:146-150, 2013.
66. de Villarreal, ESS, Requena, B, and Newton, RU. Does plyometric training improve strength performance? A meta-analysis. *J Sci Med Sport* 13:513-522, 2010.
67. Desmedt, JE, and Godaux, E. Ballistic contractions in man: Characteristic recruitment pattern of single motor units of the tibialis anterior muscle. *J Physiol* 264:673-693, 1977.
68. DeWeese, BH, and Nimphius, S. Speed and agility program design and technique. In *Essentials of Strength Training and Conditioning*. 4th ed. Haff, GG, and Triplett, NT, eds. Champaign, IL: Human Kinetics, 521-557, 2016.
69. DeWeese, BH, Bellon, CR, Magrum, E, Taber, CB, and Suchomel, TJ. Strengthening the springs: Improving sprint performance via strength training. *Techniques* 9:8-20, 2016.
70. DeWeese, BH, Sams, ML, and Serrano, AJ. Sliding toward Sochi—Part 1: A review of programming tactics used during the 2010-2014 quadrennial. *Natl Strength Cond Assoc Coach* 1:30-42, 2014.

71. DeWeese, BH, Sams, ML, and Serrano, AJ. Sliding toward Sochi—Part 2: A review of programming tactics used during the 2010-2014 quadrennial. *Natl Strength Cond Assoc Coach* 1:4-7, 2014.

72. DeWeese, BH, Sams, ML, Williams, JH, and Bellon, CR. The nature of speed: Enhancing sprint abilities through a short to long training approach. *Techniques* 8:8-22, 2015.

73. Dos'Santos, T, McBurnie, A, Comfort, P, and Jones, PA. The effects of six-weeks change of direction speed and technique modification training on cutting performance and movement quality in male youth soccer players. *Sports* 7:205, 2019.

74. Dos'Santos, T, Thomas, C, and Jones, PA. The effect of angle on change of direction biomechanics: Comparison and inter-task relationships. *J Sports Sci* 39:2618-2631, 2021.

75. Dos'Santos, T, Thomas, C, and Jones, PA. How early should you brake during a 180° turn? A kinetic comparison of the antepenultimate, penultimate, and final foot contacts during a 505 change of direction speed test. *J Sports Sci* 39:395-405, 2021.

76. Douglas, J, Pearson, S, Ross, A, and McGuigan, MR. Chronic adaptations to eccentric training: A systematic review. *Sports Med* 47:917-941, 2017.

77. Douglas, J, Pearson, S, Ross, A, and McGuigan, MR. Eccentric exercise: Physiological characteristics and acute responses. *Sports Med* 47:663-675, 2017.

78. Duchateau, J, and Hainaut, K. Mechanisms of muscle and motor unit adaptation to explosive power training. In *Strength and Power in Sport*. Paavov, VK, ed. Oxford, UK: Blackwell Science, 315-330, 2003.

79. Duehring, MD, Feldmann, CR, and Ebben, WP. Strength and conditioning practices of United States high school strength and conditioning coaches. *J Strength Cond Res* 23:2188-2203, 2009.

80. Dufour, SP, Lampert, E, Doutreleau, S, Lonsdorfer-Wolf, E, Billat, VL, Piquard, F, and Richard, R. Eccentric cycle exercise: Training application of specific circulatory adjustments. *Med Sci Sports Exerc* 36:1900-1906, 2004.

81. Dvir, Z, and Müller, S. Multiple-joint isokinetic dynamometry: A critical review. *J Strength Cond Res* 34:587-601, 2020.

82. Ebben, WP, and Blackard, DO. Strength and conditioning practices of National Football League strength and conditioning coaches. *J Strength Cond Res* 15:48-58, 2001.

83. Ebben, WP, Carroll, RM, and Simenz, CJ. Strength and conditioning practices of National Hockey League strength and conditioning coaches. *J Strength Cond Res* 18:889-897, 2004.

84. Ebben, WP, Fauth, ML, Garceau, LR, and Petushek, EJ. Kinetic quantification of plyometric exercise intensity. *J Strength Cond Res* 25:3288-3298, 2011.

85. Ebben, WP, Feldmann, CR, Vanderzanden, TL, Fauth, ML, and Petushek, EJ. Periodized plyometric training is effective for women, and performance is not influenced by the length of post-training recovery. *J Strength Cond Res* 24:1-7, 2010.

86. Ebben, WP, Hintz, MJ, and Simenz, CJ. Strength and conditioning practices of Major League Baseball strength and conditioning coaches. *J Strength Cond Res* 19:538-546, 2005.

87. Ebben, WP, and Jensen, RL. Electromyographic and kinetic analysis of traditional, chain, and elastic band squats. *J Strength Cond Res* 16:547-550, 2002.

88. Ebben, WP, Simenz, C, and Jensen, RL. Evaluation of plyometric intensity using electromyography. *J Strength Cond Res* 22:861-868, 2008.

89. Ebben, WP, Suchomel, TJ, and Garceau, LR. The effect of plyometric training volume on jumping performance. In *Proceedings of the 32nd International Conference of Biomechanics in Sports, Johnson City, TN, USA, 12-16 July, 2014*. Holland, MA: International Society of Biomechanics in Sports, 566-569, 2014.

90. Ebben, WP, Vanderzanden, T, Wurm, BJ, and Petushek, EJ. Evaluating plyometric exercises using time to stabilization. *J Strength Cond Res* 24:300-306, 2010.

91. Ellenbecker, TS, and Davies, GJ. The application of isokinetics in testing and rehabilitation of the shoulder complex. *J Athl Train* 35:338-350, 2000.

92. Esformes, JI, Keenan, M, Moody, J, and Bampouras, TM. Effect of different types of conditioning contraction on upper body postactivation potentiation. *J Strength Cond Res* 25:143-148, 2011.

93. Farup, J, and Sorensen, H. Postactivation potentiation: Upper body force development changes after maximal force intervention. *J Strength Cond Res* 24:1874-1879, 2010.

94. Fleck, SJ, and Kraemer, WJ. *Designing Resistance Training Programs*. 4th ed. Champaign, IL: Human Kinetics, 2014.

95. Flores, FJ, Sedano, S, and Redondo, JC. Optimal load and power spectrum during jerk and back jerk in competitive weightlifters. *J Strength Cond Res* 31:809-816, 2017.

96. French, DN, Kraemer, WJ, and Cooke, CB. Changes in dynamic exercise performance following a sequence of preconditioning isometric muscle actions. *J Strength Cond Res* 17:678-685, 2003.

97. Gentil, P, Fisher, J, and Steele, J. A review of the acute effects and long-term adaptations of single- and multi-joint exercises during resistance training. *Sports Med* 47:843-855, 2017.

98. Gepfert, M, Trybulski, R, Stastny, P, and Wilk, M. Fast eccentric movement tempo elicits higher physiological responses than medium eccentric tempo in ice-hockey players. *Int J Environ Res Public Health* 18:7694, 2021.

99. Gillies, EM, Putman, CT, and Bell, GJ. The effect of varying the time of concentric and eccentric muscle actions during resistance training on skeletal muscle adaptations in women. *Eur J Appl Physiol* 97:443-453, 2006.

100. Gilmore, SL, Brilla, LR, Suprak, DN, Chalmers, GR, and Dahlquist, DT. Effect of a high-intensity isometric potentiating warm-up on bat velocity. *J Strength Cond Res* 33:152-158, 2019.

101. Gioftsidou, A, Beneka, A, Malliou, P, Pafis, G, and Godolias, G. Soccer players' muscular imbalances: Restoration with an isokinetic strength training program. *Percept Mot Skills* 103:151-159, 2006.

102. Gioftsidou, A, Ispirlidis, I, Pafis, G, Malliou, P, Bikos, C, and Godolias, G. Isokinetic strength training program for muscular imbalances in professional soccer players. *Sport Sci Health* 2:101-105, 2008.

103. Golik-Peric, D, Drapsin, M, Obradovic, B, and Drid, P. Short-term isokinetic training versus isotonic training: Effects on asymmetry in strength of thigh muscles. *J Hum Kinet* 30:29-35, 2011.

104. Gottschall, JS, and Kram, R. Ground reaction forces during downhill and uphill running. *J Biomech* 38:445-452, 2005.

105. Grimby, G. Isokinetic training. *Int J Sports Med* 3:S61-S64, 1982.

106. Grimby, G. Progressive resistance exercise for injury rehabilitation: Special emphasis on isokinetic training. *Sports Med* 2:309-315, 1985.

107. Grønfeldt, BM, Lindberg Nielsen, J, Mieritz, RM, Lund, H, and Aagaard, P. Effect of blood-flow restricted vs heavy-load strength training on muscle strength: Systematic review and meta-analysis. *Scand J Med Sci Sports* 30:837-848, 2020.

108. Guilhem, G, Cornu, C, and Guével, A. Neuromuscular and muscle-tendon system adaptations to isotonic and isokinetic eccentric exercise. *Ann Phys Rehabil Med* 53:319-341, 2010.

109. Guilhem, G, Cornu, C, Maffiuletti, NA, and Guével, A. Neuromuscular adaptations to isoload versus isokinetic eccentric resistance training. *Med Sci Sports Exerc* 45:326-335, 2013.

110. Gullich, A, and Schmidtbleicher, D. MVC-induced short-term potentiation of explosive force. *New Stud Athletics* 11:67-81, 1996.

111. Haff, GG. Roundtable discussion: Machines versus free weights. *Strength Cond J* 22:18-30, 2000.

112. Haff, GG, Whitley, A, McCoy, LB, O'Bryant, HS, Kilgore, JL, Haff, EE, Pierce, K, and Stone, MH. Effects of different set configurations on barbell velocity and displacement during a clean pull. *J Strength Cond Res* 17:95-103, 2003.

113. Handford, MJ, Bright, TE, Mundy, P, Lake, J, Theis, N, and Hughes, JD. A conceptual framework of different eccentric training methods. *Strength Cond J* 46:148-158, 2024.

114. Handford, MJ, Bright, TE, Mundy, P, Lake, JP, Theis, N, and Hughes, JD. The need for eccentric speed: A narrative review of the effects of accelerated eccentric actions during resistance-based training. *Sports Med* 52:2061-2083, 2022.

115. Harden, M, Wolf, A, Haff, GG, Hicks, KM, and Howatson, G. Repeatability and specificity of eccentric force output and the implications for eccentric training load prescription. *J Strength Cond Res* 33:676-683, 2019.

116. Harper, DJ, Carling, C, and Kiely, J. High-intensity acceleration and deceleration demands in elite team sports competitive match play: A systematic review and meta-analysis of observational studies. *Sports Med* 49:1923-1947, 2019.

117. Harper, DJ, Cohen, DD, Carling, C, and Kiely, J. Can countermovement jump neuromuscular performance qualities differentiate maximal horizontal deceleration ability in team sport athletes? *Sports* 8:76, 2020.

118. Harper, DJ, Jordan, AR, and Kiely, J. Relationships between eccentric and concentric knee strength capacities and maximal linear deceleration ability in male academy soccer players. *J Strength Cond Res* 35:465-472, 2021.

119. Harper, DJ, McBurnie, AJ, Santos, TD, Eriksrud, O, Evans, M, Cohen, DD, Rhodes, D, Carling, C, and Kiely, J. Biomechanical and neuromuscular performance requirements of horizontal deceleration: A review with implications for random intermittent multi-directional sports. *Sports Med* 52:2321-2354, 2022.

120. Harper, DJ, Morin, J-B, Carling, C, and Kiely, J. Measuring maximal horizontal deceleration ability using radar technology: Reliability and sensitivity of kinematic and kinetic variables. *Sports Biomech* 22:1192-1208, 2023.

121. Harris, NK, Woulfe, CJ, Wood, MR, Dulson, DK, Gluchowski, AK, and Keogh, JB. Acute physiological responses to strongman training compared to traditional strength training. *J Strength Cond Res* 30:1397-1408, 2016.

122. Harrison, JS. Bodyweight training: A return to basics. *Strength Cond J* 32:52-55, 2010.

123. Harrison, JS, Schoenfeld, B, and Schoenfeld, ML. Applications of kettlebells in exercise program design. *Strength Cond J* 33:86-89, 2011.

124. Haven, KL, and Sigward, SM. Whole body mechanics differ among running and cutting maneuvers in skilled athletes. *Gait Posture* 42:240-245, 2015.

125. Hedrick, A. Teaching the clean. *Strength Cond J* 26:70-72, 2004.

126. Hindle, B, Lorimer, A, Winwood, P, Brimm, D, and Keogh, JWL. The biomechanical characteristics of the strongman atlas stone lift. *PeerJ* 9:e12066, 2021.

127. Hindle, BR, Lorimer, A, Winwood, P, Brimm, D, and Keogh, JWL. The biomechanical characteristics of the strongman yoke walk. *Front Sports Act Living* 3:110, 2021.

128. Hindle, BR, Lorimer, A, Winwood, P, and Keogh, JWL. The biomechanics and applications of strongman exercises: A systematic review. *Sports Med Open* 5:49, 2019.

129. Hislop, HJ, and Perrine, J. The isokinetic concept of exercise. *Phys Ther* 47:114-117, 1967.

130. Hoffman, JR, Cooper, J, Wendell, M, and Kang, J. Comparison of Olympic vs. traditional power lifting training programs in football players. *J Strength Cond Res* 18:129-135, 2004.

131. Hollander, DB, Kraemer, RR, Kilpatrick, MW, Ramadan, ZG, Reeves, GV, Francois, M, Hebert, EP, and Tryniecki, JL. Maximal eccentric and concentric strength discrepancies between young men and women for dynamic resistance exercise. *J Strength Cond Res* 21:34-40, 2007.

132. Holmstrup, ME, Jensen, BT, Evans, WS, and Marshall, EC. Eight weeks of kettlebell swing training does not improve sprint performance in recreationally active females. *Int J Exerc Sci* 9:437-444, 2016.

133. Hortobágyi, T, and Katch, FI. Eccentric and concentric torque-velocity relationships during arm flexion and extension: Influence of strength level. *Eur J Appl Physiol Occup Physiol* 60:395-401, 1990.

134. Hughes, L, Paton, B, Rosenblatt, B, Gissane, C, and Patterson, SD. Blood flow restriction training in clinical musculoskeletal rehabilitation: A systematic review and meta-analysis. *Br J Sports Med* 51:1003-1011, 2017.

135. Hunter, JP, Marshall, RN, and McNair, PJ. Relationships between ground reaction force impulse and kinematics of sprint-running acceleration. *J Appl Biomech* 21:31-43, 2005.

136. Israetel, MA, McBride, JM, Nuzzo, JL, Skinner, JW, and Dayne, AM. Kinetic and kinematic differences between squats performed with and without elastic bands. *J Strength Cond Res* 24:190-194, 2010.

137. James, LP, Haff, GG, Kelly, VG, Connick, M, Hoffman, B, and Beckman, EM. The impact of strength level on adaptations to combined weightlifting, plyometric and ballistic training. *Scand J Med Sci Sports* 28:1494-1505, 2018.

138. Jarvis, MM, Graham-Smith, P, and Comfort, P. A methodological approach to quantifying plyometric intensity. *J Strength Cond Res* 30:2522-2532, 2016.

139. Jay, K, Frisch, D, Hansen, K, Zebis, MK, Andersen, CH, Mortensen, OS, and Andersen, LL. Kettlebell training for musculoskeletal and cardiovascular health: A randomized controlled trial. *Scand J Work Environ Health* 37:196-203, 2011.

140. Jay, K, Jakobsen, MD, Sundstrup, E, Skotte, JH, Jørgensen, MB, Andersen, CH, Pedersen, MT, and Andersen, LL. Effects of kettlebell training on postural coordination and jump performance: A randomized controlled trial. *J Strength Cond Res* 27:1202-1209, 2013.

141. Jensen, RL, and Ebben, WP. Quantifying plyometric intensity via rate of force development, knee joint, and ground reaction forces. *J Strength Cond Res* 21:763-767, 2007.

142. Jo, E, Judelson, DA, Brown, LE, Coburn, JW, and Dabbs, NC. Influence of recovery duration after a potentiating stimulus on muscular power in recreationally trained individuals. *J Strength Cond Res* 24:343-347, 2010.

143. Jones, PA, Thomas, C, Dos'Santos, T, McMahon, JJ, and Graham-Smith, P. The role of eccentric strength in 180 turns in female soccer players. *Sports* 5:42, 2017.

144. Kadlec, D, Miller-Dicks, M, and Nimphius, S. Training for "worst-case" scenarios in sidestepping: Unifying strength and conditioning and perception-action approaches. *Sports Med Open* 9:22, 2023.

145. Kawamori, N, Nosaka, K, and Newton, RU. Relationships between ground reaction impulse and sprint acceleration performance in team sport athletes. *J Strength Cond Res* 27:568-573, 2013.

146. Keogh, JWL, Kattan, A, Logan, S, Bensley, J, Muller, C, and Powell, L. A preliminary kinematic gait analysis of a strongman event: The farmers walk. *Sports* 2:24-33, 2014.

147. Keogh, JWL, Newlands, C, Blewett, S, Payne, A, and Chun-Er, L. A kinematic analysis of a strongman-type event: The heavy sprint-style sled pull. *J Strength Cond Res* 24:3088-3097, 2010.

148. Khlifa, R, Aouadi, R, Hermassi, S, Chelly, MS, Jlid, MC, Hbacha, H, and Castagna, C. Effects of a plyometric training program with and without added load on jumping ability in basketball players. *J Strength Cond Res* 24:2955-2961, 2010.

149. Kilduff, LP, Bevan, HR, Kingsley, MI, Owen, NJ, Bennett, MA, Bunce, PJ, Hore, AM, Maw, JR, and Cunningham, DJ. Postactivation potentiation in professional rugby players: Optimal recovery. *J Strength Cond Res* 21:1134-1138, 2007.

150. Kipp, K, Malloy, PJ, Smith, J, Giordanelli, MD, Kiely, MT, Geiser, CF, and Suchomel, TJ. Mechanical demands of the hang power clean and jump shrug: A joint-level perspective. *J Strength Cond Res* 32:466-474, 2018.

151. Kojić, F, Ranisavljev, I, Ćosić, D, Popović, D, Stojiljković, S, and Ilić, V. Effects of resistance training on hypertrophy, strength and tensiomyography parameters of elbow flexors: Role of eccentric phase duration. *Biol Sport* 38:587-594, 2021.

152. Komi, PV, Viitasalo, JT, Rauramaa, R, and Vihko, V. Effect of isometric strength training on mechanical, electrical, and metabolic aspects of muscle function. *Eur J Appl Physiol Occup Physiol* 40:45-55, 1978.

153. Kon, M, Ikeda, T, Homma, T, Akimoto, T, Suzuki, Y, and Kawahara, T. Effects of acute hypoxia on metabolic and hormonal responses to resistance exercise. *Med Sci Sports Exerc* 42:1279-1285, 2010.

154. Kon, M, Ikeda, T, Homma, T, and Suzuki, Y. Effects of low-intensity resistance exercise under acute systemic hypoxia on hormonal responses. *J Strength Cond Res* 26:611-617, 2012.

155. Kossow, AJ, and Ebben, WP. Kinetic analysis of horizontal plyometric exercise intensity. *J Strength Cond Res* 32:1222-1229, 2018.

156. Kotarsky, CJ, Christensen, BK, Miller, JS, and Hackney, KJ. Effect of progressive calisthenic push-up training on muscle strength and thickness. *J Strength Cond Res* 32:651-659, 2018.

157. Kowalchuk, K, and Butcher, S. Eccentric overload flywheel training in older adults. *J Funct Morphol Kinesiol* 4:61, 2019.

158. Lahti, J, Jiménez-Reyes, P, Cross, MR, Samozino, P, Chassaing, P, Simond-Cote, B, Ahtiainen, JP, and Morin, J-B. Individual sprint force-velocity profile adaptations to in-season assisted and resisted velocity-based training in professional rugby. *Sports* 8:74, 2020.

159. Lake, JP, and Lauder, MA. Kettlebell swing training improves maximal and explosive strength. *J Strength Cond Res* 26:2228-2233, 2012.

160. Lake, JP, Lauder, MA, Smith, NA, and Shorter, KA. A comparison of ballistic and non-ballistic lower-body resistance exercise and the methods used to identify their positive lifting phases. *J Appl Biomech* 28:431-437, 2012.

161. Lake, JP, Mundy, PD, Comfort, P, McMahon, JJ, Suchomel, TJ, and Carden, P. The effect of barbell load on vertical jump landing force-time characteristics. *J Strength Cond Res* 35:25-32, 2021.

162. LaStayo, PC, Woolf, JM, Lewek, MD, Snyder-Mackler, L, Reich, T, and Lindstedt, SL. Eccentric muscle contractions: Their contribution to injury, prevention, rehabilitation, and sport. *J Ortho Sports Phys Ther* 33:557-571, 2003.

163. Lesmes, GR, Costill, DL, Coyle, EF, and Fink, WJ. Muscle strength and power changes during maximal isokinetic training. *Med Sci Sports* 10:266-269, 1978.

164. Lindstedt, SL, LaStayo, PC, and Reich, TE. When active muscles lengthen: Properties and consequences of eccentric contractions. *News Physiol Sci* 16:256-261, 2001.

165. Lixandrão, ME, Ugrinowitsch, C, Berton, R, Vechin, FC, Conceição, MS, Damas, F, Libardi, CA, and Roschel, H. Magnitude of muscle strength and mass adaptations between high-load resistance training versus low-load resistance training associated with blood-flow restriction: A systematic review and meta-analysis. *Sports Med* 48:361-378, 2018.

166. Loenneke, JP, Wilson, JM, Marín, PJ, Zourdos, MC, and Bemben, MG. Low intensity blood flow restriction training: A meta-analysis. *Eur J Appl Physiol* 112:1849-1859, 2012.

167. Lorenz, D, and Reiman, M. The role and implementation of eccentric training in athletic rehabilitation: Tendinopathy, hamstring strains, and ACL reconstruction. *Int J Sports Phys Ther* 6:27-44, 2011.

168. Lum, D, and Barbosa, TM. Brief review: Effects of isometric strength training on strength and dynamic performance. *Int J Sports Med* 40:363-375, 2019.

169. Lum, D, Barbosa, TM, Joseph, R, and Balasekaran, G. Effects of two isometric strength training methods on jump and sprint performances: A randomized controlled trial. *J Sci Sport Exerc* 3:115-124, 2021.

170. Lum, D, Comfort, P, Barbosa, TM, and Balasekaran, G. Comparing the effects of plyometric and isometric strength training on dynamic and isometric force-time characteristics. *Biol Sport* 39:189-197, 2022.

171. Lum, D, Haff, GG, and Barbosa, TM. The relationship between isometric force-time characteristics and dynamic performance: A systematic review. *Sports* 8:63, 2020.

172. Lum, D, Joseph, R, Ong, KY, Tang, JM, and Suchomel, TJ. Comparing the effects of long-term vs. periodic inclusion of isometric strength training on strength and dynamic performances. *J Strength Cond Res* 37:305-314, 2023.

173. MacIntosh, BR, Herzog, W, Suter, E, Wiley, JP, and Sokolosky, J. Human skeletal muscle fibre types and force: Velocity properties. *Eur J Appl Physiol Occup Physiol* 67:499-506, 1993.

174. Manocchia, P, Spierer, DK, Lufkin, AKS, Minichiello, J, and Castro, J. Transference of kettlebell training to strength, power, and endurance. *J Strength Cond Res* 27:477-484, 2013.

175. Marcinik, EJ, Hodgdon, JA, Mittleman, K, and O'Brien, JJ. Aerobic/calisthenic and aerobic/circuit weight training programs for Navy men: A comparative study. *Med Sci Sports Exerc* 17:482-487, 1985.

176. Markovic, G. Does plyometric training improve vertical jump height? A meta-analytical review. *Br J Sports Med* 41:349-355, 2007.

177. Maroto-Izquierdo, S, García-López, D, Fernandez-Gonzalo, R, Moreira, OC, González-Gallego, J, and de Paz, JA. Skeletal muscle functional and structural adaptations after eccentric overload flywheel resistance training: A systematic review and meta-analysis. *J Sci Med Sport* 20:943-951, 2017.

178. Martín-Rivera, F, Beato, M, Alepuz-Moner, V, and Maroto-Izquierdo, S. Use of concentric linear velocity to monitor flywheel exercise load. *Front Physiol* 13:1616, 2022.

179. Martínez-Hernández, D. Flywheel eccentric training: How to effectively generate eccentric overload. *Strength Cond J* 46:234-250, 2024.

180. Mattocks, KT, Jessee, MB, Mouser, JG, Dankel, SJ, Buckner, SL, Bell, ZW, Owens, JG, Abe T, and Loenneke JP. The application of blood flow restriction: Lessons from the laboratory. *Curr Sports Med Rep* 17:129-134, 2018.

181. McBurnie, AJ, Harper, DJ, Jones, PA, and Dos'Santos, T. Deceleration training in team sports: Another potential 'vaccine' for sports-related injury? *Sports Med* 52:1-12, 2022.

182. McCaw, ST, and Friday, JJ. A comparison of muscle activity between a free weight and machine bench press. *J Strength Cond Res* 8:259-264, 1994.

183. McCurdy, KW, Langford, GA, Doscher, MW, Wiley, LP, and Mallard, KG. The effects of short-term unilateral and bilateral lower-body resistance training on measures of strength and power. *J Strength Cond Res* 19:9-15, 2005.

184. McCurdy, KW, O'Kelley, E, Kutz, M, Langford, G, Ernest, J, and Torres, M. Comparison of lower extremity EMG between the 2-leg squat and modified single-leg squat in female athletes. *J Sport Rehabil* 19:57-70, 2010.

185. McQuilliam, SJ, Clark, DR, Erskine, RM, and Brownlee, TE. Mind the gap! A survey comparing current strength training methods used in men's versus women's first team and academy soccer. *Sci Med Footb* 6:597-604, 2022.

186. Mero, A. Force-time characteristics and running velocity of male sprinters during the acceleration phase of sprinting. *Res Quart Exerc Sport* 59:94-98, 1988.

187. Merrigan, JJ, Tufano, JJ, Falzone, M, and Jones, MT. Effectiveness of accentuated eccentric loading: Contingent on concentric load. *Int J Sports Physiol Perform* 16:66-72, 2020.

188. Merrigan, JJ, Tufano, JJ, and Jones, MT. Potentiating effects of accentuated eccentric loading are dependent upon relative strength. *J Strength Cond Res* 35:1208-1216, 2021.

189. Mike, J, Kerksick, CM, and Kravitz, L. How to incorporate eccentric training into a resistance training program. *Strength Cond J* 37:5-17, 2015.

190. Miyamoto, N, Wakahara, T, Ema, R, and Kawakami, Y. Further potentiation of dynamic muscle strength after resistance training. *Med Sci Sports Exerc* 45:1323-1330, 2013.

191. Moolyk, AN, Carey, JP, and Chiu, LZF. Characteristics of lower extremity work during the impact phase of jumping and weightlifting. *J Strength Cond Res* 27:3225-3232, 2013.

192. Moran, J, Ramirez-Campillo, R, Liew, B, Chaabene, H, Behm, DG, García-Hermoso, A, Izquierdo, M, and Granacher, U. Effects of bilateral and unilateral resistance training on horizontally orientated movement performance: A systematic review and meta-analysis. *Sports Med* 51:225-242, 2021.

193. Morin, J-B, Capelo-Ramirez, F, Rodriguez-Pérez, MA, Cross, MR, and Jimenez-Reyes, P. Individual adaptation kinetics following heavy resisted sprint training. *J Strength Cond Res* 36:1158-1161, 2022.

194. Morin, J-B, Slawinski, J, Dorel, S, Couturier, A, Samozino, P, Brughelli, M, and Rabita, G. Acceleration capability in elite sprinters and ground impulse: Push more, brake less? *J Biomech* 48:3149-3154, 2015.

195. Morris, SJ, Oliver, JL, Pedley, JS, Haff, GG, and Lloyd, RS. Comparison of weightlifting, traditional resistance training and plyometrics on strength, power and speed: A systematic review with meta-analysis. *Sports Med* 52:1533-1554, 2022.

196. Morriss, CJ, Tolfrey, K, and Coppack, RJ. Effects of short-term isokinetic training on standing long-jump performance in untrained men. *J Strength Cond Res* 15:498-502, 2001.

197. Muñoz-López, A, Nakamura, FY, and Beato, M. Eccentric overload differences between loads and training variables on flywheel training. *Biol Sport* 40:1151-1158, 2023.

198. Nagahara, R, Kanehisa, H, Matsuo, A, and Fukunaga, T. Are peak ground reaction forces related to better sprint acceleration performance? *Sports Biomech* 20:360-369, 2021.

199. Nagahara, R, Mizutani, M, Matsuo, A, Kanehisa, H, and Fukunaga, T. Association of sprint performance with ground reaction forces during acceleration and maximal speed phases in a single sprint. *J Appl Biomech* 34:104-110, 2018.

200. Nagahara, R, Takai, Y, Kanehisa, H, and Fukunaga, T. Vertical impulse as a determinant of combination of step length and frequency during sprinting. *Int J Sports Med* 39:282-290, 2018.

201. Nedergaard, NJ, Kersting, U, and Lake, M. Using accelerometry to quantify deceleration during a high-intensity soccer turning manoeuvre. *J Sports Sci* 32:1897-1905, 2014.

202. Newell, KM. Coordination, control, and skill. In *Differing Perspectives in Motor Learning, Memory, and Control*. Goodman, D, Wilberg, RB, and Franks, IM, eds. Amsterdam, Netherlands: North-Holland, 295-317, 1985.

203. Newton, RU, Kraemer, WJ, Häkkinen, K, Humphries, B, and Murphy, AJ. Kinematics, kinetics, and muscle activation during explosive upper body movements. *J Appl Biomech* 12:31-43, 1996.

204. Nimphius, S, Callaghan, SJ, Bezodis, NE, and Lockie, RG. Change of direction and agility tests: Challenging our current measures of performance. *Strength Cond J* 40:26-38, 2018.

205. Nimphius, S, Callaghan, SJ, Spiteri, T, and Lockie, RG. Change of direction deficit: A more isolated measure of change of direction performance than total 505 time. *J Strength Cond Res* 30:3024-3032, 2016.

206. Nimphius, S, Geib, G, Spiteri, T, and Carlisle, D. "Change of direction deficit" measurement in Division I American football players. *J Aust Strength Cond* 21:115-117, 2013.

207. Nóbrega, SR, Barroso, R, Ugrinowitsch, C, da Costa, JLF, Alvarez, IF, Barcelos, C, and Libardi, CA. Self-selected vs. fixed repetition duration: Effects on number of repetitions and muscle activation in resistance-trained men. *J Strength Cond Res* 32:2419-2424, 2018.

208. Noto, T, Hashimoto, G, Takagi, T, Awaya, T, Araki, T, Shiba, M, Iijima, R, Hara, H, Moroi, M, and Nakamura, M. Paget-Schroetter syndrome resulting from thoracic outlet syndrome and KAATSU training. *Intern Med* 56:2595-2601, 2017.

209. Núñez, FJ, Galiano, C, Muñoz-López, A, and Floria, P. Is possible an eccentric overload in a rotary inertia device? Comparison of force profile in a cylinder-shaped and a cone-shaped axis devices. *J Sports Sci* 38:1624-1628, 2020.

210. Nuñez Sanchez, FJ, and de Villarreal, ESS. Does flywheel paradigm training improve muscle volume and force? A meta-analysis. *J Strength Cond Res* 31:3177-3186, 2017.

211. Nuzzo, JL, Pinto, MD, Nosaka, K, and Steele, J. The eccentric:concentric strength ratio of human skeletal muscle in vivo: Meta-analysis of the influences of sex, age, joint action, and velocity. *Sports Med* 53:1125-1136, 2023.

212. Okudaira, M, Willwacher, S, Kawama, R, Ota, K, and Tanigawa, S. Sprinting kinematics and inter-limb coordination patterns at varying slope inclinations. *J Sports Sci* 39:2444-2453, 2021.

213. Oranchuk, DJ, Storey, AG, Nelson, AR, and Cronin, JB. Isometric training and long-term adaptations: Effects of muscle length, intensity, and intent: A systematic review. *Scand J Med Sci Sports* 29:484-503, 2019.

214. Oranchuk, DJ, Storey, AG, Nelson, AR, and Cronin, JB. Scientific basis for eccentric quasi-isometric resistance training: A narrative review. *J Strength Cond Res* 33:2846-2859, 2019.

215. Östenberg, A, Roos, E, Ekdah, C, and Roos, H. Isokinetic knee extensor strength and functional performance in healthy female soccer players. *Scand J Med Sci Sports* 8:257-264, 1998.

216. Otto, WH, III, Coburn, JW, Brown, LE, and Spiering, BA. Effects of weightlifting vs. kettlebell training on vertical jump, strength, and body composition. *J Strength Cond Res* 26:1199-1202, 2012.

217. Otzen, R. Recovering the past for the present: The history of Victorian calisthenics. *Lilith* 5:96-112, 1988.

218. Oxfeldt, M, Overgaard, K, Hvid, LG, and Dalgas, U. Effects of plyometric training on jumping, sprint performance, and lower body muscle strength in healthy adults: A systematic review and meta-analyses. *Scand J Med Sci Sports* 29:1453-1465, 2019.

219. Patterson, SD, Hughes, L, Warmington, S, Burr, J, Scott, BR, Owens, J, Abe, T, Nielsen JL, Libardi CA, and Laurentino G. Blood flow restriction exercise: Considerations of methodology, application, and safety. *Front Physiol* 10:533, 2019.

220. Pearson, J, Wadhi, T, Barakat, C, Aube, D, Schoenfeld, BJ, Andersen, JC, Barroso, R, Ugrinowitsch, C, and De Souza, EO. Does varying repetition tempo in a single-joint lower body exercise augment muscle size and strength in resistance-trained men? *J Strength Cond Res* 36:2162-2168, 2022.

221. Peñailillo, L, Blazevich, A, Numazawa, H, and Nosaka, K. Metabolic and muscle damage profiles of concentric versus repeated eccentric cycling. *Med Sci Sports Exerc* 45:1773-1781, 2013.

222. Pereira, PEA, Motoyama, YL, Esteves, GJ, Quinelato, WC, Botter, L, Tanaka, KH, and Azevedo, P. Resistance training with slow speed of movement is better for hypertrophy and muscle strength gains than fast speed of movement. *Int J Appl Exerc Physiol* 5:37-43, 2016.

223. Perera, E, Zhu, XM, Horner, NS, Bedi, A, Ayeni, OR, and Khan, M. Effects of blood flow restriction therapy for muscular strength, hypertrophy, and endurance in healthy and special populations: A systematic review and meta-analysis. *Clin J Sport Med* 32:531-545, 2022.

224. Petrakos, G, Morin, J-B, and Egan, B. Resisted sled sprint training to improve sprint performance: A systematic review. *Sports Med* 46:381-400, 2016.

225. Petré, H, Wernstål, F, and Mattsson, CM. Effects of flywheel training on strength-related variables: A meta-analysis. *Sports Med Open* 4:55, 2018.

226. Prevost, MC, Nelson, AG, and Maraj, BK. The effect of two days of velocity-specific isokinetic training on torque production. *J Strength Cond Res* 13:35-39, 1999.

227. Ramirez-Campillo, R, Sanchez-Sanchez, J, Gonzalo-Skok, O, Rodríguez-Fernandez, A, Carretero, M, and Nakamura, FY. Specific changes in young soccer player's fitness after traditional bilateral vs. unilateral combined strength and plyometric training. *Front Physiol* 9:265, 2018.

228. Ramirez-Campillo, R, Moran, J, Chaabene, H, Granacher, U, Behm, DG, García-Hermoso, A, and Izquierdo, M. Methodological characteristics and future directions for plyometric jump training research: A scoping review update. *Scand J Med Sci Sports* 30:983-997, 2020.

229. Ramírez-delaCruz, M, Bravo-Sánchez, A, Esteban-García, P, Jiménez, F, and Abián-Vicén, J. Effects of plyometric training on lower body muscle architecture, tendon structure, stiffness and physical performance: A systematic review and meta-analysis. *Sports Med Open* 8:1-29, 2022.

230. Raya-González, J, Castillo, D, de Keijzer, KL, and Beato, M. Considerations to optimize strength and muscle mass gains through flywheel resistance devices: A narrative review. *Strength Cond J* 45:111-121, 2023.

231. Raya-González, J, de Keijzer, KL, Bishop, C, and Beato, M. Effects of flywheel training on strength-related variables in female populations. A systematic review. *Res Sports Med* 30:353-370, 2022.

232. Renals, L, Lake, J, Keogh, J, and Austin, K. Strongman log push press: The effect log diameter has on force-time characteristics. *J Strength Cond Res* 32:2693-2700, 2018.

233. Rixon, KP, Lamont, HS, and Bemben, MG. Influence of type of muscle contraction, gender, and lifting experience on postactivation potentiation performance. *J Strength Cond Res* 21:500-505, 2007.

234. Roberts, BM, Nuckols, G, and Krieger, JW. Sex differences in resistance training: A systematic review and meta-analysis. *J Strength Cond Res* 34:1448-1460, 2020.

235. Ross, A, Leveritt, M, and Riek, S. Neural influences on sprint running. *Sports Med* 31:409-425, 2001.

236. Rossi, FE, Schoenfeld, BJ, Ocetnik, S, Young, J, Vigotsky, A, Contreras, B, Krieger, JW, Miller, MG, and Cholewa, J. Strength, body composition, and functional outcomes in the squat versus leg press exercises. *J Sports Med Phys Fitness* 58:263-270, 2018.

237. Schilling, BK, Falvo, MJ, and Chiu, LZ. Force-velocity, impulse-momentum relationships: Implications for efficacy of purposefully slow resistance training. *J Sports Sci Med* 7:299-304, 2008.

238. Schoenfeld, BJ, Ogborn, DI, and Krieger, JW. Effect of repetition duration during resistance training on muscle hypertrophy: A systematic review and meta-analysis. *Sports Med* 45:577-585, 2015.

239. Schot, P, Dart, J, and Schuh, M. Biomechanical analysis of two change-of-direction maneuvers while running. *J Orthop Sports Phys Ther* 22:254-258, 1995.

240. Schwanbeck, S, Chilibeck, PD, and Binsted, G. A comparison of free weight squat to Smith machine squat using electromyography. *J Strength Cond Res* 23:2588-2591, 2009.

241. Schwanbeck, SR, Cornish, SM, Barss, T, and Chilibeck, PD. Effects of training with free weights versus machines on muscle mass, strength, free testosterone, and free cortisol levels. *J Strength Cond Res* 34:1851-1859, 2020.

242. Scott, BR, Girard, O, Rolnick, N, McKee, JR, and Goods, PSR. An updated panorama of blood-flow-restriction methods. *Int J Sports Physiol Perform* 8:1461-1465, 2023.

243. Segers, N, Waldron, M, Howe, LP, Patterson, SD, Moran, J, Jones, B, Kidgell, DJ, and Tallent, J. Slow-speed compared with fast-speed eccentric muscle actions are detrimental to jump performance in elite soccer players in-season. *Int J Sports Physiol Perform* 17:1425-1431, 2022.

244. Seitz, LB, de Villarreal, ESS, and Haff, GG. The temporal profile of postactivation potentiation is related to strength level. *J Strength Cond Res* 28:706-715, 2014.

245. Seitz, LB, Mina, MA, and Haff, GG. Postactivation potentiation of horizontal jump performance across multiple sets of a contrast protocol. *J Strength Cond Res* 30:2733-2740, 2016.

246. Seitz, LB, Trajano, GS, and Haff, GG. The back squat and the power clean: Elicitation of different degrees of potentiation. *Int J Sports Physiol Perform* 9:643-649, 2014.

247. Sheppard, JM, Dingley, AA, Janssen, I, Spratford, W, Chapman, DW, and Newton, RU. The effect of assisted jumping on vertical jump height in high-performance volleyball players. *J Sci Med Sport* 14:85-89, 2011.

248. Sherman, WM, Pearson, DR, Plyley, MJ, Costill, DL, Habansky, AJ, and Vogelgesang, DA. Isokinetic rehabilitation after surgery: A review of factors which are important for developing physiotherapeutic techniques after knee surgery. *Am J Sports Med* 10:155-161, 1982.

249. Shibata, K, Takizawa, K, Nosaka, K, and Mizuno, M. Effects of prolonging eccentric phase duration in parallel back-squat training to momentary failure on muscle cross-sectional area, squat one repetition maximum, and performance tests in university soccer players. *J Strength Cond Res* 35:668-674, 2021.

250. Shvartz, E, and Tamir, D. Effect of calisthenics on strength, muscular endurance and total body reaction and movement times. *J Sports Med Phys Fitness* 11:75-79, 1971.

251. Sigward, SM, Cesar, GM, and Havens, KL. Predictors of frontal plane knee moments during side-step cutting to 45 and 110 men and women: Implications for ACL injury. *Clin J Sport Med* 25:529-534, 2015.

252. Silvester, LJ, and Bryce, GR. The effect of variable resistance and free-weight training programs on strength and vertical jump. *Strength Cond J* 3:30-33, 1981.

253. Simenz, CJ, Dugan, CA, and Ebben, WP. Strength and conditioning practices of National Basketball Association strength and conditioning coaches. *J Strength Cond Res* 19:495-504, 2005.

254. Singla, D, Hussain, ME, and Moiz, JA. Effect of upper body plyometric training on physical performance in healthy individuals: A systematic review. *Phys Ther Sport* 29:51-60, 2018.

255. Smith, MJ, and Melton, P. Isokinetic versus isotonic variable-resistance training. *Am J Sports Med* 9:275-279, 1981.

256. Soriano, MA, García-Ramos, A, Calderbank, J, Marín, PJ, Sainz de Baranda, P, and Comfort, P. Does sex impact the differences and relationships in the one repetition maximum performance across weightlifting overhead pressing exercises? *J Strength Cond Res* 36:1930-1935, 2022.

257. Soriano, MA, García-Ramos, A, Torres-González, A, Castillo-Palencia, J, Marín, PJ, de Baranda, PS, and Comfort, P. Comparison of 1-repetition-maximum performance across 3 weightlifting overhead pressing exercises and sport groups. *Int J Sports Physiol Perform* 15:862-867, 2020.

258. Soriano, MA, Lake, JP, Comfort, P, Suchomel, TJ, McMahon, JJ, Jiménez-Ormeño, E, and Sainz de Baranda, P. No differences in weightlifting overhead pressing exercises kinetics. *Sports Biomech*, 2021. [e-pub ahead of print].

259. Soriano, MA, Suchomel, TJ, and Comfort, P. Weightlifting overhead pressing derivatives: A review of the literature. *Sports Med* 49:867-885, 2019.

260. Speirs, DE, Bennett, MA, Finn, CV, and Turner, AP. Unilateral vs. bilateral squat training for strength, sprints, and agility in academy rugby players. *J Strength Cond Res* 30:386-392, 2016.

261. Spiteri, T, Cochrane, JL, Hart, NH, Haff, GG, and Nimphius, S. Effect of strength on plant foot kinetics and kinematics during a change of direction task. *Eur J Sport Sci* 13:646-652, 2013.

262. Spiteri, T, Newton, RU, Binetti, M, Hart, NH, Sheppard, JM, and Nimphius, S. Mechanical determinants of faster change of direction and agility performance in female basketball athletes. *J Strength Cond Res* 29:2205-2214, 2015.

263. Spiteri, T, Nimphius, S, Hart, NH, Specos, C, Sheppard, JM, and Newton, RU. Contribution of strength characteristics to change of direction and agility performance in female basketball athletes. *J Strength Cond Res* 28:2415-2423, 2014.

264. Spranger, MD, Krishnan, AC, Levy, PD, O'Leary, DS, and Smith, SA. Blood flow restriction training and the exercise pressor reflex: A call for concern. *Am J Physiol Heart Circ Physiol* 309:H1440-H1452, 2015.

265. Stasinaki, A-N, Zaras, N, Methenitis, S, Bogdanis, G, and Terzis, G. Rate of force development and muscle architecture after fast and slow velocity eccentric training. *Sports* 7:41, 2019.

266. Stone, MH, Collins, D, Plisk, S, Haff, GG, and Stone, ME. Training principles: Evaluation of modes and methods of resistance training. *Strength Cond J* 22:65-76, 2000.

267. Suchomel, TJ. The gray area of programming weightlifting exercises. *Natl Strength Cond Assoc Coach* 7:6-14, 2020.

268. Suchomel, TJ, Beckham, GK, and Wright, GA. Lower body kinetics during the jump shrug: Impact of load. *J Trainol* 2:19-22, 2013.

269. Suchomel, TJ, Beckham, GK, and Wright, GA. The impact of load on lower body performance variables during the hang power clean. *Sports Biomech* 13:87-95, 2014.

270. Suchomel, TJ, Beckham, GK, and Wright, GA. Effect of various loads on the force-time characteristics of the hang high pull. *J Strength Cond Res* 29:1295-1301, 2015.

271. Suchomel, TJ, Cantwell, CJ, Campbell, BA, Schroeder, ZS, Marshall, LK, and Taber, CB. Braking and propulsion phase characteristics of traditional and accentuated eccentric loaded back squats. *J Hum Kinet* 91:121-133, 2024.

272. Suchomel, TJ, and Comfort, P. Developing muscular strength and power. In *Advanced Strength and Conditioning: An Evidence-Based Approach*. Turner, A, and Comfort, P, eds. New York: Routledge, 13-39, 2022.

273. Suchomel, TJ, Comfort, P, and Lake, JP. Enhancing the force-velocity profile of athletes using weightlifting derivatives. *Strength Cond J* 39:10-20, 2017.

274. Suchomel, TJ, Comfort, P, and Stone, MH. Weightlifting pulling derivatives: Rationale for implementation and application. *Sports Med* 45:823-839, 2015.

275. Suchomel, TJ, DeWeese, BH, Beckham, GK, Serrano, AJ, and French, SM. The hang high pull: A progressive exercise into weightlifting derivatives. *Strength Cond J* 36:79-83, 2014.

276. Suchomel, TJ, DeWeese, BH, Beckham, GK, Serrano, AJ, and Sole, CJ. The jump shrug: A progressive exercise into weightlifting derivatives. *Strength Cond J* 36:43-47, 2014.

277. Suchomel, TJ, DeWeese, BH, and Serrano, AJ. The power clean and power snatch from the knee. *Strength Cond J* 38:98-105, 2016.

278. Suchomel, TJ, Garceau, LR, and Ebben, WP. Verbal instruction effect on stretch shortening cycle duration and reactive strength index-modified during plyometrics. *Med Sci Sports Exerc* 45:S428, 2013.

279. Suchomel, TJ, Garceau, LR, and Ebben, WP. The effect of verbal instruction on lower body performance variables during various plyometrics. *Med Sci Sports Exerc* 46:961, 2014.

280. Suchomel, TJ, Lake, JP, and Comfort, P. Load absorption force-time characteristics following the second pull of weightlifting derivatives. *J Strength Cond Res* 31:1644-1652, 2017.

281. Suchomel, TJ, Lamont, HS, and Moir, GL. Understanding vertical jump potentiation: A deterministic model. *Sports Med* 46:809-828, 2016.

282. Suchomel, TJ, McKeever, SM, and Comfort, P. Training with weightlifting derivatives: The effects of force and velocity overload stimuli. *J Strength Cond Res* 34:1808-1818, 2020.

283. Suchomel, TJ, McKeever, SM, McMahon, JJ, and Comfort, P. The effect of training with weightlifting catching or pulling derivatives on squat jump and countermovement jump force-time adaptations. *J Funct Morphol Kinesiol* 5:28, 2020.

284. Suchomel, TJ, McKeever, SM, Nolen, JD, and Comfort, P. Muscle architectural and force-velocity curve adaptations following 10 weeks of training with weightlifting catching and pulling derivatives. *J Sports Sci Med* 21:504-516, 2022.

285. Suchomel, TJ, Nimphius, S, Bellon, CR, and Stone, MH. The importance of muscular strength: Training considerations. *Sports Med* 48:765-785, 2018.

286. Suchomel, TJ, Nimphius, S, and Stone, MH. The importance of muscular strength in athletic performance. *Sports Med* 46:1419-1449, 2016.

287. Suchomel, TJ, and Sato, K. Baseball resistance training: Should power clean variations be incorporated? *J Athl Enhancement* 2:2, 2013.

288. Suchomel, TJ, Sato, K, DeWeese, BH, Ebben, WP, and Stone, MH. Potentiation effects of half-squats performed in a ballistic or non-ballistic manner. *J Strength Cond Res* 30:1652-1660, 2016.

289. Suchomel, TJ, Sato, K, DeWeese, BH, Ebben, WP, and Stone, MH. Potentiation following ballistic and non-ballistic complexes: The effect of strength level. *J Strength Cond Res* 30:1825-1833, 2016.

290. Suchomel, TJ, and Sole, CJ. Force-time curve comparison between weightlifting derivatives. *Int J Sports Physiol Perform* 12:431-439, 2017.

291. Suchomel, TJ, and Sole CJ. Power-time curve comparison between weightlifting derivatives. *J Sports Sci Med* 16:407-413, 2017.

292. Suchomel, TJ, Taber, CB, Sole, CJ, and Stone, MH. Force-time differences between ballistic and non-ballistic half-squats. *Sports* 6:79, 2018.

293. Suchomel, TJ, Taber, CB, and Wright, GA. Jump shrug height and landing forces across various loads. *Int J Sports Physiol Perform* 11:61-65, 2016.

294. Suchomel, TJ, Wagle, JP, Douglas, J, Taber, CB, Harden, M, Haff, GG, and Stone, MH. Implementing eccentric resistance

294. training—Part 1: A brief review of existing methods. *J Funct Morphol Kinesiol* 4:38, 2019.

295. Suchomel, TJ, Wagle, JP, Douglas, J, Taber, CB, Harden, M, Haff, GG, and Stone, MH. Implementing eccentric resistance training—Part 2: Practical recommendations. *J Funct Morphol Kinesiol* 4:55, 2019.

296. Suchomel, TJ, Wright, GA, Kernozek, TW, and Kline, DE. Kinetic comparison of the power development between power clean variations. *J Strength Cond Res* 28:350-360, 2014.

297. Suchomel, TJ, Wright, GA, and Lottig, J. Lower extremity joint velocity comparisons during the hang power clean and jump shrug at various loads. In *Proceedings of the 32nd International Conference of Biomechanics in Sports, Johnson City, TN, USA, 12-16 July, 2014*. Holland, MA: International Society of Biomechanics in Sports, 749-752, 2014.

298. Taber, CB, Bellon, CR, Abbott, H, and Bingham, GE. Roles of maximal strength and rate of force development in maximizing muscular power. *Strength Cond J* 38:71-78, 2016.

299. Takano, B. Coaching optimal technique in the snatch and the clean and jerk: Part I. *Strength Cond J* 9:50-59, 1987.

300. Takano, B. Coaching optimal technique in the snatch and the clean and jerk: Part II. *Strength Cond J* 9:52-56, 1987.

301. Takano, B. Coaching optimal techniques in the snatch and the clean and jerk: Part III. *Strength Cond J* 10:54-59, 1988.

302. Teo, SY, Newton, MJ, Newton, RU, Dempsey, AR, and Fairchild, TJ. Comparing the effectiveness of a short-term vertical jump versus weightlifting program on athletic power development. *J Strength Cond Res* 30:2741-2748, 2016.

303. Tesch, PA, Fernandez-Gonzalo, R, and Lundberg, TR. Clinical applications of iso-inertial, eccentric-overload (YoYo™) resistance exercise. *Front Physiol* 8:241, 2017.

304. Thompson, WR. Worldwide survey of fitness trends for 2020. *ACSM's Health Fit J* 23:10-18, 2019.

305. Timmons, JF, Hone, M, Duffy, O, and Egan, B. When matched for relative leg strength at baseline, male and female older adults respond similarly to concurrent aerobic and resistance exercise training. *J Strength Cond Res* 36:2927-2933, 2022.

306. Tricoli, V, Lamas, L, Carnevale, R, and Ugrinowitsch, C. Short-term effects on lower-body functional power development: Weightlifting vs. vertical jump training programs. *J Strength Cond Res* 19:433-437, 2005.

307. Tsourlou, T, Gerodimos, V, Kellis, E, Stavropoulos, N, and Kellis, S. The effects of a calisthenics and a light strength training program on lower limb muscle strength and body composition in mature women. *J Strength Cond Res* 17:590-598, 2003.

308. van Cutsem, M, Duchateau, J, and Hainaut, K. Changes in single motor unit behaviour contribute to the increase in contraction speed after dynamic training in humans. *J Physiol* 513:295-305, 1998.

309. Van den Tillaar, R. Effect of descent velocity upon muscle activation and performance in two-legged free weight back squats. *Sports* 7:15, 2019.

310. Vanrenterghem, J, Venables, E, Pataky, T, and Robinson, MA. The effect of running speed on knee mechanical loading in females during side cutting. *J Biomech* 45:2444-2449, 2012.

311. Verheul, J, Nedergaard, NJ, Pogson, M, Lisboa, P, Gregson, W, Vanrenterghem, J, and Robinson, MA. Biomechanical loading during running: Can a two mass-spring-damper model be used to evaluate ground reaction forces for high-intensity tasks? *Sports Biomech* 20:571-582, 2021.

312. Verkhoshansky, YV, and Siff, MC. *Supertraining*. Verkhoshansky, 2009.

313. Vicens-Bordas, J, Esteve, E, Fort-Vanmeerhaeghe, A, Bandholm, T, and Thorborg, K. Is inertial flywheel resistance training superior to gravity-dependent resistance training in improving muscle strength? A systematic review with meta-analyses. *J Sci Med Sport* 21:75-83, 2018.

314. Vidmar, MF, Baroni, BM, Michelin, AF, Mezzomo, M, Lugokenski, R, Pimentel, GL, and Silva, MF. Isokinetic eccentric training is more effective than constant load eccentric training for quadriceps rehabilitation following anterior cruciate ligament reconstruction: A randomized controlled trial. *Braz J Phys Ther* 24:424-432, 2020.

315. Wagle, JP, Taber, CB, Cunanan, AJ, Bingham, GE, Carroll, K, DeWeese, BH, Sato, K, and Stone, MH. Accentuated eccentric loading for training and performance: A review. *Sports Med* 47:2473-2495, 2017.

316. Walker, S, Blazevich, AJ, Haff, GG, Tufano, JJ, Newton, RU, and Häkkinen, K. Greater strength gains after training with accentuated eccentric than traditional isoinertial loads in already strength-trained men. *Front Physiol* 7:1-12, 2016.

317. Wallace, BJ, Winchester, JB, and McGuigan, MR. Effects of elastic bands on force and power characteristics during the back squat exercise. *J Strength Cond Res* 20:268-272, 2006.

318. Weyand, PG, Sandell, RF, Prime, DNL, and Bundle, MW. The biological limits to running speed are imposed from the ground up. *J Appl Physiol* 108:950-961, 2010.

319. Weyand, PG, Sternlight, DB, Bellizzi, MJ, and Wright, S. Faster top running speeds are achieved with greater ground forces not more rapid leg movements. *J Appl Physiol* 89:1991-1999, 2000.

320. Wilk M, Gepfert, M, Krzysztofik, M, Mostowik, A, Filip, A, Hajduk, G, and Zajac, A. Impact of duration of eccentric movement in the one-repetition maximum test result in the bench press among women. *J Sports Sci Med* 19:317-322, 2020.

321. Wilk, M, Golas, A, Krzysztofik, M, Nawrocka, M, and Zajac, A. The effects of eccentric cadence on power and velocity of the bar during the concentric phase of the bench press movement. *J Sports Sci Med* 18:191-197, 2019.

322. Wilk, M, Golas, A, Stastny, P, Nawrocka, M, Krzysztofik, M, and Zajac, A. Does tempo of resistance exercise impact training volume? *J Hum Kinet* 62:241-250, 2018.

323. Wilk, M, Golas, A, Zmijewski, P, Krzysztofik, M, Filip, A, Del Coso, J, and Tufano, JJ. The effects of the movement tempo on the one-repetition maximum bench press results. *J Hum Kinetic* 72:151-159, 2020.

324. Wilk, M, Stastny, P, Golas, A, Nawrocka, M, Jelen, K, Zajac, A, and Tufano, JJ. Physiological responses to different neuromuscular movement task during eccentric bench press. *Neuro Endocrinol Lett* 39:26-32, 2018.

325. Wilk, M, Tufano, JJ, and Zajac, A. The influence of movement tempo on acute neuromuscular, hormonal, and mechanical responses to resistance exercise—A mini review. *J Strength Cond Res* 34:2369-2383, 2020.

326. Wilk, M, Zajac, A, and Tufano, JJ. The influence of movement tempo during resistance training on muscular strength and hypertrophy responses: A review. *Sports Med* 51:1629-1650, 2021.

327. Winwood, PW, Cronin, JB, Brown, SR, and Keogh, JWL. A biomechanical analysis of the heavy sprint-style sled pull and comparison with the back squat. *Int J Sports Sci Coach* 10:851-868, 2015.

328. Winwood, PW, Cronin, JB, Brown, SR, and Keogh, JWL. A biomechanical analysis of the strongman log lift and comparison with weightlifting's clean and jerk. *Int J Sports Sci Coach* 10:869-886, 2015.

329. Winwood, PW, Cronin, JB, Keogh, JWL, Dudson, MK, and Gill, ND. How coaches use strongman implements in strength and conditioning practice. *Int J Sports Sci Coach* 9:1107-1125, 2014.

330. Winwood, PW, Cronin, JB, Posthumus, LR, Finlayson, SJ, Gill, ND, and Keogh, JWL. Strongman vs. traditional resistance training effects on muscular function and performance. *J Strength Cond Res* 29:429-439, 2015.

331. Winwood, PW, Keogh, JW, and Harris, NK. The strength and conditioning practices of strongman competitors. *J Strength Cond Res* 25:3118-3128, 2011.

332. Winwood, PW, Pritchard, HJ, Wilson, D, Dudson, M, and Keogh, JWL. The competition-day preparation strategies of strongman athletes. *J Strength Cond Res* 33:2308-2320, 2019.

333. Wortman, RJ, Brown, SM, Savage-Elliott, I, Finley, ZJ, and Mulcahey, MK. Blood flow restriction training for athletes: A systematic review. *Am J Sports Med* 49:1938-1944, 2021.

334. Wren, C, Beato, M, McErlain-Naylor, SA, Iacono, AD, and de Keijzer, KL. Concentric phase assistance enhances eccentric peak power during flywheel squats: Intersession reliability and the linear relationship between concentric and eccentric phases. *Int J Sports Physiol Perform* 18:428-434, 2023.

335. Young, WB, Pryor, JF, and Wilson, GJ. Effect of instructions on characteristics of countermovement and drop jump performance. *J Strength Cond Res* 9:232-236, 1995.

336. Zatsiorsky, VM. *Science and Practice of Strength Training*. Champaign, IL: Human Kinetics, 1995.

337. Zemke, B, and Wright, G. The use of strongman type implements and training to increase sport performance in collegiate athletes. *Strength Cond J* 33:1-7, 2011.

338. Zhang, Q, Léam, A, Fouré, A, Wong, DP, and Hautier, CA. Relationship between explosive strength capacity of the knee muscles and deceleration performance in female professional soccer players. *Front Physiol* 12:723041, 2021.

Chapter 6

1. Aagaard, P, and Andersen, JL. Correlation between contractile strength and myosin heavy chain isoform composition in human skeletal muscle. *Med Sci Sports Exerc* 30:1217-1222, 1998.

2. Aboodarda, SJ, Yusof, A, Osman, NAA, Thompson, MW, and Mokhtar, AH. Enhanced performance with elastic resistance during the eccentric phase of a countermovement jump. *Int J Sports Physiol Perform* 8:181-187, 2013.

3. Andersen, V, Pedersen, H, Fimland, MS, Shaw, MP, Solstad, TEJ, Stien, N, Cumming, KT, and Saeterbakken, AH. Acute effects of elastic bands as resistance or assistance on EMG, kinetics, and kinematics during deadlift in resistance-trained men. *Front Sports Act Living* 2:598284, 2020.

4. Andersen, V, Prieske, O, Stien, N, Cumming, K, Solstad, TEJ, Paulsen, G, van den Tillaar, R, Pedersen, H, and Saeterbakken, AH. Comparing the effects of variable and traditional resistance training on maximal strength and muscle power in healthy adults: A systematic review and meta-analysis. *J Sci Med Sport* 25:1023-1032, 2022.

5. Anderson, CE, Sforzo, GA, and Sigg, JA. The effects of combining elastic and free weight resistance on strength and power in athletes. *J Strength Cond Res* 22:567-574, 2008.

6. Archer, DC, Brown, LE, Coburn, JW, Galpin, AJ, Drouet, PC, Leyva, WD, Munger, CN, and Wong, MA. Effects of short-term jump squat training with and without chains on strength and power in recreational lifters. *Int J Kinesiol Sports Sci* 4:18-24, 2016.

7. Armstrong, R, Baltzopoulos, V, Langan-Evans, C, Clark, D, Jarvis, J, Stewart, C, and O'Brien, T. An investigation of movement dynamics and muscle activity during traditional and accentuated-eccentric squatting. *PLoS One* 17:e0276096, 2022.

8. Ataee, J, Koozehchian, MS, Kreider, RB, and Zuo, L. Effectiveness of accommodation and constant resistance training on maximal strength and power in trained athletes. *PeerJ* 2:e441, 2014.

9. Baker, DG, and Newton, RU. Effect of kinetically altering a repetition via the use of chain resistance on velocity during the bench press. *J Strength Cond Res* 23:1941-1946, 2009.

10. Bellar, D, Judge, LW, Turk, M, and Judge, M. Efficacy of potentiation of performance through overweight implement throws on male and female collegiate and elite weight throwers. *J Strength Cond Res* 26:1469-1474, 2012.

11. Berning, JM, Coker, CA, and Adams, KJ. Using chains for strength and conditioning. *Strength Cond J* 26:80-84, 2004.

12. Berning, JM, Coker, CA, and Briggs, D. The biomechanical and perceptual influence of chain resistance on the performance of the Olympic clean. *J Strength Cond Res* 22:390-395, 2008.

13. Bevan, HR, Cunningham, DJ, Tooley, EP, Owen, NJ, Cook, CJ, and Kilduff, LP. Influence of postactivation potentiation on sprinting performance in professional rugby players. *J Strength Cond Res* 24:701-705, 2010.

14. Blazevich, AJ, and Babault, N. Post-activation potentiation (PAP) versus post-activation performance enhancement (PAPE) in humans: Historical perspective, underlying mechanisms, and current issues. *Front Physiol* 10:1-19, 2019.

15. Boullosa, D, Del Rosso, S, Behm, DG, and Foster, C. Post-activation potentiation (PAP) in endurance sports: A review. *Eur J Sport Sci* 18:595-610, 2018.

16. Brandenburg, JE, and Docherty, D. The effects of accentuated eccentric loading on strength, muscle hypertrophy, and neural adaptations in trained individuals. *J Strength Cond Res* 16:25-32, 2002.

17. Bridgeman, LA, McGuigan, MR, Gill, ND, and Dulson, DK. The effects of accentuated eccentric loading on the drop jump exercise and the subsequent postactivation potentiation response. *J Strength Cond Res* 31:1620-1626, 2017.

18. Bullock, N, and Comfort, P. An investigation into the acute effects of depth jumps on maximal strength performance. *J Strength Cond Res* 25:3137-3141, 2011.

19. Buttifant, D, and Hrysomallis, C. Effect of various practical warm-up protocols on acute lower-body power. *J Strength Cond Res* 29:656-660, 2015.

20. Byrne, PJ, Kenny, J, and O' Rourke, B. The acute potentiating effect of depth jumps on sprint performance. *J Strength Cond Res* 28:610-615, 2014.

21. Campbell, BA, Canwell, CJ, Marshall, LK, Schroeder, ZS, Chard, JB, Sundh, AE, Taber, CB, and Suchomel, TJ. The effect of load on accentuated eccentric loaded back squat performance in resistance-trained women. *J Strength Cond Res* 37:e921-e922, 2023.

22. Carzoli, JP, Sousa, CA, Belcher, DJ, Helms, ER, Khamoui, AV, Whitehurst, M, and Zourdos, MC. The effects of eccentric phase duration on concentric outcomes in the back squat and bench press in well-trained males. *J Sports Sci* 37:2676-2684, 2019.

23. Chiu, LZF, Fry, AC, Galpin, AJ, Salem, GJ, and Cabarkapa, D. Regulatory light-chain phosphorylation during weightlifting training: Association with postactivation performance enhancement. *J Strength Cond Res* 37:e563-e568, 2023.

24. Clarkson, PM, Kroll, W, and McBride, TC. Maximal isometric strength and fiber type composition in power and endurance athletes. *Eur J Appl Physiol Occup Physiol* 44:35-42, 1980.

25. Coker, CA, Berning, JM, and Briggs, DL. A preliminary investigation of the biomechanical and perceptual influence of chain resistance on the performance of the snatch. *J Strength Cond Res* 20:887-891, 2006.

26. Comfort, P, Allen, M, and Graham-Smith, P. Comparisons of peak ground reaction force and rate of force development during variations of the power clean. *J Strength Cond Res* 25:1235-1239, 2011.

27. Comfort, P, Allen, M, and Graham-Smith, P. Kinetic comparisons during variations of the power clean. *J Strength Cond Res* 25:3269-3273, 2011.

28. Comfort, P, Dos'Santos, T, Thomas, C, McMahon, JJ, and Suchomel, TJ. An investigation into the effects of excluding the catch phase of the power clean on force-time characteristics during isometric and dynamic tasks: An intervention study. *J Strength Cond Res* 32:2116-2129, 2018.

29. Comyns, TM, Harrison, AJ, Hennessy, LK, and Jensen, RL. The optimal complex training rest interval for athletes from anaerobic sports. *J Strength Cond Res* 20:471-476, 2006.

30. Cormie, P, McGuigan, MR, and Newton, RU. Influence of strength on magnitude and mechanisms of adaptation to power training. *Med Sci Sports Exerc* 42:1566-1581, 2010.

31. Crewther, BT, Kilduff, LP, Cook, CJ, Middleton, MK, Bunce, PJ, and Yang, GZ. The acute potentiating effects of back squats on athlete performance. *J Strength Cond Res* 25:3319-3325, 2011.

32. Crum, AJ, Kawamori, N, Stone, MH, and Haff, GG. The acute effects of moderately loaded concentric-only quarter squats on vertical jump performance. *J Strength Cond Res* 26:914-925, 2012.

33. Cuenca-Fernández, F, Smith, IC, Jordan, MJ, MacIntosh, BR, López-Contreras, G, Arellano, R, and Herzog, W. Nonlocalized postactivation performance enhancement (PAPE) effects in trained athletes: A pilot study. *Appl Physiol Nutr Metab* 42:1122-1125, 2017.

34. DeWeese, BH, Hornsby, G, Stone, M, and Stone, MH. The training process: Planning for strength–power training in track and field. Part 2: Practical and applied aspects. *J Sport Health Sci* 4:318-324, 2015.

35. Doan, BK, Newton, RU, Marsit, JL, Triplett-McBride, NT, Koziris, LP, Fry, AC, and Kraemer, WJ. Effects of increased eccentric loading on bench press 1RM. *J Strength Cond Res* 16:9-13, 2002.

36. Douglas, J, Pearson, S, Ross, A, and McGuigan, M. Effects of accentuated eccentric loading on muscle properties, strength, power, and speed in resistance-trained rugby players. *J Strength Cond Res* 32:2750-2761, 2018.

37. Dubé, JJ, Broskey, NT, Despines, AA, Stefanovic-Racic, M, Toledo, FG, Goodpaster, BH, and Amati, F. Muscle characteristics and substrate energetics in lifelong endurance athletes. *Med Sci Sports Exerc* 48:472-480, 2016.

38. Duchateau, J, and Hainaut, K. Mechanisms of muscle and motor unit adaptation to explosive power training. In *Strength and Power in Sport*. Paavov, VK, ed. Oxford: Blackwell Scientific, 315-330, 2003.

39. Ebben, WP, and Jensen, RL. Electromyographic and kinetic analysis of traditional, chain, and elastic band squats. *J Strength Cond Res* 16:547-550, 2002.

40. Ebben, WP, Wurm, B, Garceau, LR, and Suchomel, TJ. Supermaximal loads potentiate subsequent high load back squat performance. *J Strength Cond Res* 27:S94-S95, 2013.

41. Farup, J, and Sorensen, H. Postactivation potentiation: Upper body force development changes after maximal force intervention. *J Strength Cond Res* 24:1874-1879, 2010.

42. Finlay, MJ, Bridge, CA, Greig, M, and Page, RM. Postactivation performance enhancement of amateur boxers' punch force and neuromuscular performance following 2 upper-body conditioning activities. *Int J Sports Physiol Perform* 17:1621-1633, 2022.

43. Finlay, MJ, Bridge, CA, Greig, M, and Page, RM. Upper-body post-activation performance enhancement for athletic performance: A systematic review with meta-analysis and recommendations for future research. *Sports Med* 52:847-871, 2022.

44. Fleck, SJ, and Kraemer, WJ. *Designing Resistance Training Programs*. Champaign, IL: Human Kinetics, 2014.

45. Franchi, MV, Atherton, PJ, Reeves, ND, Flück, M, Williams, J, Mitchell, WK, Selby, A, Beltran Valls, RM, and Narici, MV. Architectural, functional and molecular responses to concentric and eccentric loading in human skeletal muscle. *Acta Physiol* 210:642-654, 2014.

46. Galpin, AJ, Malyszek, KK, Davis, KA, Record, SM, Brown, LE, Coburn, JW, Harmon, RA, Steele, JM, and Manolovitz, AD. Acute effects of elastic bands on kinetic characteristics during the deadlift at moderate and heavy loads. *J Strength Cond Res* 29:3271-3278, 2015.

47. García-López, D, Hernández-Sánchez, S, Martín, E, Marín, PJ, Zarzosa, F, and Herrero, AJ. Free-weight augmentation with elastic bands improves bench press kinematics in professional rugby players. *J Strength Cond Res* 30:2493-2499, 2016.

48. Gepfert, M, Trybulski, R, Stastny, P, and Wilk, M. Fast eccentric movement tempo elicits higher physiological responses than medium eccentric tempo in ice-hockey players. *Int J Environ Res Public Health* 18:7694, 2021.

49. Gillies, EM, Putman, CT, and Bell, GJ. The effect of varying the time of concentric and eccentric muscle actions during resistance training on skeletal muscle adaptations in women. *Eur J Appl Physiol* 97:443-453, 2006.

50. Gilmore, SL, Brilla, LR, Suprak, DN, Chalmers, GR, and Dahlquist, DT. Effect of a high-intensity isometric potentiating warm-up on bat velocity. *J Strength Cond Res* 33:152-158, 2019.

51. Godwin, MS, Fernandes, JFT, and Twist, C. Effects of variable resistance using chains on bench throw performance in trained rugby players. *J Strength Cond Res* 32:950-954, 2018.

52. Gouvêa, AL, Fernandes, IA, César, EP, Silva, WAB, and Gomes, PSC. The effects of rest intervals on jumping performance: A meta-analysis on post-activation potentiation studies. *J Sports Sci* 31:459-467, 2013.

53. Gullich, A, and Schmidtbleicher, D. MVC-induced short-term potentiation of explosive force. *New Stud Athletics* 11:67-81, 1996.

54. Haff, GG. Strength-isometric and dynamic testing. In *Performance Assessment in Strength and Conditioning.* Comfort, P, Jones, PA, and McMahon, JJ, eds. New York: Routledge, 166-192, 2019.

55. Häkkinen, K, Komi, PV, and Tesch, PA. Effect of combined concentric and eccentric strength training and detraining on force-time, muscle fiber and metabolic characteristics of leg extensor muscles. *Scand J Med Sci Sports* 3:50-58, 1981.

56. Hamada, T, Sale, DG, MacDougall, JD, and Tarnopolsky, MA. Postactivation potentiation, fiber type, and twitch contraction time in human knee extensor muscles. *J Appl Physiol* 88:2131-2137, 2000.

57. Hamada, T, Sale, DG, MacDougall, JD, and Tarnopolsky, MA. Interaction of fibre type, potentiation and fatigue in human knee extensor muscles. *Acta Physiol Scand* 178:165-173, 2003.

58. Handford, MJ, Rivera, FM, Maroto-Izquierdo, S, and Hughes, JD. Plyo-accentuated eccentric loading methods to enhance lower limb muscle power. *Strength Cond J* 43:54-64, 2021.

59. Harden, M, Comfort, P, and Haff, GG. Eccentric training: Scientific background and practical applications. In *Advanced Strength and Conditioning: An Evidence-Based Approach.* Turner, AN, and Comfort, P, eds. New York: Routledge, 190-212, 2022.

60. Harden, M, Wolf, A, Evans, M, Hicks, KM, Thomas, K, and Howatson, G. Four weeks of augmented eccentric loading using a novel leg press device improved leg strength in well-trained athletes and professional sprint track cyclists. *PLoS One* 15:e0236663, 2020.

61. Harden, M, Wolf, A, Russell, M, Hicks, KM, French, D, and Howatson, G. An evaluation of supramaximally loaded eccentric leg press exercise. *J Strength Cond Res* 32:2708-2714, 2018.

62. Heelas, T, Theis, N, and Hughes, JD. Muscle activation patterns during variable resistance deadlift training with and without elastic bands. *J Strength Cond Res* 35:3006-3011, 2021.

63. Higuchi, T, Nagami, T, Mizuguchi, N, and Anderson, T. The acute and chronic effects of isometric contraction conditioning on baseball bat velocity. *J Strength Cond Res* 27:216-222, 2013.

64. Hollander, DB, Kraemer, RR, Kilpatrick, MW, Ramadan, ZG, Reeves, GV, Francois, M, Hebert, EP, and Tryniecki, JL. Maximal eccentric and concentric strength discrepancies between young men and women for dynamic resistance exercise. *J Strength Cond Res* 21:34-40, 2007.

65. Israetel, MA, McBride, JM, Nuzzo, JL, Skinner, JW, and Dayne, AM. Kinetic and kinematic differences between squats performed with and without elastic bands. *J Strength Cond Res* 24:190-194, 2010.

66. James, LP, Comfort, P, Suchomel, TJ, Kelly, VG, Beckman, EM, and Haff, GG. The impact of power clean ability and training age on adaptations to weightlifting-style training. *J Strength Cond Res* 36:1560-1567, 2022.

67. James, LP, Haff, GG, Kelly, VG, Connick, M, Hoffman, B, and Beckman, EM. The impact of strength level on adaptations to combined weightlifting, plyometric and ballistic training. *Scand J Med Sci Sports* 28:1494-1505, 2018.

68. Jo, E, Judelson, DA, Brown, LE, Coburn, JW, and Dabbs, NC. Influence of recovery duration after a potentiating stimulus on muscular power in recreationally trained individuals. *J Strength Cond Res* 24:343-347, 2010.

69. Judge, LW, Bellar, D, and Judge, M. Efficacy of potentiation of performance through overweight implement throws on male and female high-school weight throwers. *J Strength Cond Res* 24:1804-1809, 2010.

70. Kelekian, GK, Zaras, N, Stasinaki, AN, Spiliopoulou, P, Karampatsos, G, Bogdanis, G, and Terzis, G. Preconditioning strategies before maximum clean performance in female weightlifters. *J Strength Cond Res* 36:2318-2321, 2022.

71. Kilduff, LP, Bevan, HR, Kingsley, MI, Owen, NJ, Bennett, MA, Bunce, PJ, Hore, AM, Maw, JR, and Cunningham, DJ. Postactivation potentiation in professional rugby players: Optimal recovery. *J Strength Cond Res* 21:1134-1138, 2007.

72. Kipp, K, Malloy, PJ, Smith, J, Giordanelli, MD, Kiely, MT, Geiser, CF, and Suchomel, TJ. Mechanical demands of the hang power clean and jump shrug: A joint-level perspective. *J Strength Cond Res* 32:466-474, 2018.

73. Kipp, K, Suchomel, TJ, and Comfort, P. Correlational analysis between joint-level kinetics of countermovement jumps and weightlifting derivatives. *J Sports Sci Med* 18:663-668, 2019.

74. Kraemer, WJ, Fleck, SJ, and Evans, WJ. Strength and power training: Physiological mechanisms of adaptation. *Exerc Sport Sci Rev* 24:363-397, 1996.

75. Kraemer, WJ, and Newton, RU. Training for muscular power. *Phys Med Rehab Clin N Am* 11:341-368, 2000.

76. Kubo, T, Hirayama, K, Nakamura, N, and Higuchi, M. Effect of accommodating elastic bands on mechanical power output during back squats. *Sports* 6:151, 2018.

77. Lates, AD, Greer, BK, Wagle, JP, and Taber, CB. Accentuated eccentric loading and cluster set configurations in the bench press. *J Strength Cond Res* 36:1485-1489, 2022.

78. Lin, Y, Xu, Y, Hong, F, Li, J, Ye, W, and Korivi, M. Effects of variable-resistance training versus constant-resistance training on maximum strength: A systematic review and meta-analysis. *Int J Environ Res Public Health* 19:8559, 2022.

79. Loturco, I, Pereira, LA, Reis, VP, Zanetti, V, Bishop, C, and McGuigan, MR. Traditional free-weight vs. variable resistance training applied to elite young soccer players during a short preseason: Effects on strength, speed, and power performance. *J Strength Cond Res* 36:3432-3439, 2022.

80. Lum, D, and Chen, SE. Comparison of loaded countermovement jump with different variable resistance intensities on inducing post-activation potentiation. *J Sci Sport Exerc* 2:167-172, 2020.

81. MacIntosh, BR, Robillard, ME, and Tomaras, EK. Should postactivation potentiation be the goal of your warm-up? *Appl Physiol Nutr Metab* 37:546-550, 2012.

82. Maffiuletti, NA, Aagaard, P, Blazevich, AJ, Folland, J, Tillin, N, and Duchateau, J. Rate of force development: Physiological and methodological considerations. *Eur J Appl Physiol* 116:1091-1116, 2016.

83. Marcora, S, and Miller, MK. The effect of knee angle on the external validity of isometric measures of lower body neuromuscular function. *J Sports Sci* 18:313-319, 2000.

84. Masamoto, N, Larson, R, Gates, T, and Faigenbaum, A. Acute effects of plyometric exercise on maximum squat performance in male athletes. *J Strength Cond Res* 17:68-71, 2003.

85. Maughan, RJ, Watson, JS, and Weir, J. Strength and cross-sectional area of human skeletal muscle. *J Physiol* 338:37-49, 1983.

86. McMaster, DT, Cronin, J, and McGuigan, M. Forms of variable resistance training. *Strength Cond J* 31:50-64, 2009.

87. Merrigan, J, Borth, J, Taber, CB, Suchomel, TJ, and Jones, MT. Application of accentuated eccentric loading to elicit acute and chronic velocity and power improvements: A narrative review. *Int J Strength Cond* 2:1-16, 2022.

88. Merrigan, JJ, Tufano, JJ, Falzone, M, and Jones, MT. Effectiveness of accentuated eccentric loading: Contingent on concentric load. *Int J Sports Physiol Perform* 16:66-72, 2020.

89. Mina, MA, Blazevich, AJ, Giakas, G, and Kay, AD. The influence of variable resistance loading on subsequent free weight maximal back squat performance. *J Strength Cond Res* 28:2988-2995, 2014.

90. Mina, MA, Blazevich, AJ, Giannis, G, Seitz, LB, and Kay, AD. Chain-loaded variable resistance warm-up improves free-weight maximal back squat performance. *Eur J Sport Sci* 16:932-939, 2016.

91. Mina, MA, Blazevich, AJ, Tsatalas, T, Giakas, G, Seitz, LB, and Kay, AD. Variable, but not free-weight, resistance back squat exercise potentiates jump performance following a comprehensive task-specific warm-up. *Scand J Med Sci Sports* 29:380-392, 2019.

92. Minetti, AE. On the mechanical power of joint extensions as affected by the change in muscle force (or cross-sectional area), ceteris paribus. *Eur J Appl Physiol* 86:363-369, 2002.

93. Montalvo, S, Gruber, LD, Gonzalez, MP, Dietze-Hermosa, MS, and Dorgo, S. Effects of augmented eccentric load bench press training on one repetition maximum performance and electromyographic activity in trained powerlifters. *J Strength Cond Res* 35:1512-1519, 2021.

94. Moore, CA, and Schilling, BK. Theory and application of augmented eccentric loading. *Strength Cond J* 27:20-27, 2005.

95. Moore, CA, Weiss, LW, Schilling, BK, Fry, AC, and Li, Y. Acute effects of augmented eccentric loading on jump squat performance. *J Strength Cond Res* 21:372-377, 2007.

96. Munger, CN, Archer, DC, Leyva, WD, Wong, MA, Coburn, JW, Costa, PB, and Brown, LE. Acute effects of eccentric overload on concentric front squat performance. *J Strength Cond Res* 31:1192-1197, 2017.

97. Newton, RU, and Kraemer, WJ. Developing explosive muscular power: Implications for a mixed methods training strategy. *Strength Cond J* 16:20-31, 1994.

98. Nijem, RM, Coburn, JW, Brown, LE, Lynn, SK, and Ciccone, AB. Electromyographic and force plate analysis of the deadlift performed with and without chains. *J Strength Cond Res* 30:1177-1182, 2016.

99. Nilo Dos Santos, WD, Gentil, P, Lima de Araújo Ribeiro, A, Vieira, CA, and Martins, WR. Effects of variable resistance training on maximal strength: A meta-analysis. *J Strength Cond Res* 32:e52-e55, 2018.

100. Nuzzo, JL, Pinto, MD, Nosaka, K, and Steele, J. The eccentric:concentric strength ratio of human skeletal muscle in vivo: Meta-analysis of the influences of sex, age, joint action, and velocity. *Sports Med* 53:1125-1136, 2023.

101. O'Leary, DD, Hope, K, and Sale, DG. Posttetanic potentiation of human dorsiflexors. *J Appl Physiol* 83:2131-2138, 1997.

102. Ojasto, T, and Häkkinen, K. Effects of different accentuated eccentric load levels in eccentric-concentric actions on acute neuromuscular, maximal force, and power responses. *J Strength Cond Res* 23:996-1004, 2009.

103. Paditsaeree, K, Intiraporn, C, and Lawsirirat, C. Comparison between the effects of combining elastic and free-weight resistance and free-weight resistance on force and power production. *J Strength Cond Res* 30:2713-2722, 2016.

104. Peng, H-T, Zhan, D-W, Song, C-Y, Chen, Z-R, Gu, C-Y, Wang, I-L, and Wang, L-I. Acute effects of squats using elastic bands on postactivation potentiation. *J Strength Cond Res* 35:3334-3340, 2021.

105. Read, P, Miller, SC, and Turner, AN. The effects of post activation potentiation on golf club head speed. *J Strength Cond Res* 27:1579-1582, 2012.

106. Rhea, MR, Kenn, JG, and Dermody, BM. Alterations in speed of squat movement and the use of accommodated resistance among college athletes training for power. *J Strength Cond Res* 23:2645-2650, 2009.

107. Rønnestad, BR. Acute effects of various whole body vibra-

tion frequencies on 1RM in trained and untrained subjects. *J Strength Cond Res* 23:2068-2072, 2009.
108. Rønnestad, BR, Holden, G, Samnoy, LE, and Paulsen, G. Acute effect of whole-body vibration on power, one-repetition maximum, and muscle activation in power lifters. *J Strength Cond Res* 26:531-539, 2012.
109. Saeterbakken, AH, Andersen, V, and Van den Tillaar, R. Comparison of kinematics and muscle activation in free-weight back squat with and without elastic bands. *J Strength Cond Res* 30:945-952, 2016.
110. Sarto, F, Franchi, MV, Rigon, PA, Grigoletto, D, Zoffoli, L, Zanuso, S, and Narici, MV. Muscle activation during leg-press exercise with or without eccentric overload. *Eur J Appl Physiol* 120:1651-1656, 2020.
111. Scott, DJ, Ditroilo, M, and Marshall, P. Effect of accommodating resistance on the postactivation potentiation response in rugby league players. *J Strength Cond Res* 32:2510-2520, 2018.
112. Seitz, LB, de Villarreal, ESS, and Haff, GG. The temporal profile of postactivation potentiation is related to strength level. *J Strength Cond Res* 28:706-715, 2014.
113. Seitz, LB, and Haff, GG. Factors modulating post-activation potentiation of jump, sprint, throw, and upper-body ballistic performances: A systematic review with meta-analysis. *Sports Med* 46:231-240, 2016.
114. Seitz, LB, Mina, MA, and Haff, GG. Postactivation potentiation of horizontal jump performance across multiple sets of a contrast protocol. *J Strength Cond Res* 30:2733-2740, 2016.
115. Sheppard, J, Hobson, S, Barker M, Taylor K, Chapman D, McGuigan M, and Newton RU. The effect of training with accentuated eccentric load counter-movement jumps on strength and power characteristics of high-performance volleyball players. *Int J Sports Sci Coach* 3:355-363, 2008.
116. Sheppard, JM, and Triplett, NT. Program design for resistance training. In *Essentials of Strength Training and Conditioning*. 4th ed. Haff, GG, and Triplett, NT, eds. Champaign, IL: Human Kinetics, 439-468, 2016.
117. Shi, L, Cai, Z, Chen, S, and Han, D. Acute effects of variable resistance training on force, velocity, and power measures: A systematic review and meta-analysis. *PeerJ* 10:e13870, 2022.
118. Shoepe, T, Ramirez, D, Rovetti, R, Kohler, D, and Almstedt, H. The effects of 24 weeks of resistance training with simultaneous elastic and free weight loading on muscular performance of novice lifters. *J Hum Kinet* 29:93-106, 2011.
119. Shoepe, TC, Ramirez, DA, and Almstedt, HC. Elastic band prediction equations for combined free-weight and elastic band bench presses and squats. *J Strength Cond Res* 24:195-200, 2010.
120. Smith, IC, and MacIntosh, BR. A comment on "A new taxonomy for postactivation potentiation in sport." *Int J Sports Physiol Perform* 16:163, 2021.
121. Spiteri, T, Nimphius, S, Hart, NH, Specos, C, Sheppard, JM, and Newton, RU. Contribution of strength characteristics to change of direction and agility performance in female basketball athletes. *J Strength Cond Res* 28:2415-2423, 2014.
122. Stevenson, MW, Warpeha, JM, Dietz, CC, Giveans, RM, and Erdman, AG. Acute effects of elastic bands during the free-weight barbell back squat exercise on velocity, power, and force production. *J Strength Cond Res* 24:2944-2954, 2010.

123. Stone, MH, Collins, D, Plisk, S, Haff, GG, and Stone, ME. Training principles: Evaluation of modes and methods of resistance training. *Strength Cond J* 22:65-76, 2000.
124. Stone, MH, Stone, M, and Sands, WA. *Principles and Practice of Resistance Training*. Champaign, IL: Human Kinetics, 2007.
125. Strokosch, A, Louit, L, Seitz, L, Clarke, R, and Hughes, JD. Impact of accommodating resistance in potentiating horizontal-jump performance in professional rugby league players. *Int J Sports Physiol Perform* 13:1223-1229, 2018.
126. Suchomel, TJ, Cantwell, CJ, Campbell, BA, Schroeder, ZS, Marshall, LK, and Taber, CB. Braking and propulsion phase characteristics of traditional and accentuated eccentric loaded back squats. *J Hum Kinet* 91:121-133, 2024.
127. Suchomel, TJ, Comfort, P, and Lake, JP. Enhancing the force-velocity profile of athletes using weightlifting derivatives. *Strength Cond J* 39:10-20, 2017.
128. Suchomel, TJ, Comfort, P, and Stone, MH. Weightlifting pulling derivatives: Rationale for implementation and application. *Sports Med* 45:823-839, 2015.
129. Suchomel, TJ, Lamont, HS, and Moir, GL. Understanding vertical jump potentiation: A deterministic model. *Sports Med* 46:809-828, 2016.
130. Suchomel, TJ, McKeever, SM, and Comfort, P. Training with weightlifting derivatives: The effects of force and velocity overload stimuli. *J Strength Cond Res* 34:1808-1818, 2020.
131. Suchomel, TJ, McKeever, SM, McMahon, JJ, and Comfort, P. The effect of training with weightlifting catching or pulling derivatives on squat jump and countermovement jump force-time adaptations. *J Funct Morphol Kinesiol* 5:28, 2020.
132. Suchomel, TJ, McKeever, SM, Nolen, JD, and Comfort, P. Muscle architectural and force-velocity curve adaptations following 10 weeks of training with weightlifting catching and pulling derivatives. *J Sports Sci Med* 21:504-516, 2022.
133. Suchomel, TJ, Nimphius, S, Bellon, CR, and Stone, MH. The importance of muscular strength: Training considerations. *Sports Med* 48:765-785, 2018.
134. Suchomel, TJ, Nimphius, S, and Stone, MH. The importance of muscular strength in athletic performance. *Sports Med* 46:1419-1449, 2016.
135. Suchomel, TJ, Sato, K, DeWeese, BH, Ebben, WP, and Stone, MH. Potentiation effects of half-squats performed in a ballistic or non-ballistic manner. *J Strength Cond Res* 30:1652-1660, 2016.
136. Suchomel, TJ, Sato, K, DeWeese, BH, Ebben, WP, and Stone, MH. Potentiation following ballistic and non-ballistic complexes: The effect of strength level. *J Strength Cond Res* 30:1825-1833, 2016.
137. Suchomel, TJ, Sato, K, DeWeese, BH, Ebben, WP, and Stone, MH. Relationships between potentiation effects following ballistic half-squats and bilateral symmetry. *Int J Sports Physiol Perform* 11:448-454, 2016.
138. Suchomel, TJ, and Sole, CJ. Force-time curve comparison between weightlifting derivatives. *Int J Sports Physiol Perform* 12:431-439, 2017.
139. Suchomel, TJ, and Sole, CJ. Power-time curve comparison between weightlifting derivatives. *J Sports Sci Med* 16:407-413, 2017.

140. Suchomel, TJ, Wagle, JP, Douglas, J, Taber, CB, Harden, M, Haff, GG, and Stone, MH. Implementing eccentric resistance training—Part 2: Practical recommendations. *J Funct Morphol Kinesiol* 4:55, 2019.

141. Suchomel, TJ, Wright, GA, Kernozek, TW, and Kline, DE. Kinetic comparison of the power development between power clean variations. *J Strength Cond Res* 28:350-360, 2014.

142. Swinton, PA, Stewart, AD, Keogh, JWL, Agouris, I, and Lloyd, R. Kinematic and kinetic analysis of maximal velocity deadlifts performed with and without the inclusion of chain resistance. *J Strength Cond Res* 25:3163-3174, 2011.

143. Taber, CB, Butler, C, Dabek, V, Kochan, B, McCormick, K, Petro, E, Suchomel, TJ, and Merrigan, J. The effects of accentuated eccentric loading on barbell and trap bar countermovement jumps. *Int J Strength Cond* 3:1-15, 2023.

144. Terzis, G, Karampatsos, G, Kyriazis, T, Kavouras, SA, and Georgiadis, G. Acute effects of countermovement jumping and sprinting on shot put performance. *J Strength Cond Res* 26:684-690, 2012.

145. Terzis, G, Spengos, K, Karampatsos, G, Manta, P, and Georgiadis, G. Acute effect of drop jumping on throwing performance. *J Strength Cond Res* 23:2592-2597, 2009.

146. Tesch, PA, and Karlsson, J. Muscle fiber types and size in trained and untrained muscles of elite athletes. *J Appl Physiol* 59:1716-1720, 1985.

147. Thorstensson, A, Grimby, G, and Karlsson, J. Force-velocity relations and fiber composition in human knee extensor muscles. *J Appl Physiol* 40:12-16, 1976.

148. Tillin, NA, and Bishop, D. Factors modulating post-activation potentiation and its effect on performance of subsequent explosive activities. *Sports Med* 39:147-166, 2009.

149. van Cutsem, M, Duchateau, J, and Hainaut, K. Changes in single motor unit behaviour contribute to the increase in contraction speed after dynamic training in humans. *J Physiol* 513:295-305, 1998.

150. van den Tillaar, R, Andersen, V, and Saeterbakken, AH. The existence of a sticking region in free weight squats. *J Hum Kinet* 42:63-71, 2014.

151. van den Tillaar, R, and Kwan, K. The effects of augmented eccentric loading upon kinematics and muscle activation in bench press performance. *J Funct Morphol Kinesiol* 5:8, 2020.

152. Vandenboom, R, Grange, RW, and Houston, ME. Threshold for force potentiation associated with skeletal myosin phosphorylation. *Am J Physiol* 265:C1456-C1462, 1993.

153. Vandenboom, R, Grange, RW, and Houston, ME. Myosin phosphorylation enhances rate of force development in fast-twitch skeletal muscle. *Am J Physiol* 268:C596-603, 1995.

154. Vandervoort, AA, Quinlan, J, and McComas, AJ. Twitch potentiation after voluntary contraction. *Exp Neurol* 81:141-152, 1983.

155. Vasconcelos, GC, Brietzke, C, Silva Cesario, JC, Bento Douetts, CD, Canestri, R, Vinicius, Í, Franco-Alvarenga, PE, and Pires, FO. No evidence of post-activation performance enhancement on endurance exercises: A comprehensive systematic review and meta-analysis. *Med Sci Sports Exerc* 56:315-327, 2024.

156. Wagle, JP, Cunanan, AJ, Carroll, KM, Sams, ML, Wetmore, A, Bingham, GE, Taber, CB, DeWeese, BH, Sato, K, and Stuart, CA. Accentuated eccentric loading and cluster set configurations in the back squat: A kinetic and kinematic analysis. *J Strength Cond Res* 35:420-427, 2021.

157. Wagle, JP, Taber, CB, Carroll, KM, Cunanan, AJ, Sams, ML, Wetmore, A, Bingham, GE, DeWeese, BH, Sato, K, and Stuart, CA. Repetition-to-repetition differences using cluster and accentuated eccentric loading in the back squat. *Sports* 6:59, 2018.

158. Wagle, JP, Taber, CB, Cunanan, AJ, Bingham, GE, Carroll, K, DeWeese, BH, Sato, K, and Stone, MH. Accentuated eccentric loading for training and performance: A review. *Sports Med* 47:2473-2495, 2017.

159. Walker, S, Blazevich, AJ, Haff, GG, Tufano, JJ, Newton, RU, and Häkkinen, K. Greater strength gains after training with accentuated eccentric than traditional isoinertial loads in already strength-trained men. *Front Physiol* 7:149, 2016.

160. Walker, S, Hulmi, JJ, Wernbom, M, Nyman, K, Kraemer, WJ, Ahtiainen, JP, and Häkkinen, K. Variable resistance training promotes greater fatigue resistance but not hypertrophy versus constant resistance training. *Eur J Appl Physiol* 113:2233-2244, 2013.

161. Wallace, BJ, Bergstrom, HC, and Butterfield, TA. Muscular bases and mechanisms of variable resistance training efficacy. *Int J Sports Sci Coach* 13:1177-1188, 2018.

162. Wallace, BJ, Winchester, JB, and McGuigan, MR. Effects of elastic bands on force and power characteristics during the back squat exercise. *J Strength Cond Res* 20:268-272, 2006.

163. Wilk, M, Golas, A, Krzysztofik, M, Nawrocka, M, and Zajac, A. The effects of eccentric cadence on power and velocity of the bar during the concentric phase of the bench press movement. *J Sports Sci Med* 18:191-197, 2019.

164. Wilson, JM, Duncan, NM, Marin, PJ, Brown, LE, Loenneke, JP, Wilson, SM, Jo, E, Lowery, RP, and Ugrinowitsch, C. Meta-analysis of postactivation potentiation and power: Effects of conditioning activity, volume, gender, rest periods, and training status. *J Strength Cond Res* 27:854-859, 2013.

165. Wyland, TP, Van Dorin, JD, and Reyes, GFC. Postactivation potentation effects from accommodating resistance combined with heavy back squats on short sprint performance. *J Strength Cond Res* 29:3115-3123, 2015.

166. Zamparo, P, Minetti, A, and di Prampero, P. Interplay among the changes of muscle strength, cross-sectional area and maximal explosive power: Theory and facts. *Eur J Appl Physiol* 88:193-202, 2002.

167. Zatsiorsky, V. *Science and Practice of Strength Training*. Champaign, IL: Human Kinetics, 1995.

168. Zimmermann, HB, MacIntosh, BR, and Dal Pupo, J. Does postactivation potentiation (PAP) increase voluntary performance? *Appl Physiol Nutr Metab* 45:349-356, 2020.

Chapter 7

1. Aagaard, P, Simonsen, EB, Andersen, JL, Magnusson, P, and Dyhre-Poulsen, P. Increased rate of force development and neural drive of human skeletal muscle following resistance training. *J Appl Physiol* 93:1318-1326, 2002.

2. Adams, K, O'Shea, JP, O'Shea, KL, and Climstein, M. The effect of six weeks of squat, plyometric and squat-plyometric training on power production. *J Appl Sport Sci Res* 6:36-41, 1992.

3. Alcaraz, PE, Carlos-Vivas, J, Oponjuru, BO, and Martinez-Rodriguez, A. The effectiveness of resisted sled training (RST) for sprint performance: A systematic review and meta-analysis. *Sports Med* 48:2143-2165, 2018.

4. American College of Sports Medicine. American College of Sports Medicine position stand. Progression models in resistance training for healthy adults. *Med Sci Sports Exerc* 41:687-708, 2009.

5. Andersen, LL, Andersen, JL, Zebis, MK, and Aagaard, P. Early and late rate of force development: Differential adaptive responses to resistance training? *Scand J Med Sci Sports* 20:e162-e169, 2010.

6. Aubry, A, Hausswirth, C, Louis, J, Coutts, AJ, and Le Meur, Y. Functional overreaching: The key to peak performance during the taper? *Med Sci Sports Exerc* 46:1769-1777, 2014.

7. Augustsson, J, Esko, A, Thomeé, R, and Svantesson, U. Weight training of the thigh muscles using closed versus open kinetic chain exercises: A comparison of performance enhancement. *J Orthop Sports Phys Ther* 27:3-8, 1998.

8. Barker, M, Wyatt, TJ, Johnson, RL, Stone, MH, O'Bryant, HS, Poe, C, and Kent, M. Performance factors, psychological assessment, physical characteristics, and football playing ability. *J Strength Cond Res* 7:224-233, 1993.

9. Bazyler, CD, Mizuguchi, S, Harrison, AP, Sato, K, Kavanaugh, AA, DeWeese, BH, and Stone, MH. Changes in muscle architecture, explosive ability, and track and field throwing performance throughout a competitive season and following a taper. *J Strength Cond Res* 31:2785-2793, 2017.

10. Bazyler, CD, Mizuguchi, S, Sole, CJ, Suchomel, TJ, Sato, K, Kavanaugh, AA, DeWeese, BH, and Stone, MH. Jumping performance is preserved, but not muscle thickness in collegiate volleyball players after a taper. *J Strength Cond Res* 32:1029-1035, 2018.

11. Bazyler, CD, Sato, K, Wassinger, CA, Lamont, HS, and Stone, MH. The efficacy of incorporating partial squats in maximal strength training. *J Strength Cond Res* 28:3024-3032, 2014.

12. Behm, DG, and Anderson, KG. The role of instability with resistance training. *J Strength Cond Res* 20:716-722, 2006.

13. Behm, DG, Young, JD, Whitten, JH, Reid, JC, Quigley, PJ, Low, J, Li, Y, de Lima, C, Hodgson, DD, Chaouachi, A, Prieske, O, and Granacher, U. Effectiveness of traditional strength versus power training on muscle strength, power and speed with youth: A systematic review and meta-analysis. *Front Physiol* 8:423, 2017.

14. Blackburn, JR, and Morrissey, MC. The relationship between open and closed kinetic chain strength of the lower limb and jumping performance. *J Ortho Sports Phys Ther* 27:430-435, 1998.

15. Boffey, D, Clark, NW, and Fukuda, DH. Efficacy of rest redistribution during squats: Considerations for strength and sex. *J Strength Cond Res* 35:586-595, 2021.

16. Borst, SE, De Hoyos, DV, Garzarella, L, Vincent, K, Pollock, BH, Lowenthal, DT, and Pollock, ML. Effects of resistance training on insulin-like growth factor-I and IGF binding proteins. *Med Sci Sports Exerc* 33:648-653, 2001.

17. Bryanton, MA, Kennedy, MD, Carey, JP, and Chiu, LZF. Effect of squat depth and barbell load on relative muscular effort in squatting. *J Strength Cond Res* 26:2820-2828, 2012.

18. Buckner, SL, Dankel, SJ, Mattocks, KT, Jessee, MB, Mouser, JG, Counts, BR, and Loenneke, JP. The problem of muscle hypertrophy: Revisited. *Muscle Nerve* 54:1012-1014, 2016.

19. Burleson, MA, Jr., O'Bryant, HS, Stone, MH, Collins, MA, and Triplett-McBride, T. Effect of weight training exercise and treadmill exercise on post-exercise oxygen consumption. *Med Sci Sports Exerc* 30:518-522, 1998.

20. Byrd, R, Centry, R, and Boatwright, D. Effect of inter-repetition rest intervals in circuit weight training on PWC170 during arm-cranking exercise. *J Sports Med Phys Fitness* 28:336-340, 1988.

21. Cahill, MJ, Oliver, JL, Cronin, JB, Clark, K, Cross, MR, Lloyd, RS, and Lee, JE. Influence of resisted sled-pull training on the sprint force–velocity profile of male high-school athletes. *J Strength Cond Res* 34:2751-2759, 2020.

22. Cahill, MJ, Oliver, JL, Cronin, JB, Clark, KP, Cross, MR, and Lloyd, RS. Influence of resisted sled-push training on the sprint force-velocity profile of male high school athletes. *Scand J Med Sci Sports* 30:442-449, 2020.

23. Carroll, KM, Bazyler, CD, Bernards, JR, Taber, CB, Stuart, CA, DeWeese, BH, Sato, K, and Stone, MH. Skeletal muscle fiber adaptations following resistance training using repetition maximums or relative intensity. *Sports* 7:169, 2019.

24. Carroll, KM, Bernards, JR, Bazyler, CD, Taber, CB, Stuart, CA, DeWeese, BH, Sato, K, and Stone, MH. Divergent performance outcomes following resistance training using repetition maximums or relative intensity. *Int J Sports Physiol Perform* 14:46-54, 2019.

25. Carvajal-Espinoza, R, Talpey, S, and Salazar-Rojas, W. Effects of physical training on change of direction performance: A systematic review with meta-analysis. *Int J Sports Sci Coach* 18:1850-1866, 2023.

26. Chae, S, Bailey, CA, Hill, DW, McMullen, SM, Moses, SA, and Vingren, JL. Acute kinetic and kinematic responses to rest redistribution with heavier loads in resistance-trained men. *J Strength Cond Res* 37:987-993, 2023.

27. Chae, S, Hill, DW, Bailey, CA, Moses, SA, McMullen, SM, and Vingren, JL. Acute physiological and perceptual responses to rest redistribution with heavier loads in resistance-trained men. *J Strength Cond Res* 37:994-1000, 2023.

28. Clark, KP, and Weyand, PG. Are running speeds maximized with simple-spring stance mechanics? *J Appl Physiol* 117:604-615, 2014.

29. Collins, C. Resistance training, recovery and genetics: AMPD1 the gene for recovery. *J Athl Enhanc* 6:2, 2017.

30. Comfort, P, Haff, GG, Suchomel, TJ, Soriano, MA, Pierce, KC, Hornsby, WG, Haff, EE, Sommerfield, LM, Chavda, S, Morris, SJ, Fry, AC, and Stone, MH. National Strength and Conditioning Association position statement on weightlifting for sports performance. *J Strength Cond Res* 37:1163-1190, 2023.

31. Comfort, P, Jones, PA, Thomas, C, Dos'Santos, T, McMahon, JJ, and Suchomel, TJ. Changes in early and maximal isometric force production in response to moderate- and high-load strength and power training. *J Strength Cond Res* 36:593-599, 2022.

32. Comfort, P, Jones, PA, and Udall, R. The effect of load and sex on kinematic and kinetic variables during the mid-thigh clean pull. *Sports Biomech* 14:139-156, 2015.

33. Comfort, P, Udall, R, and Jones, PA. The effect of loading on kinematic and kinetic variables during the midthigh clean pull. *J Strength Cond Res* 26:1208-1214, 2012.

34. Comfort, P, Williams, R, Suchomel, TJ, and Lake, JP. A comparison of catch phase force–time characteristics during clean derivatives from the knee. *J Strength Cond Res* 31:1911-1918, 2017.

35. Cormie, P, McCaulley, GO, and McBride, JM. Power versus strength–power jump squat training: Influence on the load–power relationship. *Med Sci Sports Exerc* 39:996-1003, 2007.

36. Cormie, P, McCaulley, GO, Triplett, NT, and McBride, JM. Optimal loading for maximal power output during lower-body resistance exercises. *Med Sci Sports Exerc* 39:340-349, 2007.

37. Cormie, P, McGuigan, MR, and Newton, RU. Adaptations in athletic performance after ballistic power versus strength training. *Med Sci Sports Exerc* 42:1582-1598, 2010.

38. Cormier, P, Freitas, TT, Loturco, I, Turner, A, Virgile, A, Haff, GG, Blazevich, AJ, Agar-Newman, D, Henneberry, M, and Baker, DG. Within session exercise sequencing during programming for complex training: Historical perspectives, terminology, and training considerations. *Sports Med* 52:2371-2389, 2022.

39. Cormier, P, Freitas, TT, Rubio-Arias, JÁ, and Alcaraz, PE. Complex and contrast training: Does strength and power training sequence affect performance-based adaptations in team sports? A systematic review and meta-analysis. *J Strength Cond Res* 34:1461-1479, 2020.

40. Cortes, N, Onate, J, and Van Lunen, B. Pivot task increases knee frontal plane loading compared with sidestep and drop-jump. *J Sports Sci* 29:83-92, 2011.

41. Cuadrado-Peñafiel, V, Castaño-Zambudio, A, Martínez-Aranda, LM, González-Hernández, JM, Martín-Acero, R, and Jiménez-Reyes, P. Microdosing sprint distribution as an alternative to achieve better sprint performance in field hockey players. *Sensors* 23:650, 2023.

42. Cuevas-Aburto, J, Jukic, I, Chirosa-Ríos, LJ, González-Hernández, JM, Janicijevic, D, Barboza-González, P, Guede-Rojas, F, and García-Ramos, A. Effect of traditional, cluster, and rest redistribution set configurations on neuromuscular and perceptual responses during strength-oriented resistance training. *J Strength Cond Res* 36:1490-1497, 2022.

43. Cuthbert, M, Haff, GG, Arent, SM, Ripley, N, McMahon, JJ, Evans, M, and Comfort, P. Effects of variations in resistance training frequency on strength development in well-trained populations and implications for in-season athlete training: A systematic review and meta-analysis. *Sports Med* 51:1967-1982, 2021.

44. Cuthbert, M, Haff, GG, McMahon, JJ, Evans, M, and Comfort, P. Microdosing: A conceptual framework for use as programming strategy for resistance training in team sports. *Strength Cond J* 46:180-201, 2024.

45. Cuthbert, M, Ripley, NJ, McMahon, JJ, Evans, M, Haff, GG, and Comfort, P. The effect of Nordic hamstring exercise intervention volume on eccentric strength and muscle architecture adaptations: A systematic review and meta-analyses. *Sports Med* 50:83-99, 2020.

46. Dai, B, Garrett, WE, Gross, MT, Padua, DA, Queen, RM, and Yu, B. The effect of performance demands on lower extremity biomechanics during landing and cutting tasks. *J Sport Health Sci* 8:228-234, 2019.

47. Davies, T, Orr, R, Halaki, M, and Hackett, D. Effect of training leading to repetition failure on muscular strength: A systematic review and meta-analysis. *Sports Med* 46:487-502, 2016.

48. Davies, TB, Tran, DL, Hogan, CM, Haff, GG, and Latella, C. Chronic effects of altering resistance training set configurations using cluster sets: A systematic review and meta-analysis. *Sports Med* 51:707-736, 2021.

49. de Salles, BF, Simão, R, Miranda, F, da Silva Novaes, J, Lemos, A, and Willardson, JM. Rest interval between sets in strength training. *Sports Med* 39:765-777, 2009.

50. de Salles, BF, Simão, R, Miranda, H, Bottaro, M, Fontana, F, and Willardson, JM. Strength increases in upper and lower body are larger with longer inter-set rest intervals in trained men. *J Sci Med Sport* 13:429-433, 2010.

51. Denton, J, and Cronin, JB. Kinematic, kinetic, and blood lactate profiles of continuous and intraset rest loading schemes. *J Strength Cond Res* 20:528-534, 2006.

52. DeWeese, BH, Bellon, CR, Magrum, E, Taber, CB, and Suchomel, TJ. Strengthening the springs: Improving sprint performance via strength training. *Techniques* 9:8-20, 2016.

53. DeWeese, BH, Hornsby, G, Stone, M, and Stone, MH. The training process: Planning for strength–power training in track and field. Part 1: Theoretical aspects. *J Sport Health Sci* 4:308-317, 2015.

54. DeWeese, BH, Hornsby, G, Stone, M, and Stone, MH. The training process: Planning for strength–power training in track and field. Part 2: Practical and applied aspects. *J Sport Health Sci* 4:318-324, 2015.

55. DeWeese, BH, Sams, ML, and Serrano, AJ. Sliding toward Sochi—Part 1: A review of programming tactics used during the 2010-2014 quadrennial. *Natl Strength Cond Assoc Coach* 1:30-42, 2014.

56. DeWeese, BH, Sams, ML, and Serrano, AJ. Sliding toward Sochi—Part 2: A review of programming tactics used during the 2010-2014 quadrennial. *Natl Strength Cond Assoc Coach* 1:4-7, 2014.

57. Dias, I, de Salles, BF, Novaes, J, Costa, PB, and Simão, R. Influence of exercise order on maximum strength in untrained young men. *J Sci Med Sport* 13:65-69, 2010.

58. Dos'Santos, T, Thomas, C, and Jones, PA. How early should you brake during a 180° turn? A kinetic comparison of the antepenultimate, penultimate, and final foot contacts during a 505 change of direction speed test. *J Sports Sci* 39:395-405, 2021.

59. Ebben, WP, Fauth, ML, Garceau, LR, and Petushek, EJ. Kinetic quantification of plyometric exercise intensity. *J Strength Cond Res* 25:3288-3298, 2011.

60. Ebben, WP, Feldmann, CR, Vanderzanden, TL, Fauth, ML, and Petushek, EJ. Periodized plyometric training is effective for women, and performance is not influenced by the length of post-training recovery. *J Strength Cond Res* 24:1-7, 2010.

61. Ebben, WP, Suchomel, TJ, and Garceau, LR. The effect of plyometric training volume on jumping performance. In *Proceedings of the 32nd International Conference of Biomechanics in Sports, Johnson City, TN, USA, 12-16 July, 2014.* Holland, MA: International Society of Biomechanics in Sports, 566-569, 2014.

62. Eckard, TG, Padua, DA, Hearn, DW, Pexa, BS, and Frank, BS. The relationship between training load and injury in athletes: A systematic review. *Sports Med* 48:1929-1961, 2018.

63. Edman, KA. Double-hyperbolic nature of the force–velocity relation in frog skeletal muscle. *Adv Exp Med Biol* 226:643-652, 1988.

64. Fisher, JP, Carlson, L, Steele, J, and Smith, D. The effects of pre-exhaustion, exercise order, and rest intervals in a full-body resistance training intervention. *Appl Physiol Nutr Metab* 39:1265-1270, 2014.

65. Fleck, SJ, and Kraemer, WJ. *Designing Resistance Training Programs.* Champaign, IL: Human Kinetics, 2014.

66. Fowles, JR, and Green, HJ. Coexistence of potentiation and low-frequency fatigue during voluntary exercise in human skeletal muscle. *Can J Physiol Pharmacol* 81:1092-1100, 2003.

67. Gonzalez, AM, Ghigiarelli, JJ, Sell, KM, Shone, EW, Kelly, CF, and Mangine, GT. Muscle activation during resistance exercise at 70% and 90% 1-repetition maximum in resistance-trained men. *Muscle Nerve* 56:505-509, 2017.

68. Gorostiaga, EM, Navarro-Amézqueta, I, Calbet, JAL, Hellsten, Y, Cusso, R, Guerrero, M, Granados, C, González-Izal, M, Ibañez, J, and Izquierdo, M. Energy metabolism during repeated sets of leg press exercise leading to failure or not. *PLoS One* 7:e40621, 2012.

69. Gorostiaga, EM, Navarro-Amézqueta, I, Cusso, R, Hellsten, Y, Calbet, JAL, Guerrero, M, Granados, C, González-Izal, M, Ibáñez, J, and Izquierdo, M. Anaerobic energy expenditure and mechanical efficiency during exhaustive leg press exercise. *PLoS One* 5:e13486, 2010.

70. Gouvêa, AL, Fernandes, IA, César, EP, Silva, WAB, and Gomes, PSC. The effects of rest intervals on jumping performance: A meta-analysis on post-activation potentiation studies. *J Sports Sci* 31:459-467, 2013.

71. Grgic, J, Schoenfeld, BJ, Davies, TB, Lazinica, B, Krieger, JW, and Pedisic, Z. Effect of resistance training frequency on gains in muscular strength: A systematic review and meta-analysis. *Sports Med* 48:1207-1220, 2018.

72. Grgic, J, Schoenfeld, BJ, Skrepnik, M, Davies, TB, and Mikulic, P. Effects of rest interval duration in resistance training on measures of muscular strength: A systematic review. *Sports Med* 38:137-151, 2018.

73. Haff, GG. Roundtable discussion: Machines versus free weights. *Strength Cond J* 22:18-30, 2000.

74. Haff, GG. Quantifying workloads in resistance training: A brief review. *Prof Strength Cond* 19:31-40, 2010.

75. Haff, GG. Cluster sets: Current methods for introducing variation to training sets. Presented at National Strength and Conditioning Association National Conference, New Orleans, LA, 2016.

76. Haff, GG, and Harden, M. Cluster sets: Scientific background and practical application. In *Advanced Strength and Conditioning: An Evidence-Based Approach.* Turner, AN, and Comfort, P, eds. New York: Routledge, 213-231, 2022.

77. Haff, GG, Hobbs, RT, Haff, EE, Sands, WA, Pierce, KC, and Stone, MH. Cluster training: A novel method for introducing training program variation. *Strength Cond J* 30:67-76, 2008.

78. Haff, GG, and Nimphius, S. Training principles for power. *Strength Cond J* 34:2-12, 2012.

79. Haff, GG, Whitley, A, McCoy, LB, O'Bryant, HS, Kilgore, JL, Haff, EE, Pierce, K, and Stone, MH. Effects of different set configurations on barbell velocity and displacement during a clean pull. *J Strength Cond Res* 17:95-103, 2003.

80. Häkkinen, K. Neuromuscular fatigue and recovery in male and female athletes during heavy resistance exercise. *Int J Sports Med* 14:53-59, 1993.

81. Halson, SL. Monitoring training load to understand fatigue in athletes. *Sports Med* 44:139-147, 2014.

82. Hardee, JP, Lawrence, MM, Utter, AC, Triplett, NT, Zwetsloot, KA, and McBride, JM. Effect of inter-repetition rest on ratings of perceived exertion during multiple sets of the power clean. *Eur J Appl Physiol* 112:3141-3147, 2012.

83. Hardee, JP, Lawrence, MM, Zwetsloot, KA, Triplett, NT, Utter, AC, and McBride, JM. Effect of cluster set configurations on power clean technique. *J Sports Sci* 31:488-496, 2013.

84. Hardee, JP, Triplett, NT, Utter, AC, Zwetsloot, KA, and McBride, JM. Effect of interrepetition rest on power output in the power clean. *J Strength Cond Res* 26:883-889, 2012.

85. Harper, DJ, Morin, J-B, Carling, C, and Kiely, J. Measuring maximal horizontal deceleration ability using radar technology: Reliability and sensitivity of kinematic and kinetic variables. *Sports Biomech* 22:1192-1208, 2023.

86. Harris, GR, Stone, MH, O'Bryant, HS, Proulx, CM, and Johnson, RL. Short-term performance effects of high power, high force, or combined weight-training methods. *J Strength Cond Res* 14:14-20, 2000.

87. Haun, CT, Mumford, PW, Roberson, PA, Romero, MA, Mobley, CB, Kephart, WC, Anderson, RG, Colquhoun, RJ, Muddle, TWD, Luera, MJ, Mackey, CS, Pascoe, DD, Young, KC, Martin, JS, DeFreitas, JM, Jenkins, NDM, and Roberts, MD. Molecular, neuromuscular, and recovery responses to light versus heavy resistance exercise in young men. *Physiol Rep* 5:e13457, 2017.

88. Havens, KL, and Sigward, SM. Whole body mechanics differ among running and cutting maneuvers in skilled athletes. *Gait Posture* 42:240-245, 2015.

89. Hodges, NJ, Hayes, S, Horn, RR, and Williams, AM. Changes in coordination, control and outcome as a result of extended practice on a novel motor skill. *Ergonomics* 48:1672-1685, 2005.

90. Hodgson, M, Docherty, D, and Robbins, D. Post-activation potentiation: Underlying physiology and implications for motor performance *Sports Med* 35:585-595, 2005.

91. Hornsby, WG, Gentles, J, Comfort, P, Suchomel, TJ, Mizuguchi, S, and Stone, MH. Resistance training volume load with and without exercise displacement. *Sports* 6:137, 2018.

92. Hornsby, WG, Gentles, JA, MacDonald, CJ, Mizuguchi, S, Ramsey, MW, and Stone, MH. Maximum strength, rate of force development, jump height, and peak power alterations in weightlifters across five months of training. *Sports* 5:78, 2017.

93. Iglesias-Soler, E, Carballeira, E, Sanchez-Otero, T, Mayo, X, and Fernandez-del-Olmo, M. Performance of maximum number of repetitions with cluster-set configuration. *Int J Sports Physiol Perform* 9:637-642, 2014.

94. Izquierdo, M, González-Badillo, JJ, Häkkinen, K, Ibáñez, J, Kraemer, WJ, Altadill, A, Eslava, J, and Gorostiaga, EM. Effect of loading on unintentional lifting velocity declines during single sets of repetitions to failure during upper and lower extremity muscle actions. *Int J Sports Med* 27:718-724, 2006.

95. James, LP, Comfort, P, Suchomel, TJ, Kelly, VG, Beckman, EM, and Haff, GG. The impact of power clean ability and training age on adaptations to weightlifting-style training. *J Strength Cond Res* 36:1560-1567, 2022.

96. James, LP, Haff, GG, Kelly, VG, Connick, M, Hoffman, B, and Beckman, EM. The impact of strength level on adaptations to combined weightlifting, plyometric and ballistic training. *Scand J Med Sci Sports* 28:1494-1505, 2018.

97. James, LP, Haycraft, J, Pierobon, A, Suchomel, TJ, and Connick, M. Mixed versus focused resistance training during an Australian football pre-season. *J Funct Morphol Kinesiol* 5:99, 2020.

98. James, LP, Talpey, SW, Young, WB, Geneau, MC, Newton, RU, and Gastin, PB. Strength classification and diagnosis: Not all strength is created equal. *Strength Cond J* 45:333-341, 2023.

99. Janicijevic, D, González-Hernández, JM, Jiménez-Reyes, P, Márquez, G, and García-Ramos, A. Longitudinal effects of traditional and rest redistribution set configurations on explosive-strength and strength–endurance manifestations. *J Strength Cond Res* 37:980-986, 2023.

100. Jensen, RL, and Ebben, WP. Quantifying plyometric intensity via rate of force development, knee joint, and ground reaction forces. *J Strength Cond Res* 21:763-767, 2007.

101. Jensen, RL, Flanagan, EP, and Ebben, WP. Rate of force development and time to peak force during plyometric exercises. In *Proceedings of the 26th International Conference of Biomechanics in Sports, Seoul, Korea, 14-18, July, 2008*.

102. Jensen, RL, Flanagan, EP, Jensen, NL, and Ebben, WP. Kinetic responses during landings of plyometric exercises. In *Proceedings of the 26th International Conference of Biomechanics in Sports, Seoul, Korea, 14-18, July, 2008*.

103. Jo, E, Judelson, DA, Brown, LE, Coburn, JW, and Dabbs, NC. Influence of recovery duration after a potentiating stimulus on muscular power in recreationally trained individuals. *J Strength Cond Res* 24:343-347, 2010.

104. Jones, A, and Wood, J. *Nautilus Training Principles: Bulletin No. 1*. Arthur Jones, 1970.

105. Jukic, I, Ramos, AG, Helms, ER, McGuigan, MR, and Tufano, JJ. Acute effects of cluster and rest redistribution set structures on mechanical, metabolic, and perceptual fatigue during and after resistance training: A systematic review and meta-analysis. *Sports Med* 50:2209-2236, 2020.

106. Jukic, I, and Tufano, JJ. Rest Redistribution functions as a free and ad-hoc equivalent to commonly used velocity-based training thresholds during clean pulls at different loads. *J Hum Kinet* 68:5-16, 2019.

107. Jukic, I, Van Hooren, B, Ramos, AG, Helms, ER, McGuigan, MR, and Tufano, JJ. The effects of set structure manipulation on chronic adaptations to resistance training: A systematic review and meta-analysis. *Sports Med* 51:1061-1086, 2021.

108. Kaneko, M, Fuchimoto, T, Toji, H, and Suei, K. Training effect of different loads on the force–velocity relationship and mechanical power output in human muscle. *Scand J Sports Sci* 5:50-55, 1983.

109. Kilduff, LP, Bevan, HR, Kingsley, MI, Owen, NJ, Bennett, MA, Bunce, PJ, Hore, AM, Maw, JR, and Cunningham, DJ. Postactivation potentiation in professional rugby players: Optimal recovery. *J Strength Cond Res* 21:1134-1138, 2007.

110. Kilen, A, Hjelvang, LB, Dall, N, Kruse, NL, and Nordsborg, NB. Adaptations to short, frequent sessions of endurance and strength training are similar to longer, less frequent exercise sessions when the total volume is the same. *J Strength Cond Res* 29:S46-S51, 2015.

111. Kipp, K, Comfort, P, and Suchomel, TJ. Comparing biomechanical time series data during the hang-power clean and jump shrug. *J Strength Cond Res* 35:2389-2396, 2021.

112. Kipp, K, Malloy, PJ, Smith, J, Giordanelli, MD, Kiely, MT, Geiser, CF, and Suchomel, TJ. Mechanical demands of the hang power clean and jump shrug: A joint-level perspective. *J Strength Cond Res* 32:466-474, 2018.

113. Kossow, AJ, and Ebben, WP. Kinetic analysis of horizontal plyometric exercise intensity. *J Strength Cond Res* 32:1222-1229, 2018.

114. Kraemer, WJ. Exercise prescription in weight training: A needs analysis. *Natl Strength Cond Assoc J* 5:64-65, 1983.

115. Kraemer, WJ. A series of studies-The physiological basis for strength training in American Football: Fact over philosophy. *J Strength Cond Res* 11:131-142, 1997.

116. Kraemer, WJ, Adams, K, Cafarelli, E, Dudley, GA, Dooly, C, Feigenbaum, MS, Fleck, SJ, Franklin, B, Fry, AC, Hoffman, JR, Newton, RU, Potteiger, J, Stone, MH, Ratamess, NA, and Triplett-McBride, T. American College of Sports Medicine position stand. Progression models in resistance training for healthy adults. *Med Sci Sports Exerc* 34:364-380, 2002.

117. Kraemer, WJ, and Newton, RU. Training for muscular power. *Phys Med Rehab Clin N Am* 11:341-368, 2000.

118. Kraemer, WJ, Ratamess, N, Fry, AC, Triplett-McBride, T, Koziris, LP, Bauer, JA, Lynch, JM, and Fleck, SJ. Influence of resistance training volume and periodization on physiological and performance adaptations in collegiate women tennis players. *Am J Sports Med* 28:626-633, 2000.

119. Kramer, JB, Stone, MH, O'Bryant, HS, Conley, MS, Johnson, RL, Nieman, DC, Honeycutt, DR, and Hoke, TP. Effects of single vs. multiple sets of weight training: Impact of volume, intensity, and variation. *J Strength Cond Res* 11:143-147, 1997.

120. Krieger, JW. Single versus multiple sets of resistance exercise: A meta-regression. *J Strength Cond Res* 23:1890-1901, 2009.

121. Lake, JP, Mundy, PD, Comfort, P, McMahon, JJ, Suchomel, TJ, and Carden, P. The effect of barbell load on vertical jump landing force–time characteristics. *J Strength Cond Res* 35:25-32, 2021.

122. Latella, C, Teo, WP, Drinkwater, EJ, Kendall, K, and Haff, GG. The acute neuromuscular responses to cluster set resistance training: A systematic review and meta-analysis. *Sports Med* 49:1861-1877, 2019.

123. Lawton, T, Cronin, J, Drinkwater, E, Lindsell, R, and Pyne, D. The effect of continuous repetition training and intra-set rest training on bench press strength and power. *J Sports Med Phys Fitness* 44:361-367, 2004.

124. Lopes dos Santos, M, Uftring, M, Stahl, CA, Lockie, RG, Alvar, B, Mann, JB, and Dawes, JJ. Stress in academic and athletic performance in collegiate athletes: A narrative review of sources and monitoring strategies. *Front Sports Act Living* 2:1-10, 2020.

125. Lopez, P, Radaelli, R, Taaffe, DR, Newton, RU, Galvão, DA, Trajano, GS, Teodoro, JL, Kraemer, WJ, Häkkinen, K, and Pinto, RS. Resistance training load effects on muscle hypertrophy and strength gain: Systematic review and network meta-analysis. *Med Sci Sports Exerc* 53:1206-1216, 2021.

126. Lundberg, TR, and Weckström, K. Fixture congestion modulates post-match recovery kinetics in professional soccer players. *Res Sports Med* 25:408-420, 2017.

127. Lyttle, AD, Wilson, GJ, and Ostrowski, KJ. Enhancing performance, maximal power versus combined weights and plyometrics training. *J Strength Cond Res* 10:173-179, 1996.

128. Mangine, GT, Hoffman, JR, Wang, R, Gonzalez, AM, Townsend, JR, Wells, AJ, Jajtner, AR, Beyer, KS, Boone, CH, and Miramonti, AA. Resistance training intensity and volume affect changes in rate of force development in resistance-trained men. *Eur J Appl Physiol* 116:2367-2374, 2016.

129. Marshall, J, Bishop, C, Turner, A, and Haff, GG. Optimal training sequences to develop lower body force, velocity, power, and jump height: A systematic review with meta-analysis. *Sports Med* 51:1245-1271, 2021.

130. Marshall, PWM, McEwen, M, and Robbins, DW. Strength and neuromuscular adaptation following one, four, and eight sets of high intensity resistance exercise in trained males. *Eur J Appl Physiol* 111:3007-3016, 2011.

131. Martínez-Cava, A, Hernandez-Belmonte, A, Courel-Ibanez, J, Moran-Navarro, R, Gonzalez-Badillo, JJ, and Pallarés, JG. Bench press at full range of motion produces greater neuromuscular adaptations than partial executions after prolonged resistance training. *J Strength Cond Res* 36:10-15, 2022.

132. Matveyev, LP, and Zdornyj, AP. *Fundamentals of Sports Training.* Moscow: Progress Publishers, 1981.

133. McBride, JM, Blaak, JB, and Triplett-McBride, T. Effect of resistance exercise volume and complexity on EMG, strength, and regional body composition. *Eur J Appl Physiol* 90:626-632, 2003.

134. McBride, JM, McCaulley, GO, Cormie, P, Nuzzo, JL, Cavill, MJ, and Triplett, NT. Comparison of methods to quantify volume during resistance exercise. *J Strength Cond Res* 23:106-110, 2009.

135. McCaulley, GO, McBride, JM, Cormie, P, Hudson, MB, Nuzzo, JL, Quindry, JC, and Triplett, NT. Acute hormonal and neuromuscular responses to hypertrophy, strength and power type resistance exercise. *Eur J Appl Physiol* 105:695-704, 2009.

136. McKeever, S, and Howard, R. Resistance training. In *NSCA's Guide to High School Strength and Conditioning.* McHenry, P, and Nitka, MJ, eds. Champaign, IL: Human Kinetics, 223-258, 2022.

137. McMahon, JJ, Lake, JP, Dos'Santos, T, Jones, PA, Thomasson, ML, and Comfort, P. Countermovement jump standards in rugby league: What is a "good" performance? *J Strength Cond Res* 36:1691-1698, 2022.

138. Meechan, D, McMahon, JJ, Suchomel, TJ, and Comfort, P. A comparison of kinetic and kinematic variables during the pull from the knee and hang pull, across loads. *J Strength Cond Res* 34:1819-1829, 2020.

139. Meechan, D, McMahon, JJ, Suchomel, TJ, and Comfort, P. The effect of rest redistribution on kinetic and kinematic variables during the countermovement shrug. *J Strength Cond Res* 37:1358-1366, 2023.

140. Meechan, D, Suchomel, TJ, McMahon, JJ, and Comfort, P. A comparison of kinetic and kinematic variables during the mid-thigh pull and countermovement shrug, across loads. *J Strength Cond Res* 34:1830-1841, 2020.

141. Melby, C, Scholl, C, Edwards, G, and Bullough, R. Effect of acute resistance exercise on postexercise energy expenditure and resting metabolic rate. *J Appl Physiol* 75:1847-1853, 1993.

142. Merrigan, JJ, Tufano, JJ, Fields, JB, Oliver, JM, and Jones, MT. Rest redistribution does not alter hormone responses in resistance-trained women. *J Strength Cond Res* 34:1867-1874, 2020.

143. Merrigan, JJ, Tufano, JJ, Oliver, JM, White, JB, Fields, JB, and Jones, MT. Reducing the loss of velocity and power in women athletes via rest redistribution. *Int J Sports Physiol Perform* 15:255-261, 2020.

144. Minetti, AE. On the mechanical power of joint extensions as affected by the change in muscle force (or cross-sectional area), ceteris paribus. *Eur J Appl Physiol* 86:363-369, 2002.

145. Miranda, H, Figueiredo, T, Rodrigues, B, Paz, GA, and Simão, R. Influence of exercise order on repetition performance among all possible combinations on resistance training. *Res Sports Med* 21:355-366, 2013.

146. Miyamoto, N, Wakahara, T, Ema, R, and Kawakami, Y. Further potentiation of dynamic muscle strength after resistance training. *Med Sci Sports Exerc* 45:1323-1330, 2013.

147. Moir, GL, Snyder, BW, Connaboy, C, Lamont, HS, and Davis, SE. Using drop jumps and jump squats to assess eccentric and concentric force–velocity characteristics. *Sports* 6:125, 2018.

148. Monteiro, W, Simão, R, and Farinatti, P. Manipulation of exercise order and its influence on the number of repetitions and effort subjective perception in trained women. *Rev Bras Med Esporte* 11:146-150, 2005.

149. Mookerjee, S, and Ratamess, N. Comparison of strength differences and joint action durations between full and partial

range-of-motion bench press exercise. *J Strength Cond Res* 13:76-81, 1999.

150. Moolyk, AN, Carey, JP, and Chiu, LZF. Characteristics of lower extremity work during the impact phase of jumping and weightlifting. *J Strength Cond Res* 27:3225-3232, 2013.

151. Morris, SJ, Oliver, JL, Pedley, JS, Haff, GG, and Lloyd, RS. Taking a long-term approach to the development of weightlifting ability in young athletes. *Strength Cond J* 42:71-90, 2020.

152. Naclerio, F, Faigenbaum, AD, Larumbe-Zabala, E, Perez-Bibao, T, Kang, J, Ratamess, NA, and Triplett, NT. Effects of different resistance training volumes on strength and power in team sport athletes. *J Strength Cond Res* 27:1832-1840, 2013.

153. Nagahara, R, Mizutani, M, Matsuo, A, Kanehisa, H, and Fukunaga, T. Association of sprint performance with ground reaction forces during acceleration and maximal speed phases in a single sprint. *J Appl Biomech* 34:104-110, 2018.

154. Nagatani, T, Haff, GG, Guppy, SN, and Kendall, KL. Practical application of traditional and cluster set configurations within a resistance training program. *Strength Cond J* 44:87-101, 2022.

155. Nagatani, T, Kendall, KL, Guppy, SN, and Haff, GG. Using cluster set configurations within a resistance training programme. *Prof Strength Cond* 65:7-17, 2022.

156. Nedergaard, NJ, Kersting, U, and Lake, M. Using accelerometry to quantify deceleration during a high-intensity soccer turning manoeuvre. *J Sports Sci* 32:1897-1905, 2014.

157. Newell, KM. Coordination, control, and skill. In *Advances in Psychology*. Goodman, D, Wilberg, RB, and Franks, IM, eds. Amsterdam, Netherlands: North-Holland, 295-317, 1985.

158. Newton, RU, and Kraemer, WJ. Developing explosive muscular power: Implications for a mixed methods training strategy. *Strength Cond J* 16:20-31, 1994.

159. Newton, RU, Kraemer, WJ, Häkkinen, K, Humphries, B, and Murphy, AJ. Kinematics, kinetics, and muscle activation during explosive upper body movements. *J Appl Biomech* 12:31-43, 1996.

160. Oliver, JM, Jagim, AR, Sanchez, AC, Mardock, MA, Kelly, KA, Meredith, HJ, Smith, GL, Greenwood, M, Parker, JL, and Riechman, SE. Greater gains in strength and power with intraset rest intervals in hypertrophic training. *J Strength Cond Res* 27:3116-3131, 2013.

161. Östenberg, A, Roos, E, Ekdah, C, and Roos, H. Isokinetic knee extensor strength and functional performance in healthy female soccer players. *Scand J Med Sci Sports* 8:257-264, 1998.

162. Pallarés, JG, Martínez-Cava, A, Courel-Ibáñez, J, González-Badillo, JJ, and Morán-Navarro, R. Full squat produces greater neuromuscular and functional adaptations and lower pain than partial squats after prolonged resistance training. *Eur J Sport Sci* 20:115-124, 2020.

163. Paulsen, G, Myklestad, D, and Raastad, T. The influence of volume of exercise on early adaptations to strength training. *J Strength Cond Res* 17:115-120, 2003.

164. Peterson, MD, Rhea, MR, and Alvar, BA. Applications of the dose–response for muscular strength development: A review of meta-analytic efficacy and reliability for designing training prescription. *J Strength Cond Res* 19:950-958, 2005.

165. Petrakos, G, Morin, J-B, and Egan, B. Resisted sled sprint training to improve sprint performance: A systematic review. *Sports Med* 46:381-400, 2016.

166. Phillips, MD, Mitchell, JB, Currie-Elolf, LM, Yellott, RC, and Hubing, KA. Influence of commonly employed resistance exercise protocols on circulating IL-6 and indices of insulin sensitivity. *J Strength Cond Res* 24:1091-1101, 2010.

167. Pincivero, DM, Lephart, SM, and Karunakara, RG. Effects of rest interval on isokinetic strength and functional performance after short-term high intensity training. *Br J Sports Med* 31:229-234, 1997.

168. Plisk, SS, and Jeffreys, I. Effective needs analysis and functional training principles. In *Strength and Conditioning for Sports Performance*. Jeffreys, I, and Moody, J, eds. New York: Routledge, 2021.

169. Rahimi, R. Effect of different rest intervals on the exercise volume completed during squat bouts. *J Sports Sci Med* 4:361-366, 2005.

170. Ralston, GW, Kilgore, L, Wyatt, FB, and Baker, JS. The effect of weekly set volume on strength gain: A meta-analysis. *Sports Med* 47:2585-2601, 2017.

171. Ralston, GW, Kilgore, L, Wyatt, FB, Buchan, D, and Baker, JS. Weekly training frequency effects on strength gain: A meta-analysis. *Sports Med Open* 4:1-24, 2018.

172. Rassier, DE, and Macintosh, BR. Coexistence of potentiation and fatigue in skeletal muscle. *Braz J Med Biol Res* 33:499-508, 2000.

173. Rhea, MR, Alvar, BA, Ball, SD, and Burkett, LN. Three sets of weight training superior to 1 set with equal intensity for eliciting strength. *J Strength Cond Res* 16:525-529, 2002.

174. Rhea, MR, Kenn, JG, Peterson, MD, Massey, D, Simão, R, Marin, PJ, Favero, M, Cardozo, D, and Krein, D. Joint-angle specific strength adaptations influence improvements in power in highly trained athletes. *Hum Mov* 17:43-49, 2016.

175. Robinson, JM, Stone, MH, Johnson, RL, Penland, CM, Warren, BJ, and Lewis, RD. Effects of different weight training exercise/rest intervals on strength, power, and high intensity exercise endurance. *J Strength Cond Res* 9:216-221, 1995.

176. Roll, F, and Omer, J. Football: Tulane football winter program. *Strength Cond J* 9:34-38, 1987.

177. Romano, N, Vilaça-Alves, J, Fernandes, HM, Saavedra, F, Paz, G, Miranda, H, Simão, R, Novaes, J, and Reis, V. Effects of resistance exercise order on the number of repetitions performed to failure and perceived exertion in untrained young males. *J Hum Kinet* 39:177-183, 2013.

178. Rønnestad, BR, Nymark, BS, and Raastad, T. Effects of in-season strength maintenance training frequency in professional soccer players. *J Strength Cond Res* 25:2653-2660, 2011.

179. Rooney, KJ, Herbert, RD, and Balnave, RJ. Fatigue contributes to the strength training stimulus. *Med Sci Sports Exerc* 26:1160-1164, 1994.

180. Sale, DG. Postactivation potentiation: Role in human performance. *Exerc Sport Sci Rev* 30:138-143, 2002.

181. Sanborn, K, Boros, R, Hruby, J, Schilling, B, O'Bryant, HS, Johnson, RL, Hoke, T, Stone, ME, and Stone, MH. Short-term performance effects of weight training with multiple sets not to failure vs. a single set to failure in women. *J Strength Cond Res* 14:328-331, 2000.

182. Schlumberger, A, Stec, J, and Schmidtbleicher, D. Single- vs. multiple-set strength training in women. *J Strength Cond Res* 15:284-289, 2001.

183. Schoenfeld, BJ, Contreras, B, Vigotsky, AD, and Peterson, M. Differential effects of heavy versus moderate loads on measures of strength and hypertrophy in resistance-trained men. *J Sports Sci Med* 15:715-722, 2016.

184. Schoenfeld, BJ, Grgic, J, Ogborn, D, and Krieger, JW. Strength and hypertrophy adaptations between low-versus high-load resistance training: A systematic review and meta-analysis. *J Strength Cond Res* 31:3508-3523, 2017.

185. Schoenfeld, BJ, Pope, ZK, Benik, FM, Hester, GM, Sellers, J, Nooner, JL, Schnaiter, JA, Bond-Williams, KE, Carter, AS, and Ross, CL. Longer inter-set rest periods enhance muscle strength and hypertrophy in resistance-trained men. *J Strength Cond Res* 30:1805-1812, 2016.

186. Schot, P, Dart, J, and Schuh, M. Biomechanical analysis of two change-of-direction maneuvers while running. *J Orthop Sports Phys Ther* 22:254-258, 1995.

187. Seitz, LB, de Villarreal, ESS, and Haff, GG. The temporal profile of postactivation potentiation is related to strength level. *J Strength Cond Res* 28:706-715, 2014.

188. Seitz, LB, and Haff, GG. Factors modulating post-activation potentiation of jump, sprint, throw, and upper-body ballistic performances: A systematic review with meta-analysis. *Sports Med* 46:231-240, 2016.

189. Sforzo, GA, and Touey, PR. Manipulating exercise order affects muscular performance during a resistance exercise training session. *J Strength Cond Res* 10:20-24, 1996.

190. Sheppard, JM, and Triplett, NT. Program design for resistance training. In *Essentials of Strength Training and Conditioning*. 4th ed. Haff, GG, and Triplett, NT, eds. Champaign, IL: Human Kinetics, 439-468, 2016.

191. Siff, MC, and Verkhoshansky, Y. *Supertraining*. Denver, CO: Supertraining International, 1999.

192. Sigward, SM, Cesar, GM, and Havens, KL. Predictors of frontal plane knee moments during side-step cutting to 45 and 110 men and women: Implications for ACL injury. *Clin J Sport Med* 25:529-534, 2015.

193. Simão, R, de Salles, BF, Figueiredo, T, Dias, I, and Willardson, JM. Exercise order in resistance training. *Sports Med* 42:251-265, 2012.

194. Simão, R, de Tarso Veras Farinatti, P, Polito, MD, Maior, AS, and Fleck, SJ. Influence of exercise order on the number of repetitions performed and perceived exertion during resistance exercises. *J Strength Cond Res* 19:152-156, 2005.

195. Simão, R, de Tarso Veras Farinatti, P, Polito, MD, Viveiros, L, and Fleck, SJ. Influence of exercise order on the number of repetitions performed and perceived exertion during resistance exercise in women. *J Strength Cond Res* 21:23-28, 2007.

196. Simão, R, Spineti, J, de Salles, BF, Oliveira, LF, Matta, T, Miranda, F, Miranda, H, and Costa, PB. Influence of exercise order on maximum strength and muscle thickness in untrained men. *J Sports Sci Med* 9:1-7, 2010.

197. Soares, EG, Brown, LE, Gomes, WA, Corrêa, DA, Serpa, ÉP, da Silva, JJ, de Barros Vilela, G, Jr., Fioravanti, GZ, Aoki, MS, Lopes, CR, and Marchetti, PH. Comparison between pre-exhaustion and traditional exercise order on muscle activation and performance in trained men. *J Sports Sci Med* 15:111-117, 2016.

198. Soriano, MA, Suchomel, TJ, and Marin, PJ. The optimal load for improving maximal power production during upper-body exercises: A meta-analysis. *Sports Med* 47:757-768, 2017.

199. Spineti, J, de Salles, BF, Rhea, MR, Lavigne, D, Matta, T, Miranda, F, Fernandes, L, and Simão, R. Influence of exercise order on maximum strength and muscle volume in nonlinear periodized resistance training. *J Strength Cond Res* 24:2962-2969, 2010.

200. Stølen, T, Chamari, K, Castagna, C, and Wisløff, U. Physiology of soccer: An update. *Sports Med* 35:501-536, 2005.

201. Stone, MH, Collins, D, Plisk, S, Haff, GG, and Stone, ME. Training principles: Evaluation of modes and methods of resistance training. *Strength Cond J* 22:65-76, 2000.

202. Stone, MH, Cormie, P, Lamont, H, and Stone, ME. Developing strength and power. In *Strength and Conditioning for Sports Performance*. Jeffreys, I, and Moody, J, eds. New York: Routledge, 230-260, 2016.

203. Stone, MH, Hornsby, WG, Haff, GG, Fry, AC, Suarez, DG, Liu, J, Gonzalez-Rave, JM, and Pierce, KC. Periodization and block periodization in sports: Emphasis on strength–power training—A provocative and challenging narrative. *J Strength Cond Res* 35:2351-2371, 2021.

204. Stone, MH, Moir, G, Glaister, M, and Sanders, R. How much strength is necessary? *Phys Ther Sport* 3:88-96, 2002.

205. Stone, MH, O'Bryant, H, and Garhammer, J. A hypothetical model for strength training. *J Sports Med Phys Fitness* 21:342-351, 1981.

206. Stone, MH, Sands, WA, Pierce, KC, Ramsey, MW, and Haff, GG. Power and power potentiation among strength–power athletes: Preliminary study. *Int J Sports Physiol Perform* 3:55-67, 2008.

207. Stone, MH, Stone, M, and Sands, WA. *Principles and Practice of Resistance Training*. Champaign, IL: Human Kinetics, 2007.

208. Stone, MH, Suchomel, TJ, Hornsby, WG, Wagle, JP, and Cunanan, AJ. Exercise selection. In *Strength and Conditioning in Sports: From Science to Practice*. New York: Routledge, 252-272, 2022.

209. Suarez, DG, Mizuguchi, S, Hornsby, WG, Cunanan, AJ, Marsh, DJ, and Stone, MH. Phase-specific changes in rate of force development and muscle morphology throughout a block periodized training cycle in weightlifters. *Sports* 7:129, 2019.

210. Suchomel, TJ. The gray area of programming weightlifting exercises. *Natl Strength Cond Assoc Coach* 7:6-14, 2020.

211. Suchomel, TJ. Resistance training strategies to train the force–velocity characteristics of athletes. In *Central Virginia Sport Performance: The Manual*, Vol. 7. J DeMayo, ed. Amazon Publishing, 95-118, 2022.

212. Suchomel, TJ, Beckham, GK, and Wright, GA. Lower body kinetics during the jump shrug: Impact of load. *J Trainol* 2:19-22, 2013.

213. Suchomel, TJ, Beckham, GK, and Wright, GA. Effect of various loads on the force–time characteristics of the hang high pull. *J Strength Cond Res* 29:1295-1301, 2015.

214. Suchomel, TJ, and Comfort, P. Developing muscular strength and power. In *Advanced Strength and Conditioning: An Evidence-Based Approach*. Turner, A, and Comfort, P, eds. New York: Routledge, 13-39, 2022.

215. Suchomel, TJ, and Comfort, P. Weightlifting for sports performance. In *Advanced Strength and Conditioning: An Evidence-Based Approach*. Turner, A, and Comfort, P, eds. New York: Routledge, 283-306, 2022.

216. Suchomel, TJ, Comfort, P, and Lake, JP. Enhancing the force–velocity profile of athletes using weightlifting derivatives. *Strength Cond J* 39:10-20, 2017.

217. Suchomel, TJ, Comfort, P, and Stone, MH. Weightlifting pulling derivatives: Rationale for implementation and application. *Sports Med* 45:823-839, 2015.

218. Suchomel, TJ, Giordanelli, MD, Geiser, CF, and Kipp, K. Comparison of joint work during load absorption between weightlifting derivatives. *J Strength Cond Res* 35:S127-S135, 2021.

219. Suchomel, TJ, Lake, JP, and Comfort, P. Load absorption force–time characteristics following the second pull of weightlifting derivatives. *J Strength Cond Res* 31:1644-1652, 2017.

220. Suchomel, TJ, Lamont, HS, and Moir, GL. Understanding vertical jump potentiation: A deterministic model. *Sports Med* 46:809-828, 2016.

221. Suchomel, TJ, McKeever, SM, and Comfort, P. Training with weightlifting derivatives: The effects of force and velocity overload stimuli. *J Strength Cond Res* 34:1808-1818, 2020.

222. Suchomel, TJ, McKeever, SM, McMahon, JJ, and Comfort, P. The effect of training with weightlifting catching or pulling derivatives on squat jump and countermovement jump force–time adaptations. *J Funct Morphol Kinesiol* 5:28, 2020.

223. Suchomel, TJ, McKeever, SM, Nolen, JD, and Comfort, P. Muscle architectural and force–velocity curve adaptations following 10 weeks of training with weightlifting catching and pulling derivatives. *J Sports Sci Med* 21:504-516, 2022.

224. Suchomel, TJ, McKeever, SM, Sijuwade, O, and Carpenter, L. Propulsion phase characteristics of loaded jump variations in resistance-trained women. *Sports* 11:44, 2023.

225. Suchomel, TJ, McKeever, SM, Sijuwade, O, Carpenter, L, McMahon, JJ, Loturco, I, and Comfort, P. The effect of load placement on the power production characteristics of three lower extremity jumping exercises. *J Hum Kinet* 68:109-122, 2019.

226. Suchomel, TJ, Nimphius, S, Bellon, CR, Hornsby, WG, and Stone, MH. Training for muscular strength: Methods for monitoring and adjusting training intensity. *Sports Med* 51:2051-2066, 2021.

227. Suchomel, TJ, Nimphius, S, Bellon, CR, and Stone, MH. The importance of muscular strength: Training considerations. *Sports Med* 48:765-785, 2018.

228. Suchomel, TJ, Nimphius, S, and Stone, MH. The importance of muscular strength in athletic performance. *Sports Med* 46:1419-1449, 2016.

229. Suchomel, TJ, Sands, WA, and McNeal, JR. Comparison of static, countermovement, and drop jumps of the upper and lower extremities in U.S. junior national team male gymnasts. *Sci Gymnastics J* 8:15-30, 2016.

230. Suchomel, TJ, Sato, K, DeWeese, BH, Ebben, WP, and Stone, MH. Potentiation effects of half-squats performed in a ballistic or non-ballistic manner. *J Strength Cond Res* 30:1652-1660, 2016.

231. Suchomel, TJ, Sato, K, DeWeese, BH, Ebben, WP, and Stone, MH. Potentiation following ballistic and non-ballistic complexes: The effect of strength level. *J Strength Cond Res* 30:1825-1833, 2016.

232. Suchomel, TJ, Sato, K, DeWeese, BH, Ebben, WP, and Stone, MH. Relationships between potentiation effects following ballistic half-squats and bilateral symmetry. *Int J Sports Physiol Perform* 11:448-454, 2016.

233. Suchomel, TJ, and Sole, CJ. Force–time curve comparison between weightlifting derivatives. *Int J Sports Physiol Perform* 12:431-439, 2017.

234. Suchomel, TJ, and Stone, MH. The relationships between hip and knee extensor cross-sectional area, strength, power, and potentiation characteristics. *Sports* 5:66, 2017.

235. Suchomel, TJ, Taber, CB, Sole, CJ, and Stone, MH. Force–time differences between ballistic and non-ballistic half-squats. *Sports* 6:79, 2018.

236. Suchomel, TJ, Wagle, JP, Douglas, J, Taber, CB, Harden, M, Haff, GG, and Stone MH. Implementing eccentric resistance training—Part 1: A brief review of existing methods. *J Funct Morphol Kinesiol* 4:38, 2019.

237. Suchomel, TJ, Wagle, JP, Douglas, J, Taber, CB, Harden, M, Haff, GG, and Stone, MH. Implementing eccentric resistance training - Part 2: Practical recommendations. *J Funct Morphol Kinesiol* 4:55, 2019.

238. Suchomel, TJ, Wright, GA, Kernozek, TW, and Kline, DE. Kinetic comparison of the power development between power clean variations. *J Strength Cond Res* 28:350-360, 2014.

239. Sundstrup, E, Jakobsen, MD, Andersen, CH, Zebis, MK, Mortensen, OS, and Andersen, LL. Muscle activation strategies during strength training with heavy loading vs. repetitions to failure. *J Strength Cond Res* 26:1897-1903, 2012.

240. Takei, S, Hirayama, K, and Okada, J. Comparison of the power output between the hang power clean and hang high pull across a wide range of loads in weightlifters. *J Strength Cond Res* 35:S84-S88, 2021.

241. Thompson, SW, Lake, JP, Rogerson, D, Ruddock, A, and Barnes, A. Kinetics and kinematics of the free-weight back squat and loaded jump squat. *J Strength Cond Res* 37:1-8, 2023.

242. Thompson, SW, Rogerson, D, Ruddock, A, and Barnes, A. The effectiveness of two methods of prescribing load on maximal strength development: A systematic review. *Sports Med* 50:919-938, 2020.

243. Tillin, NA, and Bishop, D. Factors modulating post-activation potentiation and its effect on performance of subsequent explosive activities. *Sports Med* 39:147-166, 2009.

244. Toji, H, and Kaneko, M. Effect of multiple-load training on the force–velocity relationship. *J Strength Cond Res* 18:792-795, 2004.

245. Toji, H, Suei, K, and Kaneko, M. Effects of combined training programs on force–velocity relation and power output in human muscle. *Jpn J Phys Fit Sports Med* 44:439-446, 1995.

246. Toji, H, Suei, K, and Kaneko, M. Effects of combined training loads on relations among force, velocity, and power development. *Can J Appl Physiol* 22:328-336, 1997.

247. Tsoukos, A, Brown, LE, Terzis, G, Veligekas, P, and Bogdanis, GC. Potentiation of bench press throw performance using a heavy load and velocity-based repetition control. *J Strength Cond Res* 35:S72-S79, 2021.

248. Tufano, JJ, Brown, LE, and Haff, GG. Theoretical and practical aspects of different cluster set structures: A systematic review. *J Strength Cond Res* 31:848-867, 2017.

249. Tufano, JJ, Conlon, JA, Nimphius, S, Brown, LE, Banyard, HG, Williamson, BD, Bishop, LG, Hopper, AJ, and Haff, GG. Cluster sets permit greater mechanical stress without decreasing relative velocity. *Int J Sports Physiol Perform* 12:463-469, 2017.

250. Tufano, JJ, Conlon, JA, Nimphius, S, Brown, LE, Seitz, LB, Williamson, BD, and Haff, GG. Maintenance of velocity and power with cluster sets during high-volume back squats. *Int J Sports Physiol Perform* 11:885-892, 2016.

251. Tufano, JJ, Omcirk, D, Malecek, J, Pisz, A, Halaj, M, and Scott, BR. Traditional sets versus rest-redistribution: A laboratory-controlled study of a specific cluster set configuration at fast and slow velocities. *Appl Physiol Nutr Metab* 45:421-430, 2020.

252. Turner, AN, Comfort, P, McMahon, JJ, Bishop, C, Chavda, S, Read, P, Mundy, P, and Lake, JP. Developing powerful athletes, Part 1: Mechanical underpinnings. *Strength Cond J* 42:30-39, 2020.

253. Turner, AN, Comfort, P, McMahon, JJ, Bishop, C, Chavda, S, Read, P, Mundy, P, and Lake, JP. Developing powerful athletes, Part 2: Practical applications. *Strength Cond J* 43:23-31, 2021.

254. Vanrenterghem, J, Venables, E, Pataky, T, and Robinson, MA. The effect of running speed on knee mechanical loading in females during side cutting. *J Biomech* 45:2444-2449, 2012.

255. Vargas-Molina, S, Romance, R, Schoenfeld, BJ, Garcia, M, Petro, JL, Bonilla, DA, Kreider, RB, Martin-Rivera, F, and Benitez-Porres, J. Effects of cluster training on body composition and strength in resistance-trained men. *Isokinet Exerc Sci* 28:391-399, 2020.

256. Verheul, J, Nedergaard, NJ, Pogson, M, Lisboa, P, Gregson, W, Vanrenterghem, J, and Robinson, MA. Biomechanical loading during running: Can a two mass-spring-damper model be used to evaluate ground reaction forces for high-intensity tasks? *Sports Biomech* 20:571-582, 2021.

257. Viitasalo, JT, and Komi, PV. Effects of fatigue on isometric force- and relaxation-time characteristics in human muscle. *Acta Physiol Scand* 111:87-95, 1981.

258. Willardson, JM, and Burkett, LN. A comparison of 3 different rest intervals on the exercise volume completed during a workout. *J Strength Cond Res* 19:23-26, 2005.

259. Willardson, JM, and Burkett, LN. The effect of rest interval length on bench press performance with heavy vs. light loads. *J Strength Cond Res* 20:396-399, 2006.

260. Willardson, JM, and Burkett, LN. The effect of rest interval length on the sustainability of squat and bench press repetitions. *J Strength Cond Res* 20:400-403, 2006.

261. Willardson, JM, and Burkett, LN. The effect of different rest intervals between sets on volume components and strength gains. *J Strength Cond Res* 22:146-152, 2008.

262. Wilson, JM, Duncan, NM, Marin, PJ, Brown, LE, Loenneke, JP, Wilson, SM, Jo, E, Lowery, RP, and Ugrinowitsch, C. Meta-analysis of postactivation potentiation and power: Effects of conditioning activity, volume, gender, rest periods, and training status. *J Strength Cond Res* 27:854-859, 2013.

263. Wisløff, U, Castagna, C, Helgerud, J, Jones, R, and Hoff, J. Strong correlation of maximal squat strength with sprint performance and vertical jump height in elite soccer players. *Br J Sports Med* 38:285-288, 2004.

264. Wolfe, BL, Lemura, LM, and Cole, PJ. Quantitative analysis of single- vs. multiple-set programs in resistance training. *J Strength Cond Res* 18:35-47, 2004.

265. Wragg, CB, Maxwell, NS, and Doust, JH. Evaluation of the reliability and validity of a soccer-specific field test of repeated sprint ability. *Eur J Appl Physiol* 83:77-83, 2000.

266. Zając, A, Chalimoniuk, M, Maszczyk, A, Gołaś, A, and Lngfort, J. Central and peripheral fatigue during resistance exercise: A critical review. *J Hum Kinet* 49:159-169, 2015.

267. Zamparo, P, Minetti, A, and di Prampero, P. Interplay among the changes of muscle strength, cross-sectional area and maximal explosive power: Theory and facts. *Eur J Appl Physiol* 88:193-202, 2002.

268. Zaras, N, Stasinaki, A-N, Spiliopoulou, P, Mpampoulis, T, Hadjicharalambous, M, and Terzis, G. Effect of inter-repetition rest vs. traditional strength training on lower body strength, rate of force development, and muscle architecture. *Appl Sciences* 11:45, 2020.

269. Zatsiorsky, V. *Science and Practice of Strength Training.* Champaign, IL: Human Kinetics, 1995.

Chapter 8

1. Aagaard, P, and Andersen, JL. Effects of strength training on endurance capacity in top-level endurance athletes. *Scand J Med Sci Sports* 20:39-47, 2010.

2. Aagaard, P, Andersen, JL, Bennekou, M, Larsson, B, Olesen, JL, Crameri, R, Magnusson, SP, and Kjær, M. Effects of resistance training on endurance capacity and muscle fiber composition in young top-level cyclists. *Scand J Med Sci Sports* 21:e298-e307, 2011.

3. Abernethy, PJ. Influence of acute endurance activity on isokinetic strength. *J Strength Cond Res* 7:141-146, 1993.

4. Baar, K. Using molecular biology to maximize concurrent training. *Sports Med* 44:117-125, 2014.

5. Baker, D. Recent trends in high-intensity aerobic training for field sports. *Prof Strength Cond* 22:3-8, 2011.

6. Baker, D. Methods and progressions to improve the aerobic fitness of teenage field and court sport athletes. Presented at the Carroll University Sport Performance Institute Long-term

Athlete Development Clinic, Virtual Conference, February 18, 2023.

7. Baquet, G, Berthoin, S, Gerbeaux, M, and Van Praagh, E. High-intensity aerobic training during a 10 week one-hour physical education cycle: Effects on physical fitness of adolescents aged 11 to 16. *Int J Sports Med* 22:295-300, 2001.

8. Bartolomei, S, Sadres, E, Church, DD, Arroyo, E, Iii, JAG, Varanoske, AN, Wang, R, Beyer, KS, Oliveira, LP, and Stout, JR. Comparison of the recovery response from high-intensity and high-volume resistance exercise in trained men. *Eur J Appl Physiol* 117:1287-1298, 2017.

9. Bell, GJ, Syrotuik, D, Martin, TP, Burnham, R, and Quinney, HA. Effect of concurrent strength and endurance training on skeletal muscle properties and hormone concentrations in humans. *Eur J Appl Physiol* 81:418-427, 2000.

10. Bentley, DJ, Smith, PA, Davie, AJ, and Zhou, S. Muscle activation of the knee extensors following high intensity endurance exercise in cyclists. *Eur J Appl Physiol* 81:297-302, 2000.

11. Bentley, DJ, Zhou, S, and Davie, AJ. The effect of endurance exercise on muscle force generating capacity of the lower limbs. *J Sci Med Sport* 1:179-188, 1998.

12. Berryman, N, Mujika, I, Arvisais, D, Roubeix, M, Binet, C, and Bosquet, L. Strength training for middle- and long-distance performance: A meta-analysis. *Int J Sports Physiol Perform* 13:57-64, 2018.

13. Berryman, N, Mujika, I, and Bosquet, L. Concurrent training for sports performance: The 2 sides of the medal. *Int J Sports Physiol Perform* 14:279-285, 2019.

14. Berthoin, S, Gerbeaux, M, Turpin, E, Guerrin, F, Lensel-Corbeil, G, and Vandendorpe, F. Comparison of two field tests to estimate maximum aerobic speed. *J Sports Sci* 12:355-362, 1994.

15. Bishop, DJ, Bartlett, J, Fyfe, J, and Lee, M. Methodological considerations for concurrent training. In *Concurrent Aerobic and Strength Training: Scientific Basics and Practical Applications*. Schumann, M, and Rønnestad, BR, eds. Cham, Switzerland: Springer, 183-196, 2019.

16. Buchheit, M, and Laursen, P. Using HIIT weapons. In *Science and Application of High-Intensity Interval Training: Solutions to the Programming Puzzle*. Laursen, P, and Buchheit, M, eds. Champaign, IL: Human Kinetics, 73-118, 2019.

17. Buchheit, M, and Laursen, PB. High-intensity interval training, solutions to the programming puzzle. Part I: Cardiopulmonary emphasis. *Sports Med* 43:313-338, 2013.

18. Buchheit, M, and Laursen, PB. High-intensity interval training, solutions to the programming puzzle. Part II: Anaerobic energy, neuromuscular load and practical applications. *Sports Med* 43:927-954, 2013.

19. Burt, DG, and Twist, C. The effects of exercise-induced muscle damage on cycling time-trial performance. *J Strength Cond Res* 25:2185-2192, 2011.

20. Castagna, C, and Alvarez, JCB. Physiological demands of an intermittent futsal-oriented high-intensity test. *J Strength Cond Res* 24:2322-2329, 2010.

21. Chiwaridzo, M, Ferguson, GD, and Smits-Engelsman, B. A systematic review protocol investigating tests for physical or physiological qualities and game-specific skills commonly used in rugby and related sports and their psychometric properties. *Syst Rev* 5:1-9, 2016.

22. Chtara, M, Chamari, K, Chaouachi, M, Chaouachi, A, Koubaa, D, Feki, Y, Millet, GP, and Amri, M. Effects of intra-session concurrent endurance and strength training sequence on aerobic performance and capacity. *Br J Sports Med* 39:555-560, 2005.

23. Coffey, VG, and Hawley, JA. Concurrent exercise training: Do opposites distract? *J Physiol* 595:2883-2896, 2017.

24. Craig, BW, Lucas, J, Pohlman, R, and Stelling, H. The effects of running, weightlifting and a combination of both on growth hormone release. *J Strength Cond Res* 5:198-203, 1991.

25. Cunanan, AJ, DeWeese, BH, Wagle, JP, Carroll, KM, Sausaman, R, Hornsby, WG, Haff, GG, Triplett, NT, Pierce, KC, and Stone, MH. The general adaptation syndrome: A foundation for the concept of periodization. *Sports Med* 48:787-797, 2018.

26. de Salles Painelli, V, Alves, VT, Ugrinowitsch, C, Benatti, FB, Artioli, GG, Lancha, AH, Gualano, B, and Roschel, H. Creatine supplementation prevents acute strength loss induced by concurrent exercise. *Eur J Appl Physiol* 114:1749-1755, 2014.

27. de Souza, EO, Tricoli, V, Franchini, E, Paulo, AC, Regazzini, M, and Ugrinowitsch, C. Acute effect of two aerobic exercise modes on maximum strength and strength endurance. *J Strength Cond Res* 21:1286-1290, 2007.

28. Denadai, BS, de Aguiar, RA, de Lima, LCR, Greco, CC, and Caputo, F. Explosive training and heavy weight training are effective for improving running economy in endurance athletes: A systematic review and meta-analysis. *Sports Med* 47:545-554, 2017.

29. DeWeese, BH, Bellon, CR, Magrum, E, Taber, CB, and Suchomel, TJ. Strengthening the springs: Improving sprint performance via strength training. *Techniques* 9:8-20, 2016.

30. Doma, K, and Deakin, GB. The effects of combined strength and endurance training on running performance the following day. *Int J Sport Health Sci* 11:1-9, 2013.

31. Doma, K, and Deakin, GB. The effects of strength training and endurance training order on running economy and performance. *Appl Physiol Nutr Metab* 38:651-656, 2013.

32. Doma, K, Deakin, GB, Schumann, M, and Bentley, DJ. Training considerations for optimising endurance development: An alternate concurrent training perspective. *Sports Med* 49:669-682, 2019.

33. Doncaster, GG, and Twist, C. Exercise-induced muscle damage from bench press exercise impairs arm cranking endurance performance. *Eur J Appl Physiol* 112:4135-4142, 2012.

34. Dupont, G, Blondel, N, Lensel, G, and Berthoin, S. Critical velocity and time spent at a high level of for short intermittent runs at supramaximal velocities. *Can J Appl Physiol* 27:103-115, 2002.

35. Eddens, L, van Someren, K, and Howatson, G. The role of intra-session exercise sequence in the interference effect: A systematic review with meta-analysis. *Sports Med* 48:177-188, 2018.

36. Eihara, Y, Takao, K, Sugiyama, T, Maeo, S, Terada, M, Kanehisa, H, and Isaka, T. Heavy resistance training versus plyometric training for improving running economy and running time trial performance: A systematic review and meta-analysis. *Sports Med Open* 8:138, 2022.

37. Eklund, D, Pulverenti, T, Bankers, S, Avela, J, Newton, R, Schumann, M, and Häkkinen, K. Neuromuscular adaptations to different modes of combined strength and endurance training. *Int J Sports Med* 36:120-129, 2014.

38. Eklund, D, Schumann, M, Kraemer, WJ, Izquierdo, M, Taipale, RS, and Häkkinen, K. Acute endocrine and force responses and long-term adaptations to same-session combined strength and endurance training in women. *J Strength Cond Res* 30:164-175, 2016.

39. Ellefsen, S, and Baar, K. Proposed mechanisms underlying the interference effect. In *Concurrent Aerobic and Strength Training: Scientific Basics and Practical Applications.* Schumann, M, and Rønnestad, BR, eds. Cham, Switzerland: Springer, 89-98, 2019.

40. Fyfe, J, Buchheit, M, and Laursen, P. Incorporating HIIT into a concurrent training program. In *Science and Application of High-Intensity Interval Training: Solutions to the Programming Puzzle.* Laursen, P, and Buchheit, M, eds. Champaign, IL: Human Kinetics, 119-136, 2019.

41. Fyfe, JJ, Bartlett, JD, Hanson, ED, Stepto, NK, and Bishop, DJ. Endurance training intensity does not mediate interference to maximal lower-body strength gain during short-term concurrent training. *Front Physiol* 7:487, 2016.

42. Fyfe, JJ, Bishop, DJ, and Stepto, NK. Interference between concurrent resistance and endurance exercise: Molecular bases and the role of individual training variables. *Sports Med* 44:743-762, 2014.

43. Fyfe, JJ, and Loenneke, JP. Interpreting adaptation to concurrent compared with single-mode exercise training: Some methodological considerations. *Sports Med* 48:289-297, 2018.

44. Gao, J, and Yu, L. Effects of concurrent training sequence on $\dot{V}O_2$max and lower limb strength performance: A systematic review and meta-analysis. *Front Physiol* 14:1072679, 2023.

45. García-Pallarés, J, García-Fernández, M, Sánchez-Medina, L, and Izquierdo, M. Performance changes in world-class kayakers following two different training periodization models. *Eur J Appl Physiol* 110:99-107, 2010.

46. García-Pallarés, J, Sánchez-Medina, L, Carrasco, L, Díaz, A, and Izquierdo, M. Endurance and neuromuscular changes in world-class level kayakers during a periodized training cycle. *Eur J Appl Physiol* 106:629-638, 2009.

47. Glowacki, SP, Martin, SE, Maurer, A, Baek, W, Green, JS, and Crouse, SF. Effects of resistance, endurance, and concurrent exercise on training outcomes in men. *Med Sci Sports Exerc* 36:2119-2127, 2004.

48. Gomez-Piqueras, P, Gonzalez-Villora, S, Castellano, J, and Teoldo, I. Relation between the physical demands and success in professional soccer players. *J Hum Sport Exerc* 14:1-11, 2019.

49. Häkkinen, K, Alen, M, Kraemer, WJ, Gorostiaga, E, Izquierdo, M, Rusko, H, Mikkola, J, Häkkinen, A, Valkeinen, H, and Kaarakainen, E. Neuromuscular adaptations during concurrent strength and endurance training versus strength training. *Eur J Appl Physiol* 89:42-52, 2003.

50. Harper, DJ, Carling, C, and Kiely, J. High-intensity acceleration and deceleration demands in elite team sports competitive match play: A systematic review and meta-analysis of observational studies. *Sports Med* 49:1923-1947, 2019.

51. Harper, DJ, Sandford, GN, Clubb, J, Young, M, Taberner, M, Rhodes, D, Carling, C, and Kiely, J. Elite football of 2030 will not be the same as that of 2020: What has evolved and what needs to evolve? *Scand J Med Sci Sports* 31:493-494, 2021.

52. Hickson, RC. Interference of strength development by simultaneously training for strength and endurance. *Eur J Appl Physiol Occup Physiol* 45:255-263, 1980.

53. Hoff, J, Gran, A, and Helgerud, J. Maximal strength training improves aerobic endurance performance. *Scand J Med Sci Sports* 12:288-295, 2002.

54. Inoue, DS, Panissa, VLG, Monteiro, PA, Gerosa-Neto, J, Rossi, FE, Antunes, BMM, Franchini, E, Cholewa, JM, Gobbo, LA, and Lira, FS. Immunometabolic responses to concurrent training: The effects of exercise order in recreational weightlifters. *J Strength Cond Res* 30:1960-1967, 2016.

55. Jacobs, I, Esbjörnsson, M, Sylvén, C, Holm, I, and Jansson, E. Sprint training effects on muscle myoglobin, enzymes, fiber types, and blood lactate. *Med Sci Sports Exerc* 19:368-374, 1987.

56. Jacobs, I, Kaiser, P, and Tesch, P. Muscle strength and fatigue after selective glycogen depletion in human skeletal muscle fibers. *Eur J Appl Physiol* 46:47-53, 1981.

57. Jones, TW, and Howatson, G. Immediate effects of endurance exercise on subsequent strength performance. In *Concurrent Aerobic and Strength Training: Scientific Basics and Practical Applications.* Schumann, M, and Rønnestad, BR, eds. Cham, Switzerland: Springer, 139-154, 2019.

58. Jones, TW, Howatson, G, Russell, M, and French, DN. Performance and neuromuscular adaptations following differing ratios of concurrent strength and endurance training. *J Strength Cond Res* 27:3342-3351, 2013.

59. Jones, TW, Howatson, G, Russell, M, and French, DN. Effects of strength and endurance exercise order on endocrine responses to concurrent training. *Eur J Sport Sci* 17:326-334, 2017.

60. Kraemer, WJ, Patton, JF, Gordon, SE, Harman, EA, Deschenes, MR, Reynolds, K, Newton, RU, Triplett, NT, and Dziados, JE. Compatibility of high-intensity strength and endurance training on hormonal and skeletal muscle adaptations. *J Appl Physiol* 78:976-989, 1995.

61. Laursen, P, and Buchheit, M, eds. *Science and Application of High-Intensity Interval Training: Solutions to the Programming Puzzle.* Champaign, IL: Human Kinetics, 2019.

62. Lee, MJC, Ballantyne, JK, Chagolla, J, Hopkins, WG, Fyfe, JJ, Phillips, SM, Bishop, DJ, and Bartlett, JD. Order of same-day concurrent training influences some indices of power development, but not strength, lean mass, or aerobic fitness in healthy, moderately-active men after 9 weeks of training. *PLoS One* 15:e0233134, 2020.

63. Leveritt, M, and Abernethy, PJ. Acute effects of high-intensity endurance exercise on subsequent resistance activity. *J Strength Cond Res* 13:47-51, 1999.

64. Leveritt, M, Abernethy, PJ, Barry, BK, and Logan, PA. Concurrent strength and endurance training: A review. *Sports Med* 28:413-427, 1999.

65. Leveritt, M, MacLaughlin, H, and Abernethy, PJ. Changes in leg strength 8 and 32 h after endurance exercise. *J Sports Sci* 18:865-871, 2000.

66. Llanos-Lagos, C, Ramirez-Campillo, R, Moran, J, and Sáez de Villarreal, E. Effect of strength training programs in middle- and long-distance runners' economy at different running speeds: A systematic review with meta-analysis. *Sports Med* 54:895-932, 2024.

67. Makhlouf, I, Castagna, C, Manzi, V, Laurencelle, L, Behm, DG, and Chaouachi, A. Effect of sequencing strength and endurance training in young male soccer players. *J Strength Cond Res* 30:841-850, 2016.

68. Mangine, GT, Hoffman, JR, Wang, R, Gonzalez, AM, Townsend, JR, Wells, AJ, Jajtner, AR, Beyer, KS, Boone, CH, and Miramonti, AA. Resistance training intensity and volume affect changes in rate of force development in resistance-trained men. *Eur J Appl Physiol* 116:2367-2374, 2016.

69. McCarthy, JP, Agre, JC, Graf, BK, Pozniak, MA, and Vailas, AC. Compatibility of adaptive responses with combining strength and endurance training. *Med Sci Sports Exerc* 27:429-436, 1995.

70. McCarthy, JP, Pozniak, MA, and Agre, JC. Neuromuscular adaptations to concurrent strength and endurance training. *Med Sci Sports Exerc* 34:511-519, 2002.

71. McGawley, K, and Andersson, P-I. The order of concurrent training does not affect soccer-related performance adaptations. *Int J Sports Med* 34:983-990, 2013.

72. Methenitis, S. A brief review on concurrent training: From laboratory to the field. *Sports* 6:127, 2018.

73. Mikkola, J, Rusko, H, Izquierdo, M, Gorostiaga, EM, and Häkkinen, K. Neuromuscular and cardiovascular adaptations during concurrent strength and endurance training in untrained men. *Int J Sports Med* 33:702-710, 2012.

74. Millet, GY, and Lepers, R. Alterations of neuromuscular function after prolonged running, cycling and skiing exercises. *Sports Med* 34:105-116, 2004.

75. Murlasits, Z, Kneffel, Z, and Thalib, L. The physiological effects of concurrent strength and endurance training sequence: A systematic review and meta-analysis. *J Sports Sci* 36:1212-1219, 2018.

76. Nader, GA. Concurrent strength and endurance training: From molecules to man. *Med Sci Sports Exerc* 38:1965-1970, 2006.

77. Nassis, GP, Massey, A, Jacobsen, P, Brito, J, Randers, MB, Castagna, C, Mohr, M, and Krustrup, P. Elite football of 2030 will not be the same as that of 2020: Preparing players, coaches, and support staff for the evolution. *Scand J Med Sci Sports* 30:962-964, 2020.

78. Paavolainen, L, Häkkinen, K, Hämäläinen, I, Nummela, A, and Rusko, H. Explosive-strength training improves 5-km running time by improving running economy and muscle power. *J Appl Physiol* 86:1527-1533, 1999.

79. Panissa, VLG, Julio, UF, Pinto e Silva, CM, Andreato, LV, Hardt, F, and Franchini, E. Effects of interval time between high-intensity intermittent aerobic exercise on strength performance: Analysis in individuals with different training background. *J Hum Sport Exerc* 7:815-825, 2012.

80. Panissa, VLG, Tricoli, VAA, Julio, UF, Ribeiro, N, de Azevedo Neto, RMA, Carmo, EC, and Franchini, E. Acute effect of high-intensity aerobic exercise performed on treadmill and cycle ergometer on strength performance. *J Strength Cond Res* 29:1077-1082, 2015.

81. Parra, J, Cadefau, JA, Rodas, G, Amigó, N, and Cussó, R. The distribution of rest periods affects performance and adaptations of energy metabolism induced by high-intensity training in human muscle. *Acta Physiol Scand* 169:157-165, 2000.

82. Petré, H, Hemmingsson, E, Rosdahl, H, and Psilander, N. Development of maximal dynamic strength during concurrent resistance and endurance training in untrained, moderately trained, and trained individuals: A systematic review and meta-analysis. *Sports Med* 51:991-1010, 2021.

83. Ratamess, NA, Kang, J, Porfido, TM, Ismaili, CP, Selamie, SN, Williams, BD, Kuper, JD, Bush, JA, and Faigenbaum, AD. Acute resistance exercise performance is negatively impacted by prior aerobic endurance exercise. *J Strength Cond Res* 30:2667-2681, 2016.

84. Reed, JP, Schilling, BK, and Murlasits, Z. Acute neuromuscular and metabolic responses to concurrent endurance and resistance exercise. *J Strength Cond Res* 27:793-801, 2013.

85. Reynolds, J, Connor, M, Jamil, M, and Beato, M. Quantifying and comparing the match demands of U18, U23, and 1st team English professional soccer players. *Front Physiol* 12:706451, 2021.

86. Robineau, J, Babault, N, Piscione, J, Lacome, M, and Bigard, AX. Specific training effects of concurrent aerobic and strength exercises depend on recovery duration. *J Strength Cond Res* 30:672-683, 2016.

87. Ross, A, and Leveritt, M. Long-term metabolic and skeletal muscle adaptations to short-sprint training: Implications for sprint training and tapering. *Sports Med* 31:1063-1082, 2001.

88. Sabag, A, Najafi, A, Michael, S, Esgin, T, Halaki, M, and Hackett, D. The compatibility of concurrent high intensity interval training and resistance training for muscular strength and hypertrophy: A systematic review and meta-analysis. *J Sports Sci* 36:2472-2483, 2018.

89. Sale, DG, Jacobs, I, MacDougall, JD, and Garner, S. Comparison of two regimens of concurrent strength and endurance training. *Med Sci Sports Exerc* 22:348-356, 1990.

90. Schumann, M, Feuerbacher, JF, Sünkeler, M, Freitag, N, Rønnestad, BR, Doma, K, and Lundberg, TR. Compatibility of concurrent aerobic and strength training for skeletal muscle size and function: An updated systematic review and meta-analysis. *Sports Med* 52:601-612, 2022.

91. Schumann, M, Küüsmaa, M, Newton, RU, Sirparanta, A-I, Syväoja, H, Häkkinen, A, and Häkkinen, K. Fitness and lean mass increases during combined training independent of loading order. *Med Sci Sports Exerc* 46:1758-1768, 2014.

92. Schumann, M, and Rønnestad, BR, eds. *Concurrent Aerobic and Strength Training: Scientific Basics and Practical Applications.* Cham, Switzerland: Springer, 2019.

93. Schumann, M, Walker, S, Izquierdo, M, Newton, RU, Kraemer, WJ, and Häkkinen, K. The order effect of combined endurance and strength loadings on force and hormone responses: Effects of prolonged training. *Eur J Appl Physiol* 114:867-880, 2014.

94. Seipp, D, Quittmann, OJ, Fasold, F, and Klatt, S. Concurrent training in team sports: A systematic review. *Int J Sports Sci Coach* 18:1342-1364, 2023.

95. Sporer, BC, and Wenger, HA. Effects of aerobic exercise on strength performance following various periods of recovery. *J Strength Cond Res* 17:638-644, 2003.

96. Stølen, T, Chamari, K, Castagna, C, and Wisløff, U. Physiology of soccer: An update. *Sports Med* 35:501-536, 2005.

97. Stone, MH, Hornsby, WG, Haff, GG, Fry, AC, Suarez, DG, Liu, J, Gonzalez-Rave, JM, and Pierce, KC. Periodization and block periodization in sports: Emphasis on strength-power training—A provocative and challenging narrative. *J Strength Cond Res* 35:2351-2371, 2021.

98. Suarez, DG, Mizuguchi, S, Hornsby, WG, Cunanan, AJ, Marsh, DJ, and Stone, MH. Phase-specific changes in rate of force development and muscle morphology throughout a block periodized training cycle in weightlifters. *Sports* 7:129, 2019.

99. Suchomel, TJ, and Comfort, P. Weightlifting for sports performance. In *Advanced Strength and Conditioning: An Evidence-Based Approach*. Turner, A, and Comfort, P, eds. New York: Routledge, 283-306, 2022.

100. Suchomel, TJ, Nimphius, S, Bellon, CR, Hornsby, WG, and Stone, MH. Training for muscular strength: Methods for monitoring and adjusting training intensity. *Sports Med* 51:2051-2066, 2021.

101. Suchomel, TJ, Nimphius, S, Bellon, CR, and Stone, MH. The importance of muscular strength: Training considerations. *Sports Med* 48:765-785, 2018.

102. Tabata, I, Nishimura, K, Kouzaki, M, Hirai, Y, Ogita, F, and Miyachi, M. Effects of moderate intensity-endurance and high intensity-intermittent training on anaerobic capacity and $\dot{V}O_2$max. Presented at the First Annual Congress of the European College of Sport Science: Frontiers in Sport Science, the European Perspective, Nice, France, May 28-31, 1996.

103. Taipale, RS, and Häkkinen, K. Acute hormonal and force responses to combined strength and endurance loadings in men and women: The "order effect." *PLoS One* 8:e55051, 2013.

104. Taipale, RS, Schumann, M, Mikkola, J, Nyman, K, Kyröläinen, H, Nummela, A, and Häkkinen, K. Acute neuromuscular and metabolic responses to combined strength and endurance loadings: The "order effect" in recreationally endurance trained runners. *J Sports Sci* 32:1155-1164, 2014.

105. Tan, JG, Coburn, JW, Brown, LE, and Judelson, DA. Effects of a single bout of lower-body aerobic exercise on muscle activation and performance during subsequent lower- and upper-body resistance exercise workouts. *J Strength Cond Res* 28:1235-1240, 2014.

106. Taylor, JL, and Gandevia, SC. A comparison of central aspects of fatigue in submaximal and maximal voluntary contractions. *J Appl Physiol* 104:542-550, 2008.

107. Thomas, K, Goodall, S, Stone, M, Howatson, G, Gibson, ASC, and Ansley, L. Central and peripheral fatigue in male cyclists after 4-, 20-, and 40-km time trials. *Med Sci Sports Exerc* 47:537-546, 2015.

108. Volpe, SL, Walberg-Rankin, J, Rodman, KW, and Sebolt, DR. The effect of endurance running on training adaptations in women participating in a weight lifting program. *J Strength Cond Res* 7:101-107, 1993.

109. Wilson, A, and Lichtwark, G. The anatomical arrangement of muscle and tendon enhances limb versatility and locomotor performance. *Philos Trans R Soc B Biol Sci* 366:1540-1553, 2011.

110. Wilson, JM, Marin, PJ, Rhea, MR, Wilson, SM, Loenneke, JP, and Anderson, JC. Concurrent training: A meta-analysis examining interference of aerobic and resistance exercises. *J Strength Cond Res* 26:2293-2307, 2012.

111. Wong, P-l, Chaouachi, A, Chamari, K, Dellal, A, and Wisloff, U. Effect of preseason concurrent muscular strength and high-intensity interval training in professional soccer players. *J Strength Cond Res* 24:653-660, 2010.

Chapter 9

1. Adams, GM. *Exercise Physiology Laboratory Manual*. Boston: McGraw Hill, 1998.

2. Ashby, BM, and Heegaard, JH. Role of arm motion in the standing long jump. *J Biomech* 35:1631-1637, 2002.

3. Baca, A. A comparison of methods for analyzing drop jump performance. *Med Sci Sports Exerc* 31:437-442, 1999.

4. Badby, AJ, Comfort, P, Mundy, PD, Lake, JP, and McMahon, JJ. Agreement among countermovement jump force-time variables obtained from a wireless dual force plate system and an industry gold standard system. Presented at the 40th International Society of Biomechanics in Sport Conference, Liverpool, UK, July 19-22, 2022.

5. Bailey, CA, Sato, K, Alexander, R, Chiang, C-Y, and Stone, MH. Isometric force production symmetry and jumping performance in collegiate athletes. *J Trainology* 2:1-5, 2013.

6. Bailey, CA, Sato, K, Burnett, A, and Stone, MH. Carry-over of force production symmetry in athletes of differing strength levels. *J Strength Cond Res* 29:3188-3196, 2015.

7. Bailey, CA, Sato, K, Burnett, A, and Stone, MH. Force production asymmetry in male and female athletes of differing strength levels. *Int J Sports Physiol Perform* 10:504-508, 2015.

8. Bailey, CA, Sato, K, and McInnis, TC. A technical report on reliability measurement in asymmetry studies. *J Strength Cond Res* 35:1779-1783, 2021.

9. Bailey, CA, Suchomel, TJ, Beckham, GK, Sole, CJ, and Grazer, JL. Reactive strength index-modified differences between baseball position players and pitchers. In *Proceedings of the 32nd International Conference of Biomechanics in Sports, Johnson City, TN, USA, 12-16 July, 2014*. Holland, MA: International Society of Biomechanics in Sports, 562-565, 2014.

10. Baker, D. Comparison of upper-body strength and power between professional and college-aged rugby league players. *J Strength Cond Res* 15:30-35, 2001.

11. Baker, D, Nance, S, and Moore, M. The load that maximizes the average mechanical power output during explosive bench press throws in highly trained athletes. *J Strength Cond Res* 15:20-24, 2001.

12. Baker, D, Nance, S, and Moore, M. The load that maximizes the average mechanical power output during jump squats in power-trained athletes. *J Strength Cond Res* 15:92-97, 2001.

13. Balsalobre-Fernández, C, García-Ramos, A, and Jiménez-Reyes, P. Load–velocity profiling in the military press exercise: Effects of gender and training. *Int J Sports Sci Coach* 13:743-750. 2018.

14. Banyard, HG, Nosaka, K, and Haff, GG. Reliability and validity of the load–velocity relationship to predict the 1RM back squat. *J Strength Cond Res* 31:1897-1904, 2017.

15. Barker, L, Siedlik, J, Magrini, M, Uesato, S, Wang, H, Sjovold, A, Ewing, G, and Harry, JR. Eccentric force velocity profiling: Motor control strategy considerations and relationships to strength and jump performance. *J Strength Cond Res* 37:574-580, 2023.

16. Bassa, E, Adamopoulos, I, Panoutsakopoulos, V, Xenofondos, A, Yannakos, A, Galazoulas, C, and Patikas, DA. Optimal drop height in prepubertal boys is revealed by the performance in squat jump. *Sports* 11:1, 2022.

17. Batterham, AM, and Hopkins, WG. Making meaningful inferences about magnitudes. *Int J Sports Physiol Perform* 1:50-57, 2006.

18. Bazyler, CD, Bailey, CA, Chiang, C-Y, Sato, K, and Stone, MH. The effects of strength training on isometric force production symmetry in recreationally trained males. *J Trainology* 3:6-10, 2014.

19. Bazyler, CD, Beckham, GK, and Sato, K. The use of the isometric squat as a measure of strength and explosiveness. *J Strength Cond Res* 29:1386-1392, 2015.

20. Beckham, GK, Mizuguchi, S, Carter, C, Sato, K, Ramsey, M, Lamont, H, Hornsby, G, Haff, G, and Stone, M. Relationships of isometric mid-thigh pull variables to weightlifting performance. *J Sports Med Phys Fitness* 53:573-581, 2013.

21. Beckham, GK, Sato, K, Santana, HAP, Mizuguchi, S, Haff, GG, and Stone, MH. Effect of body position on force production during the isometric midthigh pull. *J Strength Cond Res* 32:48-56, 2018.

22. Beckham, GK, Suchomel, TJ, Bailey, CA, Sole, CJ, and Grazer, JL. The relationship of the reactive strength index-modified and measures of force development in the isometric mid-thigh pull. In *Proceedings of the 32nd International Conference of Biomechanics in Sports, Johnson City, TN, USA, 12-16 July, 2014*. Holland, MA: International Society of Biomechanics in Sports, 501-504, 2014.

23. Beckham, GK, Suchomel, TJ, and Mizuguchi, S. Force plate use in performance monitoring and sport science testing. *New Stud Athl* 29:25-37, 2014.

24. Beckham, GK, Suchomel, TJ, Sole, CJ, Bailey, CA, Grazer, JL, Kim, SB, Talbot, KB, and Stone, MH. Influence of sex and maximum strength on reactive strength index-modified. *J Sports Sci Med* 18:65-72, 2019.

25. Benton, MJ, Raab, S, and Waggener, GT. Effect of training status on reliability of one repetition maximum testing in women. *J Strength Cond Res* 27:1885-1890, 2013.

26. Bevan, HR, Bunce, PJ, Owen, NJ, Bennett, MA, Cook, CJ, Cunningham, DJ, Newton, RU, and Kilduff, LP. Optimal loading for the development of peak power output in professional rugby players. *J Strength Cond Res* 24:43-47, 2010.

27. Bishop, C, Read, P, Lake, J, Loturco, I, Dawes, J, Madruga, M, Romero-Rodrigues, D, Chavda, S, and Turner, A. Unilateral isometric squat: Test reliability, interlimb asymmetries, and relationships with limb dominance. *J Strength Cond Res* 35:S144-S151, 2021.

28. Bishop, C, Read, P, McCubbine, J, and Turner, A. Vertical and horizontal asymmetries are related to slower sprinting and jump performance in elite youth female soccer players. *J Strength Cond Res* 35:56-63, 2021.

29. Bishop, C, Turner, A, Maloney, S, Lake, J, Loturco, I, Bromley, T, and Read, P. Drop jump asymmetry is associated with reduced sprint and change-of-direction speed performance in adult female soccer players. *Sports* 7:29, 2019.

30. Bishop, C, Turner, A, and Read, P. Effects of inter-limb asymmetries on physical and sports performance: A systematic review. *J Sports Sci* 36:1135-1144, 2018.

31. Blackburn, JR, and Morrissey, MC. The relationship between open and closed kinetic chain strength of the lower limb and jumping performance. *J Ortho Sports Phys Ther* 27:430-435, 1998.

32. Blazevich, AJ, Gill, N, and Newton, RU. Reliability and validity of two isometric squat tests. *J Strength Cond Res* 16:298-304, 2002.

33. Blazevich, AJ, and Gill, ND. Reliability of unfamiliar, multijoint, uni- and bilateral strength tests: Effects of load and laterality. *J Strength Cond Res* 20:226-230, 2006.

34. Bobbert, MF, and Casius, LJ. Is the effect of a countermovement on jump height due to active state development? *Med Sci Sports Exerc* 37:440-446, 2005.

35. Bobbert, MF, Gerritsen, KGM, Litjens, MCA, and Van Soest, AJ. Why is countermovement jump height greater than squat jump height? *Med Sci Sports Exerc* 28:1402-1412, 1996.

36. Brady, CJ, Harrison, AJ, and Comyns, TM. A review of the reliability of biomechanical variables produced during the isometric mid-thigh pull and isometric squat and the reporting of normative data. *Sports Biomech* 19:1-25, 2020.

37. Brady, CJ, Harrison, AJ, Flanagan, EP, Haff, GG, and Comyns, TM. A comparison of the isometric midthigh pull and isometric squat: Intraday reliability, usefulness, and the magnitude of difference between tests. *Int J Sports Physiol Perform* 13:844-852, 2018.

38. Buchheit, M, and Brown, M. Pre-season fitness testing in elite soccer: Integrating the 30–15 Intermittent Fitness Test into the weekly microcycle. *Sport Perform Sci Rep* 111:1-3, 2020.

39. Buckner, SL, Jessee, MB, Mattocks, KT, Mouser, JG, Counts, BR, Dankel, SJ, and Loenneke, JP. Determining strength: A case for multiple methods of measurement. *Sports Med* 47:193-195, 2017.

40. Burns, GT, Deneweth Zendler, J, and Zernicke, RF. Validation of a wireless shoe insole for ground reaction force measurement. *J Sports Sci* 37:1129-1138, 2019.

41. Byrne, PJ, Moran, K, Rankin, P, and Kinsella, S. A comparison of methods used to identify 'optimal' drop height for early phase adaptations in depth jump training. *J Strength Cond Res* 24:2050-2055, 2010.

42. Cabarkapa, D, Fry, AC, Cabarkapa, DV, Myers, CA, Jones, GT, and Deane, MA. Kinetic and kinematic characteristics of proficient and non-proficient 2-point and 3-point basketball shooters. *Sports* 10:2, 2021.
43. Carroll, KM, Bazyler, CD, Bernards, JR, Taber, CB, Stuart, CA, DeWeese, BH, Sato, K, and Stone, MH. Skeletal muscle fiber adaptations following resistance training using repetition maximums or relative intensity. *Sports* 7:169, 2019.
44. Carroll, KM, Bernards, JR, Bazyler, CD, Taber, CB, Stuart, CA, DeWeese, BH, Sato, K, and Stone, MH. Divergent performance outcomes following resistance training using repetition maximums or relative intensity. *Int J Sports Physiol Perform* 14:46-54, 2019.
45. Clark, KP, and Weyand, PG. Are running speeds maximized with simple-spring stance mechanics? *J Appl Physiol* 117:604-615, 2014.
46. Cometti, G, Maffiuletti, NA, Pousson, M, Chatard, JC, and Maffulli, N. Isokinetic strength and anaerobic power of elite, subelite and amateur French soccer players. *Int J Sports Med* 22:45-51, 2001.
47. Comfort, P, Dos'Santos, T, Beckham, GK, Stone, MH, Guppy, SN, and Haff, GG. Standardization and methodological considerations for the isometric midthigh pull. *Strength Cond J* 41:57-79, 2019.
48. Comfort, P, Fletcher, C, and McMahon, JJ. Determination of optimal loading during the power clean, in collegiate athletes. *J Strength Cond Res* 26:2970-2974, 2012.
49. Comfort, P, Jones, PA, and Udall, R. The effect of load and sex on kinematic and kinetic variables during the mid-thigh clean pull. *Sports Biomech* 14:139-156, 2015.
50. Comfort, P, and McMahon, JJ. Reliability of maximal back squat and power clean performances in inexperienced athletes. *J Strength Cond Res* 29:3089-3096, 2015.
51. Comfort, P, and Pearson, SJ. Scaling—Which methods best predict performance? *J Strength Cond Res* 28:1565-1572, 2014.
52. Comfort, P, Suchomel, TJ, and Stone, MH. Normalisation of early isometric force production as a percentage of peak force, during multi-joint isometric assessment. *Int J Sports Physiol Perform* 15:478-482, 2019.
53. Comfort, P, Thomas, C, Dos'Santos, T, Jones, PA, Suchomel, TJ, and McMahon, JJ. Comparison of methods of calculating dynamic strength index. *Int J Sports Physiol Perform* 13:320-325, 2018.
54. Comfort, P, Thomas, C, Dos'Santos, T, Suchomel, TJ, Jones, PA, and McMahon, JJ. Changes in dynamic strength index in response to strength training. *Sports* 6:176, 2018.
55. Comfort, P, Udall, R, and Jones, PA. The effect of loading on kinematic and kinetic variables during the midthigh clean pull. *J Strength Cond Res* 26:1208-1214, 2012.
56. Comyns, TM, Flanagan, EP, Fleming, S, Fitzgerald, E, and Harper, DJ. Inter-day reliability and usefulness of reactive strength index derived from two maximal rebound jump tests. *Int J Sports Physiol Perform* 14:1200-1204, 2019.
57. Cormie, P, McBride, JM, and McCaulley, GO. Power-time, force-time, and velocity-time curve analysis during the jump squat: Impact of load. *J Appl Biomech* 24:112-120, 2008.
58. Cormie, P, McCaulley, GO, Triplett, NT, and McBride, JM. Optimal loading for maximal power output during lower-body resistance exercises. *Med Sci Sports Exerc* 39:340-349, 2007.
59. Cormie, P, McGuigan, MR, and Newton, RU. Adaptations in athletic performance after ballistic power versus strength training. *Med Sci Sports Exerc* 42:1582-1598, 2010.
60. Cormie, P, McGuigan MR, and Newton, RU. Influence of strength on magnitude and mechanisms of adaptation to power training. *Med Sci Sports Exerc* 42:1566-1581, 2010.
61. Costley, L, Wallace, E, Johnston, M, and Kennedy, R. Reliability of bounce drop jump parameters within elite male rugby players. *J Sports Med Phys Fitness* 58:1390-1397, 2017.
62. Dai, B, Garrett, WE, Gross, MT, Padua, DA, Queen, RM, and Yu, B. The effect of performance demands on lower extremity biomechanics during landing and cutting tasks. *J Sport Health Sci* 8:228-234, 2019.
63. De Witt, JK, English, KL, Crowell, JB, Kalogera, KL, Guilliams, ME, Nieschwitz, BE, Hanson, AM, and Ploutz-Snyder, LL. Isometric midthigh pull reliability and relationship to deadlift one repetition maximum. *J Strength Cond Res* 32:528-533, 2018.
64. Donahue, PT, Hill, CM, Wilson, SJ, Williams, CC, and Garner, JC. Squat jump movement onset thresholds influence on kinetics and kinematics. *Int J Kinesiol Sports Sci* 9:1-7, 2021.
65. Dos'Santos, T, Jones, PA, Comfort, P, and Thomas, C. Effect of different onset thresholds on isometric midthigh pull force-time variables. *J Strength Cond Res* 31:3463-3473, 2017.
66. Dos'Santos, T, Thomas, C, and Jones, PA. The effect of angle on change of direction biomechanics: Comparison and inter-task relationships. *J Sports Sci* 39:2618-2631, 2021.
67. Dos'Santos, T, Thomas, C, Jones, PA, McMahon, JJ, and Comfort, P. The effect of hip joint angle on isometric midthigh pull kinetics. *J Strength Cond Res* 31:2748-2757, 2017.
68. Dos'Santos, T, Thomas, C, McBurnie, A, Comfort, P, and Jones, PA. Change of direction speed and technique modification training improves 180° turning performance, kinetics, and kinematics. *Sports (Basel)* 9:73, 2021.
69. Dos'Santos, T, Jones, PA, Kelly, J, McMahon, JJ, Comfort, P, and Thomas, C. Effect of sampling frequency on isometric midthigh-pull kinetics. *Int J Sports Physiol Perform* 14:525-530, 2019.
70. Dos'Santos, T, Thomas, C, Jones, PA, and Comfort, P. Assessing muscle-strength asymmetry via a unilateral-stance isometric midthigh pull. *Int J Sports Physiol Perform* 12:505-511, 2017.
71. Doyle, TLA. How effectively is the stretch-shortening cycle being used by athletes? *Strength Cond Coach* 13:7-12, 2005.
72. Drake, D, Kennedy, RA, and Wallace, ES. Measuring what matters in isometric multi-joint rate of force development. *J Sports Sci* 37:2667-2675, 2019.
73. Drake, D, Kennedy, RA, and Wallace, ES. Multi-joint rate of force development testing protocol affects reliability and the smallest detectible difference. *J Sports Sci* 37:1570-1581, 2019.
74. Ebben, WP, and Petushek, EJ. Using the reactive strength index modified to evaluate plyometric performance. *J Strength Cond Res* 24:1983-1987, 2010.

75. Edwards, T, Weakley, J, Woods, CT, Breed, R, Benson, AC, Suchomel, TJ, and Banyard, HG. Comparison of countermovement jump and squat jump performance between 627 state and non-state representative junior Australian football players. *J Strength Cond Res* 37:641-645, 2023.

76. Exell, T, Irwin, G, Gittoes, M, and Kerwin, D. Strength and performance asymmetry during maximal velocity sprint running. *Scand J Med Sci Sports* 27:1273-1282, 2017.

77. Faigenbaum, AD, McFarland, JE, Herman, RE, Naclerio, F, Ratamess, NA, Kang, J, and Myer, GD. Reliability of the one-repetition-maximum power clean test in adolescent athletes. *J Strength Cond Res* 26:432-437, 2012.

78. Fernandes, JFT, Dingley, AF, Garcia-Ramos, A, Perez-Castilla, A, Tufano, JJ, and Twist, C. Prediction of one repetition maximum using reference minimum velocity threshold values in young and middle-aged resistance-trained males. *Behav Sci* 11:71, 2021.

79. Flanagan, EP, and Comyns, TM. The use of contact time and the reactive strength index to optimize fast stretch-shortening cycle training. *Strength Cond J* 30:32-38, 2008.

80. Flanagan, EP, Ebben, WP, and Jensen, RL. Reliability of the reactive strength index and time to stabilization during depth jumps. *J Strength Cond Res* 22:1677-1682, 2008.

81. Gabbett, TJ, Kelly, J, Ralph, S, and Driscoll, D. Physiological and anthropometric characteristics of junior elite and sub-elite rugby league players, with special reference to starters and non-starters. *J Sci Med Sport* 12:215-222, 2009.

82. Gabbett, TJ, Nassis, GP, Oetter, E, Pretorius, J, Johnston, N, Medina, D, Rodas, G, Myslinski T, Howells D, and Beard A. The athlete monitoring cycle: A practical guide to interpreting and applying training monitoring data. *Br J Sports Med* 51:1451-1452, 2017.

83. García-López, J, Morante, JC, Ogueta-Alday, A, and Rodríguez-Marroyo, JA. The type of mat (Contact vs. Photocell) affects vertical jump height estimated from flight time. *J Strength Cond Res* 27:1162-1167, 2013.

84. Gathercole, R, Sporer, B, Stellingwerff, T, and Sleivert, G. Alternative countermovement-jump analysis to quantify acute neuromuscular fatigue. *Int J Sports Physiol Perform* 10:84-92, 2015.

85. Gehri, DJ, Ricard, MD, Kleiner, DM, and Kirkendall, DT. A comparison of plyometric training techniques for improving vertical jump ability and energy production. *J Strength Cond Res* 12:85-89, 1998.

86. Gillies, EM, Putman, CT, and Bell, GJ. The effect of varying the time of concentric and eccentric muscle actions during resistance training on skeletal muscle adaptations in women. *Eur J Appl Physiol* 97:443-453, 2006.

87. Gleason, BH, Suchomel, TJ, Brewer, C, McMahon, E, Lis, RP, and Stone, MH. Defining the sport scientist: Common specialties and subspecialties. *Strength Cond J* 46:18-27, 2024.

88. Gleason, BH, Suchomel, TJ, Brewer, C, McMahon, EL, Lis, RP, and Stone, MH. Defining the sport scientist. *Strength Cond J* 46:2-17, 2024.

89. González-Badillo, JJ, and Sánchez-Medina, L. Movement velocity as a measure of loading intensity in resistance training. *Int J Sports Med* 31:347-352, 2010.

90. Gonzalo-Skok, O, Tous-Fajardo, J, Suarez-Arrones, L, Arjol-Serrano, JL, Casajús, JA, and Mendez-Villanueva, A. Single-leg power output and between-limbs imbalances in team-sport players: Unilateral versus bilateral combined resistance training. *Int J Sports Physiol Perform* 12:106-114, 2017.

91. Greig, L, Aspe, RR, Hall, A, Comfort, P, Cooper, K, and Swinton, PA. The predictive validity of individualised load-velocity relationships for predicting 1RM: A systematic review and individual participant data meta-analysis. *Sports Med* 53:1693-1708, 2023.

92. Grgic, J, Lazinica, B, Schoenfeld, BJ, and Pedisic, Z. Test–retest reliability of the one-repetition maximum (1RM) strength assessment: A systematic review. *Sports Med Open* 6:1-16, 2020.

93. Haff, GG, Ruben, RP, Lider, J, Twine, C, and Cormie, P. A comparison of methods for determining the rate of force development during isometric midthigh clean pulls. *J Strength Cond Res* 29:386-395, 2015.

94. Haff, GG, Stone, MH, O'Bryant, HS, Harman, E, Dinan, C, Johnson, R, and Han, K-H. Force-time dependent characteristics of dynamic and isometric muscle actions. *J Strength Cond Res* 11:269-272, 1997.

95. Haff, GG, Whitley, A, McCoy, LB, O'Bryant, HS, Kilgore, JL, Haff, EE, Pierce, K, and Stone, MH. Effects of different set configurations on barbell velocity and displacement during a clean pull. *J Strength Cond Res* 17:95-103, 2003.

96. Haischer, MH, Krzyszkowski, J, Roche, S, and Kipp, K. Impulse-based dynamic strength index: Considering time-dependent force expression. *J Strength Cond Res* 35:1177-1181, 2021.

97. Häkkinen, K, Newton, RU, Walker, S, Häkkinen, A, Krapi, S, Rekola, R, Koponen, P, Kraemer, WJ, Haff, GG, and Blazevich, AJ. Effects of upper body eccentric versus concentric strength training and detraining on maximal force, muscle activation, hypertrophy and serum hormones in women. *J Sports Sci Med* 21:200-213, 2022.

98. Halperin, I, Williams, KJ, Martin, DT, and Chapman, DW. The effects of attentional focusing instructions on force production during the isometric midthigh pull. *J Strength Cond Res* 30:919-923, 2016.

99. Halson, SL. Monitoring training load to understand fatigue in athletes. *Sports Med* 44:139-147, 2014.

100. Hamilton, D. Drop jumps as an indicator of neuromuscular fatigue and recovery in elite youth soccer athletes following tournament match play. *J Aust Strength Cond* 17:3-8, 2009.

101. Hara, M, Shibayama, A, Takeshita, D, and Fukashiro, S. The effect of arm swing on lower extremities in vertical jumping. *J Biomech* 39:2503-2511, 2006.

102. Hara, M, Shibayama, A, Takeshita, D, Hay, DC, and Fukashiro, S. A comparison of the mechanical effect of arm swing and countermovement on the lower extremities in vertical jumping. *Hum Mov Sci* 27:636-648, 2008.

103. Harden, M, Wolf, A, Evans, M, Hicks, KM, Thomas, K, and Howatson, G. Four weeks of augmented eccentric loading using a novel leg press device improved leg strength in well-trained athletes and professional sprint track cyclists. *PLoS One* 15:e0236663, 2020.

104. Harper, DJ, Morin, J-B, Carling, C, and Kiely, J. Measuring maximal horizontal deceleration ability using radar technology: Reliability and sensitivity of kinematic and kinetic variables. *Sports Biomech* 22:1192-1208, 2023.

105. Harry, JR, Krzyszkowski, J, Chowning, LD, and Kipp, K. Phase-specific force and time predictors of standing long jump distance. *J Appl Biomech* 37:400-407, 2021.

106. Harry, JR, Paquette, MR, Schilling, BK, Barker, LA, James, CR, and Dufek, JS. Kinetic and electromyographic subphase characteristics with relation to countermovement vertical jump performance. *J Appl Biomech* 34:291-297, 2018.

107. Hawkins, SB, Doyle, TLA, and McGuigan, MR. The effect of different training programs on eccentric energy utilization in college-aged males. *J Strength Cond Res* 23:1996-2002, 2009.

108. Hoeger, WWK, Hopkins, DR, Barette, SL, and Hale, DF. Relationship between repetitions and selected percentages of one repetition maximum: A comparison between untrained and trained males and females. *J Strength Cond Res* 4:47-54, 1990.

109. Hopkins, WG. Measures of reliability in sports medicine and science. *Sports Med* 30:1-15, 2000.

110. Hornsby, WG, Gentles, JA, MacDonald, CJ, Mizuguchi, S, Ramsey, MW, and Stone, MH. Maximum strength, rate of force development, jump height, and peak power alterations in weightlifters across five months of training. *Sports* 5:78, 2017.

111. Impellizzeri, FM, and Marcora, SM. Test validation in sport physiology: Lessons learned from clinimetrics. *Int J Sports Physiol Perform* 4:269-277, 2009.

112. James, LP, and Comfort, P. The reliability of novel, temporal-based dynamic strength index metrics. *Sports Biomech*, 2022. [e-pub ahead of print].

113. James, LP, Haff, GG, Kelly, VG, Connick, M, Hoffman, B, and Beckman, EM. The impact of strength level on adaptations to combined weightlifting, plyometric and ballistic training. *Scand J Med Sci Sports* 28:1494-1505, 2018.

114. James, LP, Roberts, LA, Haff, GG, Kelly, VG, and Beckman, EM. Validity and reliability of a portable isometric mid-thigh clean pull. *J Strength Cond Res* 31:1378-1386, 2017.

115. James, LP, Talpey, SW, Young, WB, Geneau, MC, Newton, RU, and Gastin, PB. Strength classification and diagnosis: Not all strength is created equal. *Strength Cond J* 45:333-341, 2023.

116. James, LP, Weakley, J, Comfort, P, and Huynh, M. The relationship between isometric and dynamic strength following resistance training: A systematic review, meta-analysis, and level of agreement. *Int J Sports Physiol Perform* 19:2-12, 2023.

117. Jidovtseff, B, Harris, NK, Crielaard, J-M, and Cronin, JB. Using the load-velocity relationship for 1RM prediction. *J Strength Cond Res* 25:267-270, 2011.

118. Jiménez-Reyes, P, Samozino, P, Brughelli, M, and Morin, J-B. Effectiveness of an individualized training based on force-velocity profiling during jumping. *Front Physiol* 7:677, 2017.

119. Jiménez-Reyes, P, Samozino, P, Cuadrado-Peñafiel, V, Conceição, F, González-Badillo, JJ, and Morin, J-B. Effect of countermovement on power–force–velocity profile. *Eur J Appl Physiol* 114:2281-2288, 2014.

120. Jiménez-Reyes, P, Samozino, P, García-Ramos, A, Cuadrado-Peñafiel, V, Brughelli, M, and Morin, J-B. Relationship between vertical and horizontal force-velocity-power profiles in various sports and levels of practice. *PeerJ* 6:e5937, 2018.

121. Jiménez-Reyes, P, Samozino, P, Pareja-Blanco, F, Conceição, F, Cuadrado-Peñafiel, V, González-Badillo, JJ, and Morin, J-B. Validity of a simple method for measuring force-velocity-power profile in countermovement jump. *Int J Sports Physiol Perform* 12:36-43, 2017.

122. Julio, UF, Panissa, VLG, and Franchini, E. Prediction of one repetition maximum from the maximum number of repetitions with submaximal loads in recreationally strength-trained men. *Sci Sport* 27:e69-e76, 2012.

123. Kawamori, N, Crum, AJ, Blumert, PA, Kulik, JR, Childers, JT, Wood, JA, Stone, MH, and Haff, GG. Influence of different relative intensities on power output during the hang power clean: Identification of the optimal load. *J Strength Cond Res* 19:698-708, 2005.

124. Kenny, IC, Caireallàin, AÓ, and Comyns, TM. Validation of an electronic jump mat to assess stretch-shortening cycle function. *J Strength Cond Res* 26:1601-1608, 2012.

125. Khuu, S, Musalem, L, and Beach, TAC. Verbal instructions acutely affect drop vertical jump biomechanics-implications for athletic performance and injury risk assessments. *J Strength Cond Res* 29:2816-2826, 2015.

126. Kilduff, LP, Bevan, H, Owen, N, Kingsley, MI, Bunce, P, Bennett, M, and Cunningham, D. Optimal loading for peak power output during the hang power clean in professional rugby players. *Int J Sports Physiol Perform* 2:260-269, 2007.

127. Kipp, K, Kiely, MT, and Geiser, CF. The reactive strength index modified is a valid measure of explosiveness in collegiate female volleyball players. *J Strength Cond Res* 30:1341-1347, 2016.

128. Kipp, K, Kiely, MT, Giordanelli, MD, Malloy, PJ, and Geiser, CF. Biomechanical determinants of the reactive strength index during drop jumps. *Int J Sports Physiol Perform* 13:44-49, 2018.

129. Kipp, K, Malloy, PJ, Smith, J, Giordanelli, MD, Kiely, MT, Geiser, CF, and Suchomel, TJ. Mechanical demands of the hang power clean and jump shrug: A joint-level perspective. *J Strength Cond Res* 32:466-474, 2018.

130. Kotani, Y, Lake, J, Guppy, SN, Poon, W, Nosaka, K, and Haff, GG. Agreement in squat jump force-time characteristics between Smith machine and free-weight squat jump force-time characteristics. *J Strength Cond Res* 37:1955-1962, 2023.

131. Kotani, Y, Lake, JP, Guppy, SN, Poon, W, Nosaka, K, Hori, N, and Haff, GG. Reliability of the squat jump force-velocity and load-velocity profiles. *J Strength Cond Res* 36:3000-3007, 2022.

132. Kozinc, Ž, Pleša, J, and Šarabon, N. Questionable utility of the eccentric utilization ratio in relation to the performance of volleyball players. *Int J Environ Res Public Health* 18:11754, 2021.

133. Kozinc, Ž, Žitnik, J, Smajla, D, and Šarabon, N. The difference between squat jump and countermovement jump in 770 male and female participants from different sports. *Eur J Sport Sci* 22:985-993, 2022.

134. Kraemer, WJ, Ratamess, NA, Fry, AC, and French, DN. Strength training: Development and evaluation of methodology. In *Physiological Assessment of Human Fitness*. Maud, P, and Foster, C, eds. Champaign, IL: Human Kinetics, 119-150, 2006.

135. Krzyszkowski, J, Chowning, LD, and Harry, JR. Phase-specific predictors of countermovement jump performance that distinguish good from poor jumpers. *J Strength Cond Res* 36:1257-1263, 2022.

136. Lake, JP, Mundy, PD, Comfort, P, McMahon, JJ, Suchomel, TJ, and Carden, P. Concurrent validity of a portable force plate using vertical jump force-time characteristics. *J Appl Biomech* 34:410-413, 2018.

137. Lake, JP, Mundy, PD, Comfort, P, and Suchomel, TJ. Do the peak and mean force methods of assessing vertical jump force asymmetry agree? *Sports Biomech* 19:227-234, 2020.

138. Layer, JS, Grenz, C, Hinshaw, TJ, Smith, DT, Barrett, SF, and Dai, B. Kinetic analysis of isometric back squats and isometric belt squats. *J Strength Cond Res* 32:3301-3309, 2018.

139. Lees, A, Vanrenterghem, J, and Clercq, DD. Understanding how an arm swing enhances performance in the vertical jump. *J Biomech* 37:1929-1940, 2004.

140. Lees, A, Vanrenterghem, J, and De Clercq, D. The energetics and benefit of an arm swing in submaximal and maximal vertical jump performance. *J Sports Sci* 24:51-57, 2006.

141. Lindberg, K, Solberg, P, Bjørnsen, T, Helland, C, Rønnestad, B, Thorsen Frank, M, Haugen, T, Østerås, S, Kristoffersen, M, and Midttun, M. Force-velocity profiling in athletes: Reliability and agreement across methods. *PLoS One* 16:e0245791, 2021.

142. Lockie, RG, Callaghan, SJ, Berry, SP, Cooke, ERA, Jordan, CA, Luczo, TM, and Jeffriess, MD. Relationship between unilateral jumping ability and asymmetry on multidirectional speed in team-sport athletes. *J Strength Cond Res* 28:3557-3566, 2014.

143. Lockie, RG, Callaghan, SJ, Moreno, MR, Risso, FG, Liu, TM, Stage, AA, Birmingham-Babauta, SA, Stokes, JJ, Giuliano, DV, and Lazar, A. An investigation of the mechanics and sticking region of a one-repetition maximum close-grip bench press versus the traditional bench press. *Sports* 5:46, 2017.

144. Lockie, RG, Murphy, AJ, Knight, TJ, and de Jonge, XAKJ. Factors that differentiate acceleration ability in field sport athletes. *J Strength Cond Res* 25:2704-2714, 2011.

145. Loturco, I, Kobal, R, Moraes, JE, Kitamura, K, Cal Abad, CC, Pereira, LA, and Nakamura, FY. Predicting the maximum dynamic strength in bench press: The high precision of the bar velocity approach. *J Strength Cond Res* 31:1127-1131, 2017.

146. Loturco, I, McGuigan, M, Freitas, TT, Nakamura, F, Boullosa, D, Valenzuela, P, Pereira, LA, and Pareja-Blanco, F. Squat and countermovement jump performance across a range of loads: A comparison between Smith machine and free weight execution modes in elite sprinters. *Biol Sport* 39:1043-1048, 2022.

147. Loturco, I, Pereira, L, Kobal, R, Cal Abad, C, Fernandes, V, Ramirez-Campillo, R, and Suchomel, TJ. Portable force plates: A viable and practical alternative to rapidly and accurately monitor elite sprint performance. *Sports* 6:61, 2018.

148. Loturco, I, Pereira, LA, Abad, CCC, Gil, S, Kitamura, K, Kobal, R, and Nakamura, FY. Using bar velocity to predict maximum dynamic strength in the half-squat exercise. *Int J Sports Physiol Perform* 11:697-700, 2016.

149. Louder, T, Bressel, M, and Bressel, E. The kinetic specificity of plyometric training: Verbal cues revisited. *J Hum Kinet* 49:201-208, 2015.

150. Louder, T, Thompson, BJ, and Bressel, E. Association and agreement between reactive strength index and reactive strength index-modified scores. *Sports* 9:97, 2021.

151. Lum, D, and Aziz, AR. Relationship between isometric force-time characteristics and sprint kayaking performance. *Int J Sports Physiol Perform* 16:474-479, 2020.

152. Mackala, K, Stodólka, J, Siemienski, A, and Coh, M. Biomechanical analysis of standing long jump from varying starting positions. *J Strength Cond Res* 27:2674-2684, 2013.

153. Maloney, SJ. The relationship between asymmetry and athletic performance: A critical review. *J Strength Cond Res* 33:2579-2593, 2019.

154. Mann, JB, Bird, M, Signorile, JF, Brechue, WF, and Mayhew, JL. Prediction of anaerobic power from standing long jump in NCAA Division IA football players. *J Strength Cond Res* 35:1542-1546, 2021.

155. Marcora, S, and Miller, MK. The effect of knee angle on the external validity of isometric measures of lower body neuromuscular function. *J Sports Sci* 18:313-319, 2000.

156. Matic, MS, Pazin, NR, Mrdakovic, VD, Jankovic, NN, Ilic, DB, and Stefanovic, DLJ. Optimum drop height for maximizing power output in drop jump: The effect of maximal muscle strength. *J Strength Cond Res* 29:3300-3310, 2015.

157. Mayhew, JL, Ball, TE, and Bowen, JC. Prediction of bench press lifting ability from submaximal repetitions before and after training. *Sports Med Train Rehabil* 3:195-201, 1992.

158. Mayhew, JL, Johnson, BD, LaMonte, MJ, Lauber, D, and Kemmler, W. Accuracy of prediction equations for determining one repetition maximum bench press in women before and after resistance training. *J Strength Cond Res* 22:1570-1577, 2008.

159. McBride, JM, Haines, TL, and Kirby, TJ. Effect of loading on peak power of the bar, body, and system during power cleans, squats, and jump squats. *J Sports Sci* 29:1215-1221, 2011.

160. McClymont, D. Use of the reactive strength index (RSI) as an indicator of plyometric training conditions. In *Science and Football V: The Proceedings of the Fifth World Congress on Sports Science and Football*. Reilly, T, Cabri, J, and Araújo D, eds. New York: Routledge, 16, 2003.

161. McGuigan, M. Principles of test selection and administration. In *Essentials of Strength Training and Conditioning*. 4th ed. Haff, GG, and Triplett, NT, eds. Champaign, IL: Human Kinetics, 249-258, 2016.

162. McGuigan, MR, Cormack, S, and Newton, RU. Long-term power performance of elite Australian rules football players. *J Strength Cond Res* 23:26-32, 2009.

163. McGuigan, MR, Cormack, SJ, and Gill, ND. Strength and power profiling of athletes: Selecting tests and how to use the information for program design. *Strength Cond J* 35:7-14, 2013.

164. McGuigan, MR, La Doyle, T, Newton, M, Edwards, DJ, Nimphius, S, and Newton, RU. Eccentric utilization ratio: Effect of sport and phase of training. *J Strength Cond Res* 20:992-995, 2006.

165. McGuigan, MR, Newton, MJ, Winchester, JB, and Nelson, AG. Relationship between isometric and dynamic strength in recreationally trained men. *J Strength Cond Res* 24:2570-2573, 2010.

166. McMahon, JJ, Jones, PA, and Comfort, P. Comparison of countermovement jump–derived reactive strength index modified and underpinning force-time variables between super league and championship rugby league players. *J Strengt Cond Res* 36:226-231, 2022.

167. McMahon, JJ, Jones, PA, Dos'Santos, T, and Comfort, P. Influence of dynamic strength index on countermovement jump force-, power-, velocity-, and displacement-time curves. *Sports* 5:72, 2017.

168. McMahon, JJ, Jones, PA, Suchomel, TJ, Lake, JP, and Comfort, P. Influence of reactive strength index modified on force- and power-time curves. *Int J Sports Physiol Perform* 13:220-227, 2018.

169. McMahon, JJ, Lake, JP, and Comfort, P. Identifying and reporting position-specific countermovement jump outcome and phase characteristics within rugby league. *PLoS One* 17:e0265999, 2022.

170. McMahon, JJ, Lake, JP, Dos'Santos, T, Jones, PA, Thomasson, ML, and Comfort, P. Countermovement jump standards in rugby league: What is a "good" performance? *J Strength Cond Res* 36:1691-1698, 2022.

171. McMahon, JJ, Lake, JP, and Suchomel, TJ. Vertical jump testing. In *Performance Assessment in Strength and Conditioning*. Comfort, P, Jones, PA, and McMahon, JJ, eds. New York: Routledge, 96-116, 2019.

172. McMahon, JJ, Murphy, S, Rej, SJE, and Comfort, P. Countermovement-jump-phase characteristics of senior and academy rugby league players. *Int J Sports Physiol Perform* 12:803-811, 2017.

173. McMahon, JJ, Ripley, NJ, and Comfort, P. Force plate-derived countermovement jump normative data and benchmarks for professional rugby league players. *Sensors* 22:8669, 2022.

174. McMahon, JJ, Suchomel, TJ, Lake, JP, and Comfort, P. Understanding the key phases of the countermovement jump force-time curve. *Strength Cond J* 40:96-106, 2018.

175. McMahon, JJ, Suchomel, TJ, Lake, JP, and Comfort, P. Relationship between reactive strength index variants in rugby league players. *J Strength Cond Res* 35:280-285, 2021.

176. McMaster, DT, Gill, N, Cronin, J, and McGuigan, MR. A brief review of strength and ballistic assessment methodologies in sport. *Sports Med* 44:603-623, 2014.

177. Meechan, D, McMahon, JJ, Suchomel, TJ, and Comfort, P. A comparison of kinetic and kinematic variables during the pull from the knee and hang pull, across loads. *J Strength Cond Res* 34:1819-1829, 2020.

178. Meechan, D, McMahon, JJ, Suchomel, TJ, and Comfort, P. The effect of rest redistribution on kinetic and kinematic variables during the countermovement shrug. *J Strength Cond Res* 37:1358-1366, 2023.

179. Meechan, D, Suchomel, TJ, McMahon, JJ, and Comfort, P. A comparison of kinetic and kinematic variables during the mid-thigh pull and countermovement shrug, across loads. *J Strength Cond Res* 34:1830-1841, 2020.

180. Mero, A, Komi, PV, and Gregor, RJ. Biomechanics of sprint running. *Sports Med* 13:376-392, 1992.

181. Meyers, RW, Oliver, JL, Hughes, MG, Lloyd, RS, and Cronin, JB. Asymmetry during maximal sprint performance in 11- to 16-year-old boys. *Pediatr Exerc Sci* 29:94-102, 2017.

182. Miller, J, Comfort, P, and McMahon, JJ. *Laboratory Manual for Strength and Conditioning*. New York: Routledge, 2023.

183. Minetti, AE. On the mechanical power of joint extensions as affected by the change in muscle force (or cross-sectional area), ceteris paribus. *Eur J Appl Physiol* 86:363-369, 2002.

184. Moir, G, Sanders, R, Button, C, and Glaister, M. The influence of familiarization on the reliability of force variables measured during unloaded and loaded vertical jumps. *J Strength Cond Res* 19:140-145, 2005.

185. Moir, GL. Three different methods of calculating vertical jump height from force platform data in men and women. *Meas Phys Educ Exerc Sci* 12:207-218, 2008.

186. Moir, GL, Snyder, BW, Connaboy, C, Lamont, HS, and Davis, SE. Using drop jumps and jump squats to assess eccentric and concentric force-velocity characteristics. *Sports* 6:125, 2018.

187. Morales, J, and Sobonya, S. Use of submaximal repetition tests for predicting 1-RM strength in class athletes. *J Strength Cond Res* 10:186-189, 1996.

188. Moresi, MP, Bradshaw, EJ, Greene, D, and Naughton, G. The assessment of adolescent female athletes using standing and reactive long jumps. *Sports Biomech* 10:73-84, 2011.

189. Morin, J-B, and Samozino, P. Interpreting power-force-velocity profiles for individualized and specific training. *Int J Sports Physiol Perform* 11:267-272, 2016.

190. Morin, J-B, Samozino, P, Murata, M, Cross, MR, and Nagahara, R. A simple method for computing sprint acceleration kinetics from running velocity data: Replication study with improved design. *J Biomech* 94:82-87, 2019.

191. Morin, J-B, Slawinski, J, Dorel, S, Couturier, A, Samozino, P, Brughelli, M, and Rabita, G. Acceleration capability in elite sprinters and ground impulse: Push more, brake less? *J Biomech* 48:3149-3154, 2015.

192. Müller, S, Baur, H, König, T, Hirschmüller, A, and Mayer, F. Reproducibility of isokinetic single- and multi-joint strength measurements in healthy and injured athletes. *Isokinet Exerc Sci* 15:295-302, 2007.

193. Muñoz-López, M, Marchante, D, Cano-Ruiz, MA, Chicharro, JL, and Balsalobre-Fernández, C. Load, force and power-velocity relationships in the prone pull-up exercise. *Int J Sports Physiol Perform* 12:1249-1255, 2017.

194. Murphy, AJ, and Wilson, GJ. Poor correlations between isometric tests and dynamic performance: Relationship to muscle activation. *Eur J Appl Physiol Occup Physiol* 73:353-357, 1996.

195. Murphy, AJ, Wilson, GJ, Pryor, JF, and Newton, RU. Isometric assessment of muscular function: The effect of joint angle. *J Appl Biomech* 11:205-215, 1995.

196. Nagahara, R, Mizutani, M, Matsuo, A, Kanehisa, H, and Fukunaga, T. Association of sprint performance with ground reaction forces during acceleration and maximal speed phases in a single sprint. *J Appl Biomech* 34:104-110, 2018.

197. Nagahara, R, and Morin, J-B. Sensor insole for measuring temporal variables and vertical force during sprinting. *Proc Inst Mech Eng P: J Sports Eng Technol* 232:369-374, 2018.

198. Newton, RU, Cormie, P, and Cardinale, M. Principles of athlete testing. In *Strength and Conditioning: Biological and Practical*

Applications. Cardinale, M, Newton, RU, and Nosaka, K, eds. Chichester: Wiley-Blackwell, 2011.

199. Newton, RU, and Dugan, E. Application of strength diagnosis. *Strength Cond J* 24:50-59, 2002.

200. Newton, RU, Häkkinen, K, Häkkinen, A, McCormick, M, Volek, J, and Kraemer, WJ. Mixed-methods resistance training increases power and strength of young and older men. *Med Sci Sports Exerc* 34:1367-1375, 2002.

201. Nibali, ML, Tombleson, T, Brady, PH, and Wagner, P. Influence of familiarization and competitive level on the reliability of countermovement vertical jump kinetic and kinematic variables. *J Strength Cond Res* 29:2827-2835, 2015.

202. Nimphius, S. Re-evaluating what we "know" about female athletes in biomechanics research: Across the continuum from capacity to skill. *ISBS Proc Arch* 36:1059, 2018.

203. Nimphius, S. Exercise and sport science failing by design in understanding female athletes. *Int J Sports Physiol Perform* 14:1157-1158, 2019.

204. Nimphius, S, McGuigan, MR, Suchomel, TJ, and Newton, RU. Variability of a "force signature" during windmill softball pitching and relationship between discrete force variables and pitch velocity. *Hum Mov Sci* 47:151-158, 2016.

205. Nuzzo, JL, McBride, JM, Cormie, P, and McCaulley, GO. Relationship between countermovement jump performance and multijoint isometric and dynamic tests of strength. *J Strength Cond Res* 22:699-707, 2008.

206. O'Connor, B, Simmons, J, and O'Shea, P. *Weight Training Today*. St. Paul, MN: West Publishing, 1989.

207. Owen, NJ, Watkins, J, Kilduff, LP, Bevan, HR, and Bennett, MA. Development of a criterion method to determine peak mechanical power output in a countermovement jump. *J Strength Cond Res* 28:1552-1558, 2014.

208. Pallarés, JG, Sánchez-Medina, L, Pérez, CE, De La Cruz-Sánchez, E, and Mora-Rodriguez, R. Imposing a pause between the eccentric and concentric phases increases the reliability of isoinertial strength assessments. *J Sports Sci* 32:1165-1175, 2014.

209. Parsonage, J, Secomb, J, Dowse, R, Ferrier, B, Sheppard, J, and Nimphius, S. The assessment of isometric, dynamic, and sports-specific upper-body strength in male and female competitive surfers. *Sports* 6:53, 2018.

210. Pedley, JS, Radnor, JM, Lloyd, RS, and Oliver, JL. Analyzing drop jump ground reaction forces in Microsoft Excel. *Strength Cond J* 45:683-697, 2023.

211. Peng, H-T, Khuat, CT, Kernozek, TW, Wallace, BJ, Lo, S-L, and Song, C-Y. Optimum drop jump height in Division III athletes: Under 75% of vertical jump height. *Int J Sports Med* 38:842-846, 2017.

212. Peng, H-T, Song, C-Y, Wallace, BJ, Kernozek, TW, Wang, M-H, and Wang, Y-H. Effects of relative drop heights of drop jump biomechanics in male volleyball players. *Int J Sports Med* 40:863-870, 2019.

213. Petridis, L, Tróznai, Z, Pálinkás, G, Kalabiska, I, and Szabó, T. Modified reactive strength index in adolescent athletes competing in different sports and its relationship with force production. *Am J Sports Sci Med* 5:21-26, 2017.

214. Porter, JM, Ostrowski, EJ, Nolan, RP, and Wu, WFW. Standing long-jump performance is enhanced when using an external focus of attention. *J Strength Cond Res* 24:1746-1750, 2010.

215. Renner, KE, Williams, DSB, and Queen, RM. The reliability and validity of the Loadsol® under various walking and running conditions. *Sensors* 19:265, 2019.

216. Reynolds, JM, Gordon, TJ, and Robergs, RA. Prediction of one repetition maximum strength from multiple repetition maximum testing and anthropometry. *J Strength Cond Res* 20:584-592, 2006.

217. Ritti-Dias, RM, Avelar, A, Salvador, EP, and Cyrino, ES. Influence of previous experience on resistance training on reliability of one-repetition maximum test. *J Strength Cond Res* 25:1418-1422, 2011.

218. Rouis, M, Coudrat, L, Jaafar, H, Filliard, JR, Vandewalle, H, Barthelemy, Y, and Driss, T. Assessment of isokinetic knee strength in elite young female basketball players: Correlation with vertical jump. *J Sports Med Phys Fitness* 55:1502-1508, 2015.

219. Ruf, L, Chéry, C, and Taylor, K-L. Validity and reliability of the load-velocity relationship to predict the one-repetition maximum in deadlift. *J Strength Cond Res* 32:681-689, 2018.

220. Ryman Augustsson, S, and Svantesson, U. Reliability of the 1 RM bench press and squat in young women. *Eur J Physiother* 15:118-126, 2013.

221. Samozino, P, Edouard, P, Sangnier, S, Brughelli, M, Gimenez, P, and Morin, J-B. Force-velocity profile: Imbalance determination and effect on lower limb ballistic performance. *Int J Sports Med* 35:505-510, 2014.

222. Samozino, P, Morin, J-B, Hintzy, F, and Belli, A. A simple method for measuring force, velocity and power output during squat jump. *J Biomech* 41:2940-2945, 2008.

223. Samozino, P, Rabita, G, Dorel, S, Slawinski, J, Peyrot, N, de Villarreal, ESS, and Morin, JB. A simple method for measuring power, force, velocity properties, and mechanical effectiveness in sprint running. *Scand J Med Sci Sports* 26:648-658, 2016.

224. Samozino, P, Rejc, E, Di Prampero, PE, Belli, A, and Morin, J-B. Optimal force–velocity profile in ballistic movements—Altius: Citius or Fortius? *Med Sci Sports Exerc* 44:313-322, 2012.

225. Sanchez-Medina, L, Perez, CE, and Gonzalez-Badillo, JJ. Importance of the propulsive phase in strength assessment. *Int J Sports Med* 31:123-129, 2010.

226. Sánchez-Sixto, A, McMahon, JJ, and Floría, P. Verbal instructions affect reactive strength index modified and time-series waveforms in basketball players. *Sports Biomech* 23:211-221, 2024.

227. Sands, W, Cardinale, M, McNeal, J, Murray, S, Sole, C, Reed, J, Apostolopoulos, N, and Stone, M. Recommendations for measurement and management of an elite athlete. *Sports* 7:105, 2019.

228. Sands, WA, Kavanaugh, AA, Murray, SR, McNeal, JR, and Jemni, M. Modern techniques and technologies applied to training and performance monitoring. *Int J Sports Physiol Perform* 12:63-72, 2017.

229. Sannicandro, I, Cofano, G, Rosa, RA, and Piccinno, A. Balance training exercises decrease lower-limb strength asymmetry in young tennis players. *J Sports Sci Med* 13:397-402, 2014.

230. Šarabon, N, Kozinc, Ž, and Perman, M. Establishing reference values for isometric knee extension and flexion strength. *Front Physiol* 12:767941, 2021.

231. Sayers, MGL, Schlaeppi, M, Hitz, M, and Lorenzetti, S. The impact of test loads on the accuracy of 1RM prediction using the load-velocity relationship. *BMC Sports Sci Med Rehabil* 10:9, 2018.

232. Schmidtbleicher, D. Training for power events. In *Strength and Power in Sport*. Komi, PV, ed. London: Blackwell Scientific, 381-395, 1992.

233. Secomb, JL, Nimphius, S, Farley, ORL, Lundgren, LE, Tran, TT, and Sheppard, JM. Relationships between lower-body muscle structure and, lower-body strength, explosiveness and eccentric leg stiffness in adolescent athletes. *J Sports Sci Med* 14:691, 2015.

234. Seitz, LB, de Villarreal, ESS, and Haff, GG. The temporal profile of postactivation potentiation is related to strength level. *J Strength Cond Res* 28:706-715, 2014.

235. Seo, DI, Kim, E, Fahs, CA, Rossow, L, Young, K, Ferguson, SL, Thiebaud, R, Sherk, VD, Loenneke, JP, Kim, D, Lee, MK, Choi, KH, Bemben, DA, Bemben, MG, and So, WY. Reliability of the one-repetition maximum test based on muscle group and gender. *J Sports Sci Med* 11:221-225, 2012.

236. Sharp, AP, Cronin, JB, and Neville, J. Using smartphones for jump diagnostics: A brief review of the validity and reliability of the My Jump app. *Strength Cond J* 41:96-107, 2019.

237. Sheppard, JM, Chapman, D, and Taylor, K-L. An evaluation of a strength qualities assessment method for the lower body. *J Aust Strength Cond* 19:4-10, 2011.

238. Sheppard, JM, Cormack, S, Taylor, K-L, McGuigan, MR, and Newton, RU. Assessing the force-velocity characteristics of the leg extensors in well-trained athletes: The incremental load power profile. *J Strength Cond Res* 22:1320-1326, 2008.

239. Shimano, T, Kraemer, WJ, Spiering, BA, Volek, JS, Hatfield, DL, Silvestre, R, Vingren, JL, Fragala, MS, Maresh, CM, and Fleck, SJ. Relationship between the number of repetitions and selected percentages of one repetition maximum in free weight exercises in trained and untrained men. *J Strength Cond Res* 20:819-823, 2006.

240. Soares-Caldeira, LF, Ritti-Dias, RM, Okuno, NM, Cyrino, ES, Gurjão, ALD, and Ploutz-Snyder, LL. Familiarization indexes in sessions of 1-RM tests in adult women. *J Strength Cond Res* 23:2039-2045, 2009.

241. Sole, CJ, Mizuguchi, S, Sato, K, Moir, GL, and Stone, MH. Phase characteristics of the countermovement jump force-time curve: A comparison of athletes by jumping ability. *J Strength Cond Res* 32:1155-1165, 2018.

242. Sole, CJ, Suchomel, TJ, and Stone, MH. Preliminary scale of reference values for evaluating reactive strength index-modified in male and female NCAA Division I athletes. *Sports* 6:133, 2018.

243. Soriano, MA, Jiménez-Reyes, P, Rhea, MR, and Marín, PJ. The optimal load for maximal power production during lower-body resistance exercises: A meta-analysis. *Sports Med* 45:1191-1205, 2015.

244. Soriano, MA, Suchomel, TJ, and Marin, PJ. The optimal load for improving maximal power production during upper-body exercises: A meta-analysis. *Sports Med* 47:757-768, 2017.

245. Spiteri, T, Nimphius, S, Hart, NH, Specos, C, Sheppard, JM, and Newton, RU. Contribution of strength characteristics to change of direction and agility performance in female basketball athletes. *J Strength Cond Res* 28:2415-2423, 2014.

246. Stone, MH, O'Bryant, H, and Garhammer, J. A hypothetical model for strength training. *J Sports Med Phys Fitness* 21:342-351, 1981.

247. Stone, MH, and O'Bryant, HS. *Weight Training: A Scientific Approach*. Minneapolis, MN: Burgess International, 1987.

248. Stone, MH, Sanborn, K, O'Bryant, HS, Hartman, M, Stone, ME, Proulx, C, Ward, B, and Hruby, J. Maximum strength-power-performance relationships in collegiate throwers. *J Strength Cond Res* 17:739-745, 2003.

249. Stratford, C, Dos'Santos, T, and McMahon, JJ. Comparing drop jumps with 10/5 repeated jumps to measure reactive strength index. *Prof Strength Cond* 57:23-28, 2020.

250. Street, G, McMillan, S, Board, W, Rasmussen, M, and Heneghan, JM. Sources of error in determining countermovement jump height with the impulse method. *J Appl Biomech* 17:43-54, 2001.

251. Suarez, DG, Mizuguchi, S, Hornsby, WG, Cunanan, AJ, Marsh, DJ, and Stone, MH. Phase-specific changes in rate of force development and muscle morphology throughout a block periodized training cycle in weightlifters. *Sports* 7:129, 2019.

252. Suchomel, TJ, Bailey, CA, Sole, CJ, Grazer, JL, and Beckham, GK. Using reactive strength index-modified as an explosive performance measurement tool in Division I athletes. *J Strength Cond Res* 29:899-904, 2015.

253. Suchomel, TJ, Beckham, GK, and Wright, GA. Lower body kinetics during the jump shrug: Impact of load. *J Trainol* 2:19-22, 2013.

254. Suchomel, TJ, Beckham, GK, and Wright, GA. The impact of load on lower body performance variables during the hang power clean. *Sports Biomech* 13:87-95, 2014.

255. Suchomel, TJ, Beckham, GK, and Wright, GA. Effect of various loads on the force-time characteristics of the hang high pull. *J Strength Cond Res* 29:1295-1301, 2015.

256. Suchomel, TJ, Comfort, P, and Lake, JP. Enhancing the force-velocity profile of athletes using weightlifting derivatives. *Strength Cond J* 39:10-20, 2017.

257. Suchomel, TJ, Comfort, P, and Stone, MH. Weightlifting pulling derivatives: Rationale for implementation and application. *Sports Med* 45:823-839, 2015.

258. Suchomel, TJ, Garceau, LR, and Ebben, WP. Verbal instruction effect on stretch shortening cycle duration and reactive strength index-modified during plyometrics. *Med Sci Sports Exerc* 45:S428, 2013.

259. Suchomel, TJ, Garceau, LR, and Ebben, WP. The effect of verbal instruction on lower body performance variables during various plyometrics. *Med Sci Sports Exerc* 46:961, 2014.

260. Suchomel, TJ, McKeever, SM, and Comfort, P. Training with weightlifting derivatives: The effects of force and velocity overload stimuli. *J Strength Cond Res* 34:1808-1818, 2020.

261. Suchomel, TJ, McKeever, SM, McMahon, JJ, and Comfort, P. The effect of training with weightlifting catching or pulling derivatives on squat jump and countermovement jump force-time adaptations. *J Funct Morphol Kinesiol* 5:28, 2020.

262. Suchomel, TJ, McKeever, SM, Nolen, JD, and Comfort, P. Muscle architectural and force-velocity curve adaptations following 10 weeks of training with weightlifting catching and pulling derivatives. *J Sports Sci Med* 21:504-516, 2022.

263. Suchomel, TJ, McKeever, SM, Sijuwade, O, and Carpenter, L. Propulsion phase characteristics of loaded jump variations in resistance-trained women. *Sports* 11:44, 2023.

264. Suchomel, TJ, McKeever, SM, Sijuwade, O, Carpenter, L, McMahon, JJ, Loturco, I, and Comfort, P. The effect of load placement on the power production characteristics of three lower extremity jumping exercises. *J Hum Kinet* 68:109-122, 2019.

265. Suchomel, TJ, McMahon, JJ, and Lake, JP. Combined assessment methods. In *Performance Assessment in Strength and Conditioning*. Comfort, P, Jones, PA, and McMahon, JJ, eds. New York: Routledge, 275-290, 2019.

266. Suchomel, TJ, Nimphius, S, Bellon, CR, Hornsby, WG, and Stone, MH. Training for muscular strength: Methods for monitoring and adjusting training intensity. *Sports Med* 51:2051-2066, 2021.

267. Suchomel, TJ, Nimphius, S, Bellon, CR, and Stone, MH. The importance of muscular strength: Training considerations. *Sports Med* 48:765-785, 2018.

268. Suchomel, TJ, Nimphius, S, and Stone, MH. The importance of muscular strength in athletic performance. *Sports Med* 46:1419-1449, 2016.

269. Suchomel, TJ, Nimphius, S, and Stone, MH. Scaling isometric mid-thigh pull maximum strength in Division I athletes: Are we meeting the assumptions? *Sports Biomech* 19:532-546, 2020.

270. Suchomel, TJ, Sato, K, DeWeese, BH, Ebben, WP, and Stone, MH. Potentiation effects of half-squats performed in a ballistic or non-ballistic manner. *J Strength Cond Res* 30:1652-1660, 2016.

271. Suchomel, TJ, Sato, K, DeWeese, BH, Ebben, WP, and Stone, MH. Potentiation following ballistic and non-ballistic complexes: The effect of strength level. *J Strength Cond Res* 30:1825-1833, 2016.

272. Suchomel, TJ, Sato, K, DeWeese, BH, Ebben, WP, and Stone, MH. Relationships between potentiation effects following ballistic half-squats and bilateral symmetry. *Int J Sports Physiol Perform* 11:448-454, 2016.

273. Suchomel, TJ, Sole, CJ, Bailey, CA, Grazer, JL, and Beckham, GK. A comparison of reactive strength index-modified between six U.S. collegiate athletic teams. *J Strength Cond Res* 29:1310-1316, 2015.

274. Suchomel, TJ, Sole, CJ, Bellon, CR, and Stone, MH. Dynamic strength index: Relationships with common performance variables and contextualization of training recommendations. *J Hum Kinet* 74:59-70, 2020.

275. Suchomel, TJ, Sole, CJ, Sams, ML, Hollins, JE, Griggs, CV, and Stone, MH. The effect of a competitive season on the explosive performance characteristics of collegiate male soccer players. Presented at the Ninth Annual Sport Science and Coaches College, Johnson City, TN, December 5-6, 2014.

276. Suchomel, TJ, Sole, CJ, and Stone, MH. Comparison of methods that assess lower body stretch-shortening cycle utilization. *J Strength Cond Res* 30:547-554, 2016.

277. Suchomel, TJ, and Stone, MH. The relationships between hip and knee extensor cross-sectional area, strength, power, and potentiation characteristics. *Sports* 5:66, 2017.

278. Suchomel, TJ, Wright, GA, Kernozek, TW, and Kline, DE. Kinetic comparison of the power development between power clean variations. *J Strength Cond Res* 28:350-360, 2014.

279. Swinton, PA, Stewart, AD, Lloyd, R, Agouris, I, and Keogh, JW. Effect of load positioning on the kinematics and kinetics of weighted vertical jumps. *J Strength Cond Res* 26:906-913, 2012.

280. Thomas, C, Dos'Santos, T, and Jones, PA. A Comparison of dynamic strength index between team-sport athletes. *Sports* 5:71, 2017.

281. Thomas, C, Jones, PA, and Comfort, P. Reliability of the dynamic strength index in college athletes. *Int J Sports Physiol Perform* 10:542-545, 2015.

282. Thomas, GA, Kraemer, WJ, Spiering, BA, Volek, JS, Anderson, JM, and Maresh, CM. Maximal power at different percentages of one repetition maximum: Influence of resistance and gender. *J Strength Cond Res* 21:336-342, 2007.

283. Turner, TS, Tobin, DP, and Delahunt, E. Optimal loading range for the development of peak power output in the hexagonal barbell jump squat. *J Strength Cond Res* 29:1627-1632, 2015.

284. Valenzuela, PL, Sánchez-Martínez, G, Torrontegi, E, Vázquez-Carrión, J, Montalvo, Z, and Haff, GG. Should we base training prescription on the force–velocity profile? Exploratory study of its between-day reliability and differences between methods. *Int J Sports Physiol Perform* 16:1001-1007, 2021.

285. van den Tillaar, R, and Ettema, G. A comparison of muscle activity in concentric and counter movement maximum bench press. *J Hum Kinet* 38:63-71, 2013.

286. Van Hooren, B, and Bosch, F. Influence of muscle slack on high-intensity sport performance: A review. *Strength Cond J* 38:75-87, 2016.

287. Van Hooren, B, and Zolotarjova, J. The difference between countermovement and squat jump performances: A review of underlying mechanisms with practical applications. *J Strength Cond Res* 31:2011-2020, 2017.

288. Vanrenterghem, J, Venables, E, Pataky, T, and Robinson, MA. The effect of running speed on knee mechanical loading in females during side cutting. *J Biomech* 45:2444-2449, 2012.

289. Verheul, J, Nedergaard, NJ, Pogson, M, Lisboa, P, Gregson, W, Vanrenterghem, J, and Robinson, MA. Biomechanical loading during running: Can a two mass-spring-damper model be used to evaluate ground reaction forces for high-intensity tasks? *Sports Biomech* 20:571-582, 2021.

290. Walshe, AD, Wilson, GJ, and Murphy, AJ. The validity and reliability of a test of lower body musculotendinous stiffness. *Eur J Appl Physiol* 73:332-339, 1996.

291. Wang, R, Hoffman, JR, Tanigawa, S, Miramonti, AA, La Monica, MB, Beyer, KS, Church, DD, Fukuda, DH, and Stout, JR. Isometric mid-thigh pull correlates with strength, sprint, and agility performance in collegiate rugby union players. *J Strength Cond Res* 30:3051-3056, 2016.

292. Weakley, J, Black, G, McLaren, S, Scantlebury, S, Suchomel, TJ, McMahon, E, Watts, D, and Read, DB. Testing and profiling athletes: Recommendations for test selection, implementation, and maximizing information. *Strength Cond J* 46:159-179, 2023.

293. Weyand, PG, Sternlight, DB, Bellizzi, MJ, and Wright, S. Faster top running speeds are achieved with greater ground forces not more rapid leg movements. *J Appl Physiol* 89:1991-1999, 2000.

294. Wilson, JM, and Flanagan, EP. The role of elastic energy in activities with high force and power requirements: A brief review. *J Strength Cond Res* 22:1705-1715, 2008.

295. Woolford, S, Polglaze, T, Rowsell, G, and Spencer, M. Field testing principles and protocols. In *Physiological Tests for Elite Athletes*. Tanner, R, and Gore, C, eds. Champaign, IL: Human Kinetics, 2012.

296. Wu, WFW, Porter, JM, and Brown, LE. Effect of attentional focus strategies on peak force and performance in the standing long jump. *J Strength Cond Res* 26:1226-1231, 2012.

297. Young, KP, Haff, GG, Newton, RU, Gabbett, TJ, and Sheppard, JM. Assessment and monitoring of ballistic and maximal upper-body strength qualities in athletes. *Int J Sports Physiol Perform* 10:232-237, 2015.

298. Young, KP, Haff, GG, Newton, RU, and Sheppard, JM. Reliability of a novel testing protocol to assess upper-body strength qualities in elite athletes. *Int J Sports Physiol Perform* 9:871-875, 2014.

299. Young, WB. Laboratory strength assessment of athletes. *New Stud Athl* 10:89-96, 1995.

300. Young, WB, Pryor, JF, and Wilson, GJ. Effect of instructions on characteristics of countermovement and drop jump performance. *J Strength Cond Res* 9:232-236, 1995.

301. Zamparo, P, Minetti, A, and di Prampero, P. Interplay among the changes of muscle strength, cross-sectional area and maximal explosive power: Theory and facts. *Eur J Appl Physiol* 88:193-202, 2002.

302. Zaras, ND, Stasinaki, AN, Methenitis, SK, Krase, AA, Karampatsos, GP, Georgiadis, GV, Spengos, KM, Terzis, GD, and Zaras, N. Rate of force development, muscle architecture, and performance in young competitive track and field throwers. *J Strength Cond Res* 30:81-92, 2016.

Chapter 10

1. Andrade, DC, Manzo, O, Beltrán, AR, Alvares, C, Del Rio, R, Toledo, C, Moran, J, and Ramirez-Campillo, R. Kinematic and neuromuscular measures of intensity during plyometric jumps. *J Strength Cond Res* 34:3395-3402, 2020.

2. Arabatzi, F, and Kellis, E. Olympic weightlifting training causes different knee muscle-coactivation adaptations compared with traditional weight training. *J Strength Cond Res* 26:2192-2201, 2012.

3. Arazi, H, and Asadi, A. The relationship between the selected percentages of one repetition maximum and the number of repetitions in trained and untrained males. *FU Phys Ed Sport* 9:25-33, 2011.

4. Arede, J, Vaz, R, Gonzalo-Skok, O, Balsalobre-Fernandéz, C, Varela-Olalla, D, Madruga-Parera, M, and Leite, N. Repetitions in reserve vs maximum effort resistance training programs in youth female athletes. *J Sports Med Phys Fitness* 60:1231-1239, 2020.

5. Baechle, TR, and Earle, RW. Learning how to manipulate training variables to maximize results. In *Weight Training: Steps to Success*. Champaign, IL: Human Kinetics, 177-188, 2011.

6. Balsalobre-Fernández, C, García-Ramos, A, and Jiménez-Reyes, P. Load–velocity profiling in the military press exercise: Effects of gender and training. *Int J Sports Sci Coach* 13:743-750, 2018.

7. Balsalobre-Fernández, C, Marchante, D, Baz-Valle, E, Alonso-Molero, I, Jiménez, SL, and Muñóz-López, M. Analysis of wearable and smartphone-based technologies for the measurement of barbell velocity in different resistance training exercises. *Front Physiol* 8:649, 2017.

8. Banyard, HG, Nosaka, K, and Haff, GG. Reliability and validity of the load–velocity relationship to predict the 1RM back squat. *J Strength Cond Res* 31:1897-1904, 2017.

9. Barroso, R, Cardoso, RK, Carmo, EC, and Tricoli, V. Perceived exertion in coaches and young swimmers with different training experience. *Int J Sports Physiol Perform* 9:212-216, 2014.

10. Bartholomew, JB, Stults-Kolehmainen, MA, Elrod, CC, and Todd, JS. Strength gains after resistance training: The effect of stressful, negative life events. *J Strength Cond Res* 22:1215-1221, 2008.

11. Bompa, TO, and Buzzichelli, CA. Principles of training. In *Periodization*. Champaign, IL: Human Kinetics, 29-49, 2019.

12. Bompa, TO, and Buzzichelli, CA. Strength and power development. In *Periodization*. Champaign, IL: Human Kinetics, 229-263, 2019.

13. Borg, GAV. Perceived exertion as an indicator of somatic stress. *Scand J Rehab Med* 2:92-98, 1970.

14. Borg, GAV. Psychophysical bases of perceived exertion. *Med Sci Sports Exerc* 14:377-381, 1982.

15. Campos, GE, Luecke, TJ, Wendeln, HK, Toma, K, Hagerman, FC, Murray, TF, Ragg, KE, Ratamess NA, Kraemer WJ, and Staron RS. Muscular adaptations in response to three different resistance-training regimens: Specificity of repetition maximum training zones. *Eur J Appl Physiol* 88:50-60, 2002.

16. Carroll, KM, Bazyler, CD, Bernards, JR, Taber, CB, Stuart, CA, DeWeese, BH, Sato, K, and Stone, MH. Skeletal muscle fiber adaptations following resistance training using repetition maximums or relative intensity. *Sports* 7:169, 2019.

17. Carroll, KM, Bernards, JR, Bazyler, CD, Taber, CB, Stuart, CA, DeWeese, BH, Sato, K, and Stone, MH. Divergent performance outcomes following resistance training using repetition maximums or relative intensity. *Int J Sports Physiol Perform* 14:46-54, 2019.

18. Colquhoun, RJ, Gai, CM, Walters, J, Brannon, AR, Kilpatrick, MW, D'Agostino, DP, and Campbell, WI. Comparison of powerlifting performance in trained men using traditional and flexible daily undulating periodization. *J Strength Cond Res* 31:283-291, 2017.

19. Conceição, F, Fernandes, J, Lewis, M, Gonzaléz-Badillo, JJ, and Jimenéz-Reyes, P. Movement velocity as a measure of exercise intensity in three lower limb exercises. *J Sports Sci* 34:1099-1106, 2016.

20. Cunanan, AJ, DeWeese, BH, Wagle, JP, Carroll, KM, Sausaman, R, Hornsby, WG, Haff, GG, Triplett, NT, Pierce, KC, and Stone, MH. The general adaptation syndrome: A foundation for the concept of periodization. *Sports Med* 48:787-797, 2018.

21. Day, ML, McGuigan, MR, Brice, G, and Foster, C. Monitoring exercise intensity during resistance training using the session RPE scale. *J Strength Cond Res* 18:353-358, 2004.

22. DeLorme, TL. Restoration of muscle power by heavy-resistance exercises. *J Bone Joint Surg* 27:645-667, 1945.

23. DeLorme, TL, Ferris, BG, and Gallagher, JR. Effect of progressive resistance exercise on muscle contraction time. *Arch Phys Med Rehabil* 33:86-92, 1952.

24. DeLorme, TL, West, FE, and Shriber, WJ. Influence of progressive-resistance exercises on knee function following femoral fractures. *J Bone Joint Surg* 32:910-924, 1950.

25. DeWeese, BH, Hornsby, G, Stone, M, and Stone, MH. The training process: Planning for strength–power training in track and field. Part 1: Theoretical aspects. *J Sport Health Sci* 4:308-317, 2015.

26. DeWeese, BH, Hornsby, G, Stone, M, and Stone, MH. The training process: Planning for strength–power training in track and field. Part 2: Practical and applied aspects. *J Sport Health Sci* 4:318-324, 2015.

27. DeWeese, BH, Sams, ML, and Serrano, AJ. Sliding toward Sochi—Part 1: A review of programming tactics used during the 2010-2014 quadrennial. *Natl Strength Cond Assoc Coach* 1:30-42, 2014.

28. Duchateau, J, and Hainaut, K. Isometric or dynamic training: Differential effects on mechanical properties of a human muscle. *J Appl Physiol* 56:296-301, 1984.

29. Ebben, WP, Fauth, ML, Garceau, LR, and Petushek, EJ. Kinetic quantification of plyometric exercise intensity. *J Strength Cond Res* 25:3288-3298, 2011.

30. Ebben, WP, Suchomel, TJ, and Garceau, LR. The effect of plyometric training volume on jumping performance. In *Proceedings of the 32nd International Conference of Biomechanics in Sports, Johnson City, TN, USA, 12-16 July, 2014.* Holland, MA: International Society of Biomechanics in Sports, 566-569, 2014.

31. Fernandes, JFT, Lamb, KL, Clark, CCT, Moran, J, Drury, B, Garcia-Ramos, A, and Twist, C. Comparison of the FitroDyne and GymAware rotary encoders for quantifying peak and mean velocity during traditional multijointed exercises. *J Strength Cond Res* 35:1760-1765, 2021.

32. Fernandes, JFT, Dingley, AF, Garcia-Ramos, A, Perez-Castilla, A, Tufano, JJ, and Twist, C. Prediction of one repetition maximum using reference minimum velocity threshold values in young and middle-aged resistance-trained males. *Behav Sci* 11:71, 2021.

33. Fernández-Valdés, B, Sampaio, J, Exel, J, González, J, Tous-Fajardo, J, Jones, B, and Moras, G. The influence of functional flywheel resistance training on movement variability and movement velocity in elite rugby players. *Front Psychol* 11:1205, 2020.

34. Garcia-Ramos, A, and Jaric, S. Two-point method: A quick and fatigue-free procedure for assessment of muscle mechanical capacities and the 1 repetition maximum. *Strength Cond J* 40:54-66, 2018.

35. García-Ramos, A, Pestana-Melero, FL, Pérez-Castilla, A, Rojas, FJ, and Haff, GG. Differences in the load–velocity profile between 4 bench-press variants. *Int J Sports Physiol Perform* 13:326-331, 2018.

36. García-Ramos, A, Suzovic, D, and Pérez-Castilla, A. The load-velocity profiles of three upper-body pushing exercises in men and women. *Sports Biomech* 20:693-705, 2021.

37. Garhammer, J, and Gregor, R. Propulsion forces as a function of intensity for weightlifting and vertical jumping. *J Strength Cond Res* 6:129-134, 1992.

38. González-Badillo, JJ, and Sánchez-Medina, L. Movement velocity as a measure of loading intensity in resistance training. *Int J Sports Med* 31:347-352, 2010.

39. González-Badillo, JJ, Yañez-García, JM, Mora-Custodio, R, and Rodríguez-Rosell, D. Velocity loss as a variable for monitoring resistance exercise. *Int J Sports Med* 38:217-225, 2017.

40. Graham, T, and Cleather, DJ. Autoregulation by "repetitions in reserve" leads to greater improvements in strength over a 12-week training program than fixed loading. *J Strength Cond Res* 35:2451-2456, 2021.

41. Greig, L, Aspe, RR, Hall, A, Comfort, P, Cooper, K, and Swinton, PA. The predictive validity of individualised load-velocity relationships for predicting 1RM: A systematic review and individual participant data meta-analysis. *Sports Med* 53:1693-1708, 2023.

42. Hackett, DA, Cobley, SP, Davies, TB, Michael, SW, and Halaki, M. Accuracy in estimating repetitions to failure during resistance exercise. *J Strength Cond Res* 31:2162-2168, 2017.

43. Hackett, DA, Cobley, SP, and Halaki, M. Estimation of repetitions to failure for monitoring resistance exercise intensity: Building a case for application. *J Strength Cond Res* 32:1352-1359, 2018.

44. Hackett, DA, Johnson, NA, Halaki, M, and Chow, C-M. A novel scale to assess resistance-exercise effort. *J Sports Sci* 30:1405-1413, 2012.

45. Haff, GG, and Nimphius, S. Training principles for power. *Strength Cond J* 34:2-12, 2012.

46. Helms, ER, Byrnes, RK, Cooke, DM, Haischer, MH, Carzoli, JP, Johnson, TK, Cross, MR, Cronin, JB, Storey, AG, and Zourdos, MC. RPE vs. percentage 1RM loading in periodized programs matched for sets and repetitions. *Front Physiol* 9:247, 2018.

47. Helms, ER, Storey, A, Cross, MR, Brown, SR, Lenetsky, S, Ramsay, H, Dillen, C, and Zourdos, MC. RPE and velocity relationships for the back squat, bench press, and deadlift in powerlifters. *J Strength Cond Res* 31:292-297, 2017.

48. Herrick, AB, and Stone, WJ. The effects of periodization versus progressive resistance exercise on upper and lower body strength in women. *J Strength Cond Res* 10:72-76, 1996.

49. Hoeger, WWK, Hopkins, DR, Barette, SL, and Hale, DF. Relationship between repetitions and selected percentages of one repetition maximum: A comparison between untrained and trained males and females. *J Strength Cond Res* 4:47-54, 1990.

50. Hoffman, JR, Ratamess, NA, Klatt, M, Faigenbaum, AD, Ross, RE, Tranchina, NM, McCurley, RC, Kang, J, and Kraemer, WJ. Comparison between different off-season resistance training programs in Division III American college football players. *J Strength Cond Res* 23:11-19, 2009.

51. Holsbeeke, L, Ketelaar, M, Schoemaker, MM, and Gorter, JW. Capacity, capability, and performance: Different constructs or three of a kind? *Arch Phys Med Rehab* 90:849-855, 2009.

52. Hornsby, WG, Fry, AC, Haff, GG, and Stone, MH. Addressing the confusion within periodization research. *J Funct Morphol Kinesiol* 5:68, 2020.

53. Hornsby, WG, Gentles, JA, MacDonald, CJ, Mizuguchi, S, Ramsey, MW, and Stone, MH. Maximum strength, rate of force development, jump height, and peak power alterations in weightlifters across five months of training. *Sports* 5:78, 2017.

54. Hotermans, C, Peigneux, P, de Noordhout, AM, Moonen, G, and Maquet, P. Early boost and slow consolidation in motor skill learning. *Learn Mem* 13:580-583, 2006.

55. Izquierdo, M, González-Badillo, JJ, Häkkinen, K, Ibáñez, J, Kraemer, WJ, Altadill, A, Eslava, J, and Gorostiaga, EM. Effect of loading on unintentional lifting velocity declines during single sets of repetitions to failure during upper and lower extremity muscle actions. *Int J Sports Med* 27:718-724, 2006.

56. Izquierdo, M, Ibanez, J, González-Badillo, JJ, Häkkinen, K, Ratamess, NA, Kraemer, WJ, French, DN, Eslava, J, Altadill, A, and Asiain, X. Differential effects of strength training leading to failure versus not to failure on hormonal responses, strength, and muscle power gains. *J Appl Physiol* 100:1647-1656, 2006.

57. Jukic, I, Prnjak, K, McGuigan, MR, and Helms, ER. One velocity loss threshold does not fit all: Consideration of sex, training status, history, and personality traits when monitoring and controlling fatigue during resistance training. *Sports Med Open* 9:80, 2023.

58. Julio, UF, Panissa, VLG, and Franchini, E. Prediction of one repetition maximum from the maximum number of repetitions with submaximal loads in recreationally strength-trained men. *Sci Sport* 27:e69-e76, 2012.

59. Knight, KL. Knee rehabilitation by the daily adjustable progressive resistive exercise technique. *Am J Sports Med* 7:336-337, 1979.

60. Knight, KL. Quadriceps strengthening with the DAPRE technique: Case studies with neurological implications. *Med Sci Sports Exerc* 17:646-650, 1985.

61. Knudson, D. Qualitative biomechanical principles for application in coaching. *Sports Biomech* 6:109-118, 2007.

62. Lagally, KM, and Robertson, RJ. Construct validity of the OMNI resistance exercise scale. *J Strength Cond Res* 20:252-256, 2006.

63. Lake, JP, Mundy, PD, Comfort, P, McMahon, JJ, Suchomel, TJ, and Carden, P. The effect of barbell load on vertical jump landing force-time characteristics. *J Strength Cond Res* 35:25-32, 2021.

64. LeSuer, DA, McCormick, JH, Mayhew, JL, Wasserstein, RL, and Arnold, MD. The accuracy of prediction equations for estimating 1-RM performance in the bench press, squat, and deadlift. *J Strength Cond Res* 11:211-213, 1997.

65. Lopes dos Santos, M, Uftring, M, Stahl, CA, Lockie, RG, Alvar, B, Mann, JB, and Dawes, JJ. Stress in academic and athletic performance in collegiate athletes: A narrative review of sources and monitoring strategies. *Front Sports Act Living* 2:42, 2020.

66. Lovegrove, S, Hughes, LJ, Mansfield, SK, Read, PJ, Price, P, and Patterson, SD. Repetitions in reserve is a reliable tool for prescribing resistance training load. *J Strength Cond Res* 36:2696-2700, 2022.

67. Lum, D, and Barbosa, TM. Brief review: Effects of isometric strength training on strength and dynamic performance. *Int J Sports Med* 40:363-375, 2019.

68. MacKenzie, SJ, Lavers, RJ, and Wallace, BB. A biomechanical comparison of the vertical jump, power clean, and jump squat. *J Sports Sci* 32:1576-1585, 2014.

69. Mann, JB. *The APRE: The Scientifically Proven Fastest Way to Get Strong.* Columbia, MO: Self-published. 2011.

70. Mann, JB. A programming comparison: The APRE vs. linear perodization in short term periods (PhD dissertation). Columbia: University of Missouri, 91, 2011.

71. Mann, JB. *Developing Explosive Athletes: Use of Velocity Based Training in Athletes.* Muskegon, MI: Ultimate Athlete Concepts, 2016.

72. Mann, JB, Thyfault, JP, Ivey, PA, and Sayers, SP. The effect of autoregulatory progressive resistance exercise vs. linear periodization on strength improvement in college athletes. *J Strength Cond Res* 24:1718-1723, 2010.

73. Mansfield, SK, Peiffer, JJ, Hughes, LJ, and Scott, BR. Estimating repetitions in reserve for resistance exercise: An analysis of factors which impact on prediction accuracy. *J Strength Cond Res*, 2020. [e-pub ahead of print].

74. Maroto-Izquierdo, S, García-López, D, Fernandez-Gonzalo, R, Moreira, OC, González-Gallego, J, and de Paz, JA. Skeletal muscle functional and structural adaptations after eccentric overload flywheel resistance training: A systematic review and meta-analysis. *J Sci Med Sport* 20:943-951, 2017.

75. McBurnie, AJ, Allen, KP, Maybanks, G, McDwyer, M, Dos'Santos, T, Jones, PA, Comfort, P, and McMahon, JJ. The benefits and limitations of predicting one repetition maximum using the load-velocity relationship. *Strength Cond J* 41:28-40, 2019.

76. McGuigan, MR, and Foster, C. A new approach to monitoring resistance training. *Strength Cond J* 26:42-47, 2004.

77. 76a. McKeever, S, and Howard, R. Resistance training, in: *NSCA's Guide to High School Strength and Conditioning.* P. McHenry, M.J. Nitka, eds. Champaign, IL: Human Kinetics, 223-258, 2022.

78. McNamara, JM, and Stearne, DJ. Flexible nonlinear periodization in a beginner college weight training class. *J Strength Cond Res* 24:2012-2017, 2010.

79. Miranda, F, Simão, R, Rhea, M, Bunker, D, Prestes, J, Leite, RD, Miranda, H, de Salles, BF, and Novaes, J. Effects of linear vs. daily undulatory periodized resistance training on

maximal and submaximal strength gains. *J Strength Cond Res* 25:1824-1830, 2011.

80. Moore, CA, and Fry, AC. Nonfunctional overreaching during off-season training for skill position players in collegiate American football. *J Strength Cond Res* 21:793-800, 2007.

81. Moras, G, Fernández-Valdés, B, Vázquez-Guerrero, J, Tous-Fajardo, J, Exel, J, and Sampaio, J. Entropy measures detect increased movement variability in resistance training when elite rugby players use the ball. *J Sci Med Sport* 21:1286-1292, 2018.

82. Morita, Y, Ogawa, K, and Uchida, S. The effect of a day-time 2-hour nap on complex motor skill learning. *Sleep Biol Rhythms* 10:302-309, 2012.

83. Mulloy, F, Irwin, G, Williams, GKR, and Mullineaux, DR. Quantifying bi-variate coordination variability during longitudinal motor learning of a complex skill. *J Biomech* 95:109295, 2019.

84. Newell, KM. Coordination, control, and skill. In *Differing Perspectives in Motor Learning, Memory, and Control.* Goodman, D, Wilberg, RB, and Franks, IM, eds. Amsterdam, Netherlands: North-Holland, 295-317, 1985.

85. Nuzzo, J, Pinto, M, Nosaka, K, and Steele, J. Maximal number of repetitions at percentages of the one repetition maximum: A meta-regression and moderator analysis of sex, age, training status, and exercise. *Sports Med* 54:303-321, 2024.

86. Oranchuk, DJ, Storey, AG, Nelson, AR, and Cronin, JB. Isometric training and long-term adaptations: Effects of muscle length, intensity, and intent: A systematic review. *Scand J Med Sci Sports* 29:484-503, 2019.

87. Ormsbee, MJ, Carzoli, JP, Klemp, A, Allman, BR, Zourdos, MC, Kim, J-S, and Panton, LB. Efficacy of the repetitions in reserve-based rating of perceived exertion for the bench press in experienced and novice benchers. *J Strength Cond Res* 33:337-345, 2019.

88. Painter, KB, Haff, GG, Ramsey, MW, McBride, J, Triplett, T, Sands, WA, Lamont, HS, Stone, ME, and Stone, MH. Strength gains: Block versus daily undulating periodization weight training among track and field athletes. *Int J Sports Physiol Perform* 7:161-169, 2012.

89. Pérez-Castilla, A, García-Ramos, A, Padial, P, Morales-Artacho, AJ, and Feriche, B. Load-velocity relationship in variations of the half-squat exercise: Influence of execution technique. *J Strength Cond Res* 34:1024-1031, 2020.

90. Pérez-Castilla, A, Jaric, S, Feriche, B, Padial, P, and García-Ramos, A. Evaluation of muscle mechanical capacities through the two-load method: Optimization of the load selection. *J Strength Cond Res* 32:1245-1253, 2018.

91. Pérez-Castilla, A, Piepoli, A, Delgado-García, G, Garrido-Blanca, G, and García-Ramos, A. Reliability and concurrent validity of seven commercially available devices for the assessment of movement velocity at different intensities during the bench press. *J Strength Cond Res* 33:1258-1265, 2019.

92. Pestaña-Melero, FL, Haff, GG, Rojas, FJ, Pérez-Castilla, A, and García-Ramos, A. Reliability of the load–velocity relationship obtained through linear and polynomial regression models to predict the 1-repetition maximum load. *J Appl Biomech* 34:184-190, 2018.

93. Peterson, MD, Rhea, MR, and Alvar, BA. Applications of the dose-response for muscular strength development: A review of meta-analytic efficacy and reliability for designing training prescription. *J Strength Cond Res* 19:950-958, 2005.

94. Preatoni, E, Hamill, J, Harrison, AJ, Hayes, K, Van Emmerik, REA, Wilson, C, and Rodano, R. Movement variability and skills monitoring in sports. *Sports Biomech* 12:69-92, 2013.

95. Rhea, MR, Ball, SD, Phillips, WT, and Burkett, LN. A comparison of linear and daily undulating periodized programs with equated volume and intensity for strength. *J Strength Cond Res* 16:250-255, 2002.

96. Rice, PE, and Nimphius, S. When task constraints delimit movement strategy: Implications for isolated joint training in dancers. *Front Sports Act Living* 2:49, 2020.

97. Richens, B, and Cleather, DJ. The relationship between the number of repetitions performed at given intensities is different in endurance and strength trained athletes. *Biol Sport* 31:157-161, 2014.

98. Robertson, RJ, Goss, FL, Rutkowski, J, Lenz, B, Dixon, C, Timmer, J, Frazee, K, Dube, J, and Andreacci, J. Concurrent validation of the OMNI perceived exertion scale for resistance exercise. *Med Sci Sports Exerc* 35:333-341, 2003.

99. Ruf, L, Chéry, C, and Taylor, K-L. Validity and reliability of the load-velocity relationship to predict the one-repetition maximum in deadlift. *J Strength Cond Res* 32:681-689, 2018.

100. Sánchez-Medina, L, and González-Badillo, JJ. Velocity loss as an indicator of neuromuscular fatigue during resistance training. *Med Sci Sports Exerc* 43:1725-1734, 2011.

101. Sánchez-Medina, L, González-Badillo, JJ, Perez, CE, and Pallarés, JG. Velocity- and power-load relationships of the bench pull vs. bench press exercises. *Int J Sports Med* 35:209-216, 2014.

102. Sánchez, CC, Moreno, FJ, Vaíllo, RR, Romero, AR, Coves, Á, and Murillo, DB. The role of motor variability in motor control and learning depends on the nature of the task and the individual's capabilities. *Eur J Hum Mov* 38:12-26, 2017.

103. Scott, BR, Duthie, GM, Thornton, HR, and Dascombe, BJ. Training monitoring for resistance exercise: Theory and applications. *Sports Med* 46:687-698, 2016.

104. Shattock, K, and Tee, JC. Autoregulation in resistance training: A comparison of subjective versus objective methods. *J Strength Cond Res* 36:641-648, 2022.

105. Sheppard, JM, and Triplett, NT. Program design for resistance training. In *Essentials of Strength Training and Conditioning.* 4th ed. Haff, GG, and Triplett, NT, eds. Champaign, IL: Human Kinetics, 439-468, 2016.

106. Shimano, T, Kraemer, WJ, Spiering, BA, Volek, JS, Hatfield, DL, Silvestre, R, Vingren, JL, Fragala, MS, Maresh, CM, and Fleck, SJ. Relationship between the number of repetitions and selected percentages of one repetition maximum in free weight exercises in trained and untrained men. *J Strength Cond Res* 20:819-823, 2006.

107. Shishov, N, Melzer, I, and Bar-Haim, S. Parameters and measures in assessment of motor learning in neurorehabilitation: A systematic review of the literature. *Front Hum Neurosci* 11:82, 2017.

108. Siff, MC. *Supertraining*. Denver, CO: Supertraining Institute, 2000.
109. Soriano, MA, Suchomel, TJ, and Comfort, P. Weightlifting overhead pressing derivatives: A review of the literature. *Sports Med* 49:867-885, 2019.
110. Stone, MH, and O'Bryant, HS. *Weight Training: A Scientific Approach*. Minneapolis, MN: Burgess International, 1987.
111. Suarez, DG, Mizuguchi, S, Hornsby, WG, Cunanan, AJ, Marsh, DJ, and Stone, MH. Phase-specific changes in rate of force development and muscle morphology throughout a block periodized training cycle in weightlifters. *Sports* 7:129, 2019.
112. Suchomel, TJ, Beckham, GK, and Wright, GA. Lower body kinetics during the jump shrug: Impact of load. *J Trainol* 2:19-22, 2013.
113. Suchomel, TJ, Beckham, GK, and Wright, GA. The impact of load on lower body performance variables during the hang power clean. *Sports Biomech* 13:87-95, 2014.
114. Suchomel, TJ, Beckham, GK, and Wright, GA. Effect of various loads on the force-time characteristics of the hang high pull. *J Strength Cond Res* 29:1295-1301, 2015.
115. Suchomel, TJ, Comfort, P, and Lake, JP. Enhancing the force-velocity profile of athletes using weightlifting derivatives. *Strength Cond J* 39:10-20, 2017.
116. Suchomel, TJ, Comfort, P, and Stone, MH. Weightlifting pulling derivatives: Rationale for implementation and application. *Sports Med* 45:823-839, 2015.
117. Suchomel, TJ, McKeever, SM, and Comfort, P. Training with weightlifting derivatives: The effects of force and velocity overload stimuli. *J Strength Cond Res* 34:1808-1818, 2020.
118. Suchomel, TJ, McKeever, SM, McMahon, JJ, and Comfort, P. The effect of training with weightlifting catching or pulling derivatives on squat jump and countermovement jump force-time adaptations. *J Funct Morphol Kinesiol* 5:28, 2020.
119. Suchomel, TJ, McKeever, SM, Sijuwade, O, and Carpenter, L. Propulsion phase characteristics of loaded jump variations in resistance-trained women. *Sports* 11:44, 2023.
120. Suchomel, TJ, McKeever, SM, Sijuwade, O, Carpenter, L, McMahon, JJ, Loturco, I, and Comfort, P. The effect of load placement on the power production characteristics of three lower extremity jumping exercises. *J Hum Kinet* 68:109-122, 2019.
121. Suchomel, TJ, Nimphius, S, Bellon, CR, Hornsby, WG, and Stone, MH. Training for muscular strength: Methods for monitoring and adjusting training intensity. *Sports Med* 51:2051-2066, 2021.
122. Suchomel, TJ, Wagle, JP, Douglas, J, Taber, CB, Harden, M, Haff, GG, and Stone, MH. Implementing eccentric resistance training—Part 1: A brief review of existing methods. *J Funct Morphol Kinesiol* 4:38, 2019.
123. Suchomel, TJ, Wagle, JP, Douglas, J, Taber, CB, Harden, M, Haff, GG, and Stone, MH. Implementing eccentric resistance training—Part 2: Practical recommendations. *J Funct Morphol Kinesiol* 4:55, 2019.
124. Sweet, TW, Foster, C, McGuigan, MR, and Brice, G. Quantitation of resistance training using the session rating of perceived exertion method. *J Strength Cond Res* 18:796-802, 2004.
125. Swinton, PA, Stewart, AD, Lloyd, R, Agouris, I, and Keogh, JW. Effect of load positioning on the kinematics and kinetics of weighted vertical jumps. *J Strength Cond Res* 26:906-913, 2012.
126. Thompson, SW, Rogerson, D, Dorrell, HF, Ruddock, A, and Barnes, A. The reliability and validity of current technologies for measuring barbell velocity in the free-weight back squat and power clean. *Sports* 8:94, 2020.
127. Thompson, SW, Rogerson, D, Ruddock, A, and Barnes, A. The effectiveness of two methods of prescribing load on maximal strength development: A systematic review. *Sports Med* 50:919-938, 2020.
128. Torrejón, A, Balsalobre-Fernández, C, Haff, GG, and García-Ramos, A. The load-velocity profile differs more between men and women than between individuals with different strength levels. *Sports Biomech* 18:245-255, 2019.
129. Tufano, JJ, Conlon, JA, Nimphius, S, Brown, LE, Seitz, LB, Williamson, BD, and Haff, GG. Maintenance of velocity and power with cluster sets during high-volume back squats. *Int J Sports Physiol Perform* 11:885-892, 2016.
130. Wagle, JP, Taber, CB, Cunanan, AJ, Bingham, GE, Carroll, K, DeWeese, BH, Sato, K, and Stone, MH. Accentuated eccentric loading for training and performance: A review. *Sports Med* 47:2473-2495, 2017.
131. Weakley, JJS, Chalkley, D, Johnston, R, García-Ramos, A, Townshend, A, Dorrell, H, Pearson, M, Morrison, M, and Cole, M. Criterion validity, and interunit and between-day reliability of the FLEX for measuring barbell velocity during commonly used resistance training exercises. *J Strength Cond Res* 34:1519-1524, 2020.
132. Weakley, JJS, Mann, JB, Banyard, H, McLaren, S, Scott, T, and Garcia-Ramos, A. Velocity-based training: From theory to application. *Strength Cond J* 43:31-49, 2021.
133. Weakley, JJS, McLaren, S, Ramirez-Lopez, C, García-Ramos, A, Dalton-Barron, N, Banyard, H, Mann, B, Weaving, D, and Jones, B. Application of velocity loss thresholds during free-weight resistance training: Responses and reproducibility of perceptual, metabolic, and neuromuscular outcomes. *J Sports Sci* 38:477-485, 2020.
134. Weakley, JJS, Ramirez-Lopez, C, McLaren, S, Dalton-Barron, N, Weaving, D, Jones, B, Till, K, and Banyard, H. The effects of 10%, 20%, and 30% velocity loss thresholds on kinetic, kinematic, and repetition characteristics during the barbell back squat. *Int J Sports Physiol Perform* 15:180-188, 2020.
135. Weakley, JJS, Till, K, Read, DB, Phibbs, PJ, Roe, G, Darrall-Jones, J, and Jones, BL. The effects of superset configuration on kinetic, kinematic, and perceived exertion in the barbell bench press. *J Strength Cond Res* 34:65-72, 2020.
136. Weakley, JJS, Wilson, KM, Till, K, Banyard, H, Dyson, J, Phibbs, PJ, Read, DB, and Jones, B. Show me, tell me, encourage me: The effect of different forms of feedback on resistance training performance. *J Strength Cond Res* 34:3157-3163, 2020.
137. Weakley, JJS, Wilson, KM, Till, K, Read, DB, Darrall-Jones, J, Roe, GAB, Phibbs, PJ, and Jones, B. Visual feedback attenuates mean concentric barbell velocity loss and improves motivation, competitiveness, and perceived workload in male adolescent athletes. *J Strength Cond Res* 33:2420-2425, 2019.

138. Weakley, JJS, Wilson, KM, Till, K, Read, DB, Scantlebury, S, Sawczuk, T, Neenan, C, and Jones, B. Visual kinematic feedback enhances velocity, power, motivation and competitiveness in adolescent female athletes. *J Aust Strength Cond* 27:16-22, 2019.

139. Weber, CJ. Effects of autoregulatory progressive resistance exercise periodization versus linear periodization on muscular strength and anaerobic power in collegiate wrestlers (master's thesis). Whitewater: University of Wisconsin, 33, 2015.

140. Winchester, JB, Erickson, TM, Blaak, JB, and McBride, JM. Changes in bar-path kinematics and kinetics after power-clean training. *J Strength Cond Res* 19:177-183, 2005.

141. Wren, TAL, O'Callahan, B, Katzel, MJ, Zaslow, TL, Edison, BR, VandenBerg, CD, Conrad-Forrest, A, and Mueske, NM. Movement variability in pre-teen and teenage athletes performing sports related tasks. *Gait Posture* 80:228-233, 2020.

142. Wulf, G, and Lewthwaite, R. Optimizing performance through intrinsic motivation and attention for learning: The OPTIMAL theory of motor learning. *Psychon Bull Rev* 23:1382-1414, 2016.

143. Zhang, X, Li, H, Bi, S, Cao, Y, and Zhang, G. Auto-regulation method vs. fixed-loading method in maximum strength training for athletes: A systematic review and meta-analysis. *Front Physiol* 12:244, 2021.

144. Zourdos, MC, Goldsmith, JA, Helms, ER, Trepeck, C, Halle, JL, Mendez, KM, Cooke, DM, Haischer, MH, Sousa, CA, and Klemp, A. Proximity to failure and total repetitions performed in a set influences accuracy of intraset repetitions in reserve-based rating of perceived exertion. *J Strength Cond Res* 35:S158-S165, 2021.

145. Zourdos, MC, Klemp, A, Dolan, C, Quiles, JM, Schau, KA, Jo, E, Helms, E, Esgro, B, Duncan, S, and Merino, SG. Novel resistance training–specific rating of perceived exertion scale measuring repetitions in reserve. *J Strength Cond Res* 30:267-275, 2016.

Chapter 11

1. Abe, H. Role of histidine-related compounds as intracellular proton buffering constituents in vertebrate muscle. *Biochemistry (Mosc)* 65:757-765, 2000.

2. Almond, CS, Shin, AY, Fortescue, EB, Mannix, RC, Wypij, D, Binstadt, BA, Duncan, CN, Olson, DP, Salerno, AE, Newburger, JW, and Greenes, DS. Hyponatremia among runners in the Boston Marathon. *N Engl J Med* 352:1550-1556, 2005.

3. Anderson, L, Naughton, RJ, Close, GL, Di Michele, R, Morgans, R, Drust, B, and Morton, JP. Daily distribution of macronutrient intakes of professional soccer players from the English Premier League. *Int J Sport Nutr Exerc Metab* 27:491-498, 2017.

4. Arent, SM, Cintineo, HP, McFadden, BA, Chandler, AJ, and Arent, MA. Nutrient timing: A garage door of opportunity? *Nutrients* 12:1948, 2020.

5. Artioli, GG, Gualano, B, Smith, A, Stout, J, and Lancha, AH, Jr. Role of beta-alanine supplementation on muscle carnosine and exercise performance. *Med Sci Sports Exerc* 42:1162-1173, 2010.

6. Atkinson, FS, Brand-Miller, JC, Foster-Powell, K, Buyken, AE, and Goletzke, J. International tables of glycemic index and glycemic load values 2021: A systematic review. *Am J Clin Nutr* 114:1625-1632, 2021.

7. Atkinson, FS, Foster-Powell, K, and Brand-Miller, JC. International tables of glycemic index and glycemic load values: 2008. *Diabetes Care* 31:2281-2283, 2008.

8. Aussieker, T, Hilkens, L, Holwerda, AM, Fuchs, CJ, Houben, LHP, Senden, JM, van Dijk, J-W, Snijders, T, and van Loon, LJC. Collagen protein ingestion during recovery from exercise does not increase muscle connective protein synthesis rates. *Med Sci Sports Exerc* 55:1792-1802, 2023.

9. Ayoama, R, Hiruma, E, and Sasaki, H. Effects of creatine loading on muscular strength and endurance of female softball players. *J Sports Med Phys Fitness* 43:481-487, 2003.

10. Baar, K. Training and nutrition to prevent soft tissue injuries and accelerate return to play. *GSSI SSE* 28:1-6, 2015.

11. Baguet, A, Bourgois, J, Vanhee, L, Achten, E, and Derave, W. Important role of muscle carnosine in rowing performance. *J Appl Physiol* 109:1096-1101, 2010.

12. Baguet, A, Reyngoudt, H, Pottier, A, Everaert, I, Callens, S, Achten, E, and Derave, W. Carnosine loading and washout in human skeletal muscles. *J Appl Physiol* 106:837-842, 2009.

13. Bailey, RL, Saldanha, LG, Gahche, JJ, and Dwyer, JT. Estimating caffeine intake from energy drinks and dietary supplements in the United States. *Nutr Rev* 72:9-13, 2014.

14. Barber, JJ, McDermott, AY, McGaughey, KJ, Olmstead, JD, and Hagobian, TA. Effects of combined creatine and sodium bicarbonate supplementation on repeated sprint performance in trained men. *J Strength Cond Res* 27:252-258, 2013.

15. Bassinello, D, de Salles Painelli, V, Dolan, E, Lixandrão, M, Cajueiro, M, de Capitani, M, Saunders, B, Sale, C, Artioli, GG, Gualano, B, and Roschel, H. Beta-alanine supplementation improves isometric, but not isotonic or isokinetic strength endurance in recreationally strength-trained young men. *Amino Acids* 51:27-37, 2019.

16. Bazzucchi, I, Felici, F, Montini, M, Figura, F, and Sacchetti, M. Caffeine improves neuromuscular function during maximal dynamic exercise. *Muscle Nerve* 43:839-844, 2011.

17. Beelen, M, Tieland, M, Gijsen, AP, Vandereyt, H, Kies, AK, Kuipers, H, Saris, WHM, Koopman, R, and van Loon, LJC. Coingestion of carbohydrate and protein hydrolysate stimulates muscle protein synthesis during exercise in young men, with no further increase during subsequent overnight recovery. *J Nutr* 138:2198-2204, 2008.

18. Bellar, DM, Kamimori, G, Judge, L, Barkley, JE, Ryan, EJ, Muller, M, and Glickman, EL. Effects of low-dose caffeine supplementation on early morning performance in the standing shot put throw. *Eur J Sport Sci* 12:57-61, 2012.

19. Bemben, MG, Bemben, DA, Loftiss, DD, and Knehans, AW. Creatine supplementation during resistance training in college football athletes. *Med Sci Sports Exerc* 33:1667-1673, 2001.

20. Bender, A, and Klopstock, T. Creatine for neuroprotection in neurodegenerative disease: End of story? *Amino Acids* 48:1929-1940, 2016.

21. Berrazaga, I, Micard, V, Gueugneau, M, and Walrand, S. The role of the anabolic properties of plant-versus animal-based protein sources in supporting muscle mass maintenance: A critical review. *Nutrients* 11:1825, 2019.

22. Bertin, M, Pomponi, SM, Kokuhuta, C, Iwasaki, N, Suzuki, T, and Ellington, WR. Origin of the genes for the isoforms of creatine kinase. *Gene* 392:273-282, 2007.

23. Bishop, D, and Claudius, B. Effects of induced metabolic alkalosis on prolonged intermittent-sprint performance. *Med Sci Sports Exerc* 37:759-767, 2005.

24. Bishop, D, Edge, J, Davis, C, and Goodman, C. Induced metabolic alkalosis affects muscle metabolism and repeated-sprint ability. *Med Sci Sports Exerc* 36:807-813, 2004.

25. Blue, MN, Trexler, ET, Hirsch, KR, and Smith-Ryan, AE. A profile of body composition, omega-3 and vitamin D in National Football League players. *J Sports Med Phys Fitness* 59:87-93, 2018.

26. Bobb, A, Pringle, D, and Ryan, AJ. A brief study of the diet of athletes. *J Sports Med Phys Fitness* 9:255-262, 1969.

27. Bowers, RW, and Fox, EL. *Sports Physiology.* New York: Saunders, 1992.

28. Brooks, GA, Fahey, TD, and Baldwin, KM. *Exercise Physiology: Human Bioenergetics and Its Applications.* New York: McGraw Hill, 2005.

29. Buford, TW, Kreider, RB, Stout, JR, Greenwood, M, Campbell, B, Spano, M, Ziegenfuss, T, Lopez, H, Landis, J, and Antonio, J. International Society of Sports Nutrition position stand: Creatine supplementation and exercise. *J Int Soc Sports Nutr* 4:1-8, 2007.

30. Burke, DG, Chilibeck, PD, Parise, G, Candow, DG, Mahoney, D, and Tarnopolsky, M. Effect of creatine and weight training on muscle creatine and performance in vegetarians. *Med Sci Sports Exerc* 35:1946-1955, 2003.

31. Burke, LE, Wang, J, and Sevick, MA. Self-monitoring in weight loss: A systematic review of the literature. *J Am Diet Assoc* 111:92-102, 2011.

32. Burke, LM, Collier, GR, and Hargreaves, M. Muscle glycogen storage after prolonged exercise: Effect of the glycemic index of carbohydrate feedings. *J Appl Physiol* 75:1019-1023, 1993.

33. Burke, LM, and Hawley, JA. Swifter, higher, stronger: What's on the menu? *Science* 362:781-787, 2018.

34. Burke, LM, Hawley, JA, Wong, SHS, and Jeukendrup, AE. Carbohydrates for training and competition. *J Sports Sci* 29:S17-S27, 2011.

35. Burke, LM, van Loon, LJC, and Hawley, JA. Postexercise muscle glycogen resynthesis in humans. *J Appl Physiol* 122:1055-1067, 2017.

36. Burke, LM, Winter, JA, Cameron-Smith D, Enslen M, Farnfield M, and Decombaz J. Effect of intake of different dietary protein sources on plasma amino acid profiles at rest and after exercise. *Int J Sport Nutr Exerc Metab* 22:452-462, 2012.

37. Burleson, MA, Jr, O'Bryant, HS, Stone, MH, Collins, MA, and Triplett-McBride, T. Effect of weight training exercise and treadmill exercise on post-exercise oxygen consumption. *Med Sci Sports Exerc* 30:518-522, 1998.

38. Cabre, HE, Moore, SR, Smith-Ryan, AE, and Hackney, AC. Relative energy deficiency in sport (RED-S): Scientific, clinical, and practical implications for the female athlete. *Dtsch Z Sportmed* 73:225-234, 2022.

39. Carr, AJ, Gore, CJ, and Dawson, B. Induced alkalosis and caffeine supplementation: Effects on 2,000-m rowing performance. *Int J Sport Nutr Exerc Metab* 21:357-364, 2011.

40. Carr, BM, Webster, MJ, Boyd, JC, Hudson, GM, and Scheett, TP. Sodium bicarbonate supplementation improves hypertrophy-type resistance exercise performance. *Eur J Appl Physiol* 113:743-752, 2013.

41. Cermak, NM, de Groot, LC, Saris, WHM, and Van Loon, LJC. Protein supplementation augments the adaptive response of skeletal muscle to resistance-type exercise training: A meta-analysis. *Am J Clin Nutr* 96:1454-1464, 2012.

42. Chiang, C-M, Ismaeel, A, Griffis, RB, and Weems, S. Effects of vitamin D supplementation on muscle strength in athletes: A systematic review. *J Strength Cond Res* 31:566-574, 2017.

43. Church, DD, Hirsch, KR, Park, S, Kim, I-Y, Gwin, JA, Pasiakos, SM, Wolfe, RR, and Ferrando, AA. Essential amino acids and protein synthesis: Insights into maximizing the muscle and whole-body response to feeding. *Nutrients* 12:3717, 2020.

44. Claudino, JG, Mezêncio, B, Amaral, S, Zanetti, V, Benatti, F, Roschel, H, Gualano, B, Amadio, AC, and Serrão, JC. Creatine monohydrate supplementation on lower-limb muscle power in Brazilian elite soccer players. *J Int Soc Sports Nutr* 11:32, 2014.

45. Close, GL, Ashton, T, McArdle, A, and Maclaren, DPM. The emerging role of free radicals in delayed onset muscle soreness and contraction-induced muscle injury. *Comp Biochem Physiol A Mol Integr Physiol* 142:257-266, 2005.

46. Close, GL, Kasper, AM, and Morton, JP. Nutrition for human performance. In *Strength and Conditioning for Sports Performance.* Jeffreys, I, and Moody, J, eds. New York: Routledge, 153-190, 2021.

47. Close, GL, Russell, J, Cobley, JN, Owens, DJ, Wilson, G, Gregson, W, Fraser, WD, and Morton, JP. Assessment of vitamin D concentration in non-supplemented professional athletes and healthy adults during the winter months in the UK: Implications for skeletal muscle function. *J Sports Sci* 31:344-353, 2013.

48. Cohen, M, and Bendich, A. Safety of pyridoxine—A review of human and animal studies. *Toxicol Lett* 34:129-139, 1986.

49. Cooke, MB, Rybalka, E, Williams, AD, Cribb, PJ, and Hayes, A. Creatine supplementation enhances muscle force recovery after eccentrically-induced muscle damage in healthy individuals. *J Int Soc Sports Nutr* 6:13, 2009.

50. Cornish, SM, Chilibeck, PD, and Burke, DG. The effect of creatine monohydrate supplementation on sprint skating in ice-hockey players. *J Sports Med Phys Fitness* 46:90-98, 2006.

51. Cox, GR, Clark, SA, Cox, AJ, Halson, SL, Hargreaves, M, Hawley, JA, Jeacocke, N, Snow, RJ, Yeo, WK, and Burke, LM. Daily training with high carbohydrate availability increases exogenous carbohydrate oxidation during endurance cycling. *J Appl Physiol* 109:126-134, 2010.

52. Cribb, PJ, and Hayes, A. Effects of supplement-timing and resistance exercise on skeletal muscle hypertrophy. *Med Sci Sports Exerc* 38:1918-1925, 2006.

53. Culbertson, JY, Kreider, RB, Greenwood, M, and Cooke, M. Effects of beta-alanine on muscle carnosine and exercise performance: A review of the current literature. *Nutrients* 2:75-98, 2010.

54. da Silva, RP, de Oliveira, LF, Saunders, B, de Andrade Kratz, C, de Salles Painelli, V, da Eira Silva, V, Marins, JCB, Franchini, E, Gualano, B, and Artioli, GG. Effects of β-alanine and sodium bicarbonate supplementation on the estimated energy system contribution during high-intensity intermittent exercise. *Amino Acids* 51:83-96, 2019.

55. Dalbo, VJ, Roberts, MD, Stout, JR, and Kerksick, CM. Putting to rest the myth of creatine supplementation leading to muscle cramps and dehydration. *Br J Sports Med* 42:567-573, 2008.

56. Dawson, B, Vladich, T, and Blanksby, BA. Effects of 4 weeks of creatine supplementation in junior swimmers on freestyle sprint and swim bench performance. *J Strength Cond Res* 16:485-490, 2002.

57. de Ataide e Silva, T, Di Cavalcanti Alves de Souza, ME, de Amorim, JF, Stathis, CG, Leandro, CG, and Lima-Silva, AE. Can carbohydrate mouth rinse improve performance during exercise? A systematic review. *Nutrients* 6:1-10, 2013.

58. de Oliveira, JJ, Crisp, AH, Barbosa, CGR, de Silva, AS, Baganha, RJ, and Verlengia, R. Influence of carbohydrate mouth rinse on sprint performance: A systematic review and meta-analysis. *J Exerc Physiol Online* 20:88-99, 2017.

59. de Salles Painelli, V, Roschel, H, De Jesus, F, Sale, C, Harris, RC, Solis, MY, Benatti, FB, Gualano, B, Lancha, AH, Jr, and Artioli, GG. The ergogenic effect of beta-alanine combined with sodium bicarbonate on high-intensity swimming performance. *Appl Physiol Nutr Metab* 38:525-532, 2013.

60. de Souza, JG, Del Coso, J, Fonseca, FS, Silva, BVC, de Souza, DB, da Silva Gianoni, RL, Filip-Stachnik, A, Serrão, JC, and Claudino, JG. Risk or benefit? Side effects of caffeine supplementation in sport: A systematic review. *Eur J Nutr* 61:3823-3834, 2022.

61. Deakin, V. Iron depletion in athletes. In *Clinical Sports Nutrition*. L Burke, V Deakin, eds. Rossville, NSW: McGraw Hill Australia, 270-310, 2000.

62. Deminice, R, Rosa, FT, Franco, GS, Jordao, AA, and de Freitas, EC. Effects of creatine supplementation on oxidative stress and inflammatory markers after repeated-sprint exercise in humans. *Nutrition* 29:1127-1132, 2013.

63. Doepker, C, Lieberman, HR, Smith, AP, Peck, JD, El-Sohemy, A, and Welsh, BT. Caffeine: Friend or foe? *Annu Rev Food Sci Technol* 7:117-137, 2016.

64. Doherty, M, and Smith, PM. Effects of caffeine ingestion on rating of perceived exertion during and after exercise: A meta-analysis. *Scand J Med Sci Sports* 15:69-78, 2005.

65. Donaldson, CM, Perry, TL, and Rose, MC. Glycemic index and endurance performance. *Int J Sport Nutr Exerc Metab* 20: 154-165, 2010.

66. Douroudos, II, Fatouros, IG, Gourgoulis, V, Jamurtas, AZ, Tsitsios, T, Hatzinikolaou, A, Margonis, K, Mavromatidis, K, and Taxildaris, K. Dose-related effects of prolonged $NaHCO_3$ ingestion during high-intensity exercise. *Med Sci Sports Exerc* 38:1746-1753, 2006.

67. Duncan, MJ, and Oxford, SW. Acute caffeine ingestion enhances performance and dampens muscle pain following resistance exercise to failure. *J Sports Med Phys Fitness* 52:280-285, 2012.

68. Duncan, MJ, Stanley, M, Parkhouse, N, Cook, K, and Smith, M. Acute caffeine ingestion enhances strength performance and reduces perceived exertion and muscle pain perception during resistance exercise. *Eur J Sport Sci* 13:392-399, 2013.

69. Duncan, MJ, Weldon, A, and Price, MJ. The effect of sodium bicarbonate ingestion on back squat and bench press exercise to failure. *J Strength Cond Res* 28:1358-1366, 2014.

70. Dutra, MT, Martins, WR, Ribeiro, ALA, and Bottaro, M. The effects of strength training combined with vitamin C and E supplementation on skeletal muscle mass and strength: A systematic review and meta-analysis. *J Sports Med* 2020:1-9, 2020.

71. Ellis, D, and Close, GL. Fueling for training and performance. In *High-Performance Training for Sports: The Authoritative Guide for Ultimate Athletic Conditioning*. Joyce, D, and Lewindon, D, eds. Champaign, IL: Human Kinetics, 2022.

72. Engell, DB, Maller, O, Sawka, MN, Francesconi, RN, Drolet, L, and Young, AJ. Thirst and fluid intake following graded hypohydration levels in humans. *Physiol Behav* 40:229-236, 1987.

73. Everaert, I, Mooyaart, A, Baguet, A, Zutinic, A, Baelde, H, Achten, E, Taes, Y, De Heer, E, and Derave, W. Vegetarianism, female gender and increasing age, but not CNDP1 genotype, are associated with reduced muscle carnosine levels in humans. *Amino Acids* 40:1221-1229, 2011.

74. Ferrando, AA, Wolfe, RR, Hirsch, KR, Church, DD, Kviatkovsky, SA, Roberts, MD, Stout, JR, Gonzalez, DE, Sowinski, RJ, and Kreider, RB. International Society of Sports Nutrition position stand: Essential amino acid supplementation on skeletal muscle and performance. *J Int Soc Sports Nutr* 20:2263409, 2023.

75. Ferrugem, LC, Martini, GL, and De Souza, CG. Influence of the glycemic index of pre-exercise meals in sports performance: A systematic review. *Int J Med Rev* 5:151-158, 2018.

76. Fredholm, BB. Astra Award Lecture: Adenosine, adenosine receptors and the actions of caffeine. *Pharmacol Toxicol* 76:93-101, 1995.

77. Freitas, MC, Cholewa, J, Panissa, V, Quizzini, G, de Oliveira, JV, Figueiredo, C, Gobbo, LA, Caperuto, E, Zanchi, NE, Lira, F, and Rossi, FE. Short-time β-alanine supplementation on the acute strength performance after high-intensity intermittent exercise in recreationally trained men. *Sports* 7:108, 2019.

78. Ftaiti, F, Grélot, L, Coudreuse, JM, Nicol, C, and Coudreuse, JM. Combined effect of heat stress, dehydration and exercise on neuromuscular function in humans. *Eur J Appl Physiol* 84:87-94, 2001.

79. Ganio, MS, Klau, JF, Casa, DJ, Armstrong, LE, and Maresh, CM. Effect of caffeine on sport-specific endurance performance: A systematic review. *J Strength Cond Res* 23:315-324, 2009.

80. Gao, J, Costill, DL, Horswill, CA, and Park, SH. Sodium bicarbonate ingestion improves performance in interval swimming. *Eur J Appl Physiol Occup Physiol* 58:171-174, 1988.

81. Gleeson, M, Nieman, DC, and Pedersen, BK. Exercise, nutrition and immune function. *J Sports Sci* 22:115-125, 2004.

82. Gomez-Cabrera, M-C, Domenech, E, and Viña, J. Moderate exercise is an antioxidant: Upregulation of antioxidant genes by training. *Free Radic Biol Med* 44:126-131, 2008.

83. Graham, TE. Caffeine and exercise: Metabolism, endurance and performance. *Sports Med* 31:785-807, 2001.
84. Green, AL, Hultman, E, Macdonald, IA, Sewell, DA, and Greenhaff, PL. Carbohydrate ingestion augments skeletal muscle creatine accumulation during creatine supplementation in humans. *Am J Physiol Endocrinol Metab* 271:E821-E826, 1996.
85. Greenwood, M, Farris, J, Kreider, R, Greenwood, L, and Byars, A. Creatine supplementation patterns and perceived effects in select Division I collegiate athletes. *Clin J Sport Med* 10:191-194, 2000.
86. Greenwood, M, Kreider, R, Earnest, C, Rasmussen, C, and Almada, A. Differences in creatine retention among three nutritional formulations of oral creatine supplements. *J Exerc Physiol Online* 6:37-43, 2003.
87. Greenwood, M, Kreider, RB, Greenwood, L, and Byars, A. Cramping and injury incidence in collegiate football players are reduced by creatine supplementation. *J Athl Train* 38:216-219, 2003.
88. Greenwood, M, Kreider, RB, Melton, C, Rasmussen, C, Lancaster, S, Cantler, E, Milnor, P, and Almada, A. Creatine supplementation during college football training does not increase the incidence of cramping or injury. *Mol Cell Biochem* 244:83-88, 2003.
89. Grgic, J, and Del Coso, J. Ergogenic effects of acute caffeine intake on muscular endurance and muscular strength in women: A meta-analysis. *Int J Environ Res Public Health* 18:5773, 2021.
90. Grgic, J, Garofolini, A, Pickering, C, Duncan, MJ, Tinsley, GM, and Del Coso, J. Isolated effects of caffeine and sodium bicarbonate ingestion on performance in the yo-yo test: A systematic review and meta-analysis. *J Sci Med Sport* 23:41-47, 2020.
91. Grgic, J, Grgic, I, Del Coso, J, Schoenfeld, BJ, and Pedisic, Z. Effects of sodium bicarbonate supplementation on exercise performance: An umbrella review. *J Int Soc Sports Nutr* 18:71, 2021.
92. Grgic, J, Grgic, I, Pickering, C, Schoenfeld, BJ, Bishop, DJ, and Pedisic, Z. Wake up and smell the coffee: Caffeine supplementation and exercise performance—An umbrella review of 21 published meta-analyses. *Br J Sports Med* 54:681-688, 2020.
93. Grgic, J, and Mikulic, P. Effects of caffeine on rate of force development: A meta-analysis. *Scand J Med Sci Sports* 32:644-653, 2022.
94. Grgic, J, Mikulic, P, Schoenfeld, BJ, Bishop, DJ, and Pedisic, Z. The influence of caffeine supplementation on resistance exercise: A review. *Sports Med* 49:17-30, 2019.
95. Grgic, J, Pedisic, Z, Saunders, B, Artioli, GG, Schoenfeld, BJ, McKenna, MJ, Bishop, DJ, Kreider, RB, Stout, JR, and Kalman, DS. International Society of Sports Nutrition position stand: Sodium bicarbonate and exercise performance. *J Int Soc Sports Nutr* 18:61, 2021.
96. Grgic, J, and Pickering, C. The effects of caffeine ingestion on isokinetic muscular strength: A meta-analysis. *J Sci Med Sport* 22:353-360, 2019.
97. Grgic, J, Rodriguez, RF, Garofolini, A, Saunders, B, Bishop, DJ, Schoenfeld, BJ, and Pedisic, Z. Effects of sodium bicarbonate supplementation on muscular strength and endurance: A systematic review and meta-analysis. *Sports Med* 50:1361-1375, 2020.
98. Grgic, J, Trexler, ET, Lazinica, B, and Pedisic, Z. Effects of caffeine intake on muscle strength and power: A systematic review and meta-analysis. *J Int Soc Sports Nutr* 15:11, 2018.
99. Griffen, C, Rogerson, D, Ranchordas, M, and Ruddock, A. Effects of creatine and sodium bicarbonate coingestion on multiple indices of mechanical power output during repeated Wingate tests in trained men. *Int J Sport Nutr Exerc Metab* 25:298-306, 2015.
100. Grindstaff, PD, Kreider, R, Bishop, R, Wilson, M, Wood, L, Alexander, C, and Almada, A. Effects of creatine supplementation on repetitive sprint performance and body composition in competitive swimmers. *Int J Sport Nutr* 7:330-346, 1997.
101. Guest, NS, VanDusseldorp, TA, Nelson, MT, Grgic, J, Schoenfeld, BJ, Jenkins, NDM, Arent, SM, Antonio, J, Stout, JR, and Trexler, ET. International Society of Sports Nutrition position stand: Caffeine and exercise performance. *J Int Soc Sports Nutr* 18:1, 2021.
102. Hammond, KM, Impey, SG, Currell, K, Mitchell, N, Shepherd, SO, Jeromson, S, Hawley, JA, Close, GL, Hamilton, LD, and Sharples, AP. Postexercise high-fat feeding suppresses p70S6K1 activity in human skeletal muscle. *Med Sci Sports Exerc* 48:2108-2117, 2016.
103. Hannah, R, Stannard, RL, Minshull, C, Artioli, GG, Harris, RC, and Sale, C. β-alanine supplementation enhances human skeletal muscle relaxation speed but not force production capacity. *J Appl Physiol* 118:604-612, 2015.
104. Hansen, AK, Fischer, CP, Plomgaard, P, Andersen, JL, Saltin, B, and Pedersen, BK. Skeletal muscle adaptation: Training twice every second day vs. training once daily. *J Appl Physiol* 98:93-99, 2005.
105. Harcourt, BA, Panagiotopoulos, M, Sardelis, S, Terzis, G, and Bogdanis, GC. The effect of dehydration on vertical jump, muscle strength and sprint performance. *MDPI Proceedings* 25:10, 2019.
106. Harris, RC, Jones, G, Hill, CH, Kendrick, IP, Boobis, L, Kim, C, Kim, H, Dang, VH, Edge, J, and Wise, JA. The carnosine content of vastus lateralis in vegetarians and omnivores. *FASEB J* 21:pA944, 2007.
107. Harris, RC, Jones, GA, Kim, HJ, Kim, CK, Price, KA, and Wise, JA. Changes in muscle carnosine of subjects with 4 weeks supplementation with a controlled release formulation of beta-alanine (Carnosyn™), and for 6 weeks post. *FASEB J* 23:599.594, 2009.
108. Harris, RC, Tallon, MJ, Dunnett, M, Boobis, L, Coakley, J, Kim, HJ, Fallowfield, JL, Hill, CA, Sale, C, and Wise, JA. The absorption of orally supplied β-alanine and its effect on muscle carnosine synthesis in human vastus lateralis. *Amino Acids* 30:279-289, 2006.
109. Hayward, S, Wilborn, CD, Taylor, LW, Urbina, SL, Outlaw, JJ, Foster, CA, and Roberts, MD. Effects of a high protein and omega-3-enriched diet with or without creatine supplementation on markers of soreness and inflammation during 5 consecutive days of high volume resistance exercise in females. *J Sports Sci Med* 15:704-714, 2016.
110. Heffernan, SM, Horner, K, De Vito, G, and Conway, GE. The role of mineral and trace element supplementation in exercise and athletic performance: A systematic review. *Nutrients* 11:696, 2019.

111. Heibel, AB, Perim, PHL, Oliveira, LF, McNaughton, LR, and Saunders, B. Time to optimize supplementation: Modifying factors influencing the individual responses to extracellular buffering agents. *Front Nutr* 5:35, 2018.

112. Heileson, JL, Machek, SB, Harris, DR, Tomek, S, de Souza, LC, Kieffer, AJ, Barringer, ND, Gallucci, A, Forsse, JS, and Funderburk, LK. The effect of fish oil supplementation on resistance training-induced adaptations. *J Int Soc Sports Nutr* 20:2174704, 2023.

113. Henselmans, M, Bjørnsen, T, Hedderman, R, and Vårvik, FT. The effect of carbohydrate intake on strength and resistance training performance: A systematic review. *Nutrients* 14:856, 2022.

114. Herbert, WG. Water and electrolytes. In *Ergogenic Aids in Sport*. Williams, MH, ed. Champaign, IL: Human Kinetics, 56-98, 1983.

115. Hespel, P, Op't Eijnde, B, Van Leemputte, M, Ursø, B, Greenhaff, PL, Labarque, V, Dymarkowski, S, Van Hecke, P, and Richter, EA. Oral creatine supplementation facilitates the rehabilitation of disuse atrophy and alters the expression of muscle myogenic factors in humans. *J Physiol* 536:625-633, 2001.

116. Hespel, P, Op't, Eijnde B, and Van Leemputte, M. Opposite actions of caffeine and creatine on muscle relaxation time in humans. *J Appl Physiol* 92:513-518, 2002.

117. Hetland, ML, Haarbo, J, and Christiansen, C. Low bone mass and high bone turnover in male long distance runners. *J Clin Endocrinol Metab* 77:770-775, 1993.

118. Heung-Sang Wong, S, Sun, F-H, Chen, Y-J, Li, C, Zhang, Y-J, and Ya-Jun Huang, W. Effect of pre-exercise carbohydrate diets with high vs low glycemic index on exercise performance: A meta-analysis. *Nutr Rev* 75:327-338, 2017.

119. Higgins, MF, Wilson, S, Hill, C, Price, MJ, Duncan, M, and Tallis, J. Evaluating the effects of caffeine and sodium bicarbonate, ingested individually or in combination, and a taste-matched placebo on high-intensity cycling capacity in healthy males. *Appl Physiol Nutr Metab* 41:354-361, 2016.

120. Higgins, S, Straight, CR, and Lewis, RD. The effects of preexercise caffeinated coffee ingestion on endurance performance: An evidence-based review. *Int J Sport Nutr Exerc Metab* 26:221-239, 2016.

121. Hilton, NP, Leach, NK, Sparks, SA, Gough, LA, Craig, MM, Deb, SK, and McNaughton, LR. A novel ingestion strategy for sodium bicarbonate supplementation in a delayed-release form: A randomised crossover study in trained males. *Sports Med Open* 5:1-8, 2019.

122. Hobson, RM, Harris, RC, Martin, D, Smith, P, Macklin, B, Gualano, B, and Sale, C. Effect of beta-alanine with and without sodium bicarbonate on 2,000-m rowing performance. *Int J Sport Nutr Exerc Metab* 23:480-487, 2013.

123. Hobson, RM, Saunders, B, Ball, G, Harris, RC, and Sale, C. Effects of β-alanine supplementation on exercise performance: A meta-analysis. *Amino Acids* 43:25-37, 2012.

124. Hoffman, J, Ratamess, N, Kang, J, Mangine, G, Faigenbaum, A, and Stout, J. Effect of creatine and ß-alanine supplementation on performance and endocrine responses in strength/power athletes. *Int J Sport Nutr Exerc Metab* 16:430-446, 2006.

125. Hoffman, J, Ratamess, NA, Ross, R, Kang, J, Magrelli, J, Neese, K, Faigenbaum, AD, and Wise, JA. Beta-alanine and the hormonal response to exercise. *Int J Sports Med* 29:952-958, 2008.

126. Hoffman, JR, Ratamess, NA, Faigenbaum, AD, Ross, R, Kang, J, Stout, JR, and Wise, JA. Short-duration beta-alanine supplementation increases training volume and reduces subjective feelings of fatigue in college football players. *Nutr Res* 28:31-35, 2008.

127. Holwerda, AM, and van Loon, LJC. The impact of collagen protein ingestion on musculoskeletal connective tissue remodeling: A narrative review. *Nutr Rev* 80:1497-1514, 2022.

128. Howarth, KR, Phillips, SM, MacDonald, MJ, Richards, D, Moreau, NA, and Gibala, MJ. Effect of glycogen availability on human skeletal muscle protein turnover during exercise and recovery. *J Appl Physiol* 109:431-438, 2010.

129. Hulston, CJ, Venables, MC, Mann, CH, Martin, C, Philp, A, Baar, K, and Jeukendrup, AE. Training with low muscle glycogen enhances fat metabolism in well-trained cyclists. *Med Sci Sports Exerc* 42:2046-2055, 2010.

130. Hultman, E, Soderlund, K, Timmons, JA, Cederblad, G, and Greenhaff, PL. Muscle creatine loading in men. *J Appl Physiol* 81:232-237, 1996.

131. Impey, SG, Hearris, MA, Hammond, KM, Bartlett, JD, Louis, J, Close, GL, and Morton, JP. Fuel for the work required: A theoretical framework for carbohydrate periodization and the glycogen threshold hypothesis. *Sports Med* 48:1031-1048, 2018.

132. Invernizzi, PL, Limonta, E, Riboli, A, Bosio, A, Scurati, R, and Esposito, F. Effects of acute carnosine and β-alanine on isometric force and jumping performance. *Int J Sports Physiol Perform* 11:344-349, 2016.

133. Ivy, JL. Regulation of muscle glycogen repletion, muscle protein synthesis and repair following exercise. *J Sports Sci Med* 3:131, 2004.

134. Ivy, JL, and Portman, R. *Nutrient Timing*. Nashville, TN: Basic Health Publishing Company, 2004.

135. Jäger, R, Purpura, M, Shao, A, Inoue, T, and Kreider, RB. Analysis of the efficacy, safety, and regulatory status of novel forms of creatine. *Amino Acids* 40:1369-1383, 2011.

136. Jentjens, R, and Jeukendrup, AE. Determinants of post-exercise glycogen synthesis during short-term recovery. *Sports Med* 33:117-144, 2003.

137. Jeukendrup, A. A step towards personalized sports nutrition: Carbohydrate intake during exercise. *Sports Med* 44:25-33, 2014.

138. Jeukendrup, AE. Periodized nutrition for athletes. *Sports Med* 47:51-63, 2017.

139. Jones, NL, Sutton, JR, Taylor, R, and Toews, CJ. Effect of pH on cardiorespiratory and metabolic responses to exercise. *J Appl Physiol* 43:959-964, 1977.

140. Jones, RL, Stellingwerff, T, Artioli, GG, Saunders, B, Cooper, S, and Sale, C. Dose-response of sodium bicarbonate ingestion highlights individuality in time course of blood analyte responses. *Int J Sport Nutr Exerc Metab* 26:445-453, 2016.

141. Jouris, KB, McDaniel, JL, and Weiss, EP. The effect of omega-3 fatty acid supplementation on the inflammatory response to eccentric strength exercise. *J Sports Sci Med* 10:432-438, 2011.

142. Judelson, DA, Maresh, CM, Anderson, JM, Armstrong, LE, Casa, DJ, Kraemer, WJ, and Volek, JS. Hydration and muscular performance: Does fluid balance affect strength, power and high-intensity endurance? *Sports Med* 37:907-921, 2007.

143. Juhász I, Györe I, Csende Z, Racz L, and Tihanyi J. Creatine supplementation improves the anaerobic performance of elite junior fin swimmers. *Acta Physiol Hung* 96:325-336, 2009.

144. Kamimori, GH, Karyekar, CS, Otterstetter, R, Cox, DS, Balkin, TJ, Belenky, GL, and Eddington, ND. The rate of absorption and relative bioavailability of caffeine administered in chewing gum versus capsules to normal healthy volunteers. *Int J Pharm* 234:159-167, 2002.

145. Kasper, AM, Cocking, S, Cockayne, M, Barnard, M, Tench, J, Parker, L, McAndrew, J, Langan-Evans, C, Close, GL, and Morton, JP. Carbohydrate mouth rinse and caffeine improves high-intensity interval running capacity when carbohydrate restricted. *Eur J Sport Sci* 16:560-568, 2016.

146. Kendrick, IP, Harris, RC, Kim, HJ, Kim, CK, Dang, VH, Lam, TQ, Bui, TT, Smith, M, and Wise, JA. The effects of 10 weeks of resistance training combined with beta-alanine supplementation on whole body strength, force production, muscular endurance and body composition. *Amino Acids* 34:547-554, 2008.

147. Kerksick, CM, Arent, S, Schoenfeld, BJ, Stout, JR, Campbell, B, Wilborn, CD, Taylor, L, Kalman, D, Smith-Ryan, AE, Kreider, RB, Willoughby, D, Arciero, PJ, VanDusseldorp, TA, Ormsbee, MJ, Wildman, R, Greenwood, M, Ziegenfuss, TN, Aragon, AA, and Antonio, J. International Society of Sports Nutrition position stand: Nutrient timing. *J Int Soc Sports Nutr* 14:33, 2017.

148. Kerksick, CM, Wilborn, CD, Campbell, WI, Harvey, TM, Marcello, BM, Roberts, MD, Parker, AG, Byars, AG, Greenwood, LD, and Almada, AL. The effects of creatine monohydrate supplementation with and without D-pinitol on resistance training adaptations. *J Strength Cond Res* 23:2673-2682, 2009.

149. Kim, HJ, Kim, CK, Carpentier, A, and Poortmans, JR. Studies on the safety of creatine supplementation. *Amino Acids* 40:1409-1418, 2011.

150. Kim, J, Lee, J, Kim, S, Yoon, D, Kim, J, and Sung, DJ. Role of creatine supplementation in exercise-induced muscle damage: A mini review. *J Exerc Rehabil* 11:244-250, 2015.

151. King, A, Helms, E, Zinn, C, and Jukic, I. The ergogenic effects of acute carbohydrate feeding on resistance exercise performance: A systematic review and meta-analysis. *Sports Med* 52:2691-2712, 2022.

152. Kovacs, EM, Schmahl, RM, Senden, JM, and Brouns, F. Effect of high and low rates of fluid intake on post-exercise rehydration. *Int J Sport Nutr Exerc Metab* 12:14-23, 2002.

153. Kreider, RB. Effects of creatine supplementation on performance and training adaptations. *Mol Cell Biochem* 244:89-94, 2003.

154. Kreider, RB, Ferreira, M, Wilson, M, Grindstaff, P, Plisk, S, Reinardy, J, Cantler, E, and Almada, AL. Effects of creatine supplementation on body composition, strength, and sprint performance. *Med Sci Sports Exerc* 30:73-82, 1998.

155. Kreider, RB, and Jung, YP. Creatine supplementation in exercise, sport, and medicine. *J Exerc Nutr Biochem* 15:53-69, 2011.

156. Kreider, RB, Kalman, DS, Antonio, J, Ziegenfuss, TN, Wildman, R, Collins, R, Candow, DG, Kleiner, SM, Almada, AL, and Lopez, HL. International Society of Sports Nutrition position stand: Safety and efficacy of creatine supplementation in exercise, sport, and medicine. *J Int Soc Sports Nutr* 14:18, 2017.

157. Kreider, RB, Melton, C, Rasmussen, CJ, Greenwood, M, Lancaster, S, Cantler, EC, Milnor, P, and Almada, AL. Long-term creatine supplementation does not significantly affect clinical markers of health in athletes. *Mol Cell Biochem* 244:95-104, 2003.

158. Krustrup, P, Ermidis, G, and Mohr, M. Sodium bicarbonate intake improves high-intensity intermittent exercise performance in trained young men. *J Int Soc Sports Nutr* 12:25, 2015.

159. Lancha, AH, Jr, de Salles Painelli, V, Saunders, B, and Artioli, GG. Nutritional strategies to modulate intracellular and extracellular buffering capacity during high-intensity exercise. *Sports Med* 45:71-81, 2015.

160. Lanhers, C, Pereira, B, Naughton, G, Trousselard, M, Lesage, F-X, and Dutheil, F. Creatine supplementation and lower limb strength performance: A systematic review and meta-analyses. *Sports Med* 45:1285-1294, 2015.

161. Lanhers, C, Pereira, B, Naughton, G, Trousselard, M, Lesage, F-X, and Dutheil, F. Creatine supplementation and upper limb strength performance: A systematic review and meta-analysis. *Sports Med* 47:163-173, 2017.

162. Lee, JK, Shirreffs, SM, and Maughan, RJ. Cold drink ingestion improves exercise endurance capacity in the heat. *Med Sci Sports Exerc* 40:1637-1644, 2008.

163. Lee, JKW, and Shirreffs, SM. The influence of drink temperature on thermoregulatory responses during prolonged exercise in a moderate environment. *J Sports Sci* 25:975-985, 2007.

164. Lemon, PW. Do athletes need more dietary protein and amino acids? *Int J Sport Nutr* 5:S39-S61, 1995.

165. Leveritt, M, and Abernethy, PJ. Effects of carbohydrate restriction on strength performance. *J Strength Cond Res* 13:52-57, 1999.

166. Lim, MT, Pan, BJ, Toh, DWK, Sutanto, CN, and Kim, JE. Animal protein versus plant protein in supporting lean mass and muscle strength: A systematic review and meta-analysis of randomized controlled trials. *Nutrients* 13:661, 2021.

167. Lis, DM, Jordan, M, Lipuma, T, Smith, T, Schaal, K, and Baar, K. Collagen and vitamin C supplementation increases lower limb rate of force development. *Int J Sport Nutr Exerc Metab* 32:65-73, 2021.

168. Lobo, V, Patil, A, Phatak, A, and Chandra, N. Free radicals, antioxidants and functional foods: Impact on human health. *Pharmacogn Rev* 4:118-126, 2010.

169. Loucks, AB, Kiens, B, and Wright, HH. Energy availability in athletes. *J Sports Sci* 29:S7-S15, 2011.

170. Manore, MM. Effect of physical activity on thiamine, riboflavin, and vitamin B-6 requirements. *Am J Clin Nutr* 72:598S-606S, 2000.

171. Manore, MM. Dietary recommendations and athletic menstrual dysfunction. *Sports Med* 32:887-901, 2002.

172. Maridakis, V, O'Connor, PJ, Dudley, GA, and McCully, KK. Caffeine attenuates delayed-onset muscle pain and force loss following eccentric exercise. *J Pain* 8:237-243, 2007.

173. Marquet, L-A, Brisswalter, J, Louis, J, Tiollier, E, Burke, L, Hawley, J, and Hausswirth, C. Enhanced endurance performance by periodization of CHO intake: "Sleep low" strategy. *Med Sci Sports Exerc* 48:663-672, 2016.

174. Marriott, M, Krustrup, P, and Mohr, M. Ergogenic effects of caffeine and sodium bicarbonate supplementation on intermittent exercise performance preceded by intense arm cranking exercise. *J Int Soc Sports Nutr* 12:1-8, 2015.

175. Maté-Muñoz, JL, Lougedo, JH, Garnacho-Castaño, MV, Veiga-Herreros, P, Lozano-Estevan, MDC, García-Fernández, P, de Jesús, F, Guodemar-Pérez, J, San Juan, AF, and Domínguez, R. Effects of β-alanine supplementation during a 5-week strength training program: A randomized, controlled study. *J Int Soc Sports Nutr* 15:19, 2018.

176. Maughan, RJ. Role of micronutrients in sport and physical activity. *Br Med Bull* 55:683-690, 1999.

177. Maughan, RJ. Impact of mild dehydration on wellness and on exercise performance. *Eur J Clin Nutr* 57:S19-S23, 2003.

178. Maughan, RJ, Burke, LM, Dvorak, J, Larson-Meyer, DE, Peeling, P, Phillips, SM, Rawson, ES, Walsh, NP, Garthe, I, and Geyer, H. IOC consensus statement: Dietary supplements and the high-performance athlete. *Int J Sport Nutr Exerc Metab* 28:104-125, 2018.

179. Maughan, RJ, Depiesse, F, and Geyer, H. The use of dietary supplements by athletes. *J Sports Sci* 25:S103-S113, 2007.

180. McNaughton, LR. Bicarbonate ingestion: Effects of dosage on 60 s cycle ergometry. *J Sports Sci* 10:415-423, 1992.

181. Mero, AA, Keskinen, KL, Malvela, MT, and Sallinen, JM. Combined creatine and sodium bicarbonate supplementation enhances interval swimming. *J Strength Cond Res* 18:306-310, 2004.

182. Mettler, S, Mitchell, N, and Tipton, KD. Increased protein intake reduces lean body mass loss during weight loss in athletes. *Med Sci Sports Exerc* 42:326-337, 2010.

183. Mielgo-Ayuso, J, Marques-Jiménez, D, Refoyo, I, Del Coso, J, León-Guereño, P, and Calleja-González, J. Effect of caffeine supplementation on sports performance based on differences between sexes: A systematic review. *Nutrients* 11:2313, 2019.

184. Mielgo-Ayuso, J, Zourdos, MC, Calleja-González, J, Urdampilleta, A, and Ostojic, S. Iron supplementation prevents a decline in iron stores and enhances strength performance in elite female volleyball players during the competitive season. *Appl Physiol Nutr Metab* 40:615-622, 2015.

185. Milioni, F, de Poli, RAB, Saunders, B, Gualano, B, da Rocha, AL, da Silva, ASR, de Tarso Guerrero Muller, P, and Zagatto, AM. Effect of β-alanine supplementation during high-intensity interval training on repeated sprint ability performance and neuromuscular fatigue. *J Appl Physiol* 127:1599-1610, 2019.

186. Miller, P, Robinson, AL, Sparks, SA, Bridge, CA, Bentley, DJ, and McNaughton, LR. The effects of novel ingestion of sodium bicarbonate on repeated sprint ability. *J Strength Cond Res* 30:561-568, 2016.

187. Morton, RW, Murphy, KT, McKellar, SR, Schoenfeld, BJ, Henselmans, M, Helms, E, Aragon, AA, Devries, MC, Banfield, L, and Krieger, JW. A systematic review, meta-analysis and meta-regression of the effect of protein supplementation on resistance training-induced gains in muscle mass and strength in healthy adults. *Br J Sports Med* 52:376-384, 2018.

188. Mozaffarian, D, Katan, MB, Ascherio, A, Stampfer, MJ, and Willett, WC. Trans fatty acids and cardiovascular disease. *N Engl J Med* 354:1601-1613, 2006.

189. Mujika, I, Halson, S, Burke, LM, Balagué, G, and Farrow, D. An integrated, multifactorial approach to periodization for optimal performance in individual and team sports. *Int J Sports Physiol Perform* 13:538-561, 2018.

190. Murphy, MJ, Rushing, BR, Sumner, SJ, and Hackney, AC. Dietary supplements for athletic performance in women: Beta-alanine, caffeine, and nitrate. *Int J Sport Nutr Exerc Metab* 32:311-323, 2022.

191. Naderi, A, de Oliveira, EP, Ziegenfuss, TN, and Willems, MET. Timing, optimal dose and intake duration of dietary supplements with evidence-based use in sports nutrition. *J Exerc Nutr Biochem* 20:1-12, 2016.

192. Oppliger, RA, and Bartok, C. Hydration testing of athletes. *Sports Med* 32: 959-971, 2002.

193. Owens, DJ, Allison, R, and Close, GL. Vitamin D and the athlete: Current perspectives and new challenges. *Sports Med* 48:3-16, 2018.

194. Owens, DJ, Sharples, AP, Polydorou, I, Alwan, N, Donovan, T, Tang, J, Fraser, WD, Cooper, RG, Morton, JP, and Stewart, C. A systems-based investigation into vitamin D and skeletal muscle repair, regeneration, and hypertrophy. *Am J Physiol Endcrinol Metab* 309:E1019-E1031, 2015.

195. Owens, DJ, Tang, JCY, Bradley, WJ, Sparks, SA, Fraser, WD, Morton, JP, and Close, GL. Efficacy of high dose vitamin D supplements for elite athletes. *Med Sci Sports Exerc* 49:349-356, 2016.

196. Pallares, JG, Fernandez-Elias, VE, Ortega, JF, Munoz, G, Munoz-Guerra, J, and Mora-Rodriguez, R. Neuromuscular responses to incremental caffeine doses: Performance and side effects. *Med Sci Sports Exerc* 45:2184-2192, 2013.

197. Paxton, JZ, Grover, LM, and Baar, K. Engineering an in vitro model of a functional ligament from bone to bone. *Tissue Eng Part A* 16:3515-3525, 2010.

198. Phillips, SM. Protein requirements and supplementation in strength sports. *Nutrition* 20:689-695, 2004.

199. Phillips, SM. The science of muscle hypertrophy: Making dietary protein count. *Proc Nutr Soc* 70:100-103, 2011.

200. Phillips, SM. The impact of protein quality on the promotion of resistance exercise-induced changes in muscle mass. *Nutr Metab* 13:1-9, 2016.

201. Phillips, SM, Tipton, KD, Ferrando, AA, and Wolfe, RR. Resistance training reduces the acute exercise-induced increase in muscle protein turnover. *Am J Physiol Endocrinol Metab* 276:E118-E124, 1999.

202. Philpott, JD, Witard, OC, and Galloway, SDR. Applications of omega-3 polyunsaturated fatty acid supplementation for sport performance. *Res Sports Med* 27:219-237, 2019.

203. Pinckaers, PJM, Trommelen, J, Snijders, T, and van Loon, LJC. The anabolic response to plant-based protein ingestion. *Sports Med* 51:59-74, 2021.

204. Polito, MD, Souza, DB, Casonatto, J, and Farinatti, P. Acute effect of caffeine consumption on isotonic muscular strength and endurance: A systematic review and meta-analysis. *Science Sports* 31:119-128, 2016.

205. Powers, SK, and Jackson, MJ. Exercise-induced oxidative stress: Cellular mechanisms and impact on muscle force production. *Physiol Rev* 88:1243-1276, 2008.

206. Pruscino, CL, Ross, MLR, Gregory, JR, Savage, B, and Flanagan, TR. Effects of sodium bicarbonate, caffeine, and their combination on repeated 200-m freestyle performance. *Int J Sport Nutr Exerc Metab* 18:116-130, 2008.

207. Quesnele, JJ, Laframboise, MA, Wong, JJ, Kim, P, and Wells, GD. The effects of beta-alanine supplementation on performance: A systematic review of the literature. *Int J Sport Nutr Exerc Metab* 24:14-27, 2014.

208. Ramírez-Campillo, R, González-Jurado, JA, Martínez, C, Nakamura, FY, Peñailillo, L, Meylan, CMP, Caniuqueo, A, Cañas-Jamet, R, Moran, J, and Alonso-Martínez, AM. Effects of plyometric training and creatine supplementation on maximal-intensity exercise and endurance in female soccer players. *J Sci Med Sport* 19:682-687, 2016.

209. Ramos-Campo, DJ, Clemente-Suárez, VJ, Cupeiro, R, Benítez-Muñoz, JA, Andreu Caravaca, L, and Rubio-Arias, JÁ. The ergogenic effects of acute carbohydrate feeding on endurance performance: A systematic review, meta-analysis and meta-regression. *Crit Rev Food Sci Nutr* 14:1-10, 2023.

210. Rasmussen, BB, Tipton, KD, Miller, SL, Wolf, SE, and Wolfe, RR. An oral essential amino acid-carbohydrate supplement enhances muscle protein anabolism after resistance exercise. *J Appl Physiol* 88:386-392, 2000.

211. Rawson, ES, Conti, MP, and Miles, MP. Creatine supplementation does not reduce muscle damage or enhance recovery from resistance exercise. *J Strength Cond Res* 21:1208-1213, 2007.

212. Raya-González, J, Rendo-Urteaga, T, Domínguez, R, Castillo, D, Rodríguez-Fernández, A, and Grgic, J. Acute effects of caffeine supplementation on movement velocity in resistance exercise: A systematic review and meta-analysis. *Sports Med* 50:717-729, 2020.

213. Robinson, TM, Sewell, DA, Casey, A, Steenge, G, and Greenhaff, PL. Dietary creatine supplementation does not affect some haematological indices, or indices of muscle damage and hepatic and renal function. *Br J Sports Med* 34:284-288, 2000.

214. Rogers, DR, Lawlor, DJ, and Moeller, JL. Vitamin C supplementation and athletic performance: A review. *Curr Sports Med Rep* 22:255-259, 2023.

215. Rosene, J, Matthews, T, Ryan, C, Belmore, K, Bergsten, A, Blaisdell, J, Gaylord, J, Love, R, Marrone, M, Ward, K, and Wilson, E. Short and longer-term effects of creatine supplementation on exercise induced muscle damage. *J Sports Sci Med* 8:89-96, 2009.

216. Roy, SR, and Irwin, W. *Sports Medicine.* Englewood Cliffs, NJ: Prentice Hall, 1983.

217. Sabol, F, Grgic, J, and Mikulic, P. The effects of 3 different doses of caffeine on jumping and throwing performance: A randomized, double-blind, crossover study. *Int J Sports Physiol Perform* 14:1170-1177, 2019.

218. Sale, C, Hill, CA, Ponte, J, and Harris, RC. β-alanine supplementation improves isometric endurance of the knee extensor muscles. *J Int Soc Sports Nutr* 9:26, 2012.

219. Sale, C, Saunders, B, and Harris, RC. Effect of beta-alanine supplementation on muscle carnosine concentrations and exercise performance. *Amino Acids* 39:321-333, 2010.

220. Sale, C, Saunders, B, Hudson, S, Wise, JA, Harris, RC, and Sunderland, CD. Effect of β-alanine plus sodium bicarbonate on high-intensity cycling capacity. *Med Sci Sports Exerc* 43:1972-1978, 2011.

221. Salinero, JJ, Lara, B, and Del Coso, J. Effects of acute ingestion of caffeine on team sports performance: A systematic review and meta-analysis. *Res Sports Med* 27:238-256, 2019.

222. Saltin, B, and Stenberg, J. Circulatory response to prolonged severe exercise. *J Appl Physiol* 19:833-838, 1964.

223. Saunders, B, de Oliveira, LF, Dolan, E, Durkalec-Michalski, K, McNaughton, L, Artioli, GG, and Swinton, PA. Sodium bicarbonate supplementation and the female athlete: A brief commentary with small scale systematic review and meta-analysis. *Eur J Sport Sci* 22:745-754, 2022.

224. Savoie, F-A, Kenefick, RW, Ely, BR, Cheuvront, SN, and Goulet, EDB. Effect of hypohydration on muscle endurance, strength, anaerobic power and capacity and vertical jumping ability: A meta-analysis. *Sports Med* 45:1207-1227, 2015.

225. Sawka, MN, Burke, LM, Eichner, ER, Maughan, RJ, Montain, SJ, and Stachenfeld, NS. American College of Sports Medicine position stand. Exercise and fluid replacement. *Med Sci Sports Exerc* 39:377-390, 2007.

226. Scheid, JL, and Lupien, SP. Fitness watches and nutrition apps: Behavioral benefits and emerging concerns. *ACSM Health Fitness J* 25:21-25, 2021.

227. Schlattner, U, Klaus, A, Ramirez Rios, S, Guzun, R, Kay, L, and Tokarska-Schlattner, M. Cellular compartmentation of energy metabolism: Creatine kinase microcompartments and recruitment of B-type creatine kinase to specific subcellular sites. *Amino Acids* 48:1751-1774, 2016.

228. Schoenfeld, BJ, Aragon, AA, and Krieger, JW. The effect of protein timing on muscle strength and hypertrophy: A meta-analysis. *J Int Soc Sports Nutr* 10:53, 2013.

229. Schoffstall, JE, Branch, JD, Leutholtz, BC, and Swain, DE. Effects of dehydration and rehydration on the one-repetition maximum bench press of weight-trained males. *J Strength Cond Res* 15:102-108, 2001.

230. Shaw, G, Lee-Barthel, A, Ross, MLR, Wang, B, and Baar, K. Vitamin C–enriched gelatin supplementation before intermittent activity augments collagen synthesis. *Am J Clin Nutr* 105:136-143, 2017.

231. Siegler, JC, and Gleadall-Siddall, DO. Sodium bicarbonate ingestion and repeated swim sprint performance. *J Strength Cond Res* 24:3105-3111, 2010.

232. Silva, AJ, Machado Reis, V, Guidetti, L, Bessone Alves, F, Mota, P, Freitas, J, and Baldari, C. Effect of creatine on swimming velocity, body composition and hydrodynamic variables. *J Sports Med Phys Fitness* 47:58-64, 2007.

233. Simopoulos, AP. The importance of the ratio of omega-6/omega-3 essential fatty acids. *Biomed Pharmacother* 56:365-379, 2002.

234. Sist, M, Galloway, SDR, and Rodriguez-Sanchez, N. Effects of vitamin D supplementation on maximal strength and power in athletes: A systematic review and meta-analysis of randomized controlled trials. *Front Nutr* 10:1163313, 2023.

235. Smith-Ryan, AE, Hirsch, KR, Saylor, HE, Gould, LM, and Blue, MNM. Nutritional considerations and strategies to facilitate injury recovery and rehabilitation. *J Athl Train* 55:918-930, 2020.

236. Smith, AE, Walter, AA, Graef, JL, Kendall, KL, Moon, JR, Lockwood, CM, Fukuda, DH, Beck, TW, Cramer, JT, and Stout, JR. Effects of beta-alanine supplementation and high-intensity interval training on endurance performance and body composition in men; A double-blind trial. *J Int Soc Sports Nutr* 6:5, 2009.

237. Snijders, T, Smeets, JSJ, van Vliet, S, van Kranenburg, J, Maase, K, Kies, AK, Verdijk, LB, and van Loon, LJC. Protein ingestion before sleep increases muscle mass and strength gains during prolonged resistance-type exercise training in healthy young men. *J Nutr* 145:1178-1184, 2015.

238. Southward, K, Rutherfurd-Markwick, K, Badenhorst, C, and Ali, A. The role of genetics in moderating the inter-individual differences in the ergogenicity of caffeine. *Nutrients* 10:1352, 2018.

239. Steenge, GR, Simpson, EJ, and Greenhaff, PL. Protein- and carbohydrate-induced augmentation of whole body creatine retention in humans. *J Appl Physiol* 89:1165-1171, 2000.

240. Stegen, S, Blancquaert, L, Everaert, I, Bex, T, Taes, Y, Calders, P, Achten, E, and Derave, W. Meal and beta-alanine coingestion enhances muscle carnosine loading. *Med Sci Sports Exerc* 45:1478-1485, 2013.

241. Stellingwerff, T, Anwander, H, Egger, A, Buehler, T, Kreis, R, Decombaz, J, and Boesch, C. Effect of two β-alanine dosing protocols on muscle carnosine synthesis and washout. *Amino Acids* 42:2461-2472, 2012.

242. Stellingwerff, T, Decombaz, J, Harris, RC, and Boesch, C. Optimizing human in vivo dosing and delivery of β-alanine supplements for muscle carnosine synthesis. *Amino Acids* 43:57-65, 2012.

243. Stellingwerff, T, Spriet, LL, Watt, MJ, Kimber, NE, Hargreaves, M, Hawley, JA, and Burke, LM. Decreased PDH activation and glycogenolysis during exercise following fat adaptation with carbohydrate restoration. *Am J Physiol Endocrinol Metab* 290:E380-E388, 2006.

244. Stockton, KA, Mengersen, K, Paratz, JD, Kandiah, D, and Bennell, KL. Effect of vitamin D supplementation on muscle strength: A systematic review and meta-analysis. *Osteoporos Int* 22:859-871, 2011.

245. Stone, MH, Fleck, SJ, Triplett, NT, and Kraemer, WJ. Health- and performance-related potential of resistance training. *Sports Med* 11:210-231, 1991.

246. Stone, MH, and Karatzeferi, C. Connective tissue (and bone) response to strength training. In *Strength and Power in Sport*. Komi, PV, ed. Oxford: Blackwell Scientific, 2002.

247. Stone, MH, Keith, RE, Kearney, JT, Fleck, SJ, Wilson, GD, and Triplett, NT. Overtraining: A review of the signs, symptoms and possible causes. *J Strength Cond Res* 5:35-50, 1991.

248. Stone, MH, Sanborn, K, Smith, LL, O'Bryant, HS, Hoke, T, Utter, AC, Johnson, RL, Boros, R, Hruby, J, Pierce, KC, Stone, ME, and Garner, B. Effects of in-season (5 weeks) creatine and pyruvate supplementation on anaerobic performance and body composition in American football players. *Int J Sport Nutr* 9:146-165, 1999.

249. Stone, MH, Stone, M, and Sands, WA. *Principles and Practice of Resistance Training*. Champaign, IL: Human Kinetics, 2007.

250. Stone, MH, Suchomel, TJ, Hornsby, WG, Wagle, JP, and Cunanan, AJ. Nutrition and metabolic factors. In *Strength and Conditioning in Sports: From Science to Practice*. New York: Routledge, 145-180, 2022.

251. Stout, JR, Cramer, JT, Mielke, M, O'Kroy, J, Torok, DJ, and Zoeller, RF. Effects of twenty-eight days of beta-alanine and creatine monohydrate supplementation on the physical working capacity at neuromuscular fatigue threshold. *J Strength Cond Res* 20:928-931, 2006.

252. Stout, JR, Cramer, JT, Zoeller, RF, Torok, D, Costa, P, Hoffman, JR, Harris, RC, and O'Kroy, J. Effects of beta-alanine supplementation on the onset of neuromuscular fatigue and ventilatory threshold in women. *Amino Acids* 32:381-386, 2007.

253. Sun, R, Sun, J, Li, J, and Li, S. Effects of caffeine ingestion on physiological indexes of human neuromuscular fatigue: A systematic review and meta-analysis. *Brain Behav* 12:e2529, 2022.

254. Suominen, H. Bone mineral density and long term exercise. An overview of cross-sectional athlete studies. *Sports Med* 16:316-330, 1993.

255. Suzuki, T, Mizuta, C, Uda, K, Ishida, K, Mizuta, K, Sona, S, Compaan, DM, and Ellington, WR. Evolution and divergence of the genes for cytoplasmic, mitochondrial, and flagellar creatine kinases. *J Mol Evol* 59:218-226, 2004.

256. Suzuki, Y, Nakao, T, Maemura, H, Sato, M, Kamahara, K, Morimatsu, F, and Takamatsu, K. Carnosine and anserine ingestion enhances contribution of nonbicarbonate buffering. *Med Sci Sports Exerc* 38:334-338, 2006.

257. Tarnopolsky, MA. Protein and amino acid needs for training and bulking up. In *Clinical Sports Nutrition*. L Burke, V Deakin, eds. Rossville, NSW: McGraw Hill Australia, 90-117, 2006s.

258. Tarnopolsky, MA, and MacLennan, DP. Creatine monohydrate supplementation enhances high-intensity exercise performance in males and females. *Int J Sport Nutr Exerc Metab* 10:452-463, 2000.

259. Thompson, D, Williams, C, McGregor, SJ, Nicholas, CW, McArdle, F, Jackson, MJ, and Powell, JR. Prolonged vitamin C supplementation and recovery from demanding exercise. *Int J Sport Nutr Exerc Metab* 11:466-481, 2001.

260. Tinsley, GM, Hamm, MA, Hurtado, AK, Cross, AG, Pineda, JG, Martin, AY, Uribe, VA, and Palmer, TB. Effects of two pre-workout supplements on concentric and eccentric force production during lower body resistance exercise in males and females: A counterbalanced, double-blind, placebo-controlled trial. *J Int Soc Sports Nutr* 14:46, 2017.

261. Tipton, KD, Rasmussen, BB, Miller, SL, Wolf, SE, Owens-Stovall, SK, Petrini, BE, and Wolfe, RR. Timing of amino acid-carbohydrate ingestion alters anabolic response of muscle to resistance exercise. *Am J Physiol* 281:E197-E206, 2001.

262. Tomlinson, PB, Joseph, C, and Angioi, M. Effects of vitamin D supplementation on upper and lower body muscle strength levels in healthy individuals. A systematic review with meta-analysis. *J Sci Med Sport* 18:575-580, 2015.

263. Trexler, ET, and Smith-Ryan, AE. Creatine and caffeine: Considerations for concurrent supplementation. *Int J Sport Nutr Exerc Metab* 25:607-623, 2015.

264. Trexler, ET, Smith-Ryan, AE, Stout, JR, Hoffman, JR, Wilborn, CD, Sale, C, Kreider, RB, Jäger, R, Earnest, CP, and Bannock, L. International Society of Sports Nutrition position stand: Beta-alanine. *J Int Soc Sports Nutr* 12:30, 2015.

265. Trommelen, J, and Van Loon, LJC. Pre-sleep protein ingestion to improve the skeletal muscle adaptive response to exercise training. *Nutrients* 8:763, 2016.

266. Tyler, TF, Nicholas, SJ, Hershman, EB, Glace, BW, Mullaney, MJ, and McHugh, MP. The effect of creatine supplementation on strength recovery after anterior cruciate ligament (ACL) reconstruction: A randomized, placebo-controlled, double-blind trial. *Am J Sports Med* 32:383-388, 2004.

267. van der Beek, EJ. Vitamin supplementation and physical exercise performance. *J Sports Sci* 9:77-90, 1991.

268. Van Gammeren, D. Vitamins and minerals. In *Essentials of Sports Nutrition and Supplements*. Antonio, J, Kalman, D, Stout, JR, Greenwood, M, Willoughby, DS, and Haff, GG, eds. New York: Springer, 313-328, 2009.

269. Van Loon, LJC, Greenhaff, PL, Constantin-Teodosiu, D, Saris, WHM, and Wagenmakers, AJM. The effects of increasing exercise intensity on muscle fuel utilisation in humans. *J Physiol* 536:295-304, 2001.

270. Van Thienen, R, Van Proeyen, K, Vanden Eynde, B, Puype, J, Lefere, T, and Hespel, P. Beta-alanine improves sprint performance in endurance cycling. *Med Sci Sports Exerc* 41:898-903, 2009.

271. Vandenberghe, K, Gillis, N, Van Leemputte, M, Van Hecke, P, Vanstapel, F, and Hespel, P. Caffeine counteracts the ergogenic action of muscle creatine loading. *J Appl Physiol* 80:452-457, 1996.

272. Vandenberghe, K, Goris, M, Van Hecke, P, Van Leemputte, M, Vangerven, L, and Hespel, P. Long-term creatine intake is beneficial to muscle performance during resistance training. *J Appl Physiol* 83:2055-2063, 1997.

273. Vandenbogaerde, TJ, and Hopkins, WG. Effects of acute carbohydrate supplementation on endurance performance: A meta-analysis. *Sports Med* 41:773-792, 2011.

274. Varanoske, AN, Hoffman, JR, Church, DD, Coker, NA, Baker, KM, Dodd, SJ, Harris, RC, Oliveira, LP, Dawson, VL, Wang, R, Fukuda, DH, and Stout, JR. Comparison of sustained-release and rapid-release β-alanine formulations on changes in skeletal muscle carnosine and histidine content and isometric performance following a muscle-damaging protocol. *Amino Acids* 51:49-60, 2019.

275. Vist, GE, and Maughan, RJ. The effect of osmolality and carbohydrate content on the rate of gastric emptying of liquids in man. *J Physiol* 486:523-531, 1995.

276. Volek, JS, Kraemer, WJ, Bush, JA, Boetes, M, Incledon, T, Clark, KL, and Lynch, JM. Creatine supplementation enhances muscular performance during high-intensity resistance exercise. *J Am Diet Assoc* 97:765-770, 1997.

277. Volek, JS, Ratamess, NA, Rubin, MR, Gomez, AL, French, DN, McGuigan, MR, Scheett, TP, Sharman, MJ, Häkkinen, K, and Kraemer, WJ. The effects of creatine supplementation on muscular performance and body composition responses to short-term resistance training overreaching. *Eur J Appl Physiol* 91:628-637, 2004.

278. Wallimann, T, Tokarska-Schlattner, M, and Schlattner, U. The creatine kinase system and pleiotropic effects of creatine. *Amino Acids* 40:1271-1296, 2011.

279. Walsh, NP. Nutrition and athlete immune health: New perspectives on an old paradigm. *Sports Med* 49:153-168, 2019.

280. Walsh, RM, Noakes, TD, Hawley, JA, and Dennis, SC. Impaired high-intensity cycling performance time at low levels of dehydration. *Int J Sports Med* 15:392-398, 1994.

281. Warren, GL, Park, ND, Maresca, RD, McKibans, KI, and Millard-Stafford, ML. Effect of caffeine ingestion on muscular strength and endurance: A meta-analysis. *Med Sci Sports Exerc* 42:1375-1387, 2010.

282. WHO (World Health Organization). Protein and amino acid requirements in human nutrition. *World Health Organ Tech Rep Ser* 935:1, 2007.

283. Wickham, KA, and Spriet, LL. Administration of caffeine in alternate forms. *Sports Med* 48:79-91, 2018.

284. Wilmore, JH and Costill, DL. *Physiology of Sport and Exercise*. Champaign, IL: Human Kinetics, 1994.

285. Wilson, MM, and Morley, JE. Impaired cognitive function and mental performance in mild dehydration. *Eur J Clin Nutr* 57:S24-S29, 2003.

286. Woolf, K, and Manore, MM. B-vitamins and exercise: Does exercise alter requirements? *Int J Sport Nutr Exerc Metab* 16:453-484, 2006.

287. Yamaguchi, GC, Nemezio, K, Schulz, ML, Natali, J, Cesar, JE, Riani, LA, Gonçalves, LS, Möller, GB, Sale, C, de Medeiros, MHG, Gualano, B, and Artioli, GG. Kinetics of muscle carnosine decay after β-alanine supplementation: A 16-wk washout study. *Med Sci Sports Exerc* 53:1079-1088, 2021.

288. Yeo, WK, Carey, AL, Burke, L, Spriet, LL, and Hawley, JA. Fat adaptation in well-trained athletes: Effects on cell metabolism. *Appl Physiol Nutr Metab* 36:12-22, 2011.

289. Yeo, WK, Paton, CD, Garnham, AP, Burke, LM, Carey, AL, and Hawley, JA. Skeletal muscle adaptation and performance responses to once a day versus twice every second day endurance training regimens. *J Appl Physiol* 105:1462-1470, 2008.

290. Zajac, A, Cholewa, J, Poprzecki, S, Waskiewicz, Z, and Langfort, J. Effects of sodium bicarbonate ingestion on swim performance in youth athletes. *J Sports Sci Med* 8:45-50, 2009.

Chapter 12

1. Abaidia, A-E, Lamblin, J, Delecroix, B, Leduc, C, Mccall, A, Nédélec, M, Dawson, B, Baquet, G, and Dupont, G. Recovery from exercise-induced muscle damage: Cold-water immersion versus whole-body cryotherapy. *Int J Sports Physiol Perform* 12:402-409, 2017.

2. Afonso, J, Clemente, FM, Nakamura, FY, Morouço, P, Sarmento, H, Inman, RA, and Ramirez-Campillo, R. The effectiveness of post-exercise stretching in short-term and delayed recovery of strength, range of motion and delayed onset muscle

soreness: A systematic review and meta-analysis of randomized controlled trials. *Front Physiol* 12:25, 2021.

3. Babault, N, Cometti, C, Maffiuletti, NA, and Deley, G. Does electrical stimulation enhance post-exercise performance recovery? *Eur J Appl Physiol* 111:2501-2507, 2011.

4. Banfi, G, Lombardi, G, Colombini, A, and Melegati, G. Whole-body cryotherapy in athletes. *Sports Med* 40:509-517, 2010.

5. Barnett, A. Using recovery modalities between training sessions in elite athletes: Does it help? *Sports Med* 36:781-796, 2006.

6. Beaven, CM, Cook, C, Gray, D, Downes, P, Murphy, I, Drawer, S, Ingram, JR, Kilduff, LP, and Gill, N. Electrostimulation's enhancement of recovery during a rugby preseason. *Int J Sports Physiol Perform* 8:92-98, 2013.

7. Beckham, GK, Suchomel, TJ, and Mizuguchi, S. Force plate use in performance monitoring and sport science testing. *New Stud Athl* 29:25-37, 2014.

8. Bieuzen, F. Water-immersion therapy. In *Recovery for Performance in Sport*. Hausswirth, C, and Mujika, I, eds. Champaign, IL: Human Kinetics, 191-199, 2013.

9. Bieuzen, F, Bleakley, CM, and Costello, JT. Contrast water therapy and exercise induced muscle damage: A systematic review and meta-analysis. *PLoS One* 8:e62356, 2013.

10. Biggins, M, Purtill, H, Fowler, P, Bender, A, Sullivan, KO, Samuels, C, and Cahalan, R. Sleep in elite multi-sport athletes: Implications for athlete health and wellbeing. *Phys Ther Sport* 39:136-142, 2019.

11. Bleakley, CM, Bieuzen, F, Davison, GW, and Costello, JT. Whole-body cryotherapy: Empirical evidence and theoretical perspectives. *Open Access J Sports Med* 5:25-36, 2014.

12. Boksem, MA, and Tops, M. Mental fatigue: Costs and benefits. *Brain Res Rev* 59:125-139, 2008.

13. Bonnar, D, Bartel, K, Kakoschke, N, and Lang, C. Sleep interventions designed to improve athletic performance and recovery: A systematic review of current approaches. *Sports Med* 48:683-703, 2018.

14. Brauer, AA. Prevalence and causes of sleep problems in athletes. *Curr Sleep Med Rep* 8:180-186, 2022.

15. Broatch, JR, Bishop, DJ, and Halson, S. Lower limb sports compression garments improve muscle blood flow and exercise performance during repeated-sprint cycling. *Int J Sports Physiol Perform* 13:882-890, 2018.

16. Brown, F, Gissane, C, Howatson, G, Van Someren, K, Pedlar, C, and Hill, J. Compression garments and recovery from exercise: A meta-analysis. *Sports Med* 47:2245-2267, 2017.

17. Brown, F, Jeffries, O, Gissane, C, Howatson, G, van Someren, K, Pedlar, C, Myers, T, and Hill, JA. Custom-fitted compression garments enhance recovery from muscle damage in rugby players. *J Strength Cond Res* 36:212-219, 2022.

18. Cafarelli, E, Sim, J, Carolan, B, and Liebesman, J. Vibratory massage and short-term recovery from muscular fatigue. *Int J Sports Med* 11:474-478, 1990.

19. Caia, J, Scott, TJ, Halson, SL, and Kelly, VG. The influence of sleep hygiene education on sleep in professional rugby league athletes. *Sleep Health* 4:364-368, 2018.

20. Capps, SG, and Brook, M. Cryotherapy and intermittent pneumatic compression for soft tissue trauma. *Int J Athl Ther Train* 14:2-4, 2009.

21. Chase, JE. The impact of a single intermittent pneumatic compression bout on performance, inflammatory markers, and myoglobin in football athletes (master's thesis). Manitoba, Canada: University of Manitoba, 129, 2017.

22. Chleboun, GS, Howell, JN, Baker, HL, Ballard, TN, Graham, JL, Hallman, HL, Perkins, LE, Schauss, JH, and Conatser, RR. Intermittent pneumatic compression effect on eccentric exercise-induced swelling, stiffness, and strength loss. *Arch Phys Med Rehab* 76:744-749, 1995.

23. Cochrane, DJ, Booker, HR, Mundel, T, and Barnes, MJ. Does intermittent pneumatic leg compression enhance muscle recovery after strenuous eccentric exercise? *Int J Sports Med* 34:969-974, 2013.

24. Cook, JD, and Charest, J. Sleep and performance in professional athletes. *Curr Sleep Med Rep* 9:56-81, 2023.

25. Córdova, A, Sureda, A, Albina, ML, Linares, V, Bellés, M, and Sánchez, DJ. Oxidative stress markers after a race in professional cyclists. *Int J Sport Nutr Exerc Metab* 25:171-178, 2015.

26. Coté, DJ, Prentice, WE, Jr, Hooker, DN, and Shields, EW. Comparison of three treatment procedures for minimizing ankle sprain swelling. *Phys Ther* 68:1072-1076, 1988.

27. Counsilman, JE, and Counsilman, BE. *The New Science of Swimming*. Englewood Cliffs, NJ: Prentice Hall, 1994.

28. Coutts, A, Reaburn, P, Piva, T, and Rowswell, G. Markers of recovery in team sport athletes. *J Sci Med Sport* 5:S6, 2002.

29. Crampton, D, Donne, B, Egaña, M, and Warmington, SA. Sprint cycling performance is maintained with short-term contrast water immersion. *Med Sci Sports Exerc* 43:2180-2188, 2011.

30. Craven, J, McCartney, D, Desbrow, B, Sabapathy, S, Bellinger, P, Roberts, L, and Irwin, C. Effects of acute sleep loss on physical performance: A systematic and meta-analytical review. *Sports Med* 52:2669-2690, 2022.

31. Cunanan, AJ, DeWeese, BH, Wagle, JP, Carroll, KM, Sausaman, R, Hornsby, WG, Haff, GG, Triplett, NT, Pierce, KC, and Stone, MH. The general adaptation syndrome: A foundation for the concept of periodization. *Sports Med* 48:787-797, 2018.

32. Dattilo, M, Antunes, HKM, Medeiros, A, Neto, MM, Souza, HS, Tufik, S, and de Mello, MT. Sleep and muscle recovery: Endocrinological and molecular basis for a new and promising hypothesis. *Med Hypotheses* 77:220-222, 2011.

33. Davis, HL, Alabed, S, and Chico, TJA. Effect of sports massage on performance and recovery: A systematic review and meta-analysis. *BMJ Open Sport Exerc Med* 6:e000614, 2020.

34. Dawson, LG, Dawson, KA, and Tiidus, PM. Evaluating the influence of massage on leg strength, swelling, and pain following a half-marathon. *J Sports Sci Med* 3:37-43, 2004.

35. Delextrat, A, Calleja-González, J, Hippocrate, A, and Clarke, ND. Effects of sports massage and intermittent cold-water immersion on recovery from matches by basketball players. *J Sports Sci* 31:11-19, 2013.

36. Draper, SN, Kullman, EL, Sparks, KE, Little, K, and Thoman, J. Effects of intermittent pneumatic compression on delayed onset muscle soreness (DOMS) in long distance runners. *Int J Exerc Sci* 13:75-86, 2020.

37. Duffield, R, Murphy, A, Kellett, A, and Reid, M. Recovery from repeated on-court tennis sessions: Combining cold-water immersion, compression, and sleep interventions. *Int J Sports Physiol Perform* 9:273-282, 2014.

38. Dupuy, O, Douzi, W, Theurot, D, Bosquet, L, and Dugué, B. An evidence-based approach for choosing post-exercise recovery techniques to reduce markers of muscle damage, soreness, fatigue, and inflammation: A systematic review with meta-analysis. *Front Physiol* 9:403, 2018.

39. Farr, T, Nottle, C, Nosaka, K, and Sacco, P. The effects of therapeutic massage on delayed onset muscle soreness and muscle function following downhill walking. *J Sci Med Sport* 5:297-306, 2002.

40. Ferguson, RA, Dodd, MJ, and Paley, VR. Neuromuscular electrical stimulation via the peroneal nerve is superior to graduated compression socks in reducing perceived muscle soreness following intense intermittent endurance exercise. *Eur J Appl Physiol* 114:2223-2232, 2014.

41. Ferreira-Junior, JB, Bottaro, M, Loenneke, JP, Vieira, A, Vieira, CA, and Bemben, MG. Could whole-body cryotherapy (below −100° C) improve muscle recovery from muscle damage? *Front Physiol* 5:247, 2014.

42. Fielding, R, Riede, L, Lugo, JP, and Bellamine, A. L-carnitine supplementation in recovery after exercise. *Nutrients* 10:349, 2018.

43. Fonda, B, and Sarabon, N. Effects of whole-body cryotherapy on recovery after hamstring damaging exercise: A crossover study. *Scand J Med Sci Sports* 23:e270-e278, 2013.

44. Fowler, PM, Knez, W, Thornton, HR, Sargent, C, Mendham, AE, Crowcroft, S, Miller, J, Halson, S, and Duffield, R. Sleep hygiene and light exposure can improve performance following long-haul air travel. *Int J Sports Physiol Perform* 16:517-526, 2020.

45. French, DN, Kraemer, WJ, Volek, JS, Spiering, BA, Judelson, DA, Hoffman, JR, and Maresh, CM. Anticipatory responses of catecholamines on muscle force production. *J Appl Physiol* 102:94-102, 2007.

46. French, DN, Thompson, KG, Garland, SW, Barnes, CA, Portas, MD, Hood, PE, and Wilkes, G. The effects of contrast bathing and compression therapy on muscular performance. *Med Sci Sports Exerc* 40:1297-1306, 2008.

47. Gill, ND, Beaven, CM, and Cook, C. Effectiveness of post-match recovery strategies in rugby players. *Br J Sports Med* 40:260-263, 2006.

48. Gilson, SF, Saunders, MJ, Moran, CW, Moore, RW, Womack, CJ, and Todd, MK. Effects of chocolate milk consumption on markers of muscle recovery following soccer training: A randomized cross-over study. *J Int Soc Sports Nutr* 7:19, 2010.

49. Gregson, W, Black, MA, Jones, H, Milson, J, Morton, J, Dawson, B, Atkinson, G, and Green, DJ. Influence of cold water immersion on limb and cutaneous blood flow at rest. *Am J Sports Med* 39:1316-1323, 2011.

50. Guilhem, G, Hug, F, Couturier, A, Regnault, S, Bournat, L, Filliard, J-R, and Dorel, S. Effects of air-pulsed cryotherapy on neuromuscular recovery subsequent to exercise-induced muscle damage. *Am J Sports Med* 41:1942-1951, 2013.

51. Habay, J, Van Cutsem, J, Verschueren, J, De Bock, S, Proost, M, De Wachter, J, Tassignon, B, Meeusen, R, and Roelands, B. Mental fatigue and sport-specific psychomotor performance: A systematic review. *Sports Med* 51:1527-1548, 2021.

52. Halson, SL. Nutrition, sleep and recovery. *Eur J Sport Sci* 8:119-126, 2008.

53. Halson, SL. Does the time frame between exercise influence the effectiveness of hydrotherapy for recovery? *Int J Sports Physiol Perform* 6:147-159, 2011.

54. Halson, SL. Sleep in elite athletes and nutritional interventions to enhance sleep. *Sports Med* 44:13-23, 2014.

55. Halson, SL, Bartram, J, West, N, Stephens, J, Argus, CK, Driller, MW, Sargent, C, Lastella, M, Hopkins, WG, and Martin, DT. Does hydrotherapy help or hinder adaptation to training in competitive cyclists? *Med Sci Sports Exerc* 46:1631-1639, 2014.

56. Halson, SL, and Martin, DT. Lying to win—Placebos and sport science. *Int J Sports Physiol Perform* 8:597-599, 2013.

57. Harre, D. *Principles of Sports Training*. Berlin: Sportverlag, 1982.

58. Hausswirth, C, Louis, J, Bieuzen, F, Pournot, H, Fournier, J, Filliard, J-R, and Brisswalter, J. Effects of whole-body cryotherapy vs. far-infrared vs. passive modalities on recovery from exercise-induced muscle damage in highly-trained runners. *PLoS One* 6:e27749, 2011.

59. Hausswirth, C, and Mujika, I, eds. *Recovery for Performance in Sport*. Champaign, IL: Human Kinetics, 2013.

60. Havas, E, Parviainen, T, Vuorela, J, Toivanen, J, Nikula, T, and Vihko, V. Lymph flow dynamics in exercising human skeletal muscle as detected by scintography. *J Physiol* 504:233-239, 1997.

61. Heapy, AM, Hoffman, MD, Verhagen, HH, Thompson, SW, Dhamija, P, Sandford, FJ, and Cooper, MC. A randomized controlled trial of manual therapy and pneumatic compression for recovery from prolonged running—An extended study. *Res Sports Med* 26:354-364, 2018.

62. Hemmings, BJ. Physiological, psychological and performance effects of massage therapy in sport: A review of the literature. *Phys Ther Sport* 2:165-170, 2001.

63. Higgins, TR, Greene, DA, and Baker, MK. Effects of cold water immersion and contrast water therapy for recovery from team sport: A systematic review and meta-analysis. *J Strength Cond Res* 31:1443-1460, 2017.

64. Hilbert, JE, Sforzo, GA, and Swensen, T. The effects of massage on delayed onset muscle soreness. *Br J Sports Med* 37:72-75, 2003.

65. Hoffman, MD, Badowski, N, Chin, J, and Stuempfle, KJ. A randomized controlled trial of massage and pneumatic compression for ultramarathon recovery. *J Orthop Sports Phys Ther* 46:320-326, 2016.

66. Hooper, SL, MacKinnon, LT, and Hanrahan, S. Mood states as an indication of staleness and recovery. *Int J Sport Psychol* 28:1-12, 1997.

67. Hooper, SL, Mackinnon, LT, Howard, A, Gordon, RD, and Bachmann, AW. Markers for monitoring overtraining and recovery. *Med Sci Sports Exerc* 27:106-112, 1995.

68. Issurin, VB. New horizons for the methodology and physiology of training periodization. *Sports Med* 40:189-206, 2010.

69. Ivy, JL, and Portman, R. *Nutrient Timing*. Nashville, TN: Basic Health Publishing Company, 2004.

70. Kellmann, M, Altenburg, D, Lormes, W, and Steinacker, JM. Assessing stress and recovery during preparation for the world championships in rowing. *Sport Psychol* 15:151-167, 2001.

71. Kellmann, M, and Kallus, K. *The Recovery-Stress Questionnaire for Athletes: User Manual*. Champaign, IL: Human Kinetics, 2001.

72. Kelly, V, Holmberg, P, and Jenkins, D. Strategies to enhance athlete recovery. In *Advanced Strength and Conditioning: An Evidence-Based Approach*. Turner, AN, and Comfort, P, eds. New York: Routledge, 133-154, 2022.

73. Khanna, A, Gougoulias, N, and Maffulli, N. Intermittent pneumatic compression in fracture and soft-tissue injuries healing. *Br Med Bull* 88:147-156, 2008.

74. Kosar, AC, Candow, DG, and Putland, JT. Potential beneficial effects of whole-body vibration for muscle recovery after exercise. *J Strength Cond Res* 26:2907-2911, 2012.

75. Kraemer, WJ, Flanagan, SD, Comstock, BA, Fragala, MS, Earp, JE, Dunn-Lewis, C, Ho, J-Y, Thomas, GA, Solomon-Hill, G, and Penwell, ZR. Effects of a whole body compression garment on markers of recovery after a heavy resistance workout in men and women. *J Strength Cond Res* 24:804-814, 2010.

76. Kraemer, WJ, Hooper, DR, Kupchak, BR, Saenz, C, Brown, LE, Vingren, JL, Luk, HY, DuPont, WH, Szivak, TK, and Flanagan, SD. The effects of a roundtrip trans-American jet travel on physiological stress, neuromuscular performance, and recovery. *J Appl Physiol* 121:438-448, 2016.

77. Lastella, M, Roach, GD, Halson, SL, and Sargent, C. Sleep/wake behaviours of elite athletes from individual and team sports. *Eur J Sport Sci* 15:94-100, 2015.

78. Lau, WY, and Nosaka, K. Effect of vibration treatment on symptoms associated with eccentric exercise-induced muscle damage. *Am J Phys Med Rehabil* 90:648-657, 2011.

79. Laurent, CM, Green, JM, Bishop, PA, Sjokvist, J, Schumacker, RE, Richardson, MT, and Curtner-Smith, M. A practical approach to monitoring recovery: Development of a perceived recovery status scale. *J Strength Cond Res* 25:620-628, 2011.

80. Lee, EC, Fragala, MS, Kavouras, SA, Queen, RM, Pryor, JL, and Casa, DJ. Biomarkers in sports and exercise: Tracking health, performance, and recovery in athletes. *J Strength Cond Res* 31:2920-2937, 2017.

81. Leeder, J, Glaister, M, Pizzoferro, K, Dawson, J, and Pedlar, C. Sleep duration and quality in elite athletes measured using wristwatch actigraphy. *J Sports Sci* 30:541-545, 2012.

82. Leeder, JDC, Godfrey, M, Gibbon, D, Gaze, D, Davison, GW, Van Someren, KA, and Howatson, G. Cold water immersion improves recovery of sprint speed following a simulated tournament. *Eur J Sport Sci* 19:1166-1174, 2019.

83. Lewis, NA, Redgrave, A, Homer, M, Burden, R, Martinson, W, Moore, B, and Pedlar, CR. Alterations in redox homeostasis during recovery from unexplained underperformance syndrome in an elite international rower. *Int J Sports Physiol Perform* 13:107-111, 2018.

84. Lombardi, G, Ziemann, E, and Banfi, G. Whole-body cryotherapy in athletes: From therapy to stimulation. An updated review of the literature. *Front Physiol* 8:258, 2017.

85. Lopes dos Santos, M, Uftring, M, Stahl, CA, Lockie, RG, Alvar, B, Mann, JB, and Dawes, JJ. Stress in academic and athletic performance in collegiate athletes: A narrative review of sources and monitoring strategies. *Front Sports Act Living* 2:42, 2020.

86. Luboshitzky, R, Zabari, Z, Shen-Orr, Z, Herer, P, and Lavie, P. Disruption of the nocturnal testosterone rhythm by sleep fragmentation in normal men. *J Clin Endocrinol Metab* 86:1134-1139, 2001.

87. Main, LC, Dawson, B, Heel, K, Grove, JR, Landers, GJ, and Goodman, C. Relationship between inflammatory cytokines and self-report measures of training overload. *Res Sports Med* 18:127-139, 2010.

88. Malone, JK, Blake, C, and Caulfield, BM. Neuromuscular electrical stimulation during recovery from exercise: A systematic review. *J Strength Cond Res* 28:2478-2506, 2014.

89. Mancinelli, CA, Davis, DS, Aboulhosn, L, Brady, M, Eisenhofer, J, and Foutty, S. The effects of massage on delayed onset muscle soreness and physical performance in female collegiate athletes. *Phys Ther Sport* 7:5-13, 2006.

90. Manimmanakorn, N, Ross, JJ, Manimmanakorn, A, Lucas, SJE, and Hamlin, MJ. Effect of whole-body vibration therapy on performance recovery. *Int J Sports Physiol Perform* 10:388-395, 2015.

91. Montgomery, PG, Pyne, DB, Hopkins, WG, Dorman, JC, Cook, K, and Minahan, CL. The effect of recovery strategies on physical performance and cumulative fatigue in competitive basketball. *J Sports Sci* 26:1135-1145, 2008.

92. Mujika, I, Halson, S, Burke, LM, Balagué, G, and Farrow, D. An integrated, multifactorial approach to periodization for optimal performance in individual and team sports. *Int J Sports Physiol Perform* 13:538-561, 2018.

93. Nagasawa, Y, Komori, S, Sato, M, Tsuboi, Y, Umetani, K, Watanabe, Y, and Tamura, K. Effects of hot bath immersion on autonomic activity and hemodynamics: Comparison of the elderly patient and the healthy young. *Jpn Circ J* 65:587-592, 2001.

94. Nédélec, M, Halson, S, Abaidia, A-E, Ahmaidi, S, and Dupont, G. Stress, sleep and recovery in elite soccer: A critical review of the literature. *Sports Med* 45:1387-1400, 2015.

95. Nédélec, M, Halson, S, Delecroix, B, Abaidia, A-E, Ahmaidi, S, and Dupont, G. Sleep hygiene and recovery strategies in elite soccer players. *Sports Med* 45:1547-1559, 2015.

96. Négyesi, J, Hortobágyi, T, Hill, J, Granacher, U, and Nagatomi, R. Can compression garments reduce the deleterious effects of physical exercise on muscle strength? A systematic review and meta-analyses. *Sports Med* 52:2159-2175, 2022.

97. Nowakowska, A, Kostrzewa-Nowak, D, Buryta, R, and Nowak, R. Blood biomarkers of recovery efficiency in soccer players. *Int J Environ Res Public Health* 16:3279, 2019.

98. O'Donnell, S, and Driller, MW. Sleep-hygiene education improves sleep indices in elite female athletes. *Int J Exerc Sci* 10:522-530, 2017.

99. Overmayer, RG, and Driller, MW. Pneumatic compression fails to improve performance recovery in trained cyclists. *Int J Sports Physiol Perform* 13:490-495, 2018.

100. Pageaux, B, and Lepers, R. The effects of mental fatigue on sport-related performance. *Prog Brain Res* 240:291-315, 2018.

101. Peake, J, McGorm, H, Roberts, L, Coombes, J, Cameron-Smith, D, and Raastad, T. Chronic and acute effects of hot water immersion on strength, recovery and hypertrophy. *FASEB J* 31 lb735, 2017.

102. Peake, JM, Markworth, JF, Nosaka, K, Raastad, T, Wadley, GD, and Coffey, VG. Modulating exercise-induced hormesis: Does less equal more? *J Appl Physiol* 119:172-189, 2015.

103. Peake, JM, Neubauer, O, Della Gatta, PA, and Nosaka, K. Muscle damage and inflammation during recovery from exercise. *J Appl Physiol* 122:559-570, 2017.

104. Pinar, S, Kaya, F, Bicer, B, Erzeybek, MS, and Cotuk, HB. Different recovery methods and muscle performance after exhausting exercise: Comparison of the effects of electrical muscle stimulation and massage. *Biol Sport* 29:269-275, 2012.

105. Poppendieck, W, Wegmann, M, Ferrauti, A, Kellmann, M, Pfeiffer, M, and Meyer, T. Massage and performance recovery: A meta-analytical review. *Sports Med* 46:183-204, 2016.

106. Pournot, H, Tindel, J, Testa, R, Mathevon, L, and Lapole, T. The acute effect of local vibration as a recovery modality from exercise-induced increased muscle stiffness. *J Sports Sci Med* 15:142-147, 2016.

107. Proost, M, Habay, J, De Wachter, J, De Pauw, K, Rattray, B, Meeusen, R, Roelands, B, and Van Cutsem, J. How to tackle mental fatigue: A systematic review of potential countermeasures and their underlying mechanisms. *Sports Med* 52:2129-2158, 2022.

108. Pyne, D. Monitoring physical training loads in swimmers. *Swim Austr* 15:5-9, 1999.

109. Reynolds, AC, and Banks, S. Total sleep deprivation, chronic sleep restriction and sleep disruption. *Prog Brain Res* 185:91-103, 2010.

110. Robson-Ansley, PJ, Gleeson, M, and Ansley, L. Fatigue management in the preparation of Olympic athletes. *J Sports Sci* 27:1409-1420, 2009.

111. Rose, C, Edwards, KM, Siegler, J, Graham, K, and Caillaud, C. Whole-body cryotherapy as a recovery technique after exercise: A review of the literature. *Int J Sports Med* 38:1049-1060, 2017.

112. Rowsell, GJ, Coutts, AJ, Reaburn, P, and Hill-Haas, S. Effects of cold-water immersion on physical performance between successive matches in high-performance junior male soccer players. *J Sports Sci* 27:565-573, 2009.

113. Russell, M, Birch, J, Love, T, Cook, CJ, Bracken, RM, Taylor, T, Swift, E, Cockburn, E, Finn, C, and Cunningham, D. The effects of a single whole-body cryotherapy exposure on physiological, performance, and perceptual responses of professional academy soccer players after repeated sprint exercise. *J Strength Cond Res* 31:415-421, 2017.

114. Russell, S, Jenkins, D, Rynne, S, Halson, SL, and Kelly, V. What is mental fatigue in elite sport? Perceptions from athletes and staff. *Eur J Sport Sci* 19:1367-1376, 2019.

115. Russell, S, Jenkins, D, Smith, M, Halson, S, and Kelly, V. The application of mental fatigue research to elite team sport performance: New perspectives. *J Sci Med Sport* 22:723-728, 2019.

116. Russell, S, Jenkins, DG, Halson, SL, Juliff, LE, and Kelly, VG. How do elite female team sport athletes experience mental fatigue? Comparison between international competition, training and preparation camps. *Eur J Sport Sci* 22:877-887, 2022.

117. Sands, WA. Thinking sensibly about recovery. In *Strength and Conditioning for Sports Performance*. Jeffreys, I, and Moody, J, eds. New York: Routledge, 451-483, 2016.

118. Sands, WA, Apostolopoulos, N, Kavanaugh, AA, and Stone, MH. Recovery-adaptation. *Strength Cond J* 38:10-26, 2016.

119. Sands, WA, McNeal, JR, Murray, SR, Ramsey, MW, Sato, K, Mizuguchi, S, and Stone, MH. Stretching and its effects on recovery: A review. *Strength Cond J* 35:30-36, 2013.

120. Sands, WA, McNeal, JR, Murray, SR, and Stone, MH. Dynamic compression enhances pressure-to-pain threshold in elite athlete recovery: Exploratory study. *J Strength Cond Res* 29:1263-1272, 2015.

121. Sands, WA, and Murray, SR. Recovery 'science.' In *Strength and Conditioning for Sports Performance*. Jeffreys, I, and Moody, J, eds. New York: Routledge, 437-462, 2021.

122. Sargent, C, Lastella, M, Halson, SL, and Roach, GD. How much sleep does an elite athlete need? *Int J Sports Physiol Perform* 16:1746-1757, 2021.

123. Selfe, J, Alexander, J, Costello, JT, May, K, Garratt, N, Atkins, S, Dillon, S, Hurst, H, Davison M, and Przybyla D. The effect of three different (−135°C) whole body cryotherapy exposure durations on elite rugby league players. *PLoS One* 9:e86420, 2014.

124. Selye, H. Experimental evidence supporting the conception of "adaptation energy." *Am J Physiol* 123:758-765, 1938.

125. Selye, H. The general adaptation syndrome and the diseases of adaptation. *J Clin Endocrinol* 6:117-230, 1946.

126. Selye, H. Stress and the general adaptation syndrome. *Br Med J* 1:1383-1392, 1950.

127. Skein, M, Duffield, R, Edge, J, Short, MJ, and Muendel, T. Intermittent-sprint performance and muscle glycogen after 30 h of sleep deprivation. *Med Sci Sports Exerc* 43:1301-1311, 2011.

128. Skein, M, Duffield, R, Minett, GM, Snape, A, and Murphy, A. The effect of overnight sleep deprivation after competitive rugby league matches on postmatch physiological and perceptual recovery. *Int J Sports Physiol Perform* 8:556-564, 2013.

129. Skorski, S, Mujika, I, Bosquet, L, Meeusen, R, Coutts, AJ, and Meyer, T. The temporal relationship between exercise, recovery processes, and changes in performance. *Int J Sports Physiol Perform* 14:1015-1021, 2019.

130. Smith, LL. Acute inflammation: The underlying mechanism in delayed onset muscle soreness? *Med Sci Sports Exerc* 23:542-551, 1991.

131. Smith, MR, Thompson, C, Marcora, SM, Skorski, S, Meyer, T, and Coutts, AJ. Mental fatigue and soccer: Current knowledge and future directions. *Sports Med* 48:1525-1532, 2018.

132. Stephens, JM, Halson, S, Miller, J, Slater, GJ, and Askew, CD. Cold-water immersion for athletic recovery: One size does not fit all. *Int J Sports Physiol Perform* 12:2-9, 2017.

133. Stephens, JM, and Halson, SL. Recovery and sleep. In *NSCA's Essentials of Sport Science*. French, D, and Torres-Ronda, L, eds. Champaign, IL: Human Kinetics, 355-364, 2022.

134. Stephens, JM, Sharpe, K, Gore, C, Miller, J, Slater, GJ, Versey, N, Peiffer, J, Duffield, R, Minett, GM, and Crampton, D. Core temperature responses to cold-water immersion recovery: A pooled-data analysis. *Int J Sports Physiol Perform* 13:917-925, 2018.

135. Sun, H, Soh, KG, Roslan, S, Wazir, MRWN, and Soh, KL. Does mental fatigue affect skilled performance in athletes? A systematic review. *PLoS One* 16:e0258307, 2021.

136. Timon, R, Tejero, J, Brazo-Sayavera, J, Crespo, C, and Olcina, G. Effects of whole-body vibration after eccentric exercise on muscle soreness and muscle strength recovery. *J Phys Ther Sci* 28:1781-1785, 2016.

137. Turner, B, Pennefather, J, and Edmonds, C. Cardiovascular effects of hot water immersion (suicide soup). *Med J Aus* 2:39-40, 1980.

138. Vaile, J, Halson, S, Gill, N, and Dawson, B. Effect of hydrotherapy on recovery from fatigue. *Int J Sports Med* 29:539-544, 2007.

139. Vaile, J, Halson, S, Gill, N, and Dawson, B. Effect of hydrotherapy on the signs and symptoms of delayed onset muscle soreness. *Eur J Appl Physiol* 102:447-455, 2008.

140. Vaile, J, Halson, S, and Graham, S. Recovery review: Science vs. practice. *J Aust Strength Cond* 18:5-21, 2010.

141. Vaile, JM, Gill, ND, and Blazevich, AJ. The effect of contrast water therapy on symptoms of delayed onset muscle soreness. *J Strength Cond Res* 21:697-702, 2007.

142. Van Cutsem, J, Marcora, S, De Pauw, K, Bailey, S, Meeusen, R, and Roelands, B. The effects of mental fatigue on physical performance: A systematic review. *Sports Med* 47:1569-1588, 2017.

143. Van Hooren, B, and Peake, JM. Do we need a cool-down after exercise? A narrative review of the psychophysiological effects and the effects on performance, injuries and the long-term adaptive response. *Sports Med* 48:1575-1595, 2018.

144. Versey, N, Halson, S, and Dawson, B. Effect of contrast water therapy duration on recovery of cycling performance: A dose–response study. *Eur J Appl Physiol* 111:37-46, 2011.

145. Versey, NG, Halson, SL, and Dawson, BT. Water immersion recovery for athletes: Effect on exercise performance and practical recommendations. *Sports Med* 43:1101-1130, 2013.

146. Vieira, A, Bottaro, M, Ferreira-Junior, JB, Vieira, C, Cleto, VA, Cadore, EL, Simões, HG, Carmo, JD, and Brown, LE. Does whole-body cryotherapy improve vertical jump recovery following a high-intensity exercise bout? *Open Access J Sports Med* 6:49-54, 2015.

147. Wahl, P, Mathes, S, Bloch, W, and Zimmer, P. Acute impact of recovery on the restoration of cellular immunological homeostasis. *Int J Sports Med* 41:12-20, 2020.

148. Weakley, J, Broatch, J, O'Riordan, S, Morrison, M, Maniar, N, and Halson, SL. Putting the squeeze on compression garments: Current evidence and recommendations for future research: A systematic scoping review. *Sports Med* 52:1141-1160, 2022.

149. Webb, NP. The use of post game recovery modalities following team contact sport: A review. *J Aust Strength Cond* 21:70-79, 2013.

150. Weeks, B, and Horan, S. Massage: Diagnosis and management. *Mod Athlete Coach* 47:16-20, 2009.

151. Wiewelhove, T, Döweling, A, Schneider, C, Hottenrott, L, Meyer, T, Kellmann, M, Pfeiffer, M, and Ferrauti, A. A meta-analysis of the effects of foam rolling on performance and recovery. *Front Physiol* 10:15, 2019.

152. Williams, CA, and Ratel, S. Definitions in muscle fatigue. In *Human Muscle Fatigue*. Williams, CA, and Ratel, S, eds. London: Routledge, 3-16, 2009.

153. Wilson, LJ, Dimitriou, L, Hills, FA, Gondek, MB, and Cockburn, E. Whole body cryotherapy, cold water immersion, or a placebo following resistance exercise: A case of mind over matter? *Eur J Appl Physiol* 119:135-147, 2019.

154. Winke, M, and Williamson, S. Comparison of a pneumatic compression device to a compression garment during recovery from DOMS. *Int J Exerc Sci* 11:375-383, 2018.

155. Zandvoort, CS, De Zwart, JR, Van Keeken, BL, Viroux, PJF, and Tiemessen, IJH. A customised cold-water immersion protocol favours one-size-fits-all protocols in improving acute performance recovery. *Eur J Sport Sci* 18:54-61, 2018.

156. Zuj, KA, Prince, CN, Hughson, R, and Peterson, SD. Enhanced muscle blood flow with intermittent pneumatic compression of the lower leg during plantar flexion exercise and recovery. *J Appl Physiol* 124:302-311, 2018.

157. Zuj, KA, Prince, CN, Hughson, RL, and Peterson, SD. Superficial femoral artery blood flow with intermittent pneumatic compression of the lower leg applied during walking exercise and recovery. *J Appl Physiol* 127:559-567, 2019.

Subject Index

Note: Page references containing an italicized *f* or *t* indicate information contained in figures and tables, respectively.

A
absolute force 130
absolute strength 2
accelerated eccentrics 66*t*, 68
accentuated eccentric loading (AEL) 66*t*, 68-69, 84-92, 86*f*, 87*f*, 88*f*, 91*t*
    back squat 186*f*
    bench press 187*f*
    dumbbell jump 187*f*
    eccentric phase duration 88-89, 88*t*
    familiarization 85-86
    force-time comparisons 88*f*
    frequency 90-92, 91*t*
    loading 89-90
    subcategories of 85
accommodated resistance training. *See* variable resistance training
acetylcholine 18
actin 15, 16*f*, 18, 164
active recovery 183*f*, 188
adaptations to strength training
    detraining 27-28
    external stressors 28
    fiber type 21-22, 21*f*
    fiber type continuum 18
    fitness-fatigue paradigm 19-20
    general adaptation syndrome 19
    motor learning 21
    motor unit 15-16, 16*f*, 17*f*
    muscle architecture 23
    neural 22
    neuroendocrine 22-23
    sliding filament theory 18
    structure and function 13-15, 14*ft*
    timeline of 20
    training stimuli influences 24-26
    training year impacts 26-27
adenosine triphosphate 18, 47, 107, 162, 176
adolescent training 6-7
AEL. *See* accentuated eccentric loading (AEL)
allometric scaling 3
amino acids 164, 164*t*, 165
amortization phase 3
anabolic window 166
annual plan 51
anthropometrics, athlete 43-44
antibodies 164
antioxidants 171
APRE (autoregulatory progressive resistance exercise) 154-155, 154*f*, 155*f*
assessment of strength. *See* measuring strength
asymmetry testing 138-139
athlete characteristics considerations
    monitoring 159
    potentiation complexes 78-79
athlete development, long-term 6-7
autoregulatory progressive resistance exercise (APRE) 154-155, 154*f*, 155*f*

B
back squat 86*f*
ballistic training 22, 61-62
bench press 87*f*
β-alanine 174-175
BFR (blood-flow restriction) training 72-73
bilateral training 60-61
Björnsson, Hafþór 2
blood-flow restriction (BFR) training 72-73
bodyweight exercise 57-58
bone mineral density 170
braking forces 96

C
caffeine 176
calcium 170, 170*t*
calisthenics 57-58
carbohydrate 162-164, 162*t*, 163*t*
carbohydrate mouth rinsing 163
cardiorespiratory endurance 4
casein protein 165
catabolic state 166
catching derivatives (weightlifting) 64*t*, 66
catecholamines 22, 23, 34
change-of-direction testing 137-138
change-of-direction training (COD) 66, 71
chloride 172
clean, motor skill perspective 42*f*
cluster sets 100-101
COD (change-of-direction training) 66, 71
cold-water immersion 183*f*, 184-185, 185*f*
collagen 165
combined assessment methods 141-144, 142*t*
combined heavy and light loading 102*f*, 103-105, 104*f*, 105*t*
comparisons between stronger and weaker individuals 7-8
competitive season phase 53*f*, 55*f*
complex carbohydrates 162*t*
complex training 99
compression 183*f*, 186-188, 187*f*
compression garments 186-187
concentrated load 53*f*
concentric strength 3
concurrent training
    conditioning methods 116-122, 117*f*, 118*f*, 119*t*
    duration and volume 121
    frequency 120
    high-intensity interval training 117-118, 119*t*
    intensity 120-121
    interference effect 113-116, 114*f*
    long, slow distance 116-117
    maximal aerobic speed 117, 117*f*, 118*f*
    practical application 122
    recovery time between sessions 122
    session order 121-122
    sprint training 119
    training status 120
contrast training 99
contrast water therapy 183*f*, 186
cortisol 23, 35-36
countermovement jump 134-135, 136*f*
creatine monohydrate 172-174
cryotherapy 188-189

D
De Arte Gymnastica Libri Sex (Mercuriale) 1
dehydration 171
density, training. *See* frequency, training
detraining 27-28
dietary supplementation
    β-alanine 174-175
    caffeine 176
    creatine monohydrate 172-174
    sodium bicarbonate 175-176
Dinnie Stones 2
drop jump 134-135
dumbbell jump 87*f*
dynamic correspondence 47
dynamic strength index 50, 141-142, 142*t*
dynamic testing 131-140

E
eccentric cycling 66
eccentric strength 3
eccentric training 65-66, 66*t*, 159
    accelerated eccentrics 66*t*, 68
    accentuated eccentric loading 66*t*, 68-69, 69*f*, 84-92, 86*f*, 87*f*, 88*f*, 91*t*
    flywheel inertial training 66*t*, 67-68, 68*f*
    plyometric training 66*t*, 69
    spectrum 96*f*
    tempo 66, 66*t*
eccentric utilization ratio 50, 51, 143-144
electric muscle stimulation 183*f*, 189, 190*f*
electrolyte balance 171
endomysium 14, 14*f*
endurance training, interference effect and 113-116, 114*f*
energy balance 161, 162*t*
enzymes 164
epigenetics 29-30
epimysium 14, 14*f*
epinephrine 176
ergogenic aids
    β-alanine 174-175
    caffeine 176
    creatine monohydrate 172-174
    sodium bicarbonate 175-176
event analysis 93-94
exercise order 24, 99
exercise selection 24, 94-97, 96*f*, 97*f*
external stressors 28

F

familiarization 82, 84, 85-86, 128
fat 166-167
fat-soluble vitamins 168, 168t
fatty acids 167
fiber type 18, 21-22, 21f, 30-31, 31t
firing frequency 32-33, 33f
fitness-fatigue paradigm 19-20
flywheel inertial training (FIT) 66t, 67-68, 68f
force plate technology 135-136
force production 9-11, 10f
force production asymmetry testing 138-139
force–velocity curve 97, 97f
force–velocity profiles 50
force-velocity profiling 142-143
frequency, training 24, 98-99

G

general adaptation syndrome (GAS) 19, 181
general preparation phase 53f, 55f
genetics 29-30
group-specific allometric scaling 3
glycemic index foods 163t

H

Hall, Eddie 2
high-intensity interval training 117-118, 119t
highland games 2
horizontal jump testing 136
hormone elevation 22
hot-water immersion 183f, 185-186
hydration 171-172, 183f
hypertrophy 36-38, 38f

I

impulse-momentum theorem 8-9, 8t, 9f, 43
individualization 45t, 50-51
injuries, detraining and 27
injury risk, strength and 5-6
in-season 26-27
in-season testing 144-145
intensity 25
interference effect 113-116, 114f
interset rest intervals 107
intraset rest intervals 107-109
iron 170, 170t
isokinetic strength 5
isokinetic testing 140-141
isokinetic training 72
isometric strength 3-4
isometric testing 128-130, 129f
    isometric bench press 131
    isometric mid-thigh pull 130, 131f
    isometric squat 130-131
isometric training 23, 58-59, 159

J

joint assessment 5
jump testing 134-137, 136f, 137f

K

keratins 164
kettlebell training 59-60

L

laws of motion 8-9, 8t, 9f
leucine 165
lever system 43, 43t
linear loading 147-148, 148f
loaded jumps 66, 159
loaded jump testing 136-137, 137f
loading and repetition ranges 101-102, 102f
    combined heavy and light loading 102f, 103-105, 104f, 106t
    training to failure 102-103
    volume-load 106-107
loads, training. *See* training loads, monitoring and adjusting
long, slow distance 116-117
long-term athlete development 6-7
lymphatic drainage 187

M

machine-based training 58
macrocycle 51-54, 52t
macronutrients
    carbohydrate 162-164, 162t, 163t
    fat 166-167
    protein 164-166, 164t
magnesium 170, 170t
massage 188
massage gun therapy 190f
maximal aerobic speed 117, 117f, 118f
maximal force production 95
maximal strength testing 132-133, 134f
maximum velocity sprint ground contact 9-11, 10f
measuring recovery 192-193
measuring strength
    combined assessment methods 141-144, 142t
    comparisons between stronger and weaker individuals 7-8
    dynamic strength index 141-142, 142t
    dynamic testing 131-140
    eccentric utilization ratio 143-144
    force production asymmetry testing 138-139
    force-velocity profiling 142-143
    horizontal jump testing 136
    isokinetic testing 140-141
    isometric testing 128-131, 129f, 131f
    loaded jump testing 136-137, 137f
    maximal strength testing 132-133, 134f
    rationale for 125
    reactive strength testing 139-140
    relative strength testing 133-134
    reliability 126
    sport science and evidence-based practice integration 145-146, 145f
    sprint and change of direction testing 137-138
    standardization of methods 127-128
    test familiarity 126-127
    test order 127
    timing of testing 144-145
    vertical jump testing 134-136, 136f
mechanical specificity 47
mechanisms of strength
    fiber type 30-31, 31t
    genetics and epigenetics 29-30
    motor learning and skill acquisition 40-44, 42f, 43t
    muscle architecture 36-40, 38f, 39f
    neuroendocrine system 34-36
    neuromuscular factors 31-33, 32f, 33f

mental fatigue and recovery 189-191
mesocycle 51-54, 52t
metabolic specificity 47
metabolic window 166
microcycle 51-54, 52t, 98
microdosing 98
minerals 170-171, 170t
momentum, impulse-momentum theorem 8-9, 8t, 9f, 43
motor learning 21, 40-44, 42f, 158-159
motor unit 15, 16f, 17f
motor unit recruitment 31-32, 32f
motor unit synchronization 33
multi-joint exercises 4, 22, 24, 35, 58, 59, 65, 67, 72, 83, 94, 97f, 128
multivitamin supplementation 171
muscle architecture 36-40, 38f, 39f
muscle architecture adaptation 23
muscle characteristics 14t
muscle connective tissues 14f
muscle fibers
    characteristics 14-18, 14f, 17f
    fiber arrangement 38-39, 39f
    and hypertrophy 36-38
    type 21-22, 21f, 30-31, 31t
muscular endurance 4
musculotendinous stiffness 39-40
myofibrillar hypertrophy 23
myosin 15, 16f, 164

N

nature and nurture 29-30
nebulin 40, 164
needs analysis 93-94
net force 130
neural adaptation to training 22
neuroendocrine adaptation to training 22-23
neuroendocrine system 34-36
neuromuscular system and strength
    factors in strength 31-33, 32f, 33f
    fiber type continuum 18
    firing frequency 32-33, 33f
    motor unit 15-16, 16f, 17f
    motor unit recruitment 31-32, 32f
    motor unit synchronization 33
    neuromuscular inhibition 33
    sliding filament theory 18
    stretch-shortening cycle 33
    structure and function 13-15, 14ft
Newton's laws of motion 8-9, 8t, 9f
NFL Combine 4
nitrogen balance 164-165
nutritional considerations 161
    antioxidants 171
    caffeine 176
    carbohydrate 162-164, 162t, 163t
    creatine monohydrate 172-174
    energy balance 161, 162t
    ergogenic aids 172-177
    fat 166-167
    hydration 171-172
    macronutrients 161-167, 162t, 163t, 164t
    micronutrients 167-171, 168t, 169t, 170t
    minerals 170-171, 170t
    monitoring nutrition 178
    periodization of nutrition 177-178, 178f
    protein 164-166, 164t
    in recovery 183f, 184

sodium bicarbonate 175-176
vitamins 168-170, 168t, 169t
β-alanine 174-175

O
off-season 26
off-season testing 144
overcompensation 181
overhead pressing derivatives (weightlifting) 64t
overload 45-46, 45t
oxidative stress 171

P
partial squat 77f
pennation 23
percentage of 1 repetition maximum 148-149
performance, strength and 5
performance analysis 94
perimysium 14, 14f
periodization of nutrition 177-178, 178f
periodization of recovery 192
periodization of training 51-54, 52t, 53f, 54t
phosphagen 47
phosphocreatine 47
placebo effect, and recovery 191
plant-based protein 165
plyometric training 23, 62-63, 63f, 66t, 69, 159
pneumatic compression 187, 187f
postactivation performance enhancement 64
post-exercise nutrition planning 161
postseason 27
postseason testing 145
potassium 172
potentiation complexes 59, 64-65
    athlete characteristics considerations 78-79
    designing 76-78
    implementing in competition 79-80
    implementing in training 79, 79t
    rest intervals 109
    variable resistance training 81
powerlifting competitions 2
preseason 26
preseason testing 144
program design
    exercise order 99
    exercise selection 94-97, 96f, 97f
    frequency 98-99
    loading and repetition ranges 101-107, 102f, 104f, 105t, 106t
    needs analysis 93-94
    overall training approach 111
    rest intervals 107-109
    set structure 100-101
    training age 109-110
    training year 110-111
    volume 99-100
progressive resistance exercise 154-155, 154f, 155f
propulsion 9-11, 10f
protein 164-166, 164t
pulling derivatives (weightlifting) 64t, 66

R
rapid force production 95-96
rating of perceived exertion (RPE) 150-151, 152t
ratio scaling 3
reactive strength 4-5
reactive strength testing 139-140
recovery–adaptation 181
recovery considerations 181-182, 183f
    active recovery 183f, 188
    cold-water immersion 184-185, 185f
    compression 183f, 186-188, 187f
    concurrent training recovery time 122
    contrast water therapy 186
    cryotherapy 188-189
    electric muscle stimulation 183f, 189, 190f
    hot-water immersion 185-186
    hydration 183f
    massage 183f, 188
    massage gun therapy 190f
    measuring recovery 192-193
    mental fatigue and recovery 189-191
    nutrition 183f, 184
    periodization of recovery 192
    placebo effect 191
    recovery methods 182-189
    recovery pyramid 183f
    sleep 183-184, 183f
    stretching 183f, 188
    vibration 183f, 189, 190f
    water immersion 183f, 184-186
regeneration 181
relative strength 3
relative strength levels 75
relative strength testing 133-134
reliability, strength testing 126
repetition 99
repetition maximum zones 149-150
repetition ranges 101-102, 102f
    combined heavy and light loading 102f, 103-105, 104f, 106t
    training to failure 102-103
    volume-load 106-107
repetitions in reserve 151-152
resistance exercise training age (RETA) 43
resistance training 22
rest. *See* recovery considerations
rest intervals 25-26, 107-109
RETA (resistance exercise training age) 43
reversibility 45t, 49-50
RPE (rating of perceived exertion) 150-151, 152t

S
sarcomeres 15, 15f, 23, 38-40, 38f
sarcoplasmic hypertrophy 23
saturated fats 167
scaled strength 3
self-massage 188
Selye, Hans 19
sequencing of training 54-55, 55f
set 99
set-repetition schemes 152-154, 153f
set structure 25, 100-101
simple carbohydrates 162t
single-joint exercises 3, 24, 58, 59, 67, 94, 99, 128
skeletal muscle system
    fiber type continuum 18
    motor unit 15-16, 16f, 17f
    sliding filament theory 18
    structure and function 13-15, 14ft
skill acquisition 40-44, 42f, 158-159
sleep 183-184, 183f
sliding filament theory 18
sodium 170, 170t, 172
sodium bicarbonate 175-176
soy-based protein 165
specificity 45t, 47-48
specific preparation phase 53f, 55f
speed-strength 4
sport, strength importance within 5-8
sport analysis 93-94
sprinting 22, 70
sprint testing 137-138
sprint training 119
squat jump 77f, 134-135, 136f
standardization of methods 127-128
sticking points 65
strength
    comparisons between stronger and weaker individuals 7-8
    defined 9-11, 10f
    definitions of types of 2-5
    importance, within sport 5-8
    influence of 6f
    injury risk 5-6
    long-term athlete development 6-7
    performance 5
    redefining 8-9, 9f
strength characteristic visualization 145f
strength competitions, history of 1-2
strength-endurance 4
strength-speed 4
strength testing. *See* measuring strength
stretching 183f, 188
stretch-shortening cycle 3, 33
stronger athletes 110
strongman competitions 2
strongman training 61
subjective loading, methods of 152f
Süleymanoğlu, Naim 2, 44
supercompensation effect 19
sweating rates 171

T
task specificity 47
tempo eccentric training 66, 66t
tendon stiffness adaptations 39
test familiarity 126-127
testing. *See* measuring strength
test order 127
testosterone 22-23, 34-35
titin 39-40, 164
train-high, sleep-low model 178f
training age 43, 50, 109-110
training day 51-54, 52t
training loads, monitoring and adjusting 147, 159
    autoregulatory progressive resistance exercise 154-155, 154f, 155f
    combined heavy and light loading 102f, 103-105, 104f, 106t
    linear loading 147-148, 148f
    percentage of 1 repetition maximum 148-149
    rating of perceived exertion 150-151
    repetition maximum zones 149-150

repetitions in reserve 151-152
set-repetition best 152-154, 153f
training to failure 102-103
two-for-two rule 148
velocity-based training 156-158, 156f
volume-load 106-107
training methods
 ballistic training 61-62
 bilateral training 60-61
 blood-flow restriction training 72-73
 bodyweight exercise 57-58
 change-of-direction training 71
 eccentric training 65-70, 66t, 68f, 69f
 isokinetic training 72
 isometric training 58-59
 kettlebell training 59-60
 machine-based training 58
 plyometric training 62-63, 63f
 potentiation complexes 64-65, 75-80, 77f, 79t, 81
 relative strength levels 75
 sprinting 70
 strongman training 61
 unilateral training 60-61
 variable resistance training 65, 80-84, 83f
 weightlifting movements and derivatives 63-64, 64t
training organization
 annual plan 51
 periodization 51-54, 52t, 53f, 54t
 sequencing 54-55, 55f
training principles 45, 45t
 individualization 50-51
 overload 45-46

reversibility 49-50
specificity 47-48
variation 48-49
training session 51-54, 52t
training stimuli
 exercise order 24
 exercise selection 24
 frequency 24
 intensity 25
 rest intervals 25-26
 set structure 25
 volume 25
training to failure 102-103
training year
 in-season 26-27
 off-season 26
 postseason 27
 preseason 26
 program design 110-111
T-tubules 18
two-for-two rule 148

U

unilateral training 60-61
unsaturated fats 167

V

validity, strength testing 126
variable resistance training (VRT) 65, 80-81, 83f
 acute effects of 81
 familiarization 82-83
 frequency 84
 loading 83-84
 longitudinal adaptations to 81-82
 potentiation complexes 81

variation 45t, 48-49
velocity-based training (VBT) 156-158, 156f
vertical jump testing 134-136, 136f
vertical net braking 9-11, 10f
vibration 183f, 189, 190f
vitamin A 168, 168t
vitamin B 168, 169t
vitamin C 168, 169t, 171
vitamin E 168, 168t, 171
vitamins 168-170, 168t, 169t
volume 25, 99-100
volume-load 106-107
VRT. See variable resistance training (VRT)

W

water immersion 183f, 184-186
water-soluble vitamins 168-170, 169t
weaker athletes 109-110
weightlifting competitions 2
weightlifting movements and derivatives 63-64, 64t, 159
whey protein 165

Y

youth training 6-7

Z

zinc 170, 170t

Author Index

A

Aagaard, P 15, 22, 23, 32, 33, 38, 39, 70, 72, 78, 79, 101, 103, 116, 122
Aakvaag, A 35
Abad, CCC 5, 133
Abaidia, A-E 184, 188
Abbott, H 54, 61
Abbott, HA 46
Abe, H 174
Abe, T 23, 37, 38, 72, 73
Abernethy, PJ 40, 115, 121, 122, 163
Abián-Vicén, J 62
Aboodarda, SJ 68, 85
Aboulhosn, L 191
Abreu, L 20
Achten, E 174, 175
Adamopoulos, I 140
Adams, GM 132
Adams, K 97, 107
Adams, KJ 65, 84
Ades, PA 21, 30
Adlercreutz, H 34
Aerts, I 69
Afonso, J 188
Agar-Newman, D 99
Agouris, I 80, 81, 82, 84, 136, 157
Agre, JC 120
Aguado, X 23, 38
Aguilar-Navarro, M 30
Ahmaidi, S 184
Ahmetov, I 30
Ahmetov, II 30
Ahtiainen, J 23, 35, 36
Ahtiainen, JP 23, 70, 82
Akimoto, T 72
Akster, HA 40
Alabed, S 188, 191
Albina, ML 193
Alcaraz, PE 70, 95, 99
Alderman, BL 51
Alegre, LM 23, 38
Alen, M 18, 21, 30, 34, 36, 113, 116, 120
Alepuz-Moner, V 68
Alexander, C 173
Alexander, J 189
Alexander, R 138, 139
Al Haddad, H 70
Ali, A 176
Allen, KP 157
Allen, M 80
Allen, SV 30
Allen, WJC 67
Allison, R 168
Allman, BR 149, 150, 152
Almada, A 173, 174
Almada, AL 172, 173, 174
Almond, CS 172
Almstedt, H 84
Almstedt, HC 84
Alonso-Martínez, AM 173

Alonso-Molero, I 157
Alsayrafi, M 30
Alshewaier, S 6
Altadill, A 102, 149, 157
Altenburg, D 193
Alvar, B 28, 30, 111, 149, 181, 189
Alvar, BA 7, 25, 46, 99, 100, 102, 158
Alvares, C 62, 69, 159
Alvarez, IF 66
Alvarez, JCB 116
Alves, VT 115
Alves de Souza, ME 163
Alwan, N 168
Am, W 23, 36
Amadio, AC 173
Amaral, S 173
Amati, F 79
Amca, AM 47
American College of Sports Medicine 24, 98
Amigó, N 119
Amri, M 113
Anastasiades, L 30
Anaya, M 53
Andersen, CH 59, 102
Andersen, JC 66
Andersen, JL 15, 22, 23, 33, 38, 79, 101, 103, 116, 122, 177
Andersen, LB 6, 27
Andersen, LL 23, 39, 59, 70, 102, 103
Andersen, M 35
Andersen, TE 6, 27
Andersen, V 80, 81, 82, 84
Anderson, CE 82
Anderson, JC 113, 115, 116, 121, 122
Anderson, JM 23, 35, 36, 142, 171
Anderson, K 58
Anderson, KG 61, 94
Anderson, L 165
Anderson, RG 101
Anderson, T 80
Andersson, P-I 121
Andrade, DC 62, 69, 159
Andreacci, J 157
Andreato, LV 115
Andreu Caravaca, L 163
Andrews, WA 7
Andronikos, M 1
Androulakis-Korakakis, P 49
Angelopoulos, TJ 30
Angioi, M 168
Anousaki, E
Ansley, L 113, 115, 182
Antonio, J 172, 173, 174, 176, 177, 178
Antunes, A 29
Antunes, BMM 121
Antunes, HKM 184
Anwander, H 175
Aouadi, R 62
Apostolopoulos, N 125, 181
Appleby, BB 60

Appleby, CL 20, 29, 37
Arabatzi, F 63, 159
Aragon, AA 165, 166, 178
Araki, T 73
Araújo, D 41
Arazi, H 62, 149
Arber, MPH 30
Archer, DC 90
Arciero, PJ 178
Arede, J 151
Arellano, R 64, 76
Arent, MA 178
Arent, S 178
Arent, SM 24, 98, 172, 176, 177, 178
Argus, CK 192
Aritan, S 47
Arjol-Serrano, JL 139
Armstrong, LE 171, 176
Armstrong, R 89
Arnaoutis, G 5
Arnold, MD 149
Arroyo, E 121
Arteta, D 30
Artieda, M 30
Artioli, GG 5, 30, 115, 172, 174, 175, 176
Arvisais, D 26, 27, 45, 116, 122
Asadi, A 62, 149
Ascherio, A 167
Ash, GI 30
Ashby, BM 136
Ashley, E 30
Ashley, EA 30
Ashton, T 171
Asiain, X 149
Askew, CD 185
Aspe, RR 133, 157
Astratenkova, IV 30
Ataee, J 80
Atalag, O 44
Atherton, PJ 39, 90
Atkins, S 189
Atkinson, FS 162
Atkinson, G 184
Atterbom, H 7
Aube, D 66
Aubry, A 98
Augustsson, J 24, 58, 94
Aune, TK
Aussieker, T 165
Austin, D 2
Austin, K 30, 61
Avela, J 113, 115, 121
Avelar, A 127
Awaya, T 73
Ayeni, OR 72
Ayoama, R 173
Azevedo, P 66
Aziz, AR 5, 128
Azizi, E 39

283

B

Baar, K 113, 121, 165, 170, 177
Babault, N 4, 76, 122, 189
Babu, GJ 40
Baca, A 140
Bach, T 19
Bachmann, AW 193
Badby, AJ 135
Baddeley, AD 41
Badenhorst, C 176
Badowski, N 187
Baechle, TR 148
Baek, W 120
Baelde, H 175
Baganha, RJ 163
Bagley, JR 18, 21, 30
Baguet, A 174, 175
Bailey, CA 4, 108, 138, 139, 140
Bailey, RL 176
Bailey, S 191
Baker, D 116, 117, 121, 123, 125, 142
Baker, DG 80, 84, 99
Baker, HL 187
Baker, JS 24, 25, 98, 100
Baker, KM 174
Baker, MK 185, 186
Bakrac, ND 72
Balagué, G 177, 192
Balagué, N 41
Balasekaran, G 59
Baldari, C 173
Baldwin, KM 13, 162
Balkin, TJ 177
Ball, G 174
Ball, SD 99, 158
Ball, TE 132
Ballantyne, JK 121
Ballard, TN 187
Balnave, RJ 101
Balsalobre-Fernandéz, C 151
Balsalobre-Fernández, C 142, 157
Balshaw, TG 20, 29, 37
Baltzopoulos, V 89
Bampouras, TM 59
Bandholm, T 67
Banfi, G 188, 189
Banfield, L 165, 166
Bang, M-L 18, 40
Banister, EW 19
Bankers, S 113, 115, 121
Banks, S 183
Bannock, L 172, 174, 175
Banyard, H 156, 157
Banyard, HG 101, 125, 132, 133, 157
Baquet, G 116, 188
Barakat, C 66
Barbé, C 35
Barber, JJ 175
Barbosa, CGR 163
Barbosa, TM 46, 58, 59, 159
Barboza-González, P 108
Barcelos, C 66
Barette, SL 132, 149
Barfield, JP 63
Bar-Haim, S 154
Barker, L 143
Barker, LA 135

Barker, M 7, 110
Barker M, Taylor K 85, 91
Barkley, JE 176
Barnard, M 163
Barnard, RJ 18, 30
Barnes, A 49, 55, 103, 105, 150, 157
Barnes, CA 186
Barnes, MJ 187
Barnett, A 182, 188
Baroni, BM 72
Barr, MJ 41
Barragán, A 30
Barrett, SF 128
Barringer, ND 167
Barrios, Y 30
Barros Coelho, EJ 30
Barroso, R 55, 66, 151
Barry, BK 121
Barss, T 58
Bartel, K 183
Barthelemy, Y 140
Bartholomew, JB 149
Bartlett, J 122
Bartlett, JD 118, 121, 164, 177
Bartlett, R 41
Bartok, C 171
Bartolomei, S 121
Bartram, J 192
Bassa, E 140
Bassinello, D 174
Batterham, AM 126
Bauer, JA 99
Baulch, J 29, 30
Baur, H 128
Baz-Valle, E 157
Bazyler, CD 4, 21, 22, 23, 26, 31, 36, 46, 49, 50, 54, 95, 103, 104, 106, 107, 128, 130, 133, 138, 139, 149, 152
Bazzucchi, I 176
Beach, TAC 140
Beardsley, C 48
Beato, M 67, 68, 119, 120
Beattie, K 4
Beaven, CM 188, 189
Beck, DT 37
Beck, TW 174
Beckham, GK 4, 50, 63, 64, 95, 127, 128, 130, 136, 140, 142, 157, 159, 193
Beckman, EM 7, 46, 62, 75, 109, 110, 128, 143
Bedi, A 72
Bedrin, NG 21, 30
Beelen, M 166
Beggs, AH 30
Behm, DG 20, 24, 58, 60, 61, 62, 63, 68, 79, 94, 109, 121
Behringer, M 6
Beiter, T 30
Bekele, Z 30
Belcastro, AN 35
Belcher, DJ 66, 88
Belenky, GL 177
Bell, GJ 66, 90, 120, 132
Bell, ZW 73
Bellamine, A 182
Bellar, D 79
Bellar, DM 176
Bellés, M 193

Belli, A 50, 142, 143
Bellinger, P 183, 184
Bellizzi, MJ 70, 137
Bellon, CR 4, 21, 26, 27, 37, 46, 48, 49, 50, 53, 54, 55, 58, 61, 65, 70, 75, 79, 94, 95, 100, 102, 103, 110, 111, 119, 121, 123, 132, 141, 142, 159
Belmonte, A 6
Belmore, K 173
Beltrame, LG 20
Beltrán, AR 62, 69, 159
Beltran Valls, RM 39, 90
Bemben, DA 127, 173
Bemben, MG 65, 72, 127, 173, 189
Benatti, F 173
Benatti, FB 115, 175
Bender, A 174, 183
Bendich, A 169
Beneka, A 72
Benik, FM 25, 107
Benítez-Muñoz, JA 163
Benitez-Porres, J 101
Bennekou, M 116, 122
Bennell, KL 168
Bennett, M 142
Bennett, MA 60, 64, 76, 105, 135, 142
Bennett, S 21, 40, 41, 42
Ben-Sira, D 7
Bensley, J 61
Benson, AC 125
Bentley, DJ 113, 115, 121, 122, 175
Bento Douetts, CD 79
Benton, MJ 127
Berg, HE 67
Bergsten, A 173
Bergstrom, HC 83
Berman, N 34
Bermon, S 35
Bernards, JR 50, 54, 103, 104, 106, 133, 149, 152
Berning, JM 65, 83, 84
Berrazaga, I 165
Berry, SP 138, 139
Berryman, N 27, 45, 116, 121, 122
Bertelsen, DM 6, 27
Berthoin, S 116, 117
Bertin, M 173
Berton, R 72
Bessone Alves, F 173
Best, TM 7
Beunen, GP 29
Bevan, H 142
Bevan, HR 64, 76, 78, 105, 135, 142
Bex, T 175
Beyer, KS 19, 25, 26, 49, 100, 121, 128
Bezodis, NE 71
Bhasin, D 34
Bhasin, S 34
Bherer, L 27, 45
Bi, S 151, 152
Bicer, B 189
Bieuzen, F 185, 186, 189, 192
Bigard, AX 122
Biggins, M 183
Bijker, K 3
Bikos, C 72
Bilby, GE 24

Billat, VL 65
Billeter, R 18, 30
Binet, C 116, 122
Binetti, M 71
Bingham, GE 54, 61, 67, 68, 84, 85, 90, 91, 92, 157
Binstadt, BA 172
Binsted, G 58
Birch, J 189
Bird, M 136
Birk, R 29, 30
Birmingham-Babauta, SA 44, 132
Biro, G 22, 23, 34, 35
Bisher, ME 18
Bishop, C 6, 24, 67, 82, 95, 97, 99, 103, 138, 139
Bishop, D 4, 20, 76, 78, 79, 84, 109, 175
Bishop, DJ 30, 113, 118, 121, 122, 172, 175, 176, 187
Bishop, LG 101
Bishop, PA 193
Bishop, R 173
Bjørnsen, T 50, 143, 163
Blaak, JB 99, 158
Black, G 126
Black, MA 184
Blackard, DO 63
Blackburn, JR 24, 58, 94, 128
Blackman, MR 35
Blaisdell, J 173
Blake, C 189
Blancquaert, L 175
Blankenstein, MA 29, 30
Blanksby, BA 173
Blazevich, A 66
Blazevich, AJ 3, 4, 22, 23, 32, 38, 39, 69, 76, 78, 81, 85, 90, 91, 99, 127, 128, 130, 132, 186
Bleakley, CM 185, 186, 189
Bleisch, W 34
Bleisch, WV 34
Blessing, D 48
Blewett, S 61
Bloch, W 182
Blondel, N
Blue, MN 168
Blue, MNM 172
Blumert, PA 142
Board, W 135
Boatwright, D 101
Bobb, A 169
Bobbert, MF 33, 58, 137, 143
Boesch, C 175
Boetes, M 173
Boffey, D 108
Bogdanis, G 66, 80
Bogdanis, GC 105, 171
Bojsen-Møller, J 23, 39
Boksem, MA 191
Bompa, TO 51, 147, 149, 152
Bondarchuk, A 54
Bond-Williams, KE 25, 107
Bonifazi, M 23, 36
Bonilla, DA 101
Bonnar, D 183
Boobis, L 174, 175
Booker, HR 187
Boomsma, DI 29, 30

Boone, CH 19, 25, 26, 49, 100, 121
Boone, J 19
Borg, GAV 150
Boros, R 99, 173
Borst, SE 99
Borth, J 85
Bosch, F 143
Bosco, C 34
Bosio, A 174
Bosquet, L 26, 27, 45, 116, 121, 122, 181, 186, 192
Bottaro, M 5, 107, 170, 189
Botter, L 66
Bottinelli, R 21, 30
Bouchard, C 30
Boullosa, D 79, 137
Boullosa, DA 20
Bourgois, J 174
Bourgois, JG 19
Bournat, L 189
Bowen, JC 132
Bowers, RW 171
Boyd, JC 175
Bracken, RM 189
Bradley, WJ 168
Bradshaw, EJ 136
Brady, CJ 128
Brady, M 191
Brady, PH 134
Brainerd, EL 39
Branch, JD 171
Brandenburg, JE 85, 90
Brand-Miller, JC 162
Brannon, AR 154
Brannon, MF 18, 21, 30, 31
Branscheidt, M 53
Brauer, AA 183
Brauner, T 7
Bravo-Sánchez, A 62
Brazo-Sayavera, J 189
Brechue, WF 4, 41, 136
Breed, R 125
Breigan, B 22, 35
Bressel, E 4, 135, 140
Bressel, M 135
Brewer, C 7, 145
Brice, G 150, 159
Bridge, CA 78, 80, 175
Bridgeman, LA 89
Brietzke, C 79
Briggs, D 83
Briggs, DL 83
Bright, TE 66, 68
Brilla, LR 59, 80
Brimm, D 61
Brisswalter, J 177, 189
Brito, J 116, 120
Britton, SL 30
Brixen, K 35
Broatch, J 187
Broatch, JR 187
Bromley, T 138, 139
Bronks, R 23, 38
Brook, M 187
Brooke, MH 30
Brooks, GA 13, 162
Broskey, NT 79

Brouns, F 172
Brown, C 22, 23, 34, 36
Brown, F 186, 187
Brown, LE 25, 59, 63, 65, 72, 75, 78, 80, 81, 82, 84, 90, 99, 100, 101, 105, 108, 109, 115, 136, 159, 187, 189
Brown, M 144
Brown, SM 72
Brown, SR 61, 70, 157
Brownlee, TE 67
Brughelli, M 9, 70, 71, 137, 142, 143
Brumitt, J 60
Brumello, E 18, 40
Bruner, H 23, 36
Brunette, M 60
Brunotte, F 23, 37
Bryant, KR 28
Bryanton, MA 46, 95
Bryce, GR 58
Buchan, D 24, 98
Buchheit, M 70, 117, 118, 121, 123, 144
Buckner, SL 29, 73, 102, 132
Buckthorpe, MW 46
Buehler, T 175
Buford, TW 174
Bui, TT 174
Bullock, N 24, 77
Bullough, R 106
Bunce, P 142
Bunce, PJ 20, 64, 76, 105, 142
Bundle, MW 70
Bunker, D 149
Burden, R 182
Burger, H 1
Burgomaster, KA 47
Burke, DG 173
Burke, JR 22, 32
Burke, L 167, 177
Burke, LE 178
Burke, LM 163, 164, 166, 167, 171, 172, 177, 192
Burke, RE 18, 30, 46
Burkett, LN 46, 99, 103, 107, 158
Burleson, MA, Jr 106, 161
Burnett, A 138, 139
Burnham, R 120
Burns, GT 138
Burr, J 72, 73
Burt, DG 113
Burt, LA 7
Burton, JD 50
Buryta, R 193
Bush, FA 23, 35
Bush, JA 35, 121, 173
Bushnell, T 48
Busso, T 36, 52
Butcher, S 67
Butler, C 89
Butterfield, TA 83
Buttifant, D 81
Button, C 21, 40, 41, 42, 137
Buxens, A 30
Buyken, AE 162
Buzzichelli, CA 147, 149, 152
Byars, A 174
Byars, AG 173
Byrd, R 48, 101
Byrd, RJ 35

Byrne, JM 68
Byrne, N 30
Byrne, PJ 80, 140
Byrnes, RK 150, 151

C

Cabarkapa, D 76, 126
Cabarkapa, DV 126
Cabre, HE 161
Cabri, JMH 72
Caccavale, F 47
Cadefau, JA 119
Cadore, EL 189
Cafarelli, E 107, 189
Cahalan, R 183
Cahill, MJ 70, 95, 96
Cai, Z 81, 82, 84
Caia, J 184
Caillaud, C 189
Caillou, N 21, 41
Caireálláin, AÓ 139
Cajueiro, M 174
Cal Abad, C 135
Cal Abad, CC 133
Calbet, JAL 25, 100, 101, 107
Calderbank, J 64
Calders, P 175
Callaghan, SJ 71, 132, 138, 139
Callan, S 7
Calleja-González, J 170, 176, 191
Callens, S 175
Calvert, TW 19
Cameron-Smith, D 186
Campbell, B 174, 178
Campbell, BA 69, 89, 90
Campbell, BI 59
Campbell, WI 154, 173
Campbell, WW 20, 23, 35
Campeiz, JM 53
Campney, HK 58
Camporesi, S 30
Campos, GE 23, 36, 149
Cañas-Jamet, R 173
Candau, R 20, 52
Candow, DG 172, 173, 174, 189
Canepari, M 21, 30
Canestri, R 79
Caniuqueo, A 173
Cannavan, D 39
Cano-Ruiz, MA 142
Cantler, E 173, 174
Cantler, EC 174
Cantwell, CJ 69, 89, 90
Canwell, CJ 90
Cao, Y 151, 152
Capelo-Ramirez, F 70
Caperuto, E 174
Capps, SG 187
Caputo, F 121
Carballeira, E 108
Carden, P 66, 96, 135, 159
Cardinale, M 36, 125, 132
Cardoso, RK 151
Cardozo, D 46, 95
Caremani, M 18, 40
Carey, AL 167, 177
Carey, JP 46, 63, 66, 95, 96

Cariolou, MA 30
Carling, C 71, 95, 116, 119, 120, 137
Carlisle, D 71
Carlock, J 7
Carlos-Vivas, J 70, 95
Carlson, L 66, 99
Carmo, EC 115, 118, 151
Carmo, JD 189
Carnevale, R 63
Carolan, B 189
Carpenter, DO 18, 32
Carpenter, L 4, 96, 136, 151, 155, 158, 159
Carpentier, A 174
Carr, AJ 175
Carr, BM 175
Carrasco, L 115, 122
Carratalá, V 30
Carretero, M 61
Carroll, K 68, 84, 85, 157
Carroll, KM 19, 30, 46, 48, 50, 51, 53, 54, 67,
        68, 90, 91, 92, 103, 104, 106, 121, 133,
        147, 149, 152, 181
Carroll, RM 63
Carson, BP 4
Carter, AS 25, 107
Carter, C 128
Caruso, O 7
Carvajal-Espinoza, R 111
Carvalho, L 55
Carvalho, MRS 30
Carzoli, JP 66, 88, 149, 150, 151, 152
Casa, DJ 171, 176, 193
Casaburi, R 34
Casáis, L 52
Casajús, JA 139
Case, MJ 6
Casey, A 174
Casillas, J-M 23, 37
Casius, LJ 143
Casonatto, J 176
Cassens, J 22, 23, 34, 35
Cassidy, J 41
Castagna, C 7, 62, 93, 110, 116, 120, 121
Castaño-Zambudio, A 98
Castellano, J 120
Castillo, A 40
Castillo, D 67, 68, 176
Castillo-Palencia, J 64
Castro, J 59
Caulfield, BM 189
Cava, A 6
Cavill, MJ 106
Cè, E 47
Ceaser, T 18
Cederblad, G 172, 173, 174
Celnik, P 53
Centry, R 101
Cermak, NM 23, 37, 166
Cerretelli, P 20, 36
César, EP 78, 109
Cesar, GM 46, 71, 96
Cesar, JE 175
Chaabene, H 60, 62
Chae, S 108
Chagolla, J 121
Chalimoniuk, M 100
Chalimoniuk M 25

Chalkley, D 156, 157
Chalmers, GR 59, 80
Chamari, K 52, 63, 93, 113, 115, 116, 120
Chandler, AJ 178
Chandra, N 171
Chang, C-C 3
Channell, BT 63
Chaouachi, A 63, 109, 113, 115, 116, 121
Chaouachi, M 113
Chapman, D 50, 141
Chapman, DW 42, 50, 69, 128
Chapman D 85, 91
Chard, JB 90
Charest, J 183
Chase, JE 187
Chassaing, P 70
Chatard, JC 140
Chatzopoulos, D 46
Chavda, S 50, 63, 95, 97, 103, 138
Chelly, MS 62
Chen, J 18, 40
Chen, S 81, 82, 84
Chen, SE 81
Chen, X 34
Chen, Y-J 163
Chen, Z-R 81
Cheng, S 22, 34, 35
Chéry, C 133, 157
Cheuvront, SN 171
Chiang, C-M 168
Chiang, C-Y 3, 138, 139
Chicharro, JL 142
Chico, TJA 188, 191
Childers, JT 142
Chilibeck, PD 58, 173
Chin, J 187
Chirosa-Ríos, LJ 108
Chiu, L 53
Chiu, LZ 66
Chiu, LZF 46, 63, 65, 66, 76, 95, 96
Chiwaridzo, M 113
Chleboun, GS 187
Chobotas, MA 22, 32
Choi, KH 127
Cholewa, J 58, 174, 175
Cholewa, JM 44, 121
Chow, C-M 151
Chowning, LD 135, 136
Christensen, BK 57
Christiansen, C 170
Chtara, M 113
Chu, DA 7
Chun-Er, L 61
Church, DD 121, 128, 165, 174
Ciccone, AB 80
Cieszczyk, P 30
Cimadoro, G 48
Cintineo, HP 178
Claessens, AL 29
Clamann, HP 31
Clark, BC 73
Clark, CCT 157
Clark, D 89
Clark, DR
Clark, K 70, 95, 96
Clark, KL 173
Clark, KP 70, 95, 96, 137, 138

Clark, MJ 34
Clark, NW 108
Clark, SA 177
Clarke, ND 191
Clarke, R 81
Clarkson, PM 30, 79
Clarys, JP 72
Claudino, JG 173, 177
Claudius, B 175
Cleather, DJ 48, 150, 151
Clemente, FM 188
Clemente-Suárez, VJ 163
Clercq, DD 135
Cleto, VA 189
Climstein, M 97
Clos, P 66
Close, GL 161, 162, 163, 164, 165, 167, 168, 170, 171, 172, 177
Clubb, J 116, 120
Coakley, J 174
Cobley, JN 168
Cobley, SP 151
Coburn, JW 59, 63, 65, 75, 80, 81, 82, 84, 90, 109, 115
Cochrane, DJ 187
Cochrane, JL 71
Cockayne, M 163
Cockburn, E 188, 189
Cocking, S 163
Coelho, DB 30
Cofano, G 139
Coffey, VG 113, 120, 192
Coglianese, R 37
Coh, M 136
Cohen, DD 71
Cohen, M 169
Coker, CA 83, 84
Coker, NA 174
Cole, M 156, 157
Cole, PJ 99
Coleman, DR 39
Colenso-Semple, LM 18, 21, 30
Collier, GR 163
Collins, C 107
Collins, D 24, 45, 47, 58, 82, 94
Collins, K 6
Collins, M 30
Collins, MA 161
Collins, R 172, 173, 174
Colombini, A 188, 189
Colquhoun, RJ 101, 154
Cometti, C 189
Cometti, G 140
Comfort, P 3, 4, 6, 8, 9, 23, 24, 33, 37, 38, 44, 46, 48, 50, 59, 61, 62, 63, 64, 66, 69, 71, 75, 77, 80, 83, 85, 94, 95, 96, 97, 98, 100, 101, 103, 105, 106, 108, 109, 111, 119, 125, 127, 130, 132, 133, 134, 135, 136, 138, 139, 140, 141, 142, 143, 146, 151, 152, 153, 155, 157, 158, 159
Compaan, DM 173
Comstock, BA 186
Comyns, TM 78, 128, 139, 140
Conatser, RR 187
Conceição, F 142, 143, 157
Conceição, MS 72
Conley, MS 99

Conlon, JA 25, 100, 101, 108, 159
Connaboy, C 96, 143
Connick, M 7, 46, 53, 62, 75, 104, 109, 110, 143
Connolly, JM 30
Connor, M 119, 120
Conrad-Forrest, A 156
Conroy, B 22, 35
Conroy, BP 7, 22, 35
Constantin-Teodosiu, D 162
Conti, MP 173
Contreras, B 23, 48, 58, 101
Conway, GE 170
Cook, C 35, 36, 188, 189
Cook, CJ 20, 64, 76, 78, 142, 189
Cook, JD 183
Cook, K 176, 184
Cooke, CB 41, 59
Cooke, DM 150, 151
Cooke, ERA 138, 139
Cooke, M 174
Cooke, MB 173
Coombes, J 186
Coons, AH 30
Cooper, J 24, 63
Cooper, K 133, 157
Cooper, MC 187
Cooper, RG 168
Cooper, S 176
Copeland, JL 23, 35
Coppack, RJ 72
Coratella, G 47, 67
Córdova, A 193
Cormack, S 136, 143
Cormack, SJ 60, 143, 145, 146
Cormie, P 23, 30, 36, 37, 46, 55, 62, 75, 97, 100, 105, 106, 109, 110, 128, 130, 132, 136, 142, 143
Cormier, P 99
Cornish, SM 58, 173
Cornu, C 72
Corrêa, DA 99
Cortes, N 71, 96
Cosgrave, C 5
Ćosić, D 66
Costa, P 174
Costa, PB 18, 21, 24, 30, 90, 99
Costello, JT 185, 186, 189
Costill, DL 18, 21, 30, 72, 161, 175
Costley, L 140
Coté, DJ 184
Cote, RW 72
Cotter, S 59
Cottin, Y 23, 37
Cotton, J 7
Cotuk, HB 189
Coudeyre, E 72
Coudrat, L 140
Coudreuse, JM 171
Counsilman, BE 193
Counsilman, JE 193
Counts, BR 102, 132
Courel-Ibanez, J 95
Courel-Ibáñez, J 95
Coutts, A 193
Coutts, AJ 98, 181, 185, 186, 191, 192
Couturier, A 9, 70, 71, 137, 189
Coves, Á 149, 150, 151
Cowley, ES

Cox, AJ 177
Cox, DS 177
Cox, GR 177
Coyle, EF 72
Craig, BW 113, 115, 116, 120
Craig, MM 176
Cramer, JT 174
Crameri, R 116, 122
Crampton, D 184, 185, 186
Craven, J 183, 184
Crespo, C 189
Crewther, B 35, 36
Crewther, BT 20, 36, 64, 76
Cribb, PJ 166, 173
Crielaard, J-M 133
Crill, MT 18
Crisp, AH 163
Cronin, J 35, 36, 48, 81, 101, 128, 132
Cronin, JB 7, 33, 43, 58, 59, 61, 62, 70, 95, 96, 108, 133, 136, 137, 139, 150, 151
Cross, AG 176
Cross, MR 70, 95, 96, 138, 150, 151, 157
Crouse, SF 120
Crowcroft, S 184
Crowell, HP 42
Crowell, JB 128
Crowther, NB 1
Crum, AJ 76, 142
Cruz, IF 5
Cruz, IR 30
Csende Z, Racz L 173
Cuadrado-Peñafiel, V 98, 137, 142, 143
Cuenca-Fernández, F 64, 76
Cuevas-Aburto, J 108
Culbertson, JY 174
Culver, BW 34
Cumming, DC 35
Cumming, K 81, 82, 84
Cumming, KT 80, 81
Cumps, E 69
Cunanan, AJ 13, 19, 25, 26, 27, 29, 30, 36, 38, 44, 46, 47, 48, 49, 51, 68, 84, 85, 90, 91, 92, 94, 100, 121, 125, 147, 157, 161, 168, 171, 172, 178, 181
Cunningham, D 142, 189
Cunningham, DJ 64, 76, 78, 105, 142
Cupeiro, R 163
Cureton, KJ 22, 35
Currell, K 167
Currie-Elolf, LM 106
Curtner-Smith, M 193
Cusso, R 25, 100, 101, 107
Cussó, R 119
Cuthbert, M 24, 94, 98
Cyrino, ES 127

D

Dabbs, NC 65, 75, 109
Dabek, V 89
da Costa, JLF 66
da Eira Silva, V 175
D'Agostino, DP 154
Dahl, HA 22, 35
Dahlquist, DT 59, 80
Dai, B 25, 71, 96, 128, 138
Dai, Y 22
Daigle, K 7

Dalbo, VJ 174
Dalen, T
Dalgas, U 62
Dall, N 98
Dal Pupo, J 76
Dalsky, GP 7
Dalton-Barron, N 157
Daly, RM 49
Damas, F 72
Dang, VH 174, 175
Daniels, J 18, 21, 30
Dankel, SJ 29, 73, 102, 132
D'Antona, G 21, 30
da Rocha, AL 174
Darrall-Jones, J 156
Dart, J 46, 71, 96
Dascombe, BJ 155, 156
da Silva, ASR 174
da Silva, JJ 99
da Silva, RP 175
da Silva Gianoni, RL 177
da Silva Novaes, J 107
Dattilo, M 184
Davids, K 21, 40, 41, 42
Davie, AJ 40, 115, 121, 122
Davies, GJ 72
Davies, J 59
Davies, KJA 19
Davies, T 25, 102
Davies, TB 24, 98, 101, 107, 151
Davis, C 175
Davis, DS 191
Davis, GW 19
Davis, HL 188, 191
Davis, IS 42
Davis, KA 80, 81, 82, 84
Davis, SE 96, 143
Davis, V 48
Davison, E 47
Davison, GW 185, 189
Dawes, J 138
Dawes, JJ 28, 30, 43, 111, 149, 181, 189
Dawson, B 173, 175, 184, 185, 186, 188
Dawson, BT 184, 185, 186
Dawson, J 183
Dawson, KA 191
Dawson, LG 191
Dawson, R 1
Dawson, TA 30
Dawson, VL 174
Day, ML 150
Dayne, AM 8, 65, 84
de Aguiar, RA 121
Deakin, GB 113, 121
Deakin, V 170
de Amorim, JF 163
de Andrade Kratz, C 175
Deane, MA 126
de Araújo Ribeiro, A 81, 82, 84
Deason, M 30
de Ataide e Silva, T 163
de Azevedo Neto, RMA 115, 118
Deb, SK 176
de Baranda, PS 64
de Barros Vilela, G 99
DeBeliso, M 65
De Blaiser, C 69

De Bleecker, C 69
De Bock, K 30
De Bock, S 191
de Capitani, M 174
De Clercq, D 135
Decombaz, J 175
de Diego, A 30
de Freitas, EC 173
DeFreitas, JM 101
de Geus, E 30
de Geus, EJ 29, 30
De Groot, G 3
de Groot, LC 166
de Haan, A 21, 31
De Heer, E 175
De Hoyos, DV 99
De Jesus, F 175
de Jesús, F 174
de Jonge, XAKJ 139
de Keijzer, KL 67, 68
delaCruz, M 62
Delahunt, E 136
de la Rosa, A 30
Del Coso, J 30, 66, 175, 176, 177
Delecroix, B 184, 188
Delextrat, A 191
Deley, G 189
Delgado-García, G 157
Delignières, D 21, 41
de Lima, C 109
de Lima, LCR 121
Della Gatta, PA 186
Dellal, A 52, 115, 116
Dello Iacono, A 68
del-Olmo, M 108
DeLong, TH 44
DeLorme, TL 154
Del Rio, R 62, 69, 159
Del Rosso, S 79
Del Vecchio, FB 5
de Medeiros, MHG 175
de Mello, MT 184
Deminice, R 173
Dempsey, AR 63
Demura, S 5
Denadai, BS 121
Denegar, CR 35
Deneweth Zendler, J 138
Dennis, SC 171
de Noordhout, AM 147
Denton, J 108
de Oliveira, EP 176
de Oliveira, JJ 163
de Oliveira, JV 174
de Oliveira, LF 175
de Oliveira, PR 53
De Pauw, K 191
de Paz, JA 67, 159
De Paz Fernández, JA 30
Depiesse, F 172
de Poli, RAB 174
Derave, W 174, 175
De Ridder, R 69
Dermody, BM 80, 82
Derom, CA 29
de Salles, BF 24, 99, 107, 149
de Salles Painelli, V 115, 174, 175

Desbrow, B 183, 184
Deschamps, T 21, 41
Deschenes, MR 34, 35, 38, 120
de Silva, AS 163
Desmedt, JE 32, 62
De Souza, CG 163
de Souza, DB 177
de Souza, EO 115, 121
De Souza, EO 55, 66
de Souza, JG 177
de Souza, LC 167
Despines, AA 79
Devaney, JM 30
De Vera, A 30
Devero, B 46
de Villareal, ESS 4, 5
de Villarreal, ESS 7, 50, 62, 63, 65, 67, 75, 78, 109, 110, 134, 138, 142, 143
De Vito, G 170
Devries, MC 165, 166
De Wachter, J 191
DeWeese, BH 3, 7, 19, 20, 24, 30, 48, 49, 50, 51, 52, 53, 54, 62, 63, 64, 65, 68, 70, 71, 75, 76, 78, 84, 85, 90, 91, 92, 103, 104, 105, 106, 109, 110, 111, 119, 121, 132, 133, 134, 147, 149, 152, 153, 157, 181
De Witt, JK 128
De Zwart, JR 185
Dhamija, P 187
Dhamrait, S 30
Dias, I 24, 99
Díaz, A 115, 122
Díaz de Durana, AL 30
Dickie, D 7
Didier, J-P 23, 37
Dietz, CC 80, 81, 84
Dietze-Hermosa, MS 85
Díez-Vega, I 30
Dillen, C 157
Dillon, S 189
Di Michele, R 165
Dimitriou, L 188
Dinan, C 128
Dingley, AA 69
Dingley, AF 133, 157
di Prampero, P 37, 79, 103, 109, 142, 143
Di Prampero, PE 50, 142
Ditroilo, M 81
Dixon, C 157
Doan, BK 85
Docherty, D 85, 90, 109
Dodd, MJ 189
Dodd, SJ 174
Doepker, C 177
Doerr, P 36
Doherty, M 176
Dohi, K 35
Dolan, C
Dolan, E 174, 175
Doma, K 113, 116, 121, 122
Domenech, E 171
Domínguez, E 52
Domínguez, R 174, 176
Donahue, PT 136
Donaldson, CM 162
Doncaster, GG 113
Donne, B 186

Donovan, T 168
Dooly, C 107
Doran, DA 6
Dorel, S 9, 50, 70, 71, 137, 138, 142, 143, 189
Dorgo, S 85
Dorman, JC 184
Dorrell, H 156, 157
Dorrell, HF 150
Doscher, MW 60
Doss, WS 3
Dos Santos, ML 43
Dos'Santos, T 64, 71, 83, 94, 95, 96, 101, 103, 130, 138, 139, 141, 142, 146, 157
Douglas, J 37, 65, 66, 67, 69, 85, 88, 90, 91, 96, 110, 150, 159
Douroudos, II 176
Doust, JH 93
Doutreleau, S 65
Douzi, W 186
Döweling, A 188
Downes, P 189
Downey, DL 6
Dowse, R 141
Doyle, TLA 51, 143, 144
Drake, D 131
Draper, SN 187
Drapsin, M 72
Drawer, S 189
Drid, P 72
Driller, MW 184, 187, 192
Drinkwater, E 101
Drinkwater, EJ 42, 50, 63, 101
Driscoll, D 125
Driss, T 140
Drolet, L 172
Drouet, PC
Drury, B 157
Drust, B 165
Druzhevskaya, AM 30
Duarte, JA 29, 30
Dube, J 157
Dubé, JJ 79
Duca, M 46, 47
Duchateau, J 22, 23, 32, 33, 39, 62, 70, 78, 159
Ducher, G 7
Dudley, GA 107, 176
Dudson, M 61
Dudson, MK 61
Duehring, MD 63
Dufek, JS 135
Duffield, R 184, 185
Duffy, O
Dufour, SP 65
Dugan, CA 63
Dugan, E 146
Dugué, B 186
Dulson, DK 61, 89
Duncan, CN 172
Duncan, M 175
Duncan, MJ 175, 176
Duncan, NM 78, 109
Duncan, S
Dunnett, M 174
Dunn-Lewis, C 186
Dupont, G 184, 188
DuPont, WH 187
Dupuy, O 27, 45, 186

Duque, A 30
Durkalec-Michalski, K 175
Dutheil, F 173
Duthie, GM 155, 156
Dutra, MT 170
Duvall, M 18, 23, 40
Dvir, Z 72
Dvorak, J 172
Dwyer, JT 176
Dyatlov, DA 30
Dyhre-Poulsen, P 22, 33, 101
Dymarkowski, S 173
Dyson, J 156
Dzekov, C 34
Dzekov, J 34
Dziados, JE 22, 34, 35, 36, 120

E

Earle, RW 148
Earnest, C 173
Earnest, CP 172, 174, 175
Earp, JE 186
Easteal, S 30
Ebben, WP 3, 4, 7, 20, 24, 62, 63, 64, 65, 69, 75, 76, 77, 78, 80, 84, 96, 105, 109, 110, 132, 134, 139, 140, 159
Eckard, TG 100
Eddens, L 116, 121
Eddington, ND 177
Edge, J 175, 184
Edgerton, VR 14, 15, 18, 23, 30, 38
Edison, BR 156
Edman, KA 97
Edmonds, C 186
Edouard, P 72, 142
Edwards, DJ 50, 143
Edwards, G 106
Edwards, KM 189
Edwards, T 125
Egan, B 70, 95
Egaña, M 186
Egger, A 175
Ehlert, T 29, 30
Eichner, ER 171, 172
Eihara, Y 116
Eisenberg, RS 18
Eisenhofer, J 191
Ekdah, C 24, 58, 94
Eklund, D 113, 115, 121
Eliakim, A 7
Ellefsen, S 121
Ellenbecker, TS 72
Ellington, WR 173
Elliott, MCCW 53
Ellis, D 161, 164, 165, 167, 177
Elrod, CC 149
El-Sohemy, A 177
Ely, BR 171
Ema, R 7, 39, 47, 65, 109
Engel, WK 18, 30
Engell, DB 172
English, KL 128
Enoka, RM 22, 32, 33, 43
Erdağ, D 47
Erdman, AG 80, 81, 84
Ereline, J 4, 5
Erickson, TM 158

Eriksrud, O 71
Ermidis, G 175
Ernest, J 27, 61
Erskine, RM 23, 39
Erzeybek, MS 189
Esbjornsson, M 30
Esbjörnsson, M 119
Esco, MR 51
Esformes, JI 59
Esgin, T 116
Esgro, B
Esko, A 24, 58, 94
Eslava, J 102, 149, 157
Esposito, F 47, 174
Esteban-García, P 62
Esteve, E 67
Esteves, GJ 66
Ettema, G 3, 132
Etxebarria, N 7
Eubank, TK 22, 35
Evans, M 24, 71, 85, 90, 91, 94, 98, 132
Evans, W 18, 21, 30
Evans, WJ 20, 32, 78
Evans, WS 59
Everaert, I 175
Ewing, G 143
Exel, J 147, 158
Exell, T 138, 139
Eynon, N 29, 30

F

Fagard, R 29
Fahey, TD 13, 22, 35, 162
Fahs, CA 127
Faigenbaum, A 77, 174
Faigenbaum, AD 7, 43, 103, 121, 132, 149, 174
Fairchild, TJ 63
Falk, B 7
Falkel, JE 22, 34
Fallowfield, JL 174
Falvey, É 5
Falvo, MJ 66
Falzone, M 69, 85
Farinatti, P 99, 176
Farley, ORL 142
Farr, T 191
Farris, J 174
Farrow, D 177, 192
Farup, J 64, 76
Fasold, F 122
Fatouros, IG 176
Fatoye, F 6
Fauth, ML 62, 69, 96, 159
Favero, M 46, 95
Fedewa, MV 51
Feher, J 18
Feigenbaum, MS 107
Feiring, DC 72
Feki, Y 113
Feldmann, CR 62, 63, 69, 96
Felici, F 176
Ferguson, GD 113
Ferguson, RA 189
Ferguson, SL 127
Feriche, B 157
Fernandes, HM 99
Fernandes, IA 78, 109

Fernandes, J 157
Fernandes, JFT 80, 133, 157
Fernandes, L 24, 99
Fernandes, V 48, 135
Fernández del Valle, M 30
Fernandez-Elias, VE 177
Fernandez-Gonzalo, R 67, 159
Fernández-Valdés, B 147, 158
Ferrando, AA 165, 166
Ferrauti, A 188
Ferreira, M 173
Ferreira-Junior, JB 189
Ferrier, B 141
Ferris, BG 154
Ferrugem, LC 163
Fethke, N 7
Feuerbacher, JF 116, 121, 122
Fielding, R 182
Fields, JB 108
Figueiredo, C 174
Figueiredo, T 24, 99
Figura, F 176
Filip, A 66
Filip-Stachnik, A 177
Filliard, JR 140
Filliard, J-R 189
Fimland, MS 80, 81
Findlay, KP 46
Fink, W 18, 21, 30
Fink, WJ 21, 72
Finlay, MJ 78, 80
Finlayson, SJ 61
Finley, ZJ 72
Finn, C 189
Finn, CV 60
Finni, T 35
Fischer, CP 177
Fisher, J 58
Fisher, JP 49, 66, 99
Fitch, HL 41
Fitzgerald, E 139
Fitzpatrick, DA 48
Flanagan, EP 96, 128, 139, 140
Flanagan, SD 186, 187
Flanagan, TR 175
Fleck, SJ 7, 22, 23, 32, 34, 35, 36, 38, 65, 78, 80, 81, 93, 99, 107, 132, 158, 161, 162
Fleming, S 139
Fletcher, C 63, 142
Fletcher, G 23, 39
Flores, FJ 64
Floria, P 67
Floría, P 140
Florini, JR 36
Floyd, RT 43
Flück, M 39, 90
Folland, J 22, 23, 32, 39, 78
Folland, JP 20, 23, 29, 37, 39, 46
Fonda, B 189
Fong, C 40
Fonseca, FS 177
Fontana, F 107
Forsse, JS 167
Forsythe, WA
Fortescue, EB 172
Fort-Vanmeerhaeghe, A 67
Foster, C 49, 79, 150, 159

Foster, CA 173
Foster-Powell, K 162
Fouré, A 71
Fournier, J 189
Foutty, S 191
Fowler, P 183
Fowler, PM 184
Fowler, VM 40
Fowles, JR 109
Fox, CD 37
Fox, EL 171
Fragala, MS 23, 35, 36, 39, 132, 158, 186, 193
Francesconi, RN 172
Franchi, MV 39, 89, 90
Franchini, E 5, 115, 118, 121, 132, 149, 175
Franco, GS 173
Franco-Alvarenga, PE 79
Francois, M 3, 65, 89, 90
Franco-Márquez, F 46
Frank, BS 100
Franklin, B 107
Franklyn-Miller, A 5
Fraser, WD 168
Frazee, K 157
Fredholm, BB 176
Freitag, N 116, 121, 122
Freitas, D 29
Freitas, J 173
Freitas, MC 174
Freitas, TT 99, 137
French, D 89
French, DN 22, 23, 34, 36, 59, 120, 121, 126, 149, 173, 186, 191
French, SM 50, 64
Friday, JJ 58
Friedl, K 35
Frisch, D 59
Frost, G 33
Fry, AC 7, 18, 21, 22, 23, 26, 27, 28, 30, 34, 35, 36, 37, 40, 46, 48, 49, 50, 51, 52, 53, 55, 63, 65, 76, 85, 89, 95, 99, 104, 107, 122, 126, 157, 158
Frykman, P 22, 35
Frykman, PN 22, 34, 35, 36
Ftaiti, F 171
Fuchimoto, T 97
Fuchs, CJ 165
Fuglevand, RJ 33
Fuhrmann, S 5
Fukashiro, S 38, 135
Fuku, N 30
Fukuda, DH 19, 39, 108, 128, 174
Fukunaga, T 23, 25, 37, 38, 39, 47, 70, 96, 137, 138
Funderburk, LK 167
Furukawa, T 18
Fyfe, J 118, 121, 122
Fyfe, JJ 49, 113, 118, 120, 121

G

Gabbett, TJ 6, 7, 41, 125, 141, 144
Gabriel, DA 33
Gage, PW 18
Gahche, JJ 176
Gai, CM 154
Gajewski, J 30
Galazoulas, C 140

Galbo, H 35
Galbraith, MA 35
Galiano, C 67
Gallagher, JR 154
Galloway, SDR 167, 168
Gallucci, A 167
Galpin, AJ 18, 21, 30, 76, 80, 81, 82, 84
Galvão, DA 101
Gambki, B 34
Gandevia, SC 121
Ganio, MS 176
Gans, C 15, 23, 38
Gao, C 30
Gao, J 116, 121, 122, 175
Gao, L 30
Gapeyeva, H 4, 5
Garatachea, N 30
Garceau, LR 62, 69, 77, 96, 140, 159
Garcia, ES 30
García, I 4, 5
Garcia, M 101
García-Fernández, M 122
García-Fernández, P 174
García-Hermoso, A 60, 62
García-López, D 67, 80, 159
García-López, J 135
García-Pallarés, J 115, 122
Garcia-Ramos, A 133, 157
García-Ramos, A 64, 108, 137, 142, 143, 149, 156, 157
Gardiner, P 22
Garhammer, J 8, 37, 53, 54, 55, 103, 142, 159
Garland, SW 186
Garnacho-Castaño, MV 174
Garner, B 173
Garner, JC 136
Garner, S 122
Garnham, AP 177
Garofolini, A 175
Garratt, N 189
Garrett, WE 25, 71, 96, 138
Garrido, ND 46
Garrido-Blanca, G 157
Garthe, I 172
Garton, F 30
Garzarella, L 99
Gastin, PB 94, 146
Gates, T 77
Gathercole, R 135
Gaunt, AS 15, 23, 38
Gaylord, J 173
Gaze, D 185
Gehri, DJ 51, 144
Geib, G 71
Geiser, CF 64, 80, 95, 96, 136, 139, 140
Geneau, MC 94, 146
Gentil, P 5, 58, 81, 82, 84
Gentles, J 44, 106
Gentles, JA 49, 106, 125, 152
Georgiades, E 29, 30
Georgiadis, G 78, 80
Georgiadis, GV 128
Georgiou, E 7
Gepfert, M 66, 88
Gerbeaux, M 116, 117
Gergley, JC 53
Gerodimos, V 58

Gerosa-Neto, J 121
Gerritsen, KGM 33, 137
Geyer, H 172
Ghigiarelli, JJ 101
Ghijs, M 19
Giakas, G 81
Giannis, G 81
Gibala, MJ 23, 37, 47, 177
Gibbon, D 185
Gibson, AL 18, 21, 30, 31
Gibson, ASC 113, 115
Gijsen, AP 166
Gil, S 133
Gilbert, C 34
Gill, N 128, 130, 132, 185, 186, 189
Gill, ND 23, 38, 61, 89, 127, 143, 145, 146, 186, 188
Gillespie, CA 30
Gillies, EM 66, 90, 132
Gillies, JD 31
Gillis, N 176
Gilmore, LA 18, 40
Gilmore, SL 59, 80
Gilpin, HE 60
Gilson, SF 193
Gimenez, P 142
Ginevičienė, V 30
Gioftsidou, A 72
Giordanelli, MD 64, 80, 95, 96, 136, 139
Girard, O 72
Gissane, C 72, 186, 187
Gittoes, M 138, 139
Gittoes, MJR 41
Giuliano, DV 44, 132
Giustanini, M 72
Giveans, RM 80, 81, 84
Glace, BW 173
Glaister, M 9, 46, 54, 110, 137, 183
Glazier, P 41
Gleadall-Siddall, DO 175
Gleason, BH 30, 145
Gleeson, M 177, 182
Glickman, EL 176
Glowacki, SP 120
Gluchowski, AK 61
Glynn, JA 70
Gobbo, LA 121, 174
Godaux, E 32, 62
Godfrey, M 185
Godolias, G 72
Godwin, MS 80
Golas, A 66, 88
Gołaś, A 100
Goldsmith, JA 151
Goletzke, J 162
Golik-Peric, D 72
Gollnick, PD 35
Gomes, PSC 78, 109
Gomes, WA 99
Gomez, AL 35, 173
Gomez-Cabrera, M-C 171
Gómez-Gallego, F 30
Gomez-Piqueras, P 120
Gonçalves, LS 175
Gondek, MB 188
González, A 30
Gonzalez, AM 19, 25, 26, 39, 49, 100, 101, 121

Gonzalez, DE 165
González, J 158
Gonzalez, JR 67
Gonzalez, MP 85
Gonzalez-Badillo, JJ 3, 63, 95, 133
Gonzaléz-Badillo, JJ 157
González-Badillo, JJ 4, 5, 46, 95, 102, 133, 142, 143, 149, 157, 159
González-Gallego, J 67, 159
González-Hernández, JM 98, 108
González-Izal, M 25, 100, 101, 107
González-Jurado, JA 173
Gonzalez-Rave, JM 27, 48, 51, 52, 53, 55, 104, 122
González-Suárez, JM 46
Gonzalez-Villora, S 120
Gonzalo-Orden, JM 23, 38
Gonzalo-Skok, O 61, 139, 151
Goodall, S 113, 115
Goodman, C 175, 184
Goodpaster, BH 79
Goods, PSR 72
Goodwin, J 41
Gordish-Dressman, H 30
Gordon, PM 30
Gordon, RD 193
Gordon, SE 20, 22, 23, 34, 35, 36, 120
Gordon, TJ 132
Gore, C 184, 185
Gore, CJ 175
Goris, M 173, 174
Gorman, A 41
Gorostiaga, E 113, 116, 120
Gorostiaga, EM 3, 7, 23, 25, 35, 100, 101, 102, 107, 120, 157
Gorter, JW 40, 158
Goss, FL 157
Gotshalk, LA 20, 23, 34, 35
Gottschall, JS 70
Gough, LA 176
Gougoulias, N 187
Gould, LM 172
Goulet, EDB 171
Gourgoulis, V 176
Gouvêa, AL 78, 109
Gouveia, É 29
Gowitzke, BA 13, 16
Gradiner, PF 14, 16, 18
Graef, JL 174
Graf, BK 120
Graham, JL 187
Graham, K 189
Graham, S 185, 186, 187
Graham, T 151
Graham, TE 176
Graham-Smith, P 62, 69, 71, 80
Gran, A 116
Granacher, U 60, 62, 109, 187
Granados, C 7, 25, 100, 101, 107
Grange, RW 78
Grant, S 30
Granzier, H 18, 40
Granzier, HL 14, 18, 40
Gras, P 23, 37
Gray, D 189
Grazer, JL 4, 140
Greco, CC 121

Green, AL 173
Green, CM 50
Green, DJ 184
Green, HJ 109
Green, JM 193
Green, JS 120
Greene, D 136
Greene, DA 7, 185, 186
Greenes, DS 172
Greenhaff, PL 162, 172, 173, 174
Greenwood, L 174
Greenwood, LD 173
Greenwood, M 101, 173, 174, 178
Greer, BK 90, 91, 92
Gregor, R 8, 159
Gregor, RJ 136
Gregorio, CC 40
Gregory, JR 175
Gregson, W 71, 95, 137, 168, 184
Greig, L 133, 157
Greig, M 78, 80
Grélot, L 171
Grenz, C 128
Grgic, I 175, 176
Grgic, J 23, 24, 37, 46, 98, 101, 107, 132, 172, 175, 176, 177
Griffen, C 175
Griffis, RB 168
Griggs, CV 140
Grigoletto, D 89
Grimby, G 72, 79
Grindstaff, P 173
Grindstaff, PD 173
Grodjinovsky, A 7
Grønfeldt, BM 72
Gross, MT 25, 71, 96, 138
Grove, JR 184
Grover, LM 170
Gruber, LD 85
Gu, C-Y 81
Gualano, B 115, 173, 174, 175
Guede-Rojas, F 108
Guerrero, M 25, 100, 101, 107
Guerrin, F 117
Guest, NS 172, 176, 177
Gueugneau, M 165
Guével, A 72
Guidetti, L 173
Guiherme, JPLF 30
Guilhem, G 72, 189
Guilliams, ME 128
Gulbin, JP 30
Gullich, A 59, 78
Günthör, W 18, 30
Guodemar-Pérez, J 174
Guppy, SN 25, 50, 101, 108, 130, 135, 136, 143
Gurjão, ALD 127
Gutiérrez-Hellín, J 30
Gutin, B 22, 35
Guzun, R 172
Gwin, JA 165

H

Haarbo, J 170
Habansky, AJ 72
Habay, J 191
Hackett, D 25, 102, 116

Hackett, DA 151
Hackney, AC 22, 23, 35, 161, 176
Hackney, KJ 57
Hadjicharalambous, M 5, 108
Haff, EE 25, 50, 63, 64, 95, 100, 101, 133
Haff, G 51, 128
Haff, GG 3, 4, 5, 7, 19, 23, 24, 25, 26, 27, 36, 37, 43, 46, 48, 50, 51, 52, 53, 54, 55, 58, 59, 62, 63, 64, 65, 66, 67, 69, 71, 75, 76, 78, 81, 82, 85, 90, 91, 94, 95, 96, 97, 98, 99, 100, 101, 103, 104, 106, 108, 109, 110, 121, 122, 127, 128, 130, 131, 132, 133, 134, 135, 136, 141, 142, 143, 147, 149, 150, 157, 159, 181
Hagerman, FC 18, 21, 22, 23, 30, 34, 36, 40, 149
Hagerman FC 30
Hagobian, TA 175
Hahn, AG 30
Hainaut, K 22, 32, 33, 62, 70, 78, 159
Haines, TL 8, 142
Hainline, B 7
Haischer, MH 142, 150, 151
Haizlip, KM 18, 30
Hajduk, G 66
Häkkinen, A 3, 20, 25, 48, 113, 116, 120, 121, 130, 132
Häkkinen, K 3, 18, 20, 21, 22, 23, 25, 30, 34, 35, 36, 37, 48, 62, 69, 82, 85, 90, 91, 95, 101, 102, 106, 113, 115, 116, 120, 121, 130, 132, 149, 157, 173
Halaj, M 108
Halaki, M 25, 102, 116, 151
Hale, DF 132, 149
Halkjaer-Kristensen, J 22, 33
Hall, A 133, 157
Hall, HK 18, 21, 30, 31
Halle, JL 151
Hallman, HL 187
Halperin, I 128
Halson, S 177, 184, 185, 186, 187, 189, 192
Halson, SL 100, 125, 177, 182, 183, 184, 185, 186, 187, 188, 189, 191, 192
Hamada, T 78
Hämäläinen, I 116
Hamill, BP 7
Hamill, J 147, 149
Hamilton, D 139
Hamilton, DL 49
Hamilton, LD 167
Hamlin, MJ 189
Hamm, MA 176
Hammami, R 63
Hammond, KM 164, 167, 177
Han, D 81, 82, 84
Han, K-H 128
Handelsman, DJ 35
Handford, MJ 66, 68, 85
Hannah, R 46, 174
Hannan, CJ 34
Hanrahan, S 193
Hansen, AK 177
Hansen, EA 53
Hansen, K 59
Hansen, KT 7
Hansen, S 34
Hanson, AM 128

Hanson, ED 30, 118, 121
Hanson, J 18
Hara, H 73
Hara, M 135
Harcourt, BA 171
Hardee, JP 23, 25, 26, 36, 100, 107
Harden, M 37, 65, 66, 67, 69, 85, 89, 90, 91, 96, 101, 110, 132, 150, 159
Hardt, F 115
Hardy, CJ
Hargreaves, M 163, 167, 177
Harman, E 22, 35, 128
Harman, EA 34, 35, 36, 120
Harman, SM 35
Harmon, RA 80, 81, 82, 84
Harper, DJ 71, 95, 116, 119, 120, 137, 139
Harre, D 181
Harrelson, A 34
Harris, C 65
Harris, DR 167
Harris, GR 37, 48, 54, 97, 104
Harris, NK 61, 133
Harris, RC 174, 175
Harrison, AJ 78, 128, 147, 149
Harrison, AP 49, 106
Harrison, BC 18, 30
Harrison, CB 7
Harrison, JS 57, 60
Harry, JR 135, 136, 143
Hart, CL 44
Hart, NH 3, 66, 71, 90, 132
Hartman, M 7, 20, 128
Hartmann, H 5
Harvey, TM 173
Hashimoto, G 73
Hatfield, DL 23, 35, 36, 132, 158
Hatzinikolaou, A 176
Haug, WB 42, 50
Haugen, T 50, 143
Haun, CT 18, 21, 31, 37, 101
Hausswirth, C 98, 177, 184, 188, 189
Hautier, CA 71
Havas, E 187
Haven, KL 71
Havens, KL 25, 46, 71, 96
Hawkins, SB 51, 144
Hawley, J 177
Hawley, JA 47, 113, 120, 164, 166, 167, 171, 177
Hay, DC 135
Haycraft, J 53, 104
Hayes, A 166, 173
Hayes, K 147, 149
Hayes, S 49, 110
Hayward, S 173
Hbacha, H 62
Heapy, AM 187
Hearn, DW 100
Hearris, MA 164, 177
Hebert, EP 3, 65, 89, 90
Heckman, CJ 22
Hedderman, R 163
Hedrick, A 63
Heegaard, JH 136
Heel, K 184
Heelas, T 80
Heffernan, SM 170
Heibel, AB 176

Heileson, JL 167
Helgerud, J 7, 110, 116
Helland, C 50, 143
Hellsten, Y 25, 100, 101, 107
Helms, E 163, 165, 166
Helms, ER 66, 88, 100, 101, 108, 150, 151, 157
Hemmings, BJ 191
Hemmingsson, E 116, 120
Heneghan, JM 135
Henneberry, M 99
Henneman, E 18, 31, 32
Hennessy, LK 78
Henselmans, M 163, 165, 166
Herbert, RD 101
Herbert, WG 172
Herer, P 183, 184
Herman, RE 132
Hermassi, S 62
Hernández, D 30
Hernandez-Belmonte, A 95
Hernández-Davó, JL 68
Hernández-Sánchez, S 80
Herrera, AA 34
Herrero, AJ 80
Herrick, AB 154
Herrington, L 7
Hershman, EB 173
Herzog, W 14, 18, 23, 39, 40, 64, 66, 76
Hespel, P 30, 173, 174, 176
Hessel, AL 18
Hester, GM 25, 107
Hetland, ML 170
Hewett, TE 7
Hicks, KM 65, 85, 89, 90, 91, 132
Hickson, RC 113, 121
Higgins, MF 175
Higgins, S 176
Higgins, TR 185, 186
Higuchi, H 23, 40
Higuchi, M 80
Higuchi, T 80
Hikida, RS 18, 21, 30, 34, 40
Hikida RS, Murray TF 30
Hilbert, JE 191
Hilkens, L 165
Hill, C 175
Hill, CA 174
Hill, CH 175
Hill, CM 136
Hill, DW 108
Hill, J 187
Hill, JA 186
Hill-Haas, S 185
Hills, FA 188
Hilton, EN 35
Hilton, NP 176
Hindle, B 61
Hindle, BR 61
Hinshaw, TJ 128
Hintz, MJ 63
Hintzy, F 142, 143
Hippocrate, A 191
Hirai, Y 116
Hirayama, K 80, 95
Hirsch, KR 165, 168, 172
Hirschberg, AL 35
Hirschmüller, A 128

Hiruma, E 173
Hislop, HJ 72
Hitz, M 133
Hjelvang, LB 98
Ho, J-Y 186
Hobbs, RT 100, 101
Hobson, RM 174, 175
Hobson, S 85, 91
Hodgdon, JA 58
Hodges, NJ 49, 110
Hodgson, DD 109
Hodgson, M 109
Hoeger, WWK 132, 149
Hoff, J 7, 110, 116
Hoffman, B 7, 46, 62, 75, 109, 110, 143
Hoffman, J 22, 35, 174
Hoffman, JR 19, 22, 24, 25, 26, 34, 39, 49, 63, 100, 107, 121, 128, 149, 172, 174, 175, 191
Hoffman, MD 187
Hogan, CM 101
Hoke, T 99, 173
Hoke, TP 99
Holden, G 77
Hollander, A 3
Hollander, DB 3, 65, 89, 90
Hollins, JE 140
Holm, I 119
Holmberg, P 185, 192
Holmstrup, ME 59
Holsbeeke, L 40, 158
Holwerda, AM 165
Homer, M 182
Homma, T 72
Hone, M
Honeycutt, DR 99
Hong, F 81, 82, 84
Hood, PE 186
Hooker, DN 184
Hooper, DR 187
Hooper, SL 193
Hope, K 78
Hopkins, DR 132, 149
Hopkins, WG 30, 121, 126, 163, 184, 192
Hoppeler, H 18, 30
Hopper, AJ 101
Horan, S 188
Hore, AM 64, 76, 105
Hori, N 7, 50, 135, 136, 143
Horn, RR 49, 110
Horne, S 39
Horner, K 170
Horner, NS 72
Hornsby, G 23, 36, 48, 49, 51, 52, 53, 54, 84, 91, 104, 105, 110, 128, 147, 152, 153
Hornsby, WG 13, 19, 21, 23, 25, 26, 27, 29, 30, 36, 38, 44, 46, 47, 48, 49, 50, 51, 52, 53, 55, 63, 94, 95, 100, 102, 103, 104, 106, 110, 121, 122, 125, 132, 147, 152, 157, 159, 161, 168, 171, 172, 178, 181
Horowits, R 18
Horswill, CA 175
Hortobágyi, T 65, 187
Horvath, SM 22, 35
Hoshikawa, Y 47
Hostler, DP 18
Hostler DP, Crill MT 30

Hotermans, C 147
Hottenrott, L 188
Houben, LHP 165
Housh, TJ 22, 35
Houston, ME 78
Houweling, PJ 30
Howard, A 193
Howard, R 7, 43, 94, 154
Howarth, KR 47, 177
Howatson, G 65, 85, 89, 90, 91, 113, 115, 116, 120, 121, 122, 132, 185, 186, 187
Howe, LP 66
Howell, JN 187
Howell, MEA 18, 21, 30, 31
Howells D 144
Hristovski, R 41
Hruby, J 20, 99, 128, 173
Hrysomallis, C 81
Huang, HD 53
Hubal, MJ 30
Hubing, KA 106
Hudson, DE 30
Hudson, GM 175
Hudson, MB 106
Hudson, S 175
Hudspeth, AJ 16
Hug, F 189
Hughes, B 6
Hughes, JD 66, 68, 80, 81, 85
Hughes, L 72, 73
Hughes, LJ 151
Hughes, MG 139
Hughson, R 187
Hughson, RL 187
Hulmi, JJ 82
Hulston, CJ 177
Hultman, E 172, 173, 174
Humphries, B 62, 95
Humphries, S 30
Hunter, G 18
Hunter, I 48
Hunter, JP 9, 70, 71
Hunter, SK 47
Hurley, BF 35
Hurst, H 189
Hurtado, AK 176
Huson, HJ 29, 30
Hussain, ME 62, 69
Huttunen, P 22, 34
Huxley, AF 18, 38
Huxley, HE 18
Huynh, M 125
Hvid, LG 62

Iacono, AD 67
Ibanez, J 3, 7, 149
Ibañez, J 101, 107
Ibáñez, J 6, 23, 25, 35, 100, 102, 107, 157
Ichinose, Y 38
Iglesias-Soler, E 108
Iida, T 47
Iii, JAG 121
Iijima, R 73
Ikebukuro, T 23
Ikeda, T 72
Ilic, DB 140
Ilić, V 66

I

Imbach, F 20
Immesberger, P 7
Impellizzeri, FM 126
Impey, SG 164, 167, 177
Incledon, T 173
Ingram, JR 189
Inman, RA 188
Inoue, DS 121
Inoue, T 174
Intiraporn, C 80
Invernizzi, PL 174
Irving, TC 14, 18, 40
Irwin, C 183, 184
Irwin, G 138, 139, 148, 158
Irwin, W 171
Isaka, T 116
Ishida, A 23, 26, 37, 46
Ishida, K 173
Ishigaki, T 23
Ismaeel, A 168
Ismaili, CP 121
Ispirlidis, I 72
Israetel, MA 65, 84
Issurin, VB 37, 48, 52, 53, 54, 55, 181
Ito, M 38
Ivey, PA 4, 28, 154
Ivy, JL 164, 166, 177, 184
Iwasaki, N 173
Izquierdo, M 3, 7, 23, 25, 35, 60, 62, 63, 100, 101, 102, 107, 113, 115, 116, 120, 121, 122, 149, 157

J

Jaafar, H 140
Jackson, MJ 171
Jacobs, I 115, 119, 122
Jacobsen, P 116, 120
Jäger, R 172, 174, 175
Jagim, AR 101
Jajtner, AR 19, 25, 26, 39, 49, 100, 121
Jakobsen, MD 59, 102
James, CR 135
James, LP 7, 46, 53, 62, 75, 94, 104, 109, 110, 125, 128, 142, 143, 146
Jamil, M 119, 120
Jamurtas, A 5
Jamurtas, AZ 176
Janicijevic, D 108
Jankovic, NN 140
Janshen, L 9
Janssen, I 69
Jansson, E 119
Jaques, R 7
Jaric, S 2, 3, 157
Jarvis, J 89
Jarvis, MM 62, 69
Jay, K 59
Jeacocke, N 177
Jeffreys, I 7, 93
Jeffries, O 186
Jeffriess, MD 138, 139
Jegu, AG 72
Jelen, K 66
Jelenkovic, A 29
Jemiolo, B 18, 21, 30
Jemni, M 125

Jenkins, D 185, 189, 191, 192
Jenkins, DG 191
Jenkins, NDM 101, 172, 176, 177
Jensen, BT 59
Jensen, J 22, 35
Jensen, NL 96
Jensen, RL 62, 65, 69, 78, 80, 84, 96, 139
Jentjens, R 164
Jeromson, S 167
Jessee, MB 29, 73, 102, 132
Jessee, TC 48
Jessell, TM 16
Jeukendrup, A 164
Jeukendrup, AE 164, 177, 178
Jezová, D 22, 35
Jidovtseff, B 133
Jiménez, F 23, 38, 62
Jiménez, SL 157
Jiménez-Ormeño, E 64
Jimenez-Reyes, P 70
Jiménéz-Reyes, P 157
Jiménez-Reyes, P 70, 98, 108, 137, 142, 143, 157
Jlid, MC 62
Jo, E 65, 75, 78, 109
Johnson, BD 132
Johnson, CA 37
Johnson, CC 35
Johnson, GO 22, 35
Johnson, NA 151
Johnson, R 128
Johnson, RL 7, 20, 25, 37, 48, 54, 97, 99, 104, 107, 110, 173
Johnson, TK 150, 151
Johnsson, A 22, 35
Johnston, K 40
Johnston, M 140
Johnston, N 144
Johnston, R 156, 157
Johnston, RD 7
Johnstone, B 28
Jones, A 99
Jones, B 66, 154, 156, 157, 158
Jones, BL 156
Jones, DA 20, 27, 59
Jones, G 175
Jones, GA 175
Jones, GT 126
Jones, H 184
Jones, MT 69, 85, 108
Jones, NL 156
Jones, PA 4, 64, 71, 94, 95, 96, 101, 103, 130, 133, 138, 140, 141, 142, 146, 157
Jones, R 7, 110
Jones, RL 176
Jones, TW 113, 115, 120, 121, 122
Jonker, B 66
Jordan, AR 71
Jordan, CA 138, 139
Jordan, M 170
Jordan, MJ 64, 76
Jordao, AA 173
Jørgensen, MB 59
Joseph, C 168
Joseph, R 59
Jostarndt-Fögen, K 18, 30
Joumaa, V 18, 40

Jouris, KB 167
Jovy, D 23, 36
Judelson, DA 22, 34, 65, 75, 109, 115, 171, 191
Judge, L 176
Judge, LW 79
Judge, M 79
Juhász I, Györe I 173
Jukic, I 100, 101, 108, 157, 163
Juliff, LE 191
Julio, UF 115, 118, 132, 149
Jung, YP 172, 173, 174
Jun Huang, W 163

K

Kaabi, S 63
Kaarakainen, E 113, 116, 120
Kaciuba-Uścilko, H 22, 35
Kadlec, D 21, 40, 41, 71
Kadlubowski, B 7
Kaiser, KK 30
Kaiser, P 115
Kakoschke, N 183
Kalabiska, I 140
Kalista, S 35
Kaliszewski, P 30
Kallus, K 161, 193
Kalman, D 178
Kalman, DS 172, 173, 174, 175, 176
Kalogera, KL 128
Kamahara, K 174
Kamen, G 22, 32, 33
Kamimori, G 176
Kamimori, GH 177
Kandell, ER 16
Kandiah, D 168
Kandoi, R 7
Kanehisa, H 23, 25, 39, 47, 70, 96, 116, 137, 138
Kaneko, M 97, 104, 105
Kang, J 24, 63, 103, 121, 132, 149, 174
Kaprio, J 29
Karamouzis, M 36
Karampatsos, G 78, 80
Karampatsos, GP 128
Karapondo, DL 22, 34
Karatzeferi, C 170
Karlsson, J 79
Karpovich, PV 3
Karunakara, RG 107
Karyekar, CS 177
Kasper, AM 162, 163, 164, 165, 167, 170, 171, 172
Kassavetis, P 53
Katan, MB 167
Katch, FI 13, 65
Katch, VI 13
Katsuta, S 5
Kattan, A 61
Katzel, MJ 156
Kauhanen, H 18, 21, 30, 36
Kavanaugh, AA 49, 106, 125, 181
Kavazis, AN 37
Kavouras, SA 80, 193
Kawahara, T 72
Kawakami, Y 7, 23, 37, 38, 39, 47, 65, 109
Kawama, R 70
Kawamori, N 7, 9, 70, 71, 76, 142
Kay, AD 81

Kay, L 172
Kaya, F 189
Kearney, JT 22, 35, 161, 162
Keenan, M 59
Keiner, M 5, 7
Keith, R 22, 23, 34, 36
Keith, RE 161, 162
Kelekian, GK 80
Kellett, A 184
Kellis, E 5, 58, 63, 159
Kellis, S 5, 58
Kellmann, M 161, 188, 193
Kelly, A 34
Kelly, CF 101
Kelly, J 125, 130
Kelly, KA 101
Kelly, V 41, 185, 189, 191, 192
Kelly, VG 7, 46, 62, 75, 109, 110, 128, 143, 184, 191
Kelso, JAS 41
Kemmler, W 132
Kempner, ES 18
Kendall, K 101
Kendall, KL 25, 101, 108, 174
Kendrick, IP 174, 175
Kenefick, RW 171
Kenn, JG 46, 80, 82, 95
Kennedy, MD 46, 95
Kennedy, R 140
Kennedy, RA 131
Kenny, IC 4, 139
Kenny, J 80
Kent, M 7, 110
Keogh, J 35, 36, 61
Keogh, JB 61
Keogh, JW 61, 136, 157
Keogh, JWL 61, 80, 81, 82, 84
Kephart, WC 101
Kerksick, CM 65, 66, 173, 174, 178
Kernozek, TW 64, 80, 95, 136, 140
Kersting, U 96
Kerwin, D 138, 139
Keskinen, KL 23, 36, 175
Ketelaar, M 40, 158
Kettler, T 34
Keul, J 23, 34
Khamoui, AV 66, 88
Khan, M 72
Khanna, A 187
Khlifa, R 62
Khoury, J 7
Khuat, CT 140
Khuu, S 140
Kidgell, DJ 66
Kieffer, AJ 167
Kiely, J 71, 95, 116, 119, 120, 137
Kiely, MT 64, 80, 95, 136, 139, 140
Kiens, B 161
Kies, AK 166
Kikuhara, N 5
Kilduff, LP 20, 64, 76, 78, 105, 135, 142, 189
Kilen, A 98
Kilgore, JL 25, 64, 95, 100, 101, 133
Kilgore, L 24, 25, 98, 100
Kilpatrick, MW 3, 65, 89, 90, 154
Kim, C 175
Kim, CK 174, 175

Kim, D 127
Kim, DH 22, 34, 35
Kim, E 127
Kim, H 175
Kim, HJ 174, 175
Kim, I-Y 165
Kim, J 173
Kim, JE 164
Kim, J-S 149, 150, 152
Kim, P 174
Kim, S 173
Kim, SB 140
Kimber, NE 167
Kinder, JE 22, 35
Kindermann, W 22, 23, 34, 35
King, A 163
King, E 5
Kingsley, MI 64, 76, 105, 142
Kinsella, S 140
Kipp, K 64, 80, 95, 96, 136, 139, 140, 142
Kirby, TJ 8, 142
Kirkendall, DT 51, 144
Kiss, B 14, 18, 40
Kitamura, K 5, 133
Kjaer, M 23, 34, 39
Kjær, M 116, 122
Klatt, M 149
Klatt, S 122
Klau, JF 176
Klaus, A 172
Klein, KE 23, 36
Kleiner, DM 51, 144
Kleiner, SM 172, 173, 174
Klemp, A 149, 150, 151, 152
Kline, DE 64, 80, 95, 136
Klissouras, V 29, 30
Klopstock, T 174
Klusiewicz, A 30
Kneffel, Z 116, 121
Knehans, AW 173
Knez, W 184
Knight, KL 154
Knight, TJ 139
Knudson, D 158
Knudson, DV 6
Knuttgen, HG 27, 34, 49, 50
Kobal, R 5, 48, 133, 135
Koceja, DM 47
Kochan, B 89
Kohler, D 84
Kojić, F 66
Kokuhuta, C 173
Kolmerer, B 40
Komi, PV 18, 20, 21, 22, 30, 34, 35, 36, 37, 59, 82, 100, 136
Komori, S 186
Kon, M 72
König, T 128
Konrad, A 7
Koopman, R 166
Koozehchian, MS 80
Koponen, P 3, 132
Korivi, M 81, 82, 84
Korkotsidis, A 7
Kosar, AC 189
Kossow, AJ 62, 69, 96
Kostrzewa-Nowak, D 193

Kotani, Y 50, 135, 136, 143
Kotarsky, CJ 57
Kotzamanidis, C 46
Koubaa, D 113
Kouvatsi, A 30
Kouzaki, M 116
Kovacs, EM 172
Kowalchuk, K 67
Kozinc, Ž 51, 128, 144
Koziris, LP 22, 27, 34, 36, 49, 50, 85, 99
Kozlowski, S 22, 35
Kraemer, RR 3, 65, 89, 90
Kraemer, WJ 3, 7, 22, 23, 25, 27, 32, 34, 35, 36, 38, 43, 48, 49, 50, 59, 62, 65, 75, 78, 80, 81, 82, 85, 93, 95, 97, 99, 101, 102, 103, 107, 110, 113, 115, 116, 120, 121, 126, 130, 132, 142, 149, 157, 158, 161, 171, 173, 186, 187, 191
Kraemer WJ 149
Krahenbuhl, G 18, 21, 30
Kram, R 70
Kramer, JB 99
Krapi, S 3, 132
Krase, AA 128
Kraska, JM 7
Kravitz, L 65, 66
Kreider, R 173, 174
Kreider, RB 80, 101, 165, 172, 173, 174, 175, 176, 178
Krein, D 46, 95
Kreis, R 175
Krieg, PA 40
Krieger, JW 23, 24, 25, 37, 46, 58, 66, 98, 99, 101, 103, 165, 166
Krishnan, AC 73
Kristensen, J 15, 23, 38
Kristoffersen, M 50, 143
Kroll, W 79
Kruger, A 51, 53
Kruse, NL 98
Krustrup, P 116, 120, 175
Krzyszkowski, J 135, 136, 142
Krzysztofik, M 66, 88
Kubo, K 23, 39
Kubo, T 80
Kugler, F 9
Kuijper, EA 29, 30
Kuipers, H 166
Kulik, JR 142
Kullman, EL 187
Kunz, M 5
Kuoppasalmi, K 34
Kupchak, BR 187
Kuper, JD 121
Kutscher, DV 22, 33
Kutz, M 27, 61
Küüsmaa, M 121
Kvetnanský, R 22, 35
Kviatkovsky, SA 165
Kvorning, T 34, 35
Kwan, K 85
Kyriazis, T 80
Kyrolainen, H 22, 34, 35
Kyröläinen, H 115, 116

L

Labarque, V 173

Labeit, S 18, 40
Lacome, M 122
Lacour, JR 36, 52
La Cruz-Sánchez, E 133
La Doyle, T 50, 143
Laframboise, MA 174
Lagally, KM 150
Lago-Ballesteros, J 52
Lago-Peñas, C 52
Lahti, J 70
Lake, J 50, 61, 68, 138, 139, 143
Lake, JP 4, 8, 50, 59, 62, 63, 64, 66, 68, 80, 83, 94, 95, 96, 97, 103, 105, 133, 135, 136, 138, 139, 140, 141, 143, 146, 159
Lake, M 96
Lam, TQ 174
Lamas, L 63
Lamb, KL 157
Lambalk, CB 29, 30
Lamblin, J 188
La Monica, MB 19, 128
Lamont, H 23, 36, 37, 100, 110, 128
Lamont, HS 4, 20, 24, 54, 64, 65, 76, 78, 95, 96, 99, 106, 107, 109, 143, 157
LaMonte, MJ 132
Lampert, E 65
Lancaster, S 174
Lancha, AH, Jr 30, 115, 174, 175
Landers, GJ 184
Landis, J 174
Landoni, L 20, 36
Lang, C 183
Langan-Evans, C 89, 163
Langford, G 27, 61
Langford, GA 60
Langfort, J. Effects 175
Lanhers, C 173
Lapole, T 189
Lara, B 176
Laroche, D 66
Larson, R 77
Larson-Meyer, DE 172
Larsson, B 116, 122
Larumbe-Zabala, E 103
LaStayo, PC 65
Lastella, M 183, 192
Latella, C 101
Lates, AD 90, 91, 92
Lau, WY 189
Lauber, D 132
Lauder, MA 59, 62
Lauersen, JB 6, 27
Laurencelle, L 121
Laurent, CM 193
Lauriot, B 21, 41
Laursen, P 117, 118, 121, 123
Laursen, PB 117, 121
Lavers, RJ 157
Lavie, P 183, 184
Lavigne, D 24, 99
Lawlor, DJ 170
Lawrence, MM 25, 26, 100, 107
Lawsirirat, C 80
Lawton, T 101
Layer, JS 128
Laynez, I 30
Lazar, A 44, 132

Lazauskus, KK 18, 21, 30
Lazinica, B 24, 98, 132, 176
Leach, NK 176
Léam, A 71
Leandro, CG 163
Leduc, C 188
Leduc, EH 30
Lee, EC 193
Lee, EJ 14, 18, 40
Lee, J 173
Lee, JE 70, 95, 96
Lee, JK 172
Lee, JKW 172
Lee, M 122
Lee, MJC 121
Lee, MK 127
Lee, ML 34
Lee, RG 33
Lee, W 72
Lee-Barthel, A 165, 170
Leeder, J 183
Leeder, JDC 185
Lees, A 135
Lefere, T 174
Lefevre, J 29
Leffers, AM 15, 23, 38
Lehmann, M 23, 34
Lehmkuhl, M 37
Leinwand, LA 18, 30
Leite, N 151
Leite, RD 149
Le Meur, Y 98
Lemon, PW
Lemos, A 107
Lemura, LM 99
Lenetsky, S 157
Lensel, G
Lensel-Corbeil, G 117
Lenz, B 157
Leonard, TR 18, 23, 40
Leong, B 22, 32
León-Guereño, P 176
Lepers, R 66, 118, 191
Lephart, SM 107
Lepley, LK 6
Leroyer, P 21, 41
Lesage, F-X 173
Leslie, KLM 50
Lesmes, GR 72
LeSuer, DA 149
Leutholtz, BC 171
Leveritt, M 22, 70, 119, 121, 122, 163
Levine, DN 18, 30
Levy, PD 73
Lewek, MD
Lewis, M 157
Lewis, NA 182
Lewis, RD 22, 23, 25, 34, 36, 107, 176
Lewthwaite, R 154
Leyva, WD 90
Li, C 163
Li, FW 14, 18, 40
Li, H 151, 152
Li, J 81, 82, 84, 176
Li, M 30
Li, S 176
Li, X 30

Li, Y 85, 89, 109
Libardi, CA 66, 72
Lichtwark, G 116
Lider, J 130
Lieber, RL 18, 40
Lieberman, HR 177
Liebesman, J 189
Liew, B 60
Lim, MT 164
Lima-Silva, AE 163
Limonta, E 174
Lin, Y 81, 82, 84
Linares, V 193
Linari, M 18, 40
Lindberg, K 50, 143
Lindberg Nielsen, J 72
Lindquist, MA 53
Lindsell, R 101
Lindstedt, SL 18, 40, 65
Linnamo, V 35
Linthorne, NP 41
Lipuma, T 170
Lira, F 174
Lira, FS 121
Lis, DM 170
Lis, RP 145
Lisboa, P 71, 95, 137
Litjens, MCA 33, 137
Little, K 187
Littlefield, KP 40
Littlefield, R 18, 40
Littlefield, RS 40
Liu, J 27, 48, 51, 52, 53, 55, 104, 122
Liu, TM 44, 132
Lixandrão, M 174
Lixandrão, ME 72
Llanos-Lagos, C
Lloyd, R 80, 81, 82, 84, 136, 157
Lloyd, RS 7, 43, 50, 63, 70, 95, 96, 110, 139
Lngfort, J 25, 100
Lo, S-L 140
Lobo, V 171
Lockie, RG 18, 21, 28, 30, 44, 71, 111, 132, 138, 139, 149, 181, 189
Lockwood, CM 174
Loeb, GE 47
Loebel, CC 34
Loenneke, JP 29, 72, 78, 102, 109, 113, 115, 116, 120, 121, 122, 127, 132, 189
Loftiss, DD 173
Logan, PA 121
Logan, S 61
Lombardi, G 188, 189
Lombardi, V 18, 40
Longman, DJA 41
Longo, S 47
Lonsdorfer-Wolf, E 65
Loos, RJ 29
Lopes dos Santos, M 28, 30, 111, 149, 181, 189
Lopez, H 174
Lopez, HL 172, 173, 174
Lopez, P 101
Lopez, SA 35
López-Contreras, G 64, 76
López Samanés, Á 67
Lorenz, D 72
Lorenzetti, S 133

Lorenzo, V 30
Lorimer, A 61
Lormes, W 193
Lottig, J 64
Loturco, I 4, 5, 48, 82, 96, 99, 133, 135, 136, 137, 138, 139, 151, 155, 158, 159
Loucks, AB 161
Louder, T 4, 135, 140
Lougedo, JH 174
Louis, J 98, 164, 177, 189
Louit, L 81
Loumaye, A 35
Love, R 173
Love, T 189
Lovegrove, S 151
Low, J 109
Lowe, T 36
Lowenthal, DT 99
Lowery, RP 78, 109
Lozano-Estevan, MDC 174
Luboshitzky, R 183, 184
Lucas, J 113, 115, 116, 120
Lucas, SJE 189
Lucia, A 29, 30
Luczo, TM 138, 139
Luden, N 18, 21, 30
Luecke, TJ 23, 36, 149
Luera, MJ 101
Lufkin, AKS 59
Lugo, JP 182
Lugokenski, R 72
Luine, VN 34
Luk, HY 187
Lum, D 5, 58, 59, 81, 128, 159
Lund, H 72
Lundberg, TR 35, 67, 98, 116, 121, 122
Lundgren, LE 142
Lupien, SP 178
Lupo, C 23, 36
Lynch, JM 22, 27, 34, 49, 50, 99, 173
Lynn, SK 80
Lyons, G 34
Lyons, M 4
Lyttle, AD 35, 97

M

Ma, F 30
Ma, W 14, 18, 40
Maase, K 166
MacArthur, DG 30
MacDonald, CJ 106, 125, 152
Macdonald, IA 173
MacDonald, JP 7
MacDonald, MJ 47, 177
MacDougall, JD 78, 122
Machado Reis, V 173
Machek, SB 167
Maciejewska-Karlowska, A 30
Macintosh, BR 109
MacIntosh, BR 14, 16, 18, 64, 66, 76
Mackala, K 136
MacKenzie, SJ 157
Mackey, CS 101
Mackinnon, LT 193
MacKinnon, LT 193
Macklin, B 175
Maclaren, DPM 171

MacLaughlin, H 121
MacLennan, DP 173
Maden-Wilkinson, TM 20, 29, 37
Madruga, M 138
Madruga-Parera, M 151
Madsen, K 35
Maemura, H 174
Maeo, S 116
Maestro, A 30
Maestroni, L 6
Maffiuletti, NA 22, 23, 32, 39, 72, 78, 140, 189
Maffulli, N 140, 187
Magill, RA 40
Magliano, L 34
Magnusson, P 22, 33, 101
Magnusson, SP 15, 22, 23, 33, 38, 39, 116, 122
Magrelli, J 174
Magrini, M 143
Magrum, E 48, 70, 111, 119
Mahoney, D 173
Maia, J 29
Main, LC 184
Maior, AS 99
Makhlouf, I 121
Malczewska-Lenczowska, J 30
Malecek, J 108
Malina, RM 7
Mallard, KG 60
Maller, O 172
Malliou, P 72
Malloy, PJ 64, 80, 95, 136, 139
Malone, JK 189
Malone, S 6
Maloney, S 138, 139
Maloney, SJ 139
Malvela, MT 175
Malyszek, KK 80, 81, 82, 84
Mancinelli, CA 191
Mangine, G 174
Mangine, GT 19, 25, 26, 39, 49, 100, 101, 121
Maniar, N 187
Manimmanakorn, A 189
Manimmanakorn, N 189
Manini, TM 73
Mann, B 157
Mann, CH 177
Mann, JB 2, 4, 28, 30, 43, 111, 136, 149, 154, 155, 157, 158, 181, 189
Mannix, RC 172
Manocchia, P 59
Manolovitz, AD 80, 81, 82, 84
Manore, MM 169
Mansfield, SK 151
Manta, P 78
Manzi, V 121
Manzo, O 62, 69, 159
Maquet, P 147
Mar, MH
Maraj, BK 72
Marcello, BM 173
Marchante, D 142, 157
Marchitelli, L 22, 35
Marchitelli, LJ 35, 36
Marcinik, EJ 58
Marcora, S 81, 128, 191
Marcora, SM 126, 191
Mardock, MA 101

Maresca, RD 176
Maresh, C 35
Maresh, CM 7, 22, 34, 35, 132, 142, 158, 171, 176, 191
Margonis, K 176
Maria Calò, C 30
Maridakis, V 176
Marin, PJ 46, 78, 95, 96, 109, 113, 115, 116, 121, 122, 142
Marín, PJ 64, 72, 80, 142
Marinho, DA 46
Marins, JCB 175
Markovic, G 62
Markworth, JF 192
Maroto-Izquierdo, S 67, 68, 85, 159
Marples, D 22, 23, 34, 36
Marques-Jiménez, D 176
Marquet, L-A 177
Márquez, G 108
Marrel, JP 72
Marriott, M 175
Marrone, M 173
Marsh, DJ 19, 25, 26, 46, 49, 100, 121, 125, 157
Marshall, B 5
Marshall, EC 59
Marshall, J 24, 99
Marshall, LK 69, 89, 90
Marshall, P 81
Marshall, PWM 103
Marshall, RN 9, 33, 70, 71
Marsit, JL 27, 49, 50, 85
Martin, AY 176
Martín, B 30
Martin, C 177
Martin, D 175
Martin, DT 128, 191, 192
Martín, E 80
Martin, JS 101
Martin, SE 120
Martin, TP 120
Martín-Acero, R 23, 38, 98
Martínez, C 173
Martínez-Aranda, LM 98
Martínez-Cava, A 95
Martínez-Hernández, D 67
Martinez-Rodriguez, A 70, 95
Martini, GL 163
Martin-Rivera, F 101
Martín-Rivera, F 68
Martins, WR 81, 82, 84, 170
Martinson, W 182
Maruyama, K 23, 40
Marx, JO 23, 35
Masamoto, N 77
Massey, A 116, 120
Massey, D 46, 95
Massey, GJ 20, 29, 37
Massidda, M 30
Mastro, AM 35
Masuda, K 5
Maszczyk, A 25, 100
Maté-Muñoz, JL 174
Mathes, S 182
Mathevon, L 189
Matic, MS 140
Matsuo, A 25, 70, 96, 137, 138
Matta, T 24, 99

Matthews, M 6
Matthews, T 173
Mattocks, KT 29, 73, 102, 132
Mattsson, CM 67
Matveyev, LP 49, 51, 104
Maughan, RJ 79, 170, 171, 172
Maulder, PS 7
Maurer, A 120
Mavromatidis, K 176
Maw, JR 64, 76, 105
Maxwell, NS 93
May, K 189
Maybanks, G 157
Mayer, F 128
Mayhew, JL 4, 41, 43, 44, 132, 136, 149
Mayo, X 108
Mazzeo, RS 34
Mazzetti, SA 35
McAndrew, J 163
McArdle, A 171
McArdle, F 171
McArdle, WD 13
McBride, J 54, 157
McBride, JM 8, 25, 26, 40, 65, 84, 97, 99, 100, 105, 106, 107, 128, 136, 142, 158
McBride, MG 7
McBride, TC 79
McBurnie, A 71, 138
McBurnie, AJ 71, 157
Mccall, A 188
McCambridge, TM 7
McCarthy, JP 120
McCartney, D 183, 184
McCaulley, GO 97, 105, 106, 128, 136, 142
McCaw, ST 58
McClymont, D 139
McComas, AJ 13, 14, 16, 18, 78
McCormick, JH 149
McCormick, K 89
McCormick, M 20, 23, 25, 35, 48, 130
McCoy, L 37
McCoy, LB 25, 64, 95, 100, 101, 133
McCubbine, J 138, 139
McCully, KK 176
McCurdy, KW 27, 60, 61
McCurley, RC 149
McCurry, D 22, 35
McDaniel, JL 167
McDermott, AY 175
McDwyer, M 157
McElhinny, AS 40
McErlain-Naylor, SA 67
McEwen, M 103
McFadden, BA 178
McFarland, JE 132
McGaughey, KJ 175
McGawley, K 121
McGee, D 48
McGee, SL 47
McGorm, H 186
McGregor, SJ 171
McGuigan, M 81, 88, 90, 91, 127, 137
McGuigan, MR 4, 7, 30, 37, 46, 50, 51, 55, 62, 65, 75, 80, 81, 82, 84, 89, 100, 101, 108, 109, 126, 128, 132, 136, 143, 144, 145, 146, 150, 157, 159, 173
McGuigan M 85, 91

McHugh, MP 173
McInnis, TC 138
McIntyre, F 41
McKay, BR 23, 37
McKee, JR 72
McKeever, S 43, 94, 154
McKeever, SM 4, 46, 64, 66, 83, 96, 105, 127, 133, 136, 143, 151, 152, 155, 158, 159
McKellar, SR 165, 166
McKenna, MJ 172, 175, 176
McKeown, A 20, 29, 37
McKibans, KI 176
McLaren, S 126, 157
McMahon, E 126, 145
McMahon, EL 145
McMahon, JJ 4, 8, 9, 24, 33, 46, 50, 63, 64, 66, 71, 83, 94, 95, 96, 97, 98, 101, 103, 105, 108, 125, 127, 130, 132, 133, 134, 135, 136, 139, 140, 141, 142, 143, 146, 151, 155, 157, 158, 159
McMaster, DT 48, 81, 128, 132
McMillan, J 37, 48, 53
McMillan, JL 22, 23, 34, 36
McMillan, S 135
McMullen, SM 108
McMurray, RG 22, 35
McNair, PJ 9, 33, 70, 71
McNamara, JM 147
McNamee, M 30
McNaughton, L 175
McNaughton, LR 175, 176
McNeal, J 125
McNeal, JR 5, 29, 96, 125, 187, 188
McNulty, KL
McQuilliam, SJ
Meckel, Y 7
Medeiros, A 184
Medina, D 144
Medlar, S 18, 21, 30
Meechan, D 33, 95, 108, 133, 142
Meen, HD 22, 35
Meeusen, R 69, 181, 186, 191, 192
Mehta, A 18
Meira, EP 60
Mekary, S 27, 45
Melby, C 106
Melegati, G 188, 189
Mello, R 22, 35, 36
Melton, C 174
Melton, P 72
Melzer, I 154
Mendez, KM 151
Mendez-Villanueva, A 70, 139
Mendham, AE 184
Mengersen, K 168
Mercuriale, G 1
Meredith, HJ 101
Merino, SG
Mero, A 70, 136
Mero, AA 175
Merrigan, J 85, 89
Merrigan, JJ 69, 85, 108
Messinis D 35
Mester, J 6
Methenitis, S 66, 122
Methenitis, SK 128
Mettler, S 165

Meyer, T 181, 186, 188, 191, 192
Meyers, RW 139
Meylan, CMP 173
Mezêncio, B 173
Mezzomo, M 72
Micaleff, J-P 21, 41
Micard, V 165
Michael, BS 70
Michael, S 116
Michael, SW 151
Michailidis C 46
Micheli, LJ 7, 43
Michelin, AF 72
Middleton, MK 20, 64, 76
Midttun, M 50, 143
Mielgo-Ayuso, J 170, 176
Mielke, M 174
Mieritz, RM 72
Mijailovich, SM 14, 18, 40
Mike, J 65, 66
Mikkola, J 113, 115, 116, 120
Mikulic, P 107, 176
Miles, MP 173
Milioni, F 174
Millard-Stafford, ML 176
Miller, IR 7
Miller, J 134, 184, 185
Miller, JS 57
Miller, MG 58
Miller, MK 81, 128
Miller, MS 21, 30
Miller, P 175
Miller, PD 7
Miller, SC 80
Miller, SL 166
Miller-Dicks, M 21, 40, 41, 71
Millet, GP 113
Millet, GY 118
Milner, M 13, 16
Milner-Brown, HS 32, 33
Milnor, P 174
Milson, J 184
Mina, MA 65, 78, 81, 82
Minahan, CL 184
Minchev, K 18, 21, 30
Minett, GM 184, 185
Minetti, A 37, 79, 103, 109, 142, 143
Minetti, AE 20, 36, 37, 55, 79, 103, 109, 142, 143
Minichiello, J 59
Minshull, C 174
Miramonti, AA 19, 25, 26, 49, 100, 121, 128
Miranda, F 24, 99, 107, 149
Miranda, H 24, 99, 107, 149
Mitchell, JB 106
Mitchell, N 165, 167
Mitchell, WK 39, 90
Mittleman, K 58
Miyachi, M 116
Miyamoto, N 7, 39, 65, 109
Mizuguchi, N 80
Mizuguchi, S 19, 25, 26, 44, 46, 49, 100, 106, 121, 125, 127, 128, 130, 135, 140, 152, 157, 188, 193
Mizuno, M 66
Mizuta, C 173
Mizuta, K 173

Mizutani, M 25, 70, 96, 137, 138
Mizutani M, Matsuo A 46
Mobley, CB 101
Moeller, JL 170
Mohr, M 116, 120, 175
Moir, G 9, 46, 54, 110, 137
Moir, GL 4, 20, 24, 28, 40, 41, 42, 64, 76, 78, 96, 99, 109, 135, 140, 143
Moiz, JA 62, 69
Mokhtar, AH 68, 85
Mokone, GG 30
Molina, JJ 30
Möller, GB 175
Monroy, JA 18, 40
Montain, SJ 171, 172
Montalvo, S 85
Montalvo, Z 51, 143
Monteiro, PA 121
Monteiro, W 99
Montgomery, H 30
Montgomery, PG 184
Montini, M 176
Montmain, J 20
Montpetit, J 26
Moody, J 59
Moody, JA 7
Mookerjee, S 95
Moolyk, AN 63, 66, 96
Moon, JR 37, 174
Moonen, G 147
Moore, B 182
Moore, CA 28, 85, 89, 158
Moore, JH 37
Moore, M 142
Moore, RW 193
Moore, SR 161
Mooyaart, A 175
Moquin, PA 46, 53
Mora-Custodio, R 46, 157
Moraes, JE 133
Morais, JE 46
Morales, J 132
Morales-Artacho, AJ 20, 29, 37, 157
Moran, CN 30
Moran, CW 193
Moran, J 60, 62, 66, 69, 157, 159, 173
Moran, K 140
Morán, M 30
Morandi, RF 30
Moran-Navarro, R 95
Morán-Navarro, R 95
Morante, JC 135
Mora-Rodriguez, R 133, 177
Moras, G 147, 158
Moreau, NA 177
Moreira, OC 67, 159
Morencos, E 30
Moreno, FJ 149, 150, 151
Moreno, MR 44, 132
Moreno Perez, V 67
Moresi, MP 136
Morgans, R 165
Morimatsu, F 174
Morin, J-B 9, 50, 70, 71, 95, 136, 137, 138, 142, 143
Morita, Y 158
Moritani, T 20
Morley, JE 171

Moroi, M 73
Morouço, P 188
Morrigi, R, Jr 55
Morris, SJ 50, 63, 95, 110
Morrison, M 156, 157, 187
Morrison, S 21, 40, 41
Morriss, CJ 72
Morrissey, MC 24, 58, 94, 128
Mortara, S 22, 35
Mortensen, OS 59, 102
Morton, J 184
Morton, JP 162, 163, 164, 165, 167, 168, 170, 171, 172, 177
Morton, RW 165, 166
Moser, D 30
Moser, DA 29, 30
Moses, SA 108
Mostowik, A 66
Mota, P 173
Motoyama, YL 66
Moungmee, P 22, 35
Moura, N 48
Mouser, JG 29, 73, 102, 132
Moyna, NM 30
Mozaffarian, D 167
Mpampoulis, T 108
Mrdakovic, VD 140
Muddle, TWD 101
Muendel, T 184
Mueske, NM 156
Mujika, I 21, 22, 26, 27, 31, 45, 46, 49, 55, 116, 121, 122, 177, 181, 184, 186, 188, 192
Mulcahey, MK 72
Mullaney, MJ 173
Muller, C 61
Muller, M 176
Müller, S 72, 128
Mullineaux, DR 148, 158
Mulloy, F 148, 158
Mumford, PW 37, 101
Mundel, T 187
Mundy, P 66, 68, 95, 97, 103
Mundy, PD 66, 96, 135, 138, 139, 159
Munger, CN 90
Muniesa, CA 30
Muñoz, A 30
Munoz, G 177
Munoz-Guerra, J 177
Muñoz-López, A 67
Muñoz-López, M 142
Muñóz-López, M 157
Munz, B 29
Muramatsu, M 47
Murata, K 39
Murata, M 138
Murillo, DB 149, 150, 151
Murlasits, Z 113, 115, 116, 121
Murphy, A 184
Murphy, AJ 4, 62, 95, 128, 131, 139, 143
Murphy, I 189
Murphy, KT 165, 166
Murphy, MJ 176
Murphy, PW 35
Murphy, S 9, 125, 135
Murray, S 125
Murray, SR 125, 181, 187, 188, 191
Murray, TF 18, 23, 36, 149

Musalem, L 140
Myer, GD 7, 43, 132
Myers, CA 126
Myers, T 186
Myklestad, D 99
Myslinski T 144
Myszka, S 41

N

Naclerio, F 103, 132
Nader, GA 113
Naderi, A 176
Nagahara, R 25, 46, 70, 96, 137, 138
Nagami, T 80
Nagasawa, Y 186
Nagatani, T 25, 101, 108
Nagatomi, R 187
Nagaya, N 34
Nagel, J 22, 35
Najafi, A 116
Nakajima, Y 47
Nakamura, F 137
Nakamura, FY 5, 61, 67, 133, 173, 188
Nakamura, M 73
Nakamura, N 80
Nakao, T 174
Nance, S 142
Narici, MV 20, 36, 39, 89, 90
Nassis, GP 116, 120, 144
Natali, J 175
Naughton, G 136, 173
Naughton, GA 7
Naughton, RJ 165
Navarro-Amézqueta, I 25, 100, 101, 107
Nawrocka, M 66, 88
Nazar, K 22, 35
Nédélec, M 184, 188
Nedergaard, NJ 71, 95, 96, 137
Neenan, C 154
Neese, K 174
Négyesi, J 187
Nein, MA 30
Nelson, AG 4, 72, 128
Nelson, AR 58, 59, 151
Nelson, MT 172, 176, 177
Nemezio, K 175
Neto, MM 184
Neubauer, O 186
Nevill, AM 2, 3
Neville, J 136, 137
Newburger, JW 172
Newell, KM 21, 40, 41, 65, 110, 149
Newlands, C 61
Newton, M 50, 143
Newton, MJ 4, 63, 128
Newton, R 113, 115, 121
Newton, RU 3, 4, 7, 9, 20, 23, 25, 30, 34, 35, 38, 40, 41, 46, 48, 50, 55, 60, 62, 63, 66, 69, 70, 71, 75, 80, 84, 85, 90, 91, 94, 95, 97, 101, 103, 107, 109, 110, 115, 120, 121, 126, 128, 130, 131, 132, 136, 141, 142, 143, 146
Nibali, ML 134
Nicholas, CW 171
Nicholas, SJ 173
Nicklas, BJ 35
Nicol, C 171

Niedergerke, R 18
Nielsen JL, Libardi CA 72, 73
Nieman, DC 99, 177
Nieschwitz, BE 128
Nieß, AM 29, 30
Nijem, RM 80
Nikula, T 187
Nilo Dos Santos, WD 81, 82, 84
Nimphius, S 3, 4, 5, 6, 7, 8, 9, 21, 25, 26, 27, 37, 40, 41, 46, 48, 49, 50, 53, 54, 55, 58, 60, 65, 66, 71, 75, 79, 90, 94, 95, 97, 98, 100, 101, 102, 103, 108, 109, 110, 111, 121, 123, 125, 126, 130, 132, 134, 141, 142, 143, 149, 159
Nindl, BC 20, 23, 34, 35
Nishikawa, K 18, 40
Nishikawa, KC 18, 40
Nishimura, K 116
Noakes, TD 30, 171
Noble, BJ 34
Nóbrega, SR 66
Nolan, RP 136
Nolen, JD 46, 64, 83, 105, 133, 143
Nooner, JL 25, 107
Nordsborg, NB 98
Nordstrom, MA
Norman, B 30
North, K 30
North, KN 30
Nosaka, K 3, 7, 9, 50, 65, 66, 70, 71, 89, 132, 133, 135, 136, 143, 157, 159, 186, 189, 191, 192
Noto, T 73
Nottebohm, F 34
Nottle, C 191
Nourrit, D 21, 41
Novaes, J 24, 99, 149
Nowak, R 40, 193
Nowakowska, A 193
Ntalles, K 7
Nuckols, G 37
Numazawa, H 66
Nummela, A 115, 116
Núñez, FJ 67
Nuñez Sanchez, FJ 67
Nuzzo, J 159
Nuzzo, JL 3, 18, 30, 65, 84, 89, 106, 128
Nygaard, H 34
Nyman, K 82, 115, 116
Nymark, BS 98

O

Obradovic, B 72
O'Brien, JJ 58
O'Brien, T 89
O'Bryant, H 37, 53, 54, 55, 103, 142
O'Bryant, HS 7, 9, 20, 25, 37, 48, 54, 64, 95, 97, 99, 100, 101, 104, 110, 128, 132, 133, 152, 161, 173
O'Callahan, B 156
Ocetnik, S 58
O'Connor, B 132
O'Connor, PJ 176
O'Donnell, S 184
Oetter, E 144
Oftebro, H 22, 35
Ogawa, K 158

Ogborn, D 23, 37, 46, 101
Ogborn, DI 66
Ogita, F 116
Ogueta-Alday, A 135
Ohlin, K 22, 35
Ojasto, T 85
Okada, J 95
O'Kelley, E 27, 61
O'Kroy, J 174
Okudaira, M 70
Okuno, NM 127
Olcina, G 189
O'Leary, DD 78
O'Leary, DS 73
Olenick, AA
Olesen, JL 116, 122
Oliveira, J 29, 30
Oliveira, LF 24, 99, 176
Oliveira, LP 121, 174
Oliveira, MGD 66
Oliver, JL 7, 43, 50, 63, 70, 95, 96, 110, 139
Oliver, JM 101, 108
Olmstead, JD 175
Olson, DP 172
Omcirk, D 108
Omer, J 100
Onate, J 71, 96
Ong, KY 59
Oponjuru, BO 70, 95
Oppliger, RA 171
Op't Eijnde, B 173
Oranchuk, DJ 58, 59, 151
Orekhov, EF 30
Orhant, E 52
O'Riordan, S 187
Orjalo, AJ 44
Ormsbee, MJ 149, 150, 152, 178
Orr, R 25, 102
Ortega, JF 177
Orysiak, J 30
Osburn, SC 37
O'Shea, JP 97
O'Shea, KL 97
O'Shea, P 132
Osman, NAA 68, 85
Östenberg, A 24, 58, 94
Österås, S 50, 143
Osterlund, T 30
Ostojic, S 170
Ostrander, EA 29, 30
Ostrander, GK 29, 30
Ostrowski, EJ 136
Ostrowski, KJ 35, 97
Ota, K 70
Otten, E 38
Ottenheijm, CAC 18, 40
Otterstetter, R 177
Otto, WH, III 59, 63
Otzen, R 57
Outlaw, JJ 173
Overgaard, K 62
Overmayer, RG 187
Owen, N 142
Owen, NJ 64, 76, 78, 105, 135, 142
Owens, DJ 168
Owens, J 72, 73
Owens, JG 73

Owens-Stovall, SK 166
Oxfeldt, M 62
Oxford, SW 176

P

Paasuke, M 4
Pääsuke, M 4, 5
Paavolainen, L 116
Padial, P 157
Padilla, S 55
Paditsaeree, K 80
Padua, DA 25, 71, 96, 100, 138
Pafis, G 72
Page, PA 68
Page, RM 78, 80
Pageaux, B 191
Painter, KB 23, 36, 54, 157
Pakarinen, A 22, 23, 34, 35, 36
Paley, VR 189
Pálinkás, G 140
Pallares, JG 177
Pallarés, JG 6, 95, 133, 159
Palmer, BM 21, 30
Palmer, TB 176
Palmieri-Smith, RM 6
Pan, BJ 164
Panagiotopoulos, M 171
Panissa, V 174
Panissa, VLG 115, 118, 121, 132, 149
Panoutsakopoulos, V 140
Panton, LB 149, 150, 152
Paoli, A 5
Papadimitriou, ID 30
Papadopoulos, K 6
Papaiakovou G 46
Pappas, CT 40
Paquette, MR 135
Paradisis, GP 41
Paratz, JD 168
Pareja-Blanco, F 46, 137, 142
Parise, G 23, 37, 173
Park, ND 176
Park, S 165
Park, SH 175
Parker, AG 173
Parker, DF 59
Parker, JL 101
Parker, L 163
Parkhouse, N 176
Parra, J 119
Parry, HA 37
Parsonage, J 141
Parviainen, T 187
Pascoe, DD 101
Pasiakos, SM 165
Pataky, T 71, 96, 138
Patikas, DA 140
Patil, A 171
Paton, B 72
Paton, CD 177
Patten, C 22, 32
Patterson, SD 66, 72, 73, 151
Pattison, NA 30
Patton, JF 34, 35, 120
Paulo, AC 115, 121
Paulsen, G 77, 81, 82, 84, 99
Paxton, JZ 170

Payne, A 61
Paz, G 99
Paz, GA 99
Pazin, NR 140
Peake, J 186
Peake, JM 186, 188, 192
Pearson, DR 72
Pearson, J 66
Pearson, M 156, 157
Pearson, S 37, 65, 88, 90, 91
Pearson, SJ 3, 130
Peck, JD 177
Pedersen, BK 177
Pedersen, H 80, 81, 82, 84
Pedersen, MT 59
Pedisic, Z 24, 98, 132, 172, 175, 176
Pedlar, C 183, 186, 187
Pedlar, CR 182
Pedley, JS 50, 63, 110, 139
Peeling, P 172
Peeters, MW 29
Peiffer, J 184, 185
Peiffer, JJ 151
Peigneux, P 147
Pellegrino, MA 21, 30
Peñailillo, L 66, 173
Peng, H-T 81, 140
Penland, CM 25, 107
Pennefather, J 186
Penwell, ZR 186
Pereira, B 72, 173
Pereira, L 135
Pereira, LA 5, 48, 82, 133, 137
Pereira, PEA 66
Pereira, S 29
Perera, E 72
Perez, CE 133, 159
Pérez, CE 133
Perez-Bibao, T 103
Pérez-Castilla, A 133, 149, 157
Periasamy, M 40
Perim, PHL 176
Perkins, LE 187
Perman, M 128
Perrey, S 20
Perrine, J 72
Perrine, JJ 14, 15, 23, 38
Perry, TL 162
Pestaña-Melero, FL 149, 157
Peter, J 18
Peter, JB 30
Peterson, M 23, 101
Peterson, MD 23, 25, 46, 95, 100, 102, 158
Peterson, SD 187
Petrakos, G 70, 95
Petré, H 67, 116, 120
Petridis, L 140
Petrini, BE 166
Petro, E 89
Petro, JL 101
Pette, D 18, 21, 30, 31
Petushek, EJ 4, 62, 69, 96, 140, 159
Pexa, BS 100
Peyrot, N 50, 138, 142, 143
Pfeiffer, M 188
Phatak, A 171
Phibbs, PJ 156

Phillips, MD 106
Phillips, SM 22, 47, 121, 165, 166, 172, 177
Phillips, WT 158
Philp, A 177
Philpott, JD 167
Piccinno, A 139
Pichardo, AW 7
Pickering, C 175, 176
Piepoli, A 157
Pierce, K 25, 64, 95, 100, 101, 133
Pierce, KC 7, 19, 20, 27, 47, 48, 50, 51, 52, 53, 55, 63, 95, 100, 101, 104, 109, 121, 122, 147, 173, 181
Pierobon, A 53, 104
Pilianidis, T 36
Pimenta, EM 30
Pimentel, GL 72
Pinar, S 189
Pincivero, DM 107
Pinckaers, PJM 165
Pineda, JG 176
Pinto, M 159
Pinto, MD 3, 65, 89
Pinto, RS 101
Pinto e Silva, CM 115
Piper, FC 41
Piper, TJ 50
Piquard, F 65
Pires, FO 79
Pirke, KM 36
Piscione, J 122
Pisz, A 108
Pitsiladis, Y 29, 30
Pitsiladis, YP 30
Piva, T 193
Pizzoferro, K 183
Pleša, J 51, 144
Plisk, S 24, 45, 47, 58, 82, 94, 173
Plisk, SS 51, 93
Plomgaard, P 177
Plotkin, DL 18, 21, 23, 31
Ploutz-Snyder, LL 127, 128
Plyley, MJ 72
Podolsky, RJ 18
Poe, C 7, 110
Pogson, M 71, 95, 137
Pohlman, R 113, 115, 116, 120
Pokrywka, A 30
Polglaze, T 126
Polito, MD 99, 176
Pollack, GH 39
Pollock, BH 99
Pollock, ML 21, 99
Polydorou, I 168
Pomponi, SM 173
Ponte, J 174
Poon, W 50, 135, 136, 143
Poortmans, JR 174
Pope, ZK 25, 107
Popović, D 66
Poppendieck, W 188
Poprzecki, S 175
Porfido, TM 121
Portas, MD 186
Porter, JM 136
Portman, R 166, 177, 184
Posthuma, D 29, 30

Posthumus, LR 61
Potteiger, J 107
Potteiger, JA 20
Pottier, A 175
Poulsen, P 15, 23, 38
Pournot, H 189
Pousson, M 140
Powell, JR 171
Powell, L 61
Powell, PL 14, 15, 23, 38
Powers, K 18, 40
Powers, KL 18, 40
Powers, SK 171
Pozniak, MA 120
Preatoni, E 147, 149
Prentice, WE, Jr 184
Prestes, J 149
Pretorius, J 144
Prevost, MC 72
Price, KA 175
Price, MJ 175
Price, P 151
Price, TB 30
Prieske, O 81, 82, 84, 109
Prime, DNL 70
Prince, CN 187
Pringle, D 169
Pritchard, HJ 61
Prnjak, K 157
Proost, M 191
Proukakis, C 7
Proulx, C 20, 128
Proulx, CM 20, 37, 48, 54, 97, 104
Pruscino, CL 175
Pryor, JF 4, 41, 69, 128, 131, 135
Pryor, JL 193
Psilander, N 116, 120
Pullinen, T 22, 34
Pulverenti, T 113, 115, 121
Purpura, M 174
Purtill, H 183
Pushkarev, VP 30
Pussieldi, G 30
Putland, JT 189
Putman, CT 66, 90, 132
Putukian, M 34
Puype, J 174
Pyne, D 101, 193
Pyne, DB 45, 184

Q

Qian, S 1
Quatman, CE 7
Queen, RM 25, 71, 96, 138, 193
Quesnele, JJ 174
Quigley, PJ 109
Quiles, JM
Quindry, JC 106
Quinelato, WC 66
Quinlan, J 78
Quinney, HA 120
Quittmann, OJ 122
Quizzini, G 174

R

Raab, S 127
Raastad, T 34, 53, 98, 99, 186, 192

Rabita, G 9, 50, 70, 71, 137, 138, 142, 143
Radaelli, R 101
Rademaker, AC 21, 31
Radnor, JM 139
Ragg, KE 18, 23, 36, 149
Rahimi, R 108
Rakobowchuk, M 47
Ralph, S 125
Ralston, GW 24, 25, 98, 100
Ramadan, ZG 3, 65, 89, 90
Ramaekers, M 30
Ramirez, D 84
Ramirez, DA 84
Ramirez-Campillo, R 60, 61, 62, 69, 135, 159, 173, 188
Ramirez-Lopez, C 157
Ramirez Rios, S 172
Ramos, AG 100, 101, 108
Ramos-Campo, DJ 163
Ramsay, H 157
Ramsbottom, R 2, 3
Ramsey, M 128
Ramsey, MW 7, 23, 36, 54, 106, 109, 125, 152, 157, 188
Ranchordas, M 175
Randers, MB 116, 120
Ranisavljev, I 66
Rankin, P 140
Rasmussen, BB 166
Rasmussen, C 173, 174
Rasmussen, CJ 174
Rasmussen, LR 23, 39
Rasmussen, M 135
Rassier, DE 109
Ratamess, N 95, 99, 174
Ratamess, NA 19, 23, 34, 35, 36, 103, 107, 121, 126, 132, 149, 173, 174
Ratel, S 181
Rattray, B 191
Raue, U 18, 21, 30
Rauramaa, R 59
Rawson, ES 172, 173
Raya-González, J 67, 68, 176
Reaburn, P 185, 193
Read, DB 126, 154, 156
Read, P 6, 80, 95, 97, 103, 138, 139
Read, PJ 151
Record, SM 80, 81, 82, 84
Redgrave, A 182
Redondo, JC 64
Reed, J 125
Reed, JP 113, 115
Reeves, GV 3, 65, 89, 90
Reeves, ND 39, 90
Refoyo, I 176
Regazzini, M 115, 121
Reggiani, C 21, 30
Regnault, S 189
Reich, T
Reich, TE 65
Reid, JC 109
Reid, M 184
Reilly, T 45
Reiman, M 72
Reinardy, J 173
Reis, V 99
Reis, VP 82
Rej, SJE 9, 125, 135

Rejc, E 50, 142
Rekola, R 3, 132
Renals, L 61
Rendo-Urteaga, T 176
Renner, KE 138
Requena, B 4, 5, 62, 63
Rey, E 52
Reyes, GFC 81
Reyneke, J 48
Reyngoudt, H 175
Reynolds, AC 183
Reynolds, J 119, 120
Reynolds, JM 132
Reynolds, K 35, 120
Rhea, M 149
Rhea, MR 24, 25, 46, 51, 80, 82, 95, 99, 100, 102, 113, 115, 116, 121, 122, 142, 158
Rhodes, D 71, 116, 120
Riani, LA 175
Ribeiro, ALA 170
Ribeiro, N 115, 118
Riboli, A 174
Ricard, MD 51, 144
Rice, PE 149
Richard, R 65
Richards, D 177
Richardson, MT 193
Richens, B 150
Richter, EA 173
Riechman, SE 101
Riede, L 182
Riek, S 22, 70
Rigon, PA 89
Ripley, N 24, 98
Ripley, NJ 94, 146
Risso, FG 44, 132
Ritti-Dias, RM 127
Rivera, FM 85
Rixon, KP 65
Roach, GD 183
Robbins, D 109
Robbins, DW 103
Robergs, RA 132
Roberson, PA 37, 101
Roberts, BM
Roberts, L 183, 184, 186
Roberts, LA 128
Roberts, MD 18, 21, 31, 37, 101, 165, 173, 174
Roberts, TJ 23, 39, 40
Robertson, CT 47
Robertson, RJ 150, 157
Robillard, ME 76
Robineau, J 122
Robinson, AL 175
Robinson, JM 25, 107
Robinson, MA 71, 95, 96, 137, 138
Robinson, TM 174
Robson-Ansley, PJ 182
Roby, FB 72
Rochau, K 46
Roche, S 142
Rodano, R 147, 149
Rodas, G 119, 144
Rodman, KW 120
Rodrigues, B 99
Rodriguez, RF 175
Rodríguez-Fernandez, A 61

Rodríguez-Fernández, A 176
Rodríguez-Marroyo, JA 135
Rodríguez-Pérez, MA 70
Rodríguez-Romo, G 30
Rodríguez-Rosell, D 46, 157
Rodriguez-Sanchez, N 168
Roe, G 156
Roe, GAB 156
Roelands, B 191
Rogers, D 53
Rogers, DR 170
Rogers, MA 35
Rogerson, D 49, 55, 103, 105, 150, 157, 175
Roi, GS 20, 36
Rojas, FJ 149, 157
Roll, F 100
Rolnick, N 72
Rolph, R 22, 35
Romance, R 101
Romano, F 5
Romano, N 99
Romero, AR 149, 150, 151
Romero, MA 37, 101
Romero-Rodrigues, D 138
Rønnestad, B 50, 143
Rønnestad, BR 34, 53, 77, 98, 116, 121, 122, 123
Rooney, KJ 101
Roos, E 24, 58, 94
Roos, H 24, 58, 94
Roosen, P 69
Rosa, FT 173
Rosa, RA 139
Roschel, H 72, 115, 173, 174, 175
Rosdahl, H 116, 120
Rose, C 189
Rose, MC 162
Rosenblatt, B 72
Rosene, J 173
Roslan, S 191
Ross, A 22, 37, 65, 70, 88, 90, 91, 119
Ross, CL 25, 107
Ross, EZ
Ross, JJ 189
Ross, MLR 165, 170, 175
Ross, R 174
Ross, RE 149
Rossi, FE 58, 121, 174
Rossi, R 21, 30
Rossow, L 127
Rotkis, TC 72
Rotstein, A 7
Roubeix, M 116, 122
Rouhier-Marcer, I 23, 37
Rouis, M 140
Rourke, B 80
Rovetti, R 84
Rowsell, G 126
Rowsell, GJ 185
Rowswell, G 193
Roy, EP 27, 49, 50
Roy, RR 14, 15, 23, 38
Roy, SR 171
Rozenek, R 37, 53
Ruben, RP 130
Rubin, MR 173
Rubinstein, N 34

Rubio-Arias, JÁ 99, 163
Ruddock, A 49, 55, 103, 105, 150, 157, 175
Ruesta, M 23, 35
Ruf, L 133, 157
Ruiz, JR 29, 30
Rundqvist, H 30
Rushing, BR 176
Rusko, H 113, 116, 120
Russell, B 29
Russell, J 168
Russell, M 89, 120, 121, 189
Russell, S 189, 191
Rutherford, OM 20, 27, 59
Rutherfurd-Markwick, K 176
Rutkowski, J 157
Ryan, AJ 35, 169
Ryan, C 173
Ryan, EJ 176
Rybalka, E 173
Ryman Augustsson, S 127
Rynne, S 191

S

Saavedra, F 99
Sabag, A 116
Sabapathy, S 183, 184
Sabol, F 176
Sacchetti, M 176
Sacco, P 191
Sadres, E 121
Saenz, C 187
Saeterbakken, AH 80, 81, 82, 84
Sáez de Villarreal, E
Sainz de Baranda, P 64
Sakaguchi, M 47
Salazar-Rojas, W 111
Saldanha, LG 176
Sale, C 172, 174, 175, 176
Sale, DG 20, 24, 47, 78, 109, 122
Salem, GJ 76
Salerno, AE 172
Salido, E 30
Salinero, JJ 176
Sallinen, JM 175
Saltin, B 18, 21, 30, 171, 177
Salvador, EP 127
Samnoy, LE 77
Samozino, P 9, 50, 70, 71, 136, 137, 138, 142, 143
Sampaio, J 147, 158
Sams, ML 50, 70, 90, 91, 92, 111, 140, 153
Samson, M 68
Samuels, C 183
Sanborn, K 20, 99, 128, 173
Sanchez, AC 101
Sánchez, CC 149, 150, 151
Sánchez, DJ 193
Sánchez-Martínez, G 51, 143
Sánchez-Medina, L 115, 122, 133, 157, 159
Sanchez-Otero, T 108
Sanchez-Sanchez, J 61
Sánchez-Sixto, A 140
Sandell, RF 70
Sander, A 7
Sander, R 55
Sanders, J 30
Sanders, R 9, 46, 54, 110, 137

Sandford, FJ 187
Sandford, GN 116, 120
Sands, W 125
Sands, WA 5, 7, 29, 38, 41, 45, 51, 52, 53, 54, 75, 96, 100, 101, 109, 110, 125, 157, 161, 181, 182, 186, 187, 188, 191, 192
Sangnier, S 142
Sang Wong, S 163
San Juan, AF 174
Sannicandro, I 139
Santana, HAP 127, 130
Sant'Ana Pereira, JA 21, 31
Santiago, C 30
Santos, TD 71
Saplinskas, JS 22, 32
Sarabon, N 189
Šarabon, N 51, 128, 144
Sardelis, S 171
Sardella, F 23, 36
Sargeant, AJ 21, 31
Sargent, C 183, 184, 192
Saris, WHM 162, 166
Sarmento, H 188
Sartin, J 22, 23, 34, 36
Sarto, F 89
Sasaki, H 173
Sato, K 3, 4, 7, 20, 24, 30, 49, 50, 54, 62, 64, 65, 67, 68, 75, 76, 78, 84, 85, 90, 91, 92, 95, 103, 104, 105, 106, 107, 109, 110, 127, 128, 130, 132, 133, 134, 135, 138, 139, 140, 142, 149, 152, 157, 188
Sato, M 174, 186
Saunders, B 172, 174, 175, 176
Saunders, MJ 193
Sausaman, R 19, 48, 51, 121, 147, 181
Sausaman, RW 47
Savage, B 175
Savage, MV 19
Savage-Elliott, I 72
Savoie, F-A 171
Sawczuk, M 30
Sawczuk, T 154
Sawka, MN 171, 172
Sayers, MGL 133
Sayers, SP 28, 154
Saylor, HE 172
Scala, D 48
Scantlebury, S 126, 154
Schaal, K 170
Schakman, O 35
Schau, KA
Schauss, JH 187
Scheett, TP 173, 175
Scheid, JL 178
Schilling, B 37, 99
Schilling, BK 18, 21, 30, 34, 36, 40, 65, 66, 85, 89, 113, 115, 135
Schjerling, P 35
Schlaeppi, M 133
Schlattner, U 172
Schlumberger, A 99
Schmahl, RM 172
Schmid, P 23, 34
Schmidtbleicher, D 7, 59, 78, 99, 140
Schmitt, WM 22, 23, 34, 35
Schnabel, A 22, 23, 34, 35
Schnaiter, JA 25, 107

Schneider, C 188
Schödl, G 1, 2
Schoemaker, MM 40, 158
Schoenfeld, B 34, 38, 60
Schoenfeld, BJ 5, 18, 21, 23, 24, 25, 31, 37, 46, 48, 58, 66, 98, 101, 107, 132, 165, 166, 172, 175, 176, 177, 178
Schoenfeld, ML 60
Schoffstall, JE 171
Scholl, C 106
Schöner, G 41
Schot, P 46, 71, 96
Schroeder, ZS 69, 89, 90
Schuh, M 46, 71, 96
Schulz, ML 175
Schumacker, RE 193
Schumann, M 113, 115, 116, 121, 122, 123
Schwab, R 22, 35
Schwanbeck, S 58
Schwanbeck, SR 58
Schwartz, JH 16
Scott, BR 72, 73, 108, 151, 155, 156
Scott, DJ 81
Scott, RA 30
Scott, T 157
Scott, TJ 184
Scruggs, SK 50
Scurati, R 174
Sebastianelli, WJ 34
Sebolt, DR 120
Secomb, J 141
Secomb, JL 142
Sedano, S 64
Sedliak, M 35
Segers, N 66
Seipp, D 122
Seitz, L 81
Seitz, LB 4, 7, 25, 62, 65, 75, 76, 78, 81, 82, 100, 101, 108, 109, 110, 134, 159
Selamie, SN 121
Selby, A 39, 90
Selfe, J 189
Sell, KM 101
Sellers, J 25, 107
Selye, H 19, 181
Semmler, JG 22, 33, 43
Senden, JM 165, 172
Seo, DI 127
Serpa, ÉP 99
Serra, A 55
Serrano, AJ 50, 63, 64, 70, 111, 153
Serrano, N 18, 21, 30
Serrão, JC 173, 177
Sevene-Adams, PG 65
Sevick, MA 178
Sewell, DA 173, 174
Sforzo, GA 82, 99, 191
Shao, A 174
Sharman, MJ 173
Sharp, AP 136, 137
Sharpe, K 184, 185
Sharples, AP 167, 168
Shattock, K 149
Shaw, G 165, 170
Shaw, MP 80, 81
Shen-Orr, Z 183, 184
Shepard, R 23, 36

Shepherd, SO 167
Sheppard, J 85, 91, 141
Sheppard, JM 3, 7, 24, 41, 50, 66, 69, 71, 89, 90, 93, 98, 100, 101, 102, 107, 128, 131, 132, 136, 141, 142, 149, 151
Sherk, VD 127
Sherman, WM 72
Sherrington, CS 15
Shi, L 81, 82, 84
Shiba, M 73
Shibata, K 66
Shibayama, A 135
Shields, EW 184
Shimano, T 132, 158
Shin, AY 172
Shirreffs, SM 172
Shishov, N 154
Shoepe, T 84
Shoepe, TC 84
Shone, EW 101
Short, MJ 184
Shorter, KA 62
Shriber, WJ 154
Shvartz, E 58
Siedlik, J 143
Sieglebaum, SA 16
Siegler, J 189
Siegler, JC 175
Siemienski, A 136
Siff, MC 47, 68, 100, 159
Sigg, JA 82
Signorile, JF 43, 136
Sigward, SM 25, 46, 71, 96
Sijuwade, O 4, 96, 136, 151, 155, 158, 159
Silva, AJ 46, 173
Silva, BVC 177
Silva, MF 72
Silva, WAB 78, 109
Silva Cesario, JC 79
Silventoinen, K 29
Silvester, LJ 58
Silvestre, R 132, 158
Sim, J 189
Simão, R 24, 46, 95, 99, 107, 149
Simenz, C 62, 69
Simenz, CJ 63
Simmons, J 132
Simões, HG 189
Simon, P 29, 30
Simond-Cote, B 70
Simonsen, EB 15, 22, 23, 33, 38, 101
Simopoulos, AP 167
Simpson, EJ 173
Singh, AB 34
Singla, D 62, 69
Sioud, R 21, 41
Siqueira, F 48
Sirparanta, A-I 121
Sist, M 168
Siu, JW 18, 21, 30
Sjøgaard, G 34
Sjokvist, J 193
Sjovold, A 143
Skein, M 184
Skinner, JW 65, 84
Skinner, RD 31
Skorski, S 181, 186, 191, 192

Skotte, JH 59
Skrepnik, M 107
Slater, GJ 184, 185
Slawinski, J 9, 50, 70, 71, 137, 138, 142, 143
Sleivert, G 135
Smajla, D 51, 144
Smeets, JSJ 166
Smilios, I 36
Smith, A 174
Smith, AE 174
Smith, AP 177
Smith, D 99
Smith, DT 128
Smith, GL 101
Smith, IC 64, 76
Smith, J 64, 80, 95, 136
Smith, LL 173, 184
Smith, M 174, 176, 189
Smith, MJ 72
Smith, MR 191
Smith, NA 62
Smith, P 175
Smith, PA 115, 121, 122
Smith, PM 176
Smith, SA 73
Smith, SL 7, 65
Smith, T 170
Smith D, Enslen M 166
Smith-Ryan, AE 161, 168, 172, 174, 175, 176, 178
Smits-Engelsman, B 113
Snape, A 184
Snijders, T 23, 37, 165, 166
Snow, RJ 177
Snyder, BW 96, 143
Snyder-Mackler, L
So, WY 127
Soares, EG 99
Soares-Caldeira, LF 127
Sobonya, S 132
Soderlund, K 172, 173, 174
Soh, KG 191
Soh, KL 191
Sokolosky, J 66
Solberg, P 50, 143
Sole, C 125
Sole, CJ 4, 24, 46, 49, 50, 62, 64, 80, 95, 106, 135, 140, 141, 142, 143, 146
Solis, MY 175
Solomon-Hill, G 186
Solstad, TEJ 80, 81, 82, 84
Somjen, G 18, 32
Sommerfield, LM 7, 50, 63, 95
Sona, S 173
Song, C-Y 81, 140
Sonmez, GT 23
Sorensen, H 64, 76
Soriano, MA 50, 63, 64, 95, 96, 142, 152, 153
Sousa, CA 66, 88, 151
Southward, K 176
Souza, DB 176
Souza, HS 184
Sowinski, RJ 165
Spano, M 174
Sparks, KE 187
Sparks, SA 168, 175, 176

Specos, C 3, 41, 66, 71, 90, 132
Speirs, DE 60
Spencer, M 126
Spengos, K 78
Spengos, KM 128
Speranza, MJA 7
Spierer, DK 59
Spiering, B 34
Spiering, BA 22, 34, 59, 63, 132, 142, 158, 191
Spiliopoulou, P 5, 80, 108
Spineti, J 24, 99
Spiteri, T 3, 41, 66, 71, 90, 132
Sporer, B 135
Sporer, BC 115, 122
Spranger, MD 73
Spratford, W 69
Spriet, LL 167, 177
Stachenfeld, NS 171, 172
Stage, AA 44, 132
Stahl, CA 28, 30, 111, 149, 181, 189
Stamford, BA 65
Stampfer, MJ 167
Stanley, M 176
Stannard, RL 174
Stapley, PJ 66
Staron, RS 18, 21, 22, 23, 30, 31, 34, 36, 40
Staron RS 30
Stasinaki, A-N 5, 66, 108
Stasinaki, AN 80, 128
Stastny, P 66, 88
Stathis, CG 163
Stavropoulos, N 58
Stearne, DJ 147
Stec, J 99
Steele, J 3, 49, 58, 65, 66, 89, 99, 159
Steele, JM 80, 81, 82, 84
Steenge, G 174
Steenge, GR 173
Stefanovic, DLJ 140
Stefanovic-Racic, M 79
Steffl, M 6
Stegen, S 175
Stein, RB 32
Steinacker, JM 193
Stelling, H 113, 115, 116, 120
Stellingwerff, T 135, 167, 175, 176
Stempel, KE 30
Stenberg, J 171
Stephens, J 192
Stephens, JM 182, 184, 185, 188, 189, 192
Stepto, NK 113, 118, 121
Stergioulas A 35
Sternlight, DB 70, 137
Stevenson, MW 80, 81, 84
Stewart, AD 80, 81, 82, 84, 136, 157
Stewart, C 89, 168
Stien, N 80, 81, 82, 84
Stockton, KA 168
Stodólka, J 136
Stojiljković, S 66
Stokes, JJ 44, 132
Stølen, T 93, 120
Stone, M 38, 41, 45, 48, 49, 51, 52, 53, 54, 75, 84, 91, 100, 104, 105, 110, 113, 115, 125, 128, 147, 152, 153, 161
Stone, ME 7, 20, 23, 24, 36, 37, 54, 58, 82, 94, 99, 100, 110, 128, 157, 173

Stone, MH 3, 4, 5, 6, 7, 8, 9, 13, 18, 19, 20, 21, 22, 23, 24, 25, 26, 27, 29, 30, 31, 34, 35, 36, 37, 38, 41, 44, 45, 46, 47, 48, 49, 50, 51, 52, 53, 54, 55, 58, 60, 62, 63, 64, 65, 66, 67, 68, 69, 75, 76, 78, 79, 82, 83, 84, 85, 91, 94, 95, 96, 97, 98, 99, 100, 101, 102, 103, 104, 105, 106, 107, 109, 110, 111, 121, 122, 123, 125, 127, 128, 130, 132, 133, 134, 135, 138, 139, 140, 141, 142, 143, 145, 146, 147, 149, 150, 152, 153, 157, 159, 161, 162, 168, 170, 171, 172, 173, 178, 181, 187, 188
Stone, WJ 154
Stone, WL 18, 21, 30, 31
Stoner, JD 4
Storer, TW 34
Storey, A 157
Storey, AG 58, 59, 150, 151
Stout, J 174
Stout, JR 19, 39, 121, 128, 165, 172, 174, 175, 176, 177, 178
Stowers, T 48
Straight, CR 176
Stratford, C 139
Street, G 135
Strike, S 5
Strokosch, A 81
Stromme, SB 35
Strømme, SB 22, 35
Stuart, C 23, 36
Stuart, CA 18, 21, 30, 31, 50, 54, 90, 91, 92, 103, 104, 106, 133, 149, 152
Stuempfle, KJ 187
Stults-Kolehmainen, MA 149
Suarez, DG 19, 23, 25, 26, 27, 30, 36, 46, 47, 48, 49, 51, 52, 53, 55, 100, 104, 121, 122, 125, 157
Suarez-Arrones, L 139
Suchomel, TJ 3, 4, 5, 6, 7, 8, 9, 13, 20, 21, 23, 24, 25, 26, 27, 29, 33, 36, 37, 38, 44, 46, 47, 48, 49, 50, 51, 53, 54, 55, 58, 59, 60, 61, 62, 63, 64, 65, 66, 67, 69, 70, 75, 76, 77, 78, 79, 80, 83, 85, 89, 90, 94, 95, 96, 97, 98, 99, 100, 101, 102, 103, 104, 105, 106, 108, 109, 110, 111, 119, 121, 123, 125, 126, 127, 128, 130, 132, 133, 134, 135, 136, 138, 139, 140, 141, 142, 143, 145, 146, 150, 151, 152, 153, 155, 157, 158, 159, 161, 168, 171, 172, 178, 193
Suei, K 97, 105
Suetta, C 35
Sugisaki, N 39
Sugiyama, T 116
Sullivan, KO 183
Sumner, SJ 176
Sun, F-H 163
Sun, H 191
Sun, J 176
Sun, R 176
Sunderland, CD 175
Sundh, AE 90
Sundstrup, E 59, 102
Sung, DJ 173
Sünkeler, M 116, 121, 122

Suominen, H 170
Suprak, DN 59, 80
Sureda, A 193
Sutanto, CN 164
Suter, E 66
Sutton, JR 156
Sutton-Charani, N 20
Suzovic, D 157
Suzuki, T 173
Suzuki, Y 72, 174
Svantesson, U 24, 58, 94, 127
Swain, DE 171
Sweet, TW 159
Swensen, T 191
Swift, E 189
Swinton, PA 80, 81, 82, 84, 133, 136, 157, 175
Sylvén, C 119
Syrotuik, D 120
Syväoja, H 121
Szabó, T 140
Szczepanowska, E 23, 35
Szivak, TK 187

T

Taaffe, DR 101
Tabata, I 116
Taber, CB 23, 24, 26, 30, 37, 46, 48, 50, 54, 61, 62, 64, 65, 66, 67, 68, 69, 70, 84, 85, 89, 90, 91, 92, 95, 96, 103, 104, 106, 110, 111, 119, 133, 149, 150, 152, 157, 159
Taberner, M 116, 120
Taes, Y 175
Taipale, RS 115, 116
Takagi, T 73
Takai, Y 70
Takamatsu, K 174
Takano, B 63
Takao, K 116
Takei, S 95
Takeshita, D 135
Takizawa, K 66
Talbot, KB 140
Tallent, J 66
Tallis, J 175
Tallon, MJ 174
Talpey, S 111
Talpey, SW 7, 94, 146
Tamir, D 58
Tamura, K 186
Tan, JG 115
Tanaka, KH 66
Tanaka, M 30
Tang, J 168
Tang, JCY 168
Tang, JM 59
Tanigawa, S 70, 128
Tarnopolsky, M 173
Tarnopolsky, MA 23, 37, 78, 165, 173
Tarso Guerrero Muller, P 174
Tarso Veras Farinatti, P 99
Tassignon, B 191
Tatár, P 22, 35
Tatyan, V 54
Taxildaris, K 176
Taylor, JL 121
Taylor, K-L 50, 133, 136, 141, 157
Taylor, L 178

Taylor, LW 173
Taylor, R 156
Taylor, T 189
Tee, JC 149
Tejero, J 189
Teka, S 30
Tench, J 163
Teo, SY 63
Teo, WP 101
Teodoro, JL 101
Teoldo, I 120
Terada, M 116
Ter Keurs, HE 40
Terzis, G 5, 66, 78, 80, 105, 108, 171
Terzis, GD 128
Tesch, A 67
Tesch, P 115
Tesch, PA 30, 37, 67, 79, 82
Testa, R 189
Thalib, L 116, 121
Tharp, GD 23, 36
Theis, N 66, 68, 80
Theurot, D 186
Thiebaud, R 127
Thissen, J-P 35
Thoman, J 187
Thomas, C 64, 71, 83, 95, 96, 101, 103, 130, 138, 141, 142
Thomas, GA 23, 35, 36, 142, 186
Thomas, K 85, 90, 91, 113, 115, 132
Thomasson, ML 94, 146
Thomeé, R 24, 58, 94
Thomis, M 29
Thomis, MA 29, 30
Thompson, BJ 4, 140
Thompson, C 191
Thompson, D 171
Thompson, FN 22, 35
Thompson, KG 186
Thompson, MW 68, 85
Thompson, PD 30
Thompson, SW 49, 55, 103, 105, 150, 157, 187
Thompson, WR 57
Thorborg, K 67
Thornton, HR 155, 156, 184
Thorsen Frank, M 50, 143
Thorstensson, A 79
Thrush, JT 18, 21, 30, 36, 40
Thyfault, JP 154
Tieland, M 166
Tiemessen, IJH 185
Tihanyi, J 34
Tiidus, PM 191
Till, K 154, 156, 157
Tillin, N 22, 23, 32, 39, 78
Tillin, NA 4, 20, 76, 78, 79, 84, 109
Timmer, J 157
Timmons, JA 172, 173, 174
Timmons, JF
Timon, R 189
Tindel, J 189
Tinsley, GM 175, 176
Tiollier, E 177
Tipton, KD 165, 166
Tobin, DP 136
Todd, JS 149
Todd, MK 193

Toews, CJ 156
Toh, DWK 164
Toivanen, J 187
Toji, H 97, 104, 105
Tokarska-Schlattner, M 172
Tokmakidis, SP 36
Toledo, C 62, 69, 159
Toledo, FG 79
Tolfrey, K 72
Tolusso, DV 51
Toma, K 18, 23, 36, 149
Tomaras, EK 76
Tombleson, T 134
Tomek, S 167
Tomlinson, PB 168
Tonino, P 14, 18, 40
Tooley, EP 78
Tops, M 191
Tornatore, G 47
Torok, D 174
Torok, DJ 174
Torrejón, A 157
Torrents, C 41
Torres, A 30
Torres, M 27, 61
Torres-González, A 64
Torres-Torrelo, J 46
Torrontegi, E 51, 143
Toth, MJ 21, 30
Touey, PR 99
Tous-Fajardo, J 139, 147, 158
Townsend, JR 19, 25, 26, 39, 49, 100, 121
Townshend, A 156, 157
Trajano, GS 62, 101
Tran, DL 101
Tran, TT 142
Tranchina, NM 149
Trappe, S 18, 21, 30
Trappe, T 18, 21, 30
Travis, SK 21, 22, 23, 26, 31, 37, 49
Tremblay, MS 23, 35
Trepeck, C 151
Treuth, MM 35
Trexler, ET 168, 172, 174, 175, 176, 177
Tricoli, V 63, 115, 121, 151
Tricoli, VAA 115, 118
Trinick, J 14, 39, 40
Triplett, NT 19, 22, 23, 24, 25, 26, 27, 34, 35, 36, 48, 49, 50, 51, 89, 93, 98, 100, 101, 102, 103, 105, 106, 107, 120, 121, 142, 147, 149, 151, 161, 162, 181
Triplett, T 54, 157
Triplett-McBride, NT 85
Triplett-McBride, T 40, 99, 107, 161
Trombitás, K 39
Trommelen, J 165, 166
Trousselard, M 173
Tróznai, Z 140
Truffi, G 55
Trybulski, R 66, 88
Tryniecki, JL 3, 65, 89, 90
Tsairis, P 18, 30
Tsatalas, T 81
Tsianos, G 30
Tsiokanos, A 5
Tsitsios, T 176
Tsolakis, C 35

Tsoukos, A 105
Tsourlou, T 58
Tsuboi, Y 186
Tsunoda, N 23
Tufano, JJ 25, 66, 69, 85, 90, 91, 100, 101, 108, 133, 157, 159
Tufik, S 184
Tuler, B 41
Turk, M 79
Turner, A 6, 24, 99, 138, 139
Turner, AN 80, 95, 97, 103
Turner, AP 60
Turner, B 186
Turner, TS 136
Turpin, E 117
Turvey, MT 41
Twine, C 130
Twist, C 80, 113, 133, 157
Tyler, TF 173

U

Uchida, S 158
Uchiyama, A 47
Uda, K 173
Udall, R 64, 95, 133, 142
Uesato, S 143
Uftring, M 28, 30, 111, 149, 181, 189
Ugrinowitsch, C 63, 66, 72, 78, 109, 115, 121
Umetani, K 186
Undheim, MB 5
Urbina, SL 173
Urdampilleta, A 170
Uribe, VA 176
Ursø, B 173
Urso, M 30
Utter, AC 25, 26, 100, 107, 173
Uyeno, TA 18, 40

V

Vaczi, M 30
Vailas, AC 120
Vaile, J 185, 186, 187
Vaile, JM 186
Vaíllo, RR 149, 150, 151
Valenzuela, P 137
Valenzuela, PL 51, 143
Valkeinen, H 113, 116, 120
van Borselen, F 22, 27, 34, 49, 50
Van Cutsem, J 191
van Cutsem, M 22, 32, 33, 62, 70, 78
VandenBerg, CD 156
Vandenberghe, K 173, 174, 176
Vandenbogaerde, TJ 163
Vandenboom, R 78
Vandendorpe, F 117
Van den Eede, E 30
Vanden Eynde, B 174
van den Tillaar, R 3, 68, 80, 81, 82, 84, 85, 132
van der Beek, EJ 169
Vandereyt, H 166
van der Sluis, S 29, 30
Vandervoort, AA 78
Vanderzanden, T 62, 69
Vanderzanden, TL 62, 69, 96
Vandewalle, H 140
van Dijk, J-W 165
Van Dorin, JD 81

VanDusseldorp, TA 172, 176, 177, 178
Van Emmerik, REA 147, 149
Van Every, DW 23
Van Gammeren, D 171
Vangerven, L 173, 174
Vangheluwe, P 40
Van Hecke, P 173, 174, 176
Vanhee, L 174
Van Helder, W 23, 35
Van Hooren, B 101, 135, 143, 188
Van Keeken, BL 185
van Kranenburg, J 166
Van Leemputte, M 30, 173, 174, 176
van Loon, LJ 23, 37
Van Loon, LJC 162, 165, 166
Van Lunen, B 71, 96
van Mechelen, M 21, 31
Vann, CG 37
Van Praagh, E 116
Van Proeyen, K 174
Vanrenterghem, J 71, 95, 96, 135, 137, 138
van Soest, AJ 33, 58, 137
van Someren, K 116, 121, 186, 187
Van Someren, KA 185
Vanstapel, F 176
Van Thienen, R 174
van Vliet, S 166
Varanoske, AN 121, 174
Varela-Olalla, D 151
Vargas-Molina, S 101
Varillas-Delgado, D 30
Vårvik, FT 163
Vasconcelos, GC 79
Vaz, R 151
Vázquez-Carrión, J 51, 143
Vázquez-Guerrero, J 147
Vechin, FC 72
Veiga-Herreros, P 174
Veligekas, P 105
Venables, E 71, 96, 138
Venables, MC 177
Veneroso, CE 30
Verde, Z 30
Verdijk, LB 23, 37, 166
Verhagen, E 69
Verhagen, HH 187
Verheul, J 71, 95, 137
Verhoeven, FM 41
Verkhoshansky, Y 54, 100
Verkhoshansky, YV 47, 52, 53, 68
Verlengia, R 163
Vermeire, K 19
Vermeulen, S 69
Verschueren, J 69, 191
Versey, N 184, 185, 186
Versey, NG 184, 185, 186
Vetrovsky, T 6
Vicens-Bordas, J 67
Vidmar, MF 72
Vieira, A 189
Vieira, C 189
Vieira, CA 81, 82, 84, 189
Vigas, M 22, 35
Vigotsky, A 37, 58
Vigotsky, AD 23, 48, 101
Vihko, V 59, 187
Viitasalo, JT 59, 100

Vilaça-Alves, J 99
Viña, J 171
Vincent, B 30
Vincent, K 99
Vingren, JL 23, 34, 35, 36, 108, 132, 158, 187
Vinicius, Í 79
Virgile, A 99
Viroux, PJF 185
Viru, A 34, 35, 36
Viru, AM 23, 35
Viru, M 34, 35, 36
Virvidakis, K 7
Vist, GE 172
Viveiros, L 99
Vladich, T 173
Vlietinck, RF 29
Vogelgesang, DA 72
Volek, J 25, 48, 130
Volek, JS 20, 22, 23, 34, 35, 36, 132, 142, 158, 171, 173, 191
Volpe, SL 120
Vom Heede, A 6
Von Walden, F 30
Vuorela, J 187

W

Wadhi, T 66
Wadley, GD 192
Wagenmakers, AJM 162
Waggener, GT 127
Wagle, JP 13, 19, 27, 29, 30, 36, 37, 38, 44, 46, 47, 48, 49, 51, 65, 66, 67, 68, 69, 84, 85, 90, 91, 92, 94, 96, 110, 121, 147, 150, 157, 159, 161, 168, 171, 172, 178, 181
Wagner, A 15, 23, 38
Wagner, CM 7
Wagner, P 134
Wagner, PP 53
Wahl, P 182
Wakahara, T 7, 39, 65, 109
Walberg-Rankin, J 120
Waldron, M 66
Walker, PM 23, 37
Walker, S 3, 69, 82, 85, 90, 91, 115, 121, 132
Wall, EJ 7
Wall, SR 35
Wallace, BB 157
Wallace, BJ 65, 80, 81, 82, 83, 84, 140
Wallace, E 140
Wallace, ES 131
Waller, MA 50
Wallimann, T 172
Walrand, S 165
Walsh, NP 169, 172
Walsh, RM 171
Walshe, AD 143
Walter, AA 174
Walters, J 154
Wang, B 165, 170
Wang, G 29, 30
Wang, H 143
Wang, I-L 81
Wang, J 178
Wang, L-I 81
Wang, M-H 140
Wang, R 19, 25, 26, 49, 100, 121, 128, 174

Wang, Y-H 140
Ward, B 20, 128
Ward, K 173
Ward, TE 44
Warmington, S 72, 73
Warmington, SA 186
Warneke, K 7
Warpeha, JM 80, 81, 84
Warren, BJ 22, 25, 35, 107
Warren, GL 176
Waskiewicz, Z 175
Wasserstein, RL 149
Wassinger, CA 95, 106, 107
Watanabe, Y 186
Watkins, J 135
Watson, JS 79
Watt, MJ 167
Watts, D 126
Wazir, MRWN 191
Wdowski, MM 41
Weakley, J 125, 126, 187
Weakley, JJS 154, 156, 157
Weatherby, R 35
Weatherby, RP 36
Weaving, D 157
Webb, NP 185
Webber, JM 30
Webborn, N 30
Weber, CJ 156
Weber, F 22, 23, 34, 35
Webster, DP 1, 2
Webster, MJ 175
Weckström, K 98
Weeks, B 188
Weems, S 168
Wegmann, M 188
Wehr, RW 58
Weidt, K 1
Weir, J 79
Weir, JP 22, 35
Weiss, EP 167
Weiss, LW 22, 30, 35, 65, 85, 89
Welde, B
Weldon, A 175
Wells, AJ 19, 25, 26, 39, 49, 100, 121
Wells, GD 174
Welsh, BT 177
Wendell, M 24, 63
Wendeln, HK 23, 36, 149
Wenger, HA 115, 122
Wernbom, M 82
Wernstål, F 67
Weseman, CA 22, 35
West, DW 22
West, FE 154
West, N 192
Westcott, WL 66
Wetmore, A 30, 90, 91, 92
Wetmore, AB 46, 53
Weyand, PG 70, 95, 96, 137, 138
Whatman, CS 7
White, JB 108
Whitehurst, M 66, 72, 88
Whitley, A 25, 64, 95, 100, 101, 133
Whitten, JH 109
WHO (World Health Organization) 165
Wickham, KA 177

Wickiewicz, TL 14, 15, 23, 38
Widmann, M 29
Wiewelhove, T 188
Wilborn, CD 172, 173, 174, 175, 178
Wildman, R 172, 173, 174, 178
Wiley, JP 66
Wiley, LP 60
Wilk, M 66, 88
Wilkerson, JE 22, 35
Wilkes, G 186
Wilk M, Gepfert 66
Willardson, JM 24, 99, 103, 107
Willems, MET 176
Willems, T 69
Willett, WC 167
Williams, A 30
Williams, AD 173
Williams, AG 30
Williams, AM 49, 110
Williams, BD 121
Williams, C 2, 3, 171
Williams, CA 181
Williams, CC 136
Williams, DSB 138
Williams, GKR 148, 158
Williams, J 39, 90
Williams, JH 70
Williams, KJ 128
Williams, R 66, 96
Williams, TD 51
Williamson, BD 25, 100, 101, 108, 159
Williamson, S 187
Willoughby, D 178
Willoughby, DS 48
Willwacher, S 70
Wilmore, JH 72, 161
Wilson, A 116
Wilson, BD 68
Wilson, C 147, 149
Wilson, D 48, 61
Wilson, E 173
Wilson, G 168
Wilson, GD 161, 162
Wilson, GJ 4, 35, 41, 69, 97, 128, 131, 135, 143
Wilson, JM 72, 78, 109, 113, 115, 116, 121, 122, 140
Wilson, KM 154, 156
Wilson, LJ 188
Wilson, M 173
Wilson, MM 171
Wilson, S 175
Wilson, SJ 136
Wilson, SM 78, 109, 113, 115, 116, 121, 122
Winchester, JB 4, 65, 80, 81, 82, 84, 128, 158
Winckler, C 48
Winke, M 187
Winter, JA 166
Winwood, P 61
Winwood, PW 61
Wirth, K 5, 7
Wise, JA 174, 175
Wisloff, U 115, 116
Wisløff, U 7, 93, 110, 120
Witard, OC 167
Witt, CC 40
Wolde, B 30
Wolf, A 65, 85, 89, 90, 91, 132

Wolf, SE 166
Wolfe, BL 99
Wolfe, RR 165, 166
Womack, CJ 193
Wong, DP 71
Wong, JJ 174
Wong, MA 90
Wong, P 115, 116
Wong, SHS 164
Wood, J 99
Wood, JA 142
Wood, L 173
Wood, MR 61
Woodhouse, L 34
Woods, CT 125
Woolf, JM
Woolf, K 169
Woolford, S 126
Wortman, RJ 72
Wortsman, J 34
Woulfe, CJ 61
Wragg, CB 93
Wren, C 67
Wren, TAL 156
Wright, G 61
Wright, GA 63, 64, 66, 80, 95, 136, 142, 157, 159
Wright, HH 161
Wright, S 70, 137
Wu, WFW 136
Wulf, G 42, 154
Wurm, B 77
Wurm, BJ 62, 69
Wuytack, F 40
Wyatt, FB 24, 25, 98, 100
Wyatt, TJ 7, 110
Wyland, TP 81
Wypij, D 172

X

Xenofondos, A 140
Xenophontos, SL 30
Xu, Y 81, 82, 84

Y

Yakovlev, NN 19
Yamaguchi, GC 175
Yamanaka, K 5
Yanai, T 39
Yanaka, T
Yañez-García, JM 46, 157
Yang, GZ 20, 64, 76
Yang, N 30
Yang, S 1
Yang, Y 30
Yannakos, A 140
Yao, W 33
Yarasheski, KE 34
Yashchaninas, II 22, 32
Yata, H 23, 39
Yavuz, HU 47
Ye, W 81, 82, 84
Yellott, RC 106
Yeo, WK 167, 177
Yeowell, G 6
Yoon, D 173
Yoshida, N 67, 68

Yoshioka, T 23, 40
Young, AJ 172
Young, J 58
Young, JD 109
Young, K 127
Young, KC 37, 101
Young, KP 128, 131, 141
Young, M 116, 120
Young, W 41
Young, WB 4, 7, 24, 41, 46, 62, 69, 94, 128, 130, 135, 136, 139, 140, 141, 146
Yu, B 25, 71, 96, 138
Yu, L 116, 121, 122
Yuen, M 18, 40
Yusof, A 68, 85
Yvert, T 30

Z

Zabari, Z 183, 184
Zagatto, AM 174
Zajac, A 66, 88, 175
Zając, A 25, 100
Zajac, FE 18, 30
Zamparo, P 37, 79, 103, 109, 142, 143
Zanchi, NE 174
Zandvoort, CS 185
Zanetti, V 82, 173
Zanuso, S 89
Zaras, N 5, 66, 80, 108, 128
Zaras, ND 128
Zarzosa, F 80
Zaslow, TL 156
Zatsiorsky, V 54, 78, 80, 109
Zatsiorsky, VM 65
Zattara, M 21, 41
Zawieja, M 7
Zdornyj, AP 49, 104
Zebis, MK 59, 102, 103
Zemke, B 61
Zernicke, RF 138
Zhan, D-W 81
Zhang, G 152
Zhang, G. Auto 151
Zhang, Q 71
Zhang, X 151, 152
Zhang, Y-J 163
Zhou, F 30
Zhou, S 22, 33, 115, 121, 122
Zhu, XM 72
Ziegenfuss, T 174
Ziegenfuss, TN 172, 173, 174, 176, 178
Ziemann, E 189
Zimmer, P 182
Zimmermann, HB 76
Zinchenko A, Johnson K 44
Zinn, C 163
Žitnik, J 51, 144
Zmijewski, P 30, 66
Zoeller, RF 174
Zoffoli, L 89
Zolotarjova, J 135, 143
Zourdos, MC 66, 72, 88, 149, 150, 151, 152, 157, 170
Zuj, KA 187
Zuo, L 80
Zutinic, A 175
Zwetsloot, KA 21, 22, 25, 26, 31, 100, 107

About the Author

Timothy J. Suchomel, PhD, CSCS,*D, RSCC,*D, mISCP, is an associate professor and the director of the sports science master's program at the University of Pittsburgh. He has published over 110 peer-reviewed journal articles on topics including strength and power development, weightlifting movements and their derivatives, eccentric training methods, postactivation potentiation, and athlete monitoring.

Suchomel has served the National Strength and Conditioning Association (NSCA) as the chair of the Sport Science Special Interest Group, Wisconsin state director, and as a research committee member, and currently serves as a senior associate editor and associate editor for the *Journal of Strength and Conditioning Research* and *Strength and Conditioning Journal*, respectively. He was honored by the NSCA with the State/Provincial Director of the Year Award in 2019, the Terry J. Housh Outstanding Young Investigator Award in 2022, and the Journal of Strength and Conditioning Research Editorial Excellence Award in 2023. Suchomel is a certified strength and conditioning specialist with distinction (CSCS,*D) and a registered strength and conditioning coach with distinction (RSCC,*D) through the NSCA and has been recognized as a master international strength and conditioning practitioner (mISCP) through the International Universities Strength and Conditioning Association (IUSCA).

Suchomel completed his bachelor's degree in Kinesiology at the University of Wisconsin-Oshkosh in 2010, master's degree in Human Performance at the University of Wisconsin-La Crosse in 2012, and doctoral degree in Sport Physiology and Performance at East Tennessee State University 2015.

HUMAN KINETICS CONTINUING EDUCATION

Now that you've read the book, complete the companion exam to **earn continuing education credit!**

Find your CE exam here:
US & International: US.HumanKinetics.com/collections/Continuing-Education
Canada: Canada.HumanKinetics.com/collections/Continuing-Education

Subscribe to our newsletters today!

US

Canada

Get exclusive offers and stay apprised of CE opportunities.